N-Ubiquityl-Lys

N-Tetradecasaccharyl-Asn

Glc₃Man₉

N-GlcNAc-Ser

Phosphatidylinositol lipid anchor

Posttranslational Modification of Proteins

Expanding Nature's Inventory

Posttranslational Modification of Proteins

Expanding Nature's Inventory

Christopher T. Walsh

Harvard Medical School

ROBERTS AND COMPANY PUBLISHERS

Englewood, Colorado

Roberts and Company Publishers
4950 South Yosemite Street, F2, #197
Greenwood Village, Colorado 80111 USA
Internet: www.roberts-publishers.com
Telephone: (303) 221-3325
Facsimile: (303) 221-3326

ORDER INFORMATION
Phone: (800) 351-1161 or (516) 422-4050
Fax: (516) 422-4097
Internet: www.roberts-publishers.com

Publisher: Ben Roberts
Production Manager: Susan Riley at Side by Side Studios
Compositor: Side by Side Studios
Copyeditor: John Murdzek
Scientific Proofreaders: Sylvie Garneau-Tsodikova and Gregory J. Gatto, Jr.
Indexer: Megan Williams
Lead Illustrators: Sylvie Garneau-Tsodikova and Gregory J. Gatto, Jr.
Cover Artist: Gregory J. Gatto, Jr.
Cover and Interior Designer: Mark Ong at Side by Side Studios
Printer and Binder: C & C Offset Printing Co., Ltd.

Three-dimensional structural representations in this book were generated using the program MolScript (Kraulis, P. J. *J. Appl. Crystallogr.* **24:**946–950 (1991)).

The cover artwork was created with MolScript and rendered with Raster3D (Merritt, E. A. and D. J. Bacon. *Meth. Enzymol.* **277:**505–524 (1997)).

Printed in China

The text of this book is set in Jansen, a typeface designed by Nicholas Kis in about 1690. Kis's original matrices were found in Germany and acquired by the Stempel foundry in 1919. This version of Janson comes directly from the Stempel foundry and was designed from the original font. It was issued by Linotype in digital form in 1985.

   The display type in this book is set in Futura, a typeface designed by Paul Renner in 1927. It is one of the classic prototypes of a geometric sans serif type. Futura was based on the Bauhaus design philosophy of form following function. The typeface's basic geometric character allows no weight stresses, serifs, or flourishes, and its long ascenders and descenders give it an elegant proportion.

ISBN 0-9747077-3-2

Library of Congress Cataloging-in-Publication Data

Walsh, Christopher.
    Posttranslational modification of proteins : expanding nature's
inventory
 / Christopher T. Walsh.
     p. ; cm.
    Includes bibliographical references and index.
     ISBN 0-9747077-3-2 (alk. paper)
     1. Post-translational modification.   I. Title.
     [DNLM: 1. Protein Processing, Post-Translational. 2.
Proteins--meta-
  bolism.   QU 475 W223p 2006]
     QH450.6.W35 2006
     572'.645--dc22

2005014060

10 9 8 7 6 5 4 3 2 1

Dedicated to Diana, Allison, and Thomas

Brief Contents

Contents

Preface

The subject of this book is the vast set of changes that occur to amino acid residues in proteins after the proteins emerge from the ribosomal biosynthetic assembly lines in both prokaryotic and eukaryotic cells. These chemical alterations of protein side chains and main-chain peptide-bond connectivity comprise a set of modifications grouped together as the posttranslational modifications of proteins: changes that occur *after* the messenger RNA code has been *translated* into the amino acid sequence code of nascent proteins.

The purpose of controlled posttranslational modifications is to increase the diversity of functional groups beyond those in the side chains of the 20–22 proteinogenic amino acids incorporated into nascent proteins. This diversity enables new chemistry, new recognition patterns for partner molecules, turns "on" and "off" enzyme activity, and controls the lifetime and location of such proteins in cells. Posttranslational modifications of proteins expand nature's repertoire by increasing the inventory of side chains available to proteins.

The diversification enabled by posttranslational modification (and also by RNA splicing, a posttranscriptional topic not covered in this book) increases the molecular variants of proteins in cells by an estimated one to two orders of magnitude over the number encoded in the genome. Thus, if there would be some 30,000 genes transcribed into RNAs and translated into proteins, there may be 300,000 to millions of protein molecular variants at any one time in cells. These protein isoforms may differ, for example, in phosphorylation, acetylation, methylation, prenylation, and/or ubiquitylation content and location at one or more amino acid residues within any given protein.

To take just one type of posttranslational modification, protein phosphorylation, there are three common amino acid side chains (serine, threonine, and tyrosine) that can be modified and up to 500 protein kinases in the human genome that can phosphorylate different proteins. The prototypic protein kinase A has over 100 identified protein substrates, while the Abl protein is known to be phosphorylated at 11 different residues. There could easily be 100,000 variants of phosphorylated proteins in the proteome of humans. The posttranslational modification of proteins is a dynamic process. There are about 150 human protein phosphatases counterbalancing the action of protein kinases, so molecular heterogeneity of the phosphoproteome is enormous and changing from moment to moment in every cell.

This book examines more than a dozen major chemical classes of posttranslational modifications that contribute to the protein diversity of structure and function. In the postgenomics era proteomics and systems biology approaches are central to biologists' efforts to provide integrated evaluations of the behavior of cells, tissues, and organisms over time. Understanding the cast of proteins with all their posttranslational variants is a fundamental prerequisite. With modern mass spectrometry methods there will doubtless be many new combinations of posttranslational modifications yet to be detected.

The approach in this book is to categorize the broad range of posttranslational modifications experienced by proteins according to the types of covalent chemical modification introduced on particular side chains of amino acid residues in proteins. The functional consequences of the covalent change are examined, including protein conformational changes, alterations in subcellular locations, modulation of enzyme activity and/or partner protein association, and change in lifetime of the modified protein species.

For each type of protein chemical modification introduced posttranslationally, there is an associated class of enzymes dedicated to those tasks. The inventory of such posttranslational modification enzymes in the proteomes of humans is quite large: 500 protein kinases; 150 phosphoprotein phosphatases; 500 proteases; >100 ubiquitylation ligases and >50 deubiquitylating enzymes; and dozens of histone acetyltransferases and deacetylases, to name a few categories. In total, between 1000 and 2000 genes, representing more than 5% of the human genome, may encode enzymes dedicated to protein posttranslational modifications (PTMs).

While proteins are one set of substrates for PTM enzymes, the complementary set of substrates that provide the covalent groups for protein side chain decorations are most often small molecules that are central players in primary metabolism, including ATP for phosphorylations, acetyl CoA for acetylations, S-adenosylmethionine for methylations, NAD for ADP-ribosylations, and O_2 for hydroxylations. Eukaryotic cells have also developed information-rich small proteins, ubiquitin and ubiquitin-like proteins, as macromolecular tags that serve to covalently mark proteins for chaperoning to different locales in cells, including proteasomes for destruction.

A major part of the flux of posttranslational covalent modifications of proteins is specific hydrolysis of particular peptide bonds in particular proteins during their life cycle. To this end there are over 500 proteases encoded in the human genome. While some have broad digestive functions, the majority probably play roles in specific stages of the maturation of subsets of proteins. This strategy is clearly exemplified in proteins such as Notch that are sent into the secretory compartments, travel to the cell surface, and then proteolytic fragments traverse back to the nucleus to act as transcription factors. Four specific protease cuts, regulated in time and space, occur to control the life cycle of Notch. The protease activities of the subcellular organelle, the proteasome, connect the covalent polyubiquitylation of client proteins to targeted proteolysis to control the lifetimes of particular proteins, such as the cyclins at particular points in the cell cycle.

I began the writing of this book because I could not find any contemporary or comprehensive treatments of protein posttranslational modification in the literature, despite the mushrooming importance of the subject to protein function, protein-based signaling cascades, chromatin regulation, and proteomics.

The organization of chapters by type of covalent modification reflects my own chemical orientation to biology and biochemistry, elaborated in the introductory chapter. While most types of modifications are discussed, the scope is purposefully not encyclopedic. Indeed, for the most prevalent types of PTMs, such as protein phosphorylation, glycosylation, and proteolysis, whole books can and have been written on each topic. Instead each chapter uses selected examples to illustrate principles relevant to each class of PTM and to provide some integrating precepts.

Because of their historic and metabolic centrality, the phosphorylations of proteins are taken up in Chapter 2, followed by cognate sulfuryl transfers in Chapter 3. Chapter 4 summarizes redox chemistry of the sulfur-containing cysteinyl and methionyl residues. Chapters 5–7 deal with the transfer of covalent organic groups, starting with the one-carbon methyl fragment, then the two-carbon acetyl fragment, and progressing to long-chain myristoyl and palmitoyl transfers to protein side chains in Chapter 7. This chapter also includes GPI anchors and prenylation reactions as part of the protein lipidation repertoire.

Chapters 8 and 9 take up proteolysis during protein life cycles and the ubiquitous ubiquitylation tagging of proteins to deliver short-lived proteins to the chambered proteases, the proteasomes.

The subject of Chapter 10 is protein glycosylation. It is estimated that as many as a third of the eukaryotic proteins that enter the sceretory protein may be glycosylated; glycosylation may aid in protein refolding and protein quality control in the endoplasmic reticulum. Both N-glycosylation at asparagine residues and O-glycosylation on serine and threonine residues are summarized. The theme of protein glycosylation is continued in Chapter 11, where the focus is on ribosyl transfer rather than the hexosyl transfers of Chapter 10. NAD is the donor substrate and ADP-ribosylation of several kinds of protein side chains is the outcome. This is chemistry practiced by several varieties of protein toxins from bacteria, including cholera toxin, diphtheria toxin, and pertussis toxin.

Chapter 12 examines protein hydroxylation. This differs from the topics in Chapters 2–11 because the protein side chains modified are not acting as electron-rich nucleophiles towards an electrophilic donor. Rather, they are unactivated CH_2 centers of proline, lysine, and asparagines side chains that get modified by nonheme iron oxygenases delivering an electrophilic oxygen. This is the underlying molecular mechanism for protein sensing of hypoxia in tissues.

Chapter 13 collects several examples of protein automodification reactions, from the conversion of tyrosine side chains to TOPA quinone, to the remarkable autoconversion of tripeptide loops in folded proteins into heterocyclized functional groups. This latter process creates electrophilic centers required for phenylalanine and histidine deaminases to function and is also the essence of chromophore creation in green fluorescent protein and its several colorful congeners. Protein autocleavage at particular peptide bonds is also examined in terms of hydrolysis, cholesterolysis, and the spectacular protein splicing reactions.

Chapter 14 treats the biotinylation, lipoylation, and phosphopantetheinylation of proteins as cognate molecular logic, converting inactive apo forms of enzymes to active holo forms as the coenzymes are covalently tethered in the enzyme active sites. Chapter 15 looks at the modifications of glutamyl and glutaminyl side chains, from the cluster of γ-carboxylations of 10–12 Glu residues in precursor forms of proteases functioning in blood coagulation cascades to the cross-linking action of the transglutaminases that build mechanical strength into the protein products. Chapter 16 notes examples of cascades of multiple types of modifications on a given protein, some antagonistic, others setting a full modification pathway in motion, illustrating the interplay and hierarchies of PTM logic for protein network control and protein-based signaling.

I am indebted to the members of my research group for inputs and discussions on many aspects of this book thoughout the course of its preparation. I thank Nathan Hillson, Wei Lu, Michelle Pacholec, Tanya Schneider, Frédéric Vaillancourt, Masashi Ueki, and Jun Yin for the preparation of artwork, with special appreciation to Michael Fischbach for the design and construction of the artwork in several chapters. I thank Greg Gatto for the bulk of the protein structure figures and for manuscript assembly. I offer special thanks and appreciation to Sylvie Garneau-Tsodikova for spearheading every aspect of manuscript preparation, including drawing much of and coordinating all of the artwork and text, and for making the book better and more accurate in myriad ways.

I also acknowledge the input of the reviewers on the following page. Their expertise and input has improved both the factual base and context of the book. Any remaining errors are purely my own.

Christopher Walsh
January 2005

Reviewers

Tadhg P. Begley, Professor of Chemistry and Chemical Biology, Cornell University

Philip A. Cole, E. K. Marshall and Thomas H. Maren Professor and Director of the Department of Pharmacology and Molecular Sciences, Johns Hopkins University School of Medicine

Benjamin F. Cravatt, Professor, The Skaggs Institute for Chemical Biology, The Scripps Research Institute

John M. Denu, Associate Professor of Biomolecular Chemistry, University of Wisconsin Medical School

Bruce Furie, Professor of Medicine, Harvard Medical School, and Chief of the Division of Hemostasis and Thrombosis at the Beth Israel Deaconess Medical Center

Michael H. Gelb, Professor of Chemistry and Biochemistry, University of Washington

Donald F. Hunt, University Professor of Chemistry and Pathology, University of Virginia

Barbara Imperiali, Class of 1922 Professor of Chemistry and Professor of Biology, Massachusetts Institute of Technology

Neil L. Kelleher, Professor of Chemistry, University of Illinois at Urbana-Champaign

Chaitan Khosla, Wells H. Rauser and Harold M. Petiprin Professor in the School of Engineering and Professor of Chemical Engineering, Chemistry, and Biochemistry, Stanford University

Min-Hao Kuo, Assistant Professor of Biochemistry and Molecular Biology, Michigan State University

Michael A. Marletta, Aldo DeBenedictis Distinguished Professor of Chemistry, University of California, Berkeley

The Author

Christopher T. Walsh is currently the Hamilton Kuhn Professor of Biological Chemistry and Molecular Pharmacology at Harvard Medical School. He is one of the leading enzymologists in the world. He has elucidated the catalytic mechanisms of a wide variety of enzymes including flavoproteins and other redox enzymes. He has also pioneered the design of mechanism-based enzyme inhibitors (or "suicide" substrates). His work has found practical application in the design of antibacterial agents, anticonvulsive agents, plant growth regulators, and antitumor drugs. His current focus is on the biosynthesis and mechanism of action of antibiotics and bacterial siderophores. He has published over 600 scientific articles and his book, *Enzymatic Reaction Mechanisms*, has educated generations of enzymologists. He has also authored *Antibiotics: Actions, Origins, Resistance*.

Professor Walsh's accomplishments have been recognized through numerous awards, which include the Eli Lilly Award in Biochemistry, the Arthur C. Cope Scholar Award in Organic Chemistry, the Repligen Award in Biological Chemistry, and the Alfred Bader Award in Bioorganic and Bioinorganic Chemistry. He is a member of the National Academy of Sciences, the Institute of Medicine, and the American Academy of Arts and Sciences.

1

Introduction

Examples of posttranslational modifications at Glu, Ser, Cys, Asn, and Pro side chains in proteins.

Protein Translation and Posttranslational Modification

Proteins are the macromolecules that execute most of the genetically instructed programs and tasks inside and outside of cells and tissues in organisms. The size range of proteins varies from about 70 amino acids [circa. 7000 daltons (7 kDa)] in molecular weight to subunits as large as 15,000 amino acids, with 1.7-megadalton molecular weight.

Proteins are informational macromolecules whose primary sequence of amino acid residues directs folding of the native proteins into three-dimensional architectures that enable specific functions (Petsko and Ringe, 2003). The sequence of the amino acid monomers found in proteins is encoded in the nucleotide triplet code of the DNA in every gene. The information is passed from DNA language to protein language through the intermediacy of messenger RNA (mRNA). The information transfer pathway, DNA to RNA to protein, worked out by molecular biologists in the 1960s, illuminated how information stored in the stable DNA sequence of genes is first transcribed into RNA molecules. Messenger RNA molecules, which can be short-lived, with half-lives of hours, are **translated** from nucleotide code (DNA and RNA) to amino acid-based code in proteins that are the major actors to execute the programs of life.

Selective transcription of DNA codons in a subset of genes by RNA polymerases into mRNAs allows regulated expression of that gene subset. The mRNA is translated via a series of adaptor RNAs, transfer RNAs (tRNAs), which read the triplet codons of mRNA and insert specific amino acids corresponding to those codons. Particular tRNAs, with specified anticodons complementary to the mRNA codons by base pairing, are recognized by aminoacyl-tRNA (aa-tRNA) synthetases, enzymes that also select particular amino acid monomers from the cellular pool of small molecules. Each aa-tRNA synthetase then activates the selected amino acid as an aminoacyl-AMP, by cleaving ATP, to create the thermodynamic activation that enables subsequent favorable transfer of this aminoacyl moiety onto the 2'- or 3'-OH of the terminal ribose of the bound, cognate tRNA molecule (Figure 1.1A). Thus, each tRNA synthetase selects one of 20 proteinogenic amino acids *and also* a cognate tRNA and joins them covalently (Figure 1.1B). The product aa-tRNAs can diffuse to the peptidyltransferase center of the ribosome and dock via the anticodon onto the triplet codon of an mRNA molecule. Polymerization of two aminoacyl moieties on adjacent (Hao et al., 2002) docked tRNAs at the peptidyltransferase center of the ribosomal RNA in the large ribosomal subunit creates the first peptide on the dipeptidyl-tRNA product. This serves as the acceptor for subsequent rounds of chain elongation as additional aminoacyl-tRNAs dock one at a time at the peptidyltransferase site, specified by the next codon of the particular mRNA being translated. The iterative selection of aminoacyl-tRNAs and polymerization of the tethered aminoacyl moieties generates the peptide bond linkages of proteins, thereby building up the protein of designated primary sequence (Figure 1.1B) and constituting the process of **protein translation**.

The inventory of side chains of amino acids incorporated into proteins during translation at the ribosomes of the hundreds of thousands of proteins produced in cells is quite limited. Canonically, only 20 amino acids were thought to be **proteino-genic**—that is, building blocks for protein biosynthesis (Figure 1.2). After some decades of exhaustive study, the inventory of genetically encoded amino acids found in proteins has been expanded to 22. The 21st amino acid, found in prokaryotes and eukaryotes, is selenocysteine (Figure 1.3A), in which the sulfur of cysteine is replaced

A.

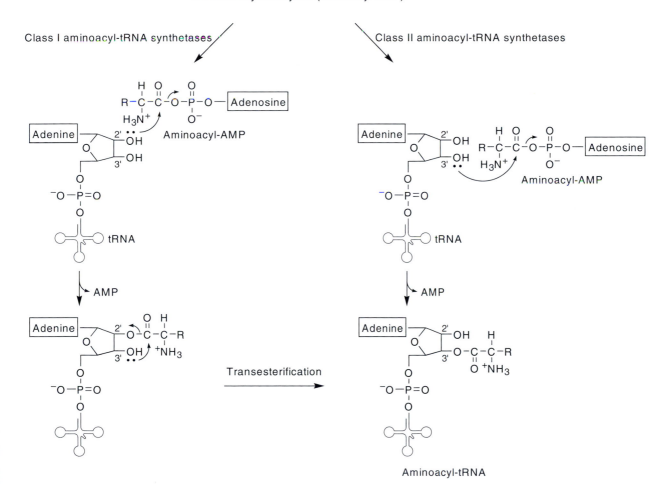

Figure 1.1 **A.** Activation of amino acids as aminoacyl-AMPs and aminoacyl-tRNAs in protein biosynthesis.

B.

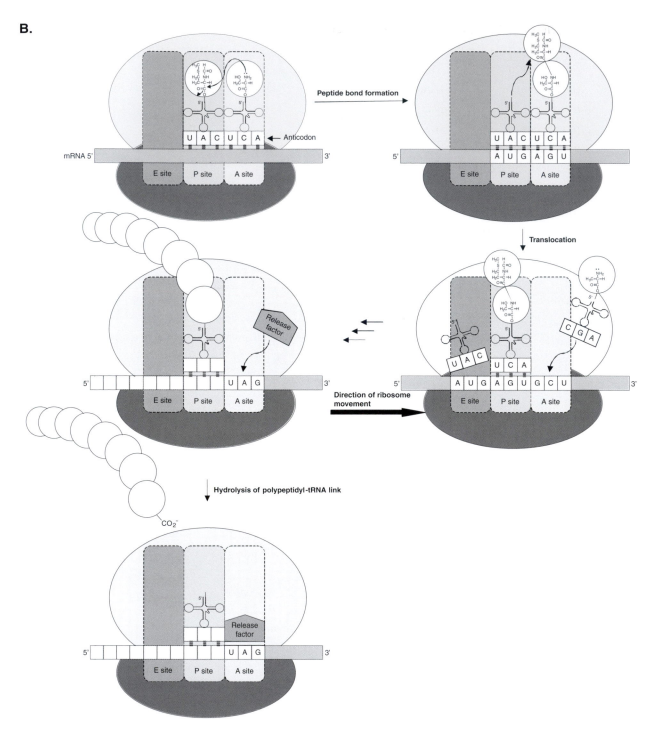

Figure 1.1 **B.** Peptide bond formation between peptidyl-tRNA as the electrophile and aminoacyl-tRNA as the nucleophile at the peptidyltransferase center of the ribosome constitutes translation of the mRNA trinucleotide code into amino acid-based protein code.

by selenium. Selenocysteine-containing enzymes and their role in peroxide detoxification catalysis are discussed in Chapter 4. The 22nd amino acid has recently been described during the x-ray structure determination of a methylamine methyltransferase (Hao et al., 2002) from an archaebacterium *Methanosarcina barkeri*, a primitive life form. Electron density indicated an unexpected and novel amino acid residue, pyrrolysine (Figure 1.3B), in the methyltransferase, where the ϵ-NH$_2$ of a lysine residue is derivatized with a β-methylpyrroline carboxylate. It appears that both the 21st and the 22nd encoded amino acids get inserted at what are usually stop codons, UGA for selenoCys and UAG for pyrroLys, by special tRNAs that read these termination codons as instructional for the suppressing 21st or 22nd amino acid. It appears that the selenoCys moiety gets constructed from a serine group as the Ser–tRNA^{SeCys} but free lysine is converted enzymatically to pyrrolysine (Py) and the

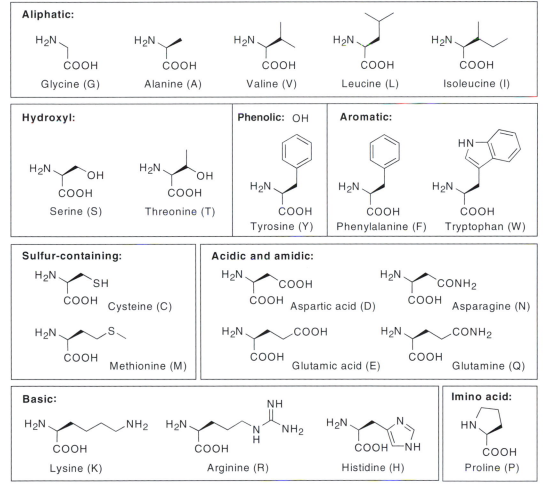

Figure 1.2 The twenty encoded proteinogenic amino acids.

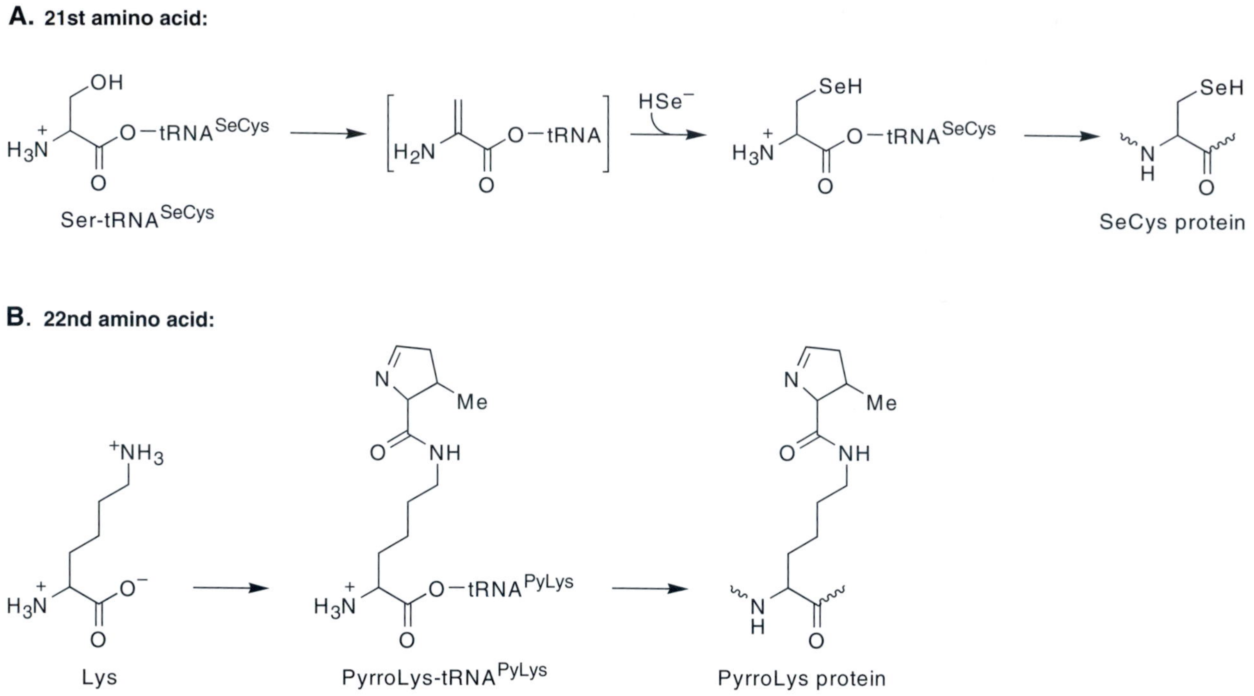

A. 21st amino acid:

B. 22nd amino acid:

Figure 1.3 **A.** Selenocysteine, the 21st amino acid found in proteins, by conversion of Ser-tRNA^{SeCys} to SeCys-tRNA^{SeCys}. **B.** Pyrrolysine, the 22nd amino acid found in proteins, by conversion of Lys-tRNA^{PyLys} in the archaeal methylamine methyltransferase enzyme.

Py is loaded onto a specialized tRNA^{PyLys} by pyrrolysyl-tRNA synthetase (Polycarpo et al., 2004) and modified as the aminoacyl-tRNA (Srinivasan et al., 2002). While it is possible that a small number of additions to the inventory of amino acids incorporated during protein translation will yet be detected, the vast majority of proteins, >99.9%, are biosynthesized from the 20 common proteinogenic amino acids.

Diversity by Posttranslational Modification

The global collection of proteins produced by organisms, unicellular to multicellular, is termed the **proteome** in analogy to the global complement of genes, the genome. While the proteome is encoded by the genome in each organism, and therefore a one-to-one correspondence of proteome to genome might be expected, proteomes are typically of higher complexity than genomes. For example, estimations of the human genome put the number of human genes at about 30,000. The human proteome is thought to have a >10–100-fold higher complexity at >300,000–3,000,000 distinct protein forms. How can this happen and what is the purpose of this increased diversity?

Proteome complexity can be built by diversification at both the mRNA level and after translation of mRNAs into proteins by covalent modification of specific proteins, a process called **posttranslational modification**. The diversification of mRNAs can happen by use of alternate promoter sequences in the 5′ upstream ends of mRNAs to start translation at different N-terminal methionine residues and so generate a few variants of the same protein of different lengths due to different starting points (a nested set).

Much more important is the alternative splicing of primary mRNA molecules in which some domains are excised and removed to generate mature mRNAs that retain different subsets of the domains of the original RNA. The domains excised are termed intervening sequences (introns), while those retained are expressed sequences (exons) (Figure 1.4). While all the introns get spliced out, one or more exons could get skipped, generating mature mRNAs with different lengths and different nucleotide sequences. The protein sequences encoded by the spliced RNAs differ if distinct combinations of exons have been spliced. mRNA splicing occurs in more than half of vertebrate RNAs and tissue-specific splicing can give rise to tissue-specific variants of the resultant proteins (Black, 2003; Maniatis and Tasic, 2002).

The third mechanism for proteome diversification occurs not at the RNA level but at the protein level by covalent modifications of the side chains of amino acid residues, and occasionally at one or more of the peptide linkages. The ensemble of covalent changes are termed posttranslational modifications to reflect the timing that these changes are introduced after the translation of mRNA nucleotide sequences into amino acid-based protein sequences at the ribosomes (Krishna and Wold, 1993). These modifications are the subject of this book.

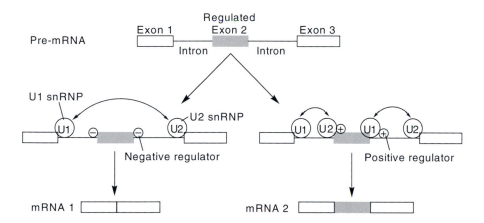

Figure 1.4 Schematic of RNA splicing: autocleavage of internal bonds at the exon–intron and intron–exon junctions to splice out the introns and ligate the exons in the frame to produce the mature, functional mRNA.

Posttranslational modifications are sometimes divided into two broad categories (Figure 1.5A). The first is the covalent addition of one or more groups, such as phosphoryl, acetyl, or glycosyl, to one or more of the amino acid side chains in a particular protein, as discussed in the next section.

The second is the hydrolytic cleavage of one or more peptide bonds in a protein by enzymes termed proteases (protein hydrolases) (Figure 1.5B). For example, most proteins secreted by cells have N-terminal signal sequences, specifying their extracellular address, which get recognized and cleaved by specific proteases during the secretion process. Hormones such as insulin are translated as a single-chain inactive prohormone that gets cleaved by a dedicated protease to the active two-chain form of insulin (Steiner, 1998). Analogously the precursor form of the leukocyte growth and differentiation protein, interleukin-1β, is cleaved at a single peptide bond to convert the inactive pro form to the active cytokine by the protease interleukin cleavage enzyme, a member of the caspase superfamily of proteases. These categories of proteolytic posttranslational modifications of primary translation products are examined in Chapter 8.

There are pathways of protein posttranslational modification, for example in the Ras family, where both covalent modification of side chains of amino acid residues and targeted proteolysis of the modified protein occur in a specific order to yield the final posttranslationally modified form of Ras. This is taken up in Chapter 7. A third form of proteolysis occurs in the phenomenon of protein splicing, in which an intervening sequence, the intein, is cleanly removed by autoproteolytic cleavage, resulting in the ligation of the two flanking external sequences, known as exons, by net transpeptidation (Paulus, 2000). This is discussed in Chapter 12 on autocatalysis. Finally, many of the proteins subjected to rapid proteolytic degradation within eukaryotic cells are degraded by proteasomes, following covalent marking by chains of the protein ubiquitin. The ubiquitylation aspects of posttranslational modification are taken up in Chapter 9. Protease action could properly occupy one or more volumes as a separate subject from all the other posttranslational modifications discussed in this book.

Almost all of the more than 200 kinds of posttranslational modifications that occur by covalent addition of groups to side chains in thousands of proteins (as well as the proteolytic cuts) are carried out by **enzymes**, proteins with catalytic activity dedicated to effecting the posttranslational modifications. The size of this catalytic posttranslational inventory is large. There are an estimated 500 proteases encoded in the human genome and a little more than 500 protein kinases for covalent phosphorylations of proteins. There are about 150 protein phosphatases opposing and balancing the action of the protein kinases. The other posttranslational modification enzymes noted in the several chapters of this book add a few hundred additional examples, such that about 5% of the human genome seems dedicated to expanding

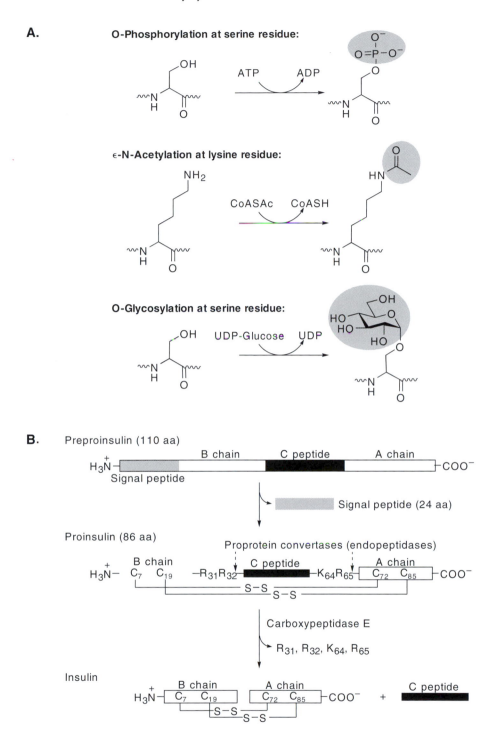

Figure 1.5 Posttranslational modification of proteins. **A.** Covalent addition of a substrate fragment to a protein side chain catalyzed by a posttranslational modification enzyme. Examples shown are generic phosphorylations, acetylations, and O-glycosylations. **B.** Covalent cleavage of one or more peptide bonds in protein substrates by proteases. Schematic processing of preproinsulin to proinsulin by signal peptidase in the endoplasmic reticulum and of proinsulin to insulin by proprotein convertases in the trans Golgi network.

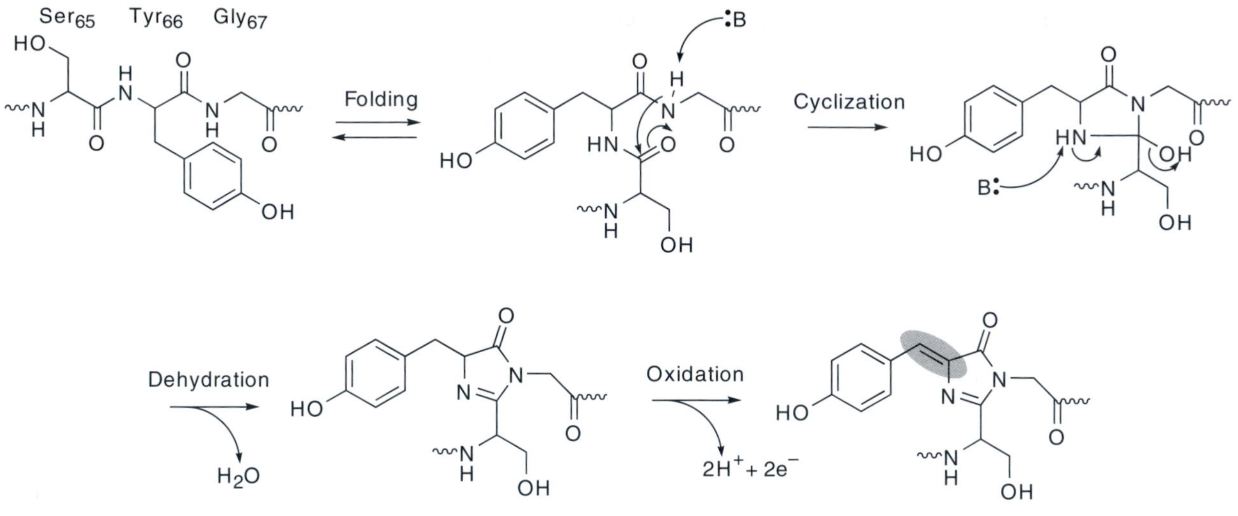

Figure 1.6 Autoconversion of the side chains of $Ser_{65}Tyr_{66}Gly_{67}$ in the proprotein form of green fluorescent protein (GFP) to the heterocyclic fluorophore.

diversity of the proteome by covalent modification of the initially translated proteome. We will examine the major categories of such modifying enzymes in the subsequent chapters of this book.

There is also a small subset of proteins that undergo automodifications—that is, modifications without the help of ancillary catalysts to effect covalent change. Among the most famous is the green fluorescent protein (GFP) (Figure 1.6) and its relatives (Tsien, 1998). We shall note the mechanism of formation of the fluorophore that requires the folded, native protein structure and rearrangement of the peptide bond connectivity in that region of GFP. The process of protein splicing, in which inteins are spliced out as extein sequences are ligated, is also an example of autocatalysis for both the cleavage and resynthesis of peptide bonds. These topics are covered in Chapter 12.

Scope of Posttranslational Modifications

The scope of the diversity introduced into the proteome by posttranslational modifications can be plotted on multiple axes. One is by the *number* of proteins modified and a corollary is the number of modified proteins thus produced. These numbers can differ for a given protein. For example, histone H3 can be covalently modified with distinct posttranslational modifications at distinct side chains. These are controlled by the specificity of posttranslational modifying enzymes for their protein substrates and the regioselectivity and sequence selectivity of the side chains modified.

The approximately 500 protein kinases, transferring phosphoryl groups from the common donor ATP (or rarely GTP), utilize three major types of side chains of specific serine, threonine, or tyrosine residues of proteins, and can transfer to different subsets depending on the sequence context surrounding a residue. Each of these 500 protein kinases could in principle be monospecific for only one protein substrate, but in practice tend to show a range of promiscuity from one or a few proteins as substrates to hundreds of proteins phosphorylated.

Protein kinases tend to be **chemospecific**, serine/threonine kinases or tyrosine kinases, but multiple serines and/or threonines in a protein substrate may be phosphorylated by a given protein serine kinase (Shabb, 2001). Analogously, up to three tyrosines may be phosphorylated on the activation loop of the insulin receptor. Many protein kinases undergo autophosphorylation on such loops as preparatory activation to populate conformers active for external protein phosphorylation. There may be sequential action of protein tyrosine kinases and protein serine kinases working on the same protein substrate, to introduce both $Ser–OPO_3^{2-}$ and $Tyr–OPO_3^{2-}$ modifications, at different side chains. The state of phosphorylation of proteins in cells is dynamic as we shall note, so the fraction of side chains of sets of protein substrates that are phosphorylated may change from moment to moment and create molecular heterogeneity both in time and in particular subregions of specific cells.

Thus, if 500 protein kinases acted on average on five proteins each and introduced on average three phosphoryl groups at particular Ser, Thr, or Tyr side chains, that alone would yield 7500 posttranslationally modified protein variants for proteome diversification. This is doubtless a dramatic underestimate of the actual scope of protein posttranslational phosphorylation. For example, it appears that the Abl tyrosine kinase can be phosphorylated at 11 distinct residues (one Ser, one Thr, and nine Tyr residues) (Hantschel and Superti-Furga, 2004). Many of the dynamic phosphorylations and dephosphorylations of proteins are involved in signal transduction pathways for selective gene expression.

More generally at any of n sites in a protein that can be posttranslationally modified, there can be 2^n variants, with each site in a modified or unmodified state. Thus, for the Abl kinase, there are 2^{11} molecular species just considering phosphorylation as one possible modification. In Chapters 5 and 6 we will note 28 sites in histone tails that can be methylated, acetylated, or phosphorylated for a possible 2^{28} variants.

A second and third axis of scope of posttranslational modifications is by the *type of amino acid side chain modified* as well as the *type of covalent chemical modification* introduced by the posttranslational modification enzymes. The generic chemical reaction in posttranslational modifications is enzyme-catalyzed transfer of an electrophilic fragment of a cosubstrate molecule onto a nucleophilic side chain of the protein undergoing modification. The attacking nucleophilic side chain of the protein to be derivatized undergoes the chemical modification by the transferring electrophile.

Therefore, side chains in proteins that can potentially act as nucleophiles are the common sites for posttranslational modification (Figure 1.7A). These are the thiolate anion of Cys; the OH groups of Ser and Thr residues and the phenolic OH of Tyr; the amine forms of Lys, His, and Arg; and the side chain carboxylates of Asp and Glu residues. The NH_2 at the N-terminus and the COO^- at the C-terminus of proteins are also potential nucleophiles that undergo posttranslational covalent modifications. The sulfur of methionine can be oxidized to the sulfoxide. This is the inventory of the most reactive side chains in the protein substrates.

An exception to this rule that nucleophilic side chains of protein substrates initiate attack on the electrophilic fragment to be transferred in the posttranslational modification occurs in protein hydroxylation (Figure 1.7B), noted in Chapter 11. In protein modifications mediated by a family of Fe^{II}-dependent hydroxylases using O_2 as cosubstrate, the activated oxygen species generated in the hydroxylase active site is capable of inserting an oxygen atom into unreactive carbon sites, such as the methylene side chains of amino acid residues. Thus, proline residues are hydroxylated at the C_4-CH_2 site and an asparagine residue at the β-CH_2 group in the modification of the HIF-1α subunit of the transcription factor controlling response to hypoxia. Analogously the C_5-CH_2 position of lysine side chains can be hydroxylated, as occurs in procollagen maturation.

A second exception to the rule that the nucleophilic side chains of amino acid residues in proteins are the sites of posttranslational modification is found in the common usage of asparagine and glutamine carboxamido groups. Although these amide nitrogens are at best weakly nucleophilic, certain sets of posttranslational modification enzyme catalysts can selectively activate them. All N-linked glycoproteins, for example, use Asn carboxamido nitrogens as nucleophiles and transglutaminases use Gln side chains for covalent connection. Analogously, in fluorophore formation in green fluorescent proteins, the isoamide form of a peptide backbone nitrogen is envisioned as the initiating nucleophile.

Table 1.1 categorizes protein posttranslational modifications according to amino acid side chain modified. Ten residues (Asp, Glu, Ser, Thr, Tyr, His, Lys, Cys, Met, and Arg) have N, O, or S atoms that can function, at appropriate ionization states, as nucleophiles in modification reactions. Two others, Asn and Gln, react via their side chain amide groups, intrinsically weak nucleophiles. The carboxamido nitrogen of Asn is the site of N-glycosylation in all eukaryotic N-glycoproteins, reflecting enzymatic enhancement of its low nucleophilicity. On the other hand, the carboxamido nitrogen of Gln side chains tends to act as an electrophile, undergoing net capture by water or by the ϵ-NH_2 group of a Lys residue in transamidations.

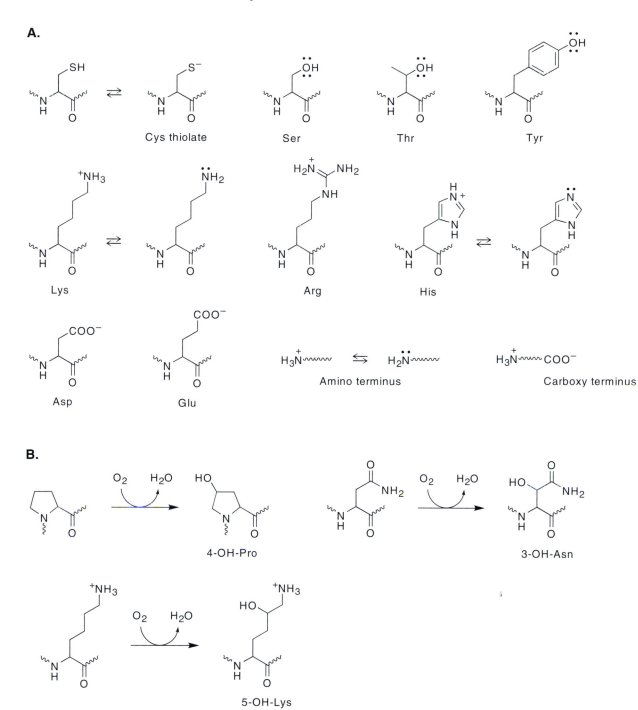

Figure 1.7 A. Electron-rich, nucleophilic side chains in proteins that undergo covalent posttranslational modification reactions. **B.** Protein side chains observed to undergo posttranslational hydroxylation.

Table 1.1 Posttranslational Modifications by Protein Residue Modified

Residue*	Reaction	Example
Asp	Phosphorylation	Protein tyrosine phosphatases; response regulators in two component systems
	Isomerization to isoAsp	
Glu	Methylation	Chemotaxis receptor proteins
	Carboxylation	Gla residues in blood coagulation
	Polyglycination	Tubulin
	Polyglutamylation	Tubulin
Ser	Phosphorylation	Protein serine kinases and phosphatases
	O-Glycosylation	Notch O-glycosylation
	Phosphopantetheinylation	Fatty acid synthase
	Autocleavages	Pyruvamidyl enzyme formation
Thr	Phosphorylation	Protein threonine kinases/phosphatases
Tyr	Phosphorylation	Tyrosine kinases/phosphatases
	Sulfation	CCR5 receptor maturation
	Ortho-nitration	Inflammatory responses
	TOPA quinone	Amine oxidase maturation
His	Phosphorylation	Sensor protein kinases in two-component regulatory systems
	Aminocarboxypropylation	Diphthamide formation
	N-Methylation	Methyl CoM reductase
Lys	N-Methylation	Histone methylation
	N-Acylation by acetyl, biotinyl, lipoyl, and ubiquityl groups	Histone acetylation; swinging arm prostetic groups; ubiquitin; SUMO (small ubiquitin-like modifier) tagging of proteins
	C-Hydroxylation	Collagen maturation
Cys	S-Hydroxylation (S-OH)	Sulfenate intermediates
	Disulfide bond formation	Protein in oxidizing environments
	Phosphorylation	PTPases
	S-Acylation	Ras
	S-Prenylation	Ras
	Protein splicing	Intein excisions
Met	Oxidation to sulfoxide	Met sulfoxide reductase
Arg	N-Methylation	Histones
	N-ADP-ribosylation	$G_{S\alpha}$
Asn	N-Glycosylation	N-Glycoproteins
	N-ADP-ribosylation	eEF2
	Protein splicing	Intein excision step
Gln	Transglutamination	Protein cross-linking
Trp	C-Mannosylation	Plasma membrane proteins
Pro	C-Hydroxylation	Collagen; HIF-1α
Gly	C-Hydroxylation	C-Terminal amide formation

* There are no known modifications on Leu, Ile, Val, Ala, and Phe side chains.

The remaining three side chains of Table 1.1, Trp, Pro, and Gly, are unlikely candidates for posttranslational modification. The eneamine reactivity of the five-membered ring in the indole side chain of tryptophan is utilized for C-mannosylation (see Chapter 10). The methylene carbons of Pro and Gly are sites for specific hydroxylation reactions, unusual modifications not requiring nucleophilic reactivity at the reacting amino acid residues. When Gly is at position 2 of proteins and the N-terminal residue is removed by aminopeptidase activity, the amino group of the newly uncovered N-terminal Gly is subject to N-acylation by the C_{14} myristoyl chain from myristoyl CoA. The remaining five amino acids (Leu, Ile, Val, Ala, and Phe) do not have obvious nucleophilic moieties and are not involved in well-defined posttranslational modifications in proteins (although they are hydroxylated in certain peptide antibiotics) (Chen et al., 2001).

Electrophilic Cosubstrates Used for Protein Posttranslational Modifications

The inventory of electrophilic fragments for group transfer reactions in protein post-translational modification is the same as that used for such group transfers in primary metabolism. The most common cosubstrates (Figure 1.8) include the small molecule substrates ATP and GTP for γ-PO_3^{2-} transfers, coenzyme A for phosphopantetheinyl transfers, the coenzyme NAD for ADP-ribosyl transfers, S-adenosylmethionine (SAM) for methyl transfers, and PAPS (phosphoadenosine phosphosulfate) for sulfuryl group transfers. Also, the biological forms of activated sugars, the nucleoside diphospho sugars (NDP sugars) are the glycosyl transfer substrates for protein glycosylations. Acyl CoA thioesters are the sources of C_2 (acetyl) acyl groups, notably in histone acetylations, while C_{14} (myristoyl) and C_{16} (palmitoyl) acyl CoA groups are the donors in acylations of proteins as part of lipidation modifications. The other sources of lipidating substrates are the biological isoprenoid donors, the C_{15} farnesyl diphosphate and the C_{20} geranylgeranyl diphosphate. The same kinds of reaction mechanisms for these electrophilic fragment transfers are utilized by posttranslational modification enzymes as for enzymes of primary metabolism. The difference is in the nature of the cosubstrate providing the attacking nucleophile: proteins for posttranslational modification enzymes versus small molecule cosubstrate metabolites for primary metabolism.

Instead of small coenzymes, some posttranslational modification enzymes use small proteins as donors of the electrophilic fragments, most notably in the ubiquitylation of proteins (Chapter 9). Ubiquitin is a 76-residue (8 kDa) protein that gets activated as the AMP derivative at its C-terminus, Gly_{76}, and then transferred via covalent thioester linkages, first to the ubiquityltransferases themselves on active site

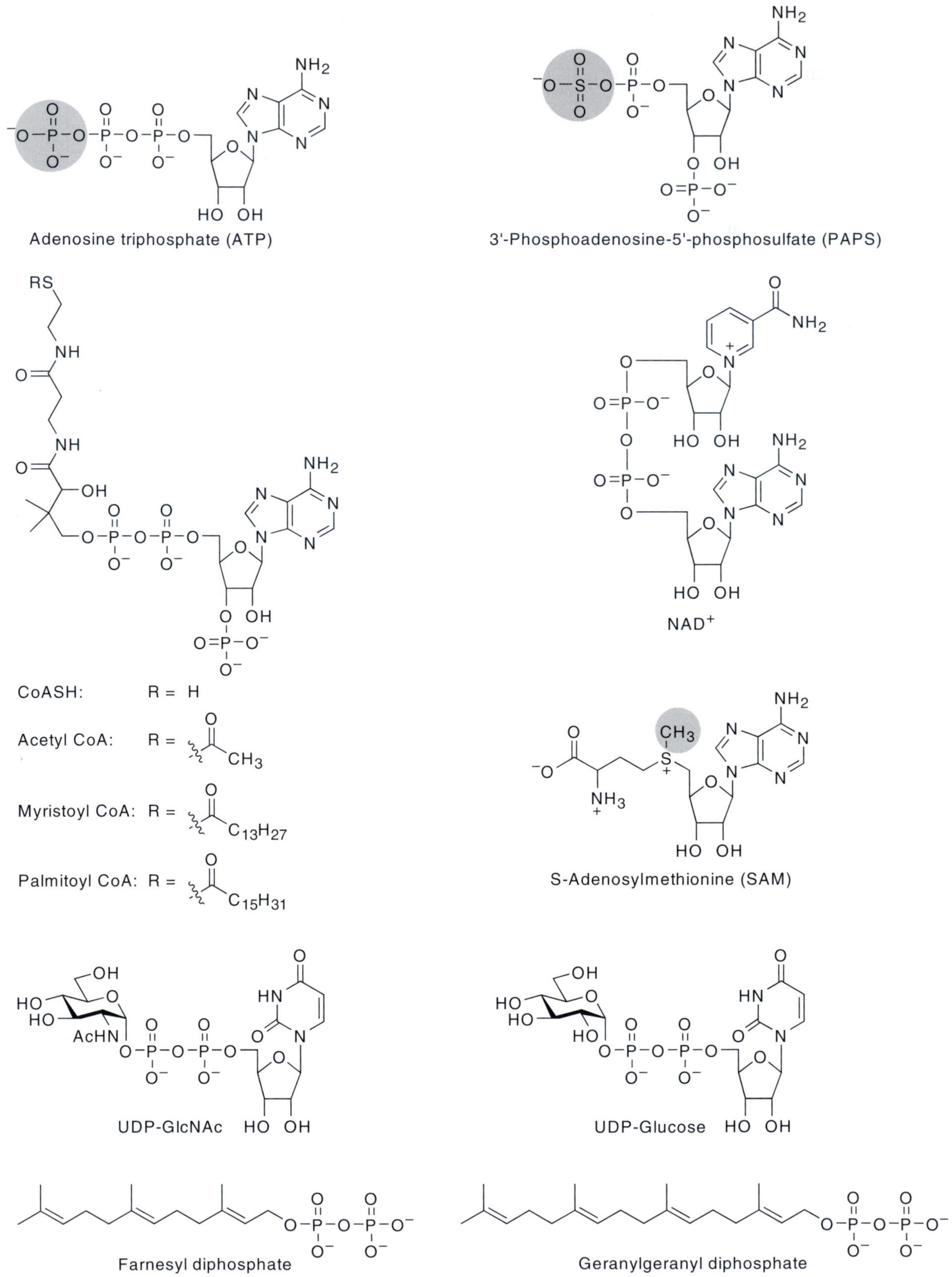

Adenosine triphosphate (ATP)

3'-Phosphoadenosine-5'-phosphosulfate (PAPS)

CoASH: R = H

Acetyl CoA: R = $\overset{O}{\underset{}{C}}$—CH$_3$

Myristoyl CoA: R = $\overset{O}{\underset{}{C}}$—C$_{13}H_{27}$

Palmitoyl CoA: R = $\overset{O}{\underset{}{C}}$—C$_{15}H_{31}$

NAD$^+$

S-Adenosylmethionine (SAM)

UDP-GlcNAc   HO OH

UDP-Glucose   HO OH

Farnesyl diphosphate

Geranylgeranyl diphosphate

Figure 1.8 Coenzymes that function as cosubstrate donors of electrophilic fragments to protein side chains catalyzed by posttranslational modification enzymes.

cysteines. Then the activated 8-kDa ubiquityl moiety, via the C-terminal Gly_{76}, is transferred to the ϵ-NH_2 groups of lysine residues in particular target proteins to generate isopeptide bonds in the covalently marked protein products. We will examine both mono- and polyubiquitylation posttranslational modifications, as well as cognate amidations of proteins by ubiquitin analogs such as the protein SUMO (small ubiquitin-like modifier), in Chapter 9.

A few posttranslational modifications remove rather than add groups to protein side chains. This occurs, for example, in the action of bacterial protein toxins that deamidate a glutamine residue in the active site of Rho proteins, inactivating the GTPase catalytic activity of these enzymes (Chapter 14). The hydrolytic deamidation of the Gln side chains is chemically analogous to the hydrolytic cleavage of main chain amide linkages by protease action.

Functional Consequences of Protein Posttranslational Modifications

Given that the multitude of enzyme-mediated posttranslational modifications increase proteome complexity to greater than 100,000 distinct proteins in a typical human cell, there are a few common functional attributes gained that allow categorization of the underlying biological logic.

Protein phosphorylations may be the most common posttranslational modification, with tens of thousands of predicted phosphorylation sites in the human proteome, and perhaps the most widely studied over the past two decades. At each protein side chain phosphorylated a polar neutral OH side chain is converted to a bulky, tetrahedral phosphate that is a mixture of monoanion ($-OPO_3^-$) and dianion ($-OPO_3^{2-}$), given the pK_{a2} values of phosphates at pH 6–7. As is detailed in Chapter 2, the introduction of one to two net negative charges has a notable effect on redistributing conformers in that microenvironment of the protein. These include conversion of unstructured regions of loops into helical regions that can drive and propagate conformational changes to other regions of the modified protein (Johnson and Lewis, 2001). These conformational changes can be intramolecular or intermolecular across subunit interfaces and create docking sites for partner proteins with motifs that can specifically recognize the tetrahedral phosphate dianionic side chains. Charge pairing of the dianionic phosphate (Figure 1.9) with a bidentate cationic arginine side chain is a common way to neutralize the charge of the introduced phosphoryl group and account for the stabilization of observed conformational changes. Protein phosphorylation events are central to protein-based signaling pathways for selective gene amplifications in both prokaryotic and eukaryotic cells.

Posttranslational modification may also expand the catalytic capacity of the modified proteins. This is exemplified in the autophosphorylation of protein kinases,

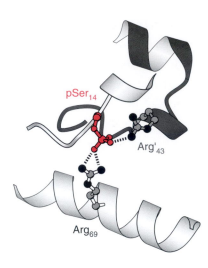

Figure 1.9 Charge pairing of dianionic phosphate side chains in proteins (shown for glycogen phosphorylase), generated by protein kinase action, with one or more cationic arginine side chains in proteins to nucleate organization or reorganization of the protein microenvironment.

such as Thr–OH to Thr–OPO$_3^{2-}$, which creates a loop movement that allows the optimization of active site geometry and is part of the switch that converts such protein kinases from the "off" state at rest to the "on" state when stimulated by some signal. Increases of 10^5 in catalytic efficiency have been observed in the double phosphorylation of the loop regions of MAP kinases. Equally or even more dramatic is the conversion of inactive apo forms of proteins to the active holo forms of proteins (Figure 1.10) by covalent installation of phosphopantetheinyl groups on serine side chains, biotinyl groups, and lipoyl groups on lysine side chains, in fatty acid synthases, carboxylases, and decarboxylases, respectively (Chapter 13). These three vitamin-derived prosthetic groups are the key functional groups in those primary metabolic enzymes and clearly enable an expanded catalytic repertoire.

Change of subcellular address is another pervasive consequence of posttranslational modification of proteins. Four examples explored in subsequent chapters are ubiquitylation, acylation, farnesylation/geranylgeranylation, and glycosyl phosphatidylinositol (GPI) anchoring (Figure 1.11). Monoubiquitylation presents an information-rich tag to send proteins from the cell membranes to recycling compartments in the trans Golgi network and on to lysosomes, while addition of four or more ubiquitins in a polyubiquitin chain marks a protein for recognition by the proteasomal machinery and ensures its proteolytic degradation. Acylation of proteins at the N-terminal glycine residue by myristoylation or at cysteine side chains by palmitoylation near the C-termini help partition the proteins from the aqueous phase to membrane phases of the cell. Analogous logic is employed for cysteinyl-S-lipidation by the C_{15} and C_{20} isoprenoid chains to enhance the membrane associative properties of proteins that have undergone farnesy-

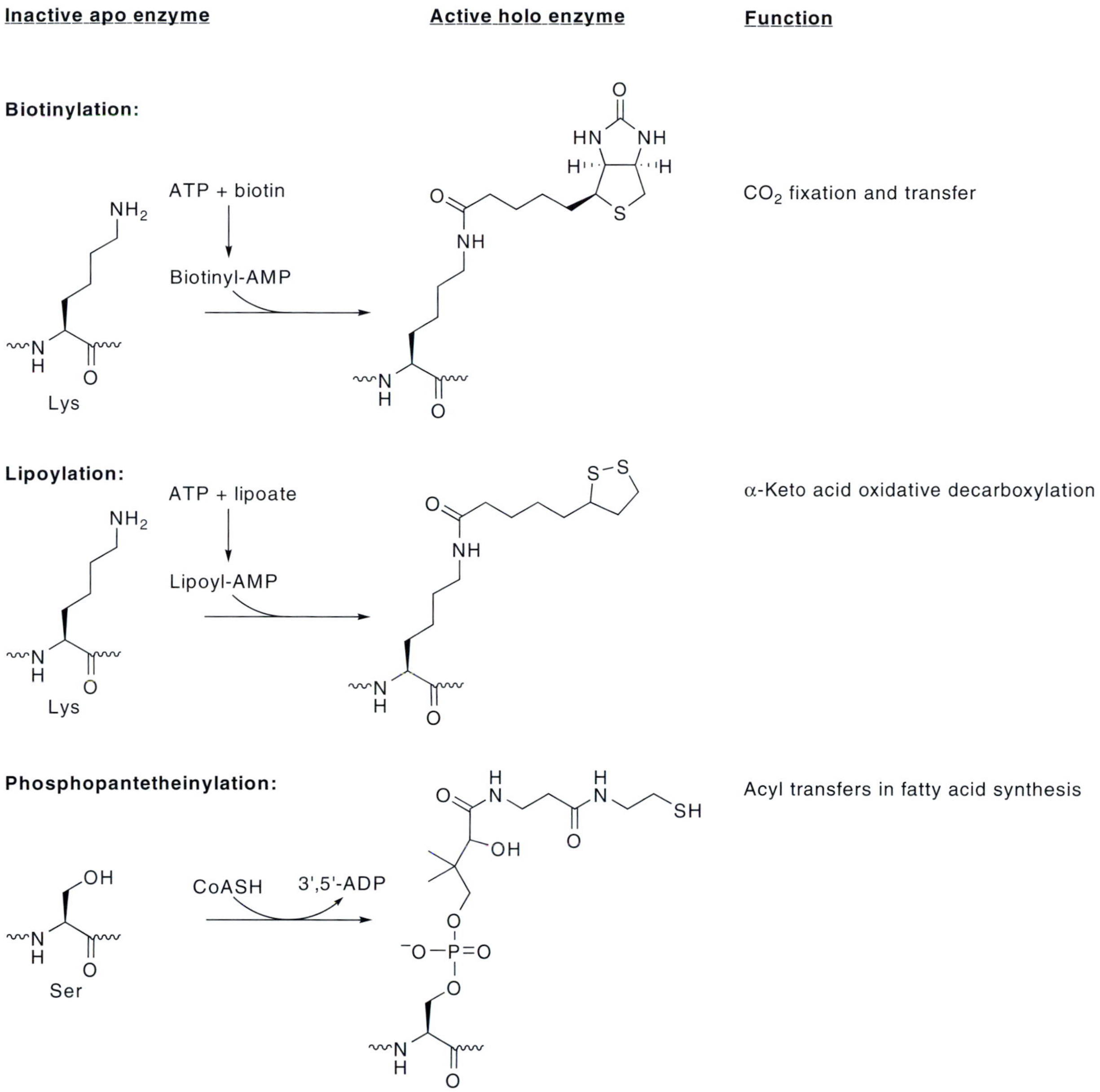

Inactive apo enzyme   **Active holo enzyme**   **Function**

Biotinylation:

CO$_2$ fixation and transfer

Lipoylation:

α-Keto acid oxidative decarboxylation

Phosphopantetheinylation:

Acyl transfers in fatty acid synthesis

Figure 1.10 Conversion of inactive apo forms of proteins into active holo forms by posttranslational biotinylation (of lysine), lipoylation (of lysine), or phosphopantetheinylation (of serine) in enzymes that carry out carboxylation, oxidative decarboxylation, and C–C bond formation, respectively.

lation and/or geranylgeranylation. A third strategy to relocate proteins is to tether them to the external face of plasma membranes of eukaryotic cells by an assembly of complex glyco phosphatidylinositol (GPI) anchors. Cleavage of the anchor at some later time can lead to sloughing of the protein into the extracellular space.

Monoubiquitylation:

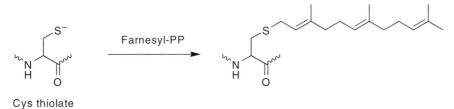

Relocation

From plasma membrane to
endocytic pathway

Ubiquityl-S-E$_2$

ϵ-Lys

Myristoylation:

Cytoplasm to membranes

Myristoyl CoA

N-Terminal Gly

$CH_3(CH_2)_{12}$

Farnesylation:

Cytoplasm to membranes

Farnesyl-PP

Cys thiolate

GPI anchoring:

Membrane protein tethering
through C-terminal GPI anchor

COO$^-$

GPI anchor

H_2NR

COO$^-$

O—P—O—Man$_3$GlcNAc

Inositol

P-lipid

Figure 1.11 Changes in the subcellular addresses of proteins enabled by covalent side chain post-translational modification: ubiquitylation of lysine, acetylation of lysine, myristoylation of the N-terminal residue, farnesylation of cysteine, and GPI anchoring via transamidation.

The posttranslational modifying enzymes that work on proteins to enable their change of subcellular localization are often found in the membrane compartments or at the membrane interfaces. Both the long-chain acyl CoA cosubstrates, myristoyl CoA and palmitoyl CoA, and the C_{15} and C_{20} isoprenyl-PP cosubstrates for the lipidations are also partitioned into the membrane interfaces and so are available to direct the compartmentalizing catalytic modifications. Protein glycosylation occurs largely in the endoplasmic reticulum (ER) and the Golgi complex. The glycosyl-transferases are ER and Golgi membrane-associated. The initial sugar substrates are nucleoside diphospho-sugars, such as UDP-glucose and GDP-mannose, but as oligosaccharide chains grow they are transferred to a C_{55} isoprenoid-PP carrier, dolichol phosphate. The undecaprenyl lipid moiety is in the ER membrane and the oligosaccharyl chain at the interface of the inner surface of the ER membrane and the ER lumenal aqueous phase.

Reversible vs Irreversible Posttranslational Modification of Proteins

Some posttranslational modifications of proteins are irreversible (Figure 1.12A). This can be due to the nature of the biological function enabled by the modification. The most pervasive irreversible posttranslational modifications are the proteolytic cleavages undergone by all proteins during their life cycles. Because the peptide bond cleavage to the acid and amine components in aqueous solution is favored thermodynamically with a K_{eq} of about 10^5, protease cleavages are irreversible biological switches. Thus, removal of the N-terminal signal sequences on all proteins passing into the endoplasmic reticulum as the first stage in eukaryotic cell secretory pathways is irreversible. So is the proprotein convertase proteolytic activity in the Golgi and trans Golgi vesicular network (e.g., for cleaving proinsulin to insulin).

The conversion of inactive apo forms of enzymes and carrier proteins to active holo forms of phosphopantetheinylated fatty acid synthases, biotinylated carboxy-lases, and lipoylated pyruvate dehydrogenases represent equivalent logic of irreversible covalent modification. The apo proteins cannot be catalytically active and the permanent addition of the prosthetic groups for the lifetime of the protein is the rationale for the posttranslational modification. Irreversibility can also be due to a lack of ready mechanism for reversal of the covalent modification. This appears to be the case for protein N-methylations of ϵ-amino groups of lysine side chains modified by SAM-dependent protein methylases. The N-methyl linkage is resistant to hydrolytic breakdown, the usual route for posttranslational removal as noted below. Recently it has been demonstrated that the CH_3 could be oxidatively removed (Chapter 5).

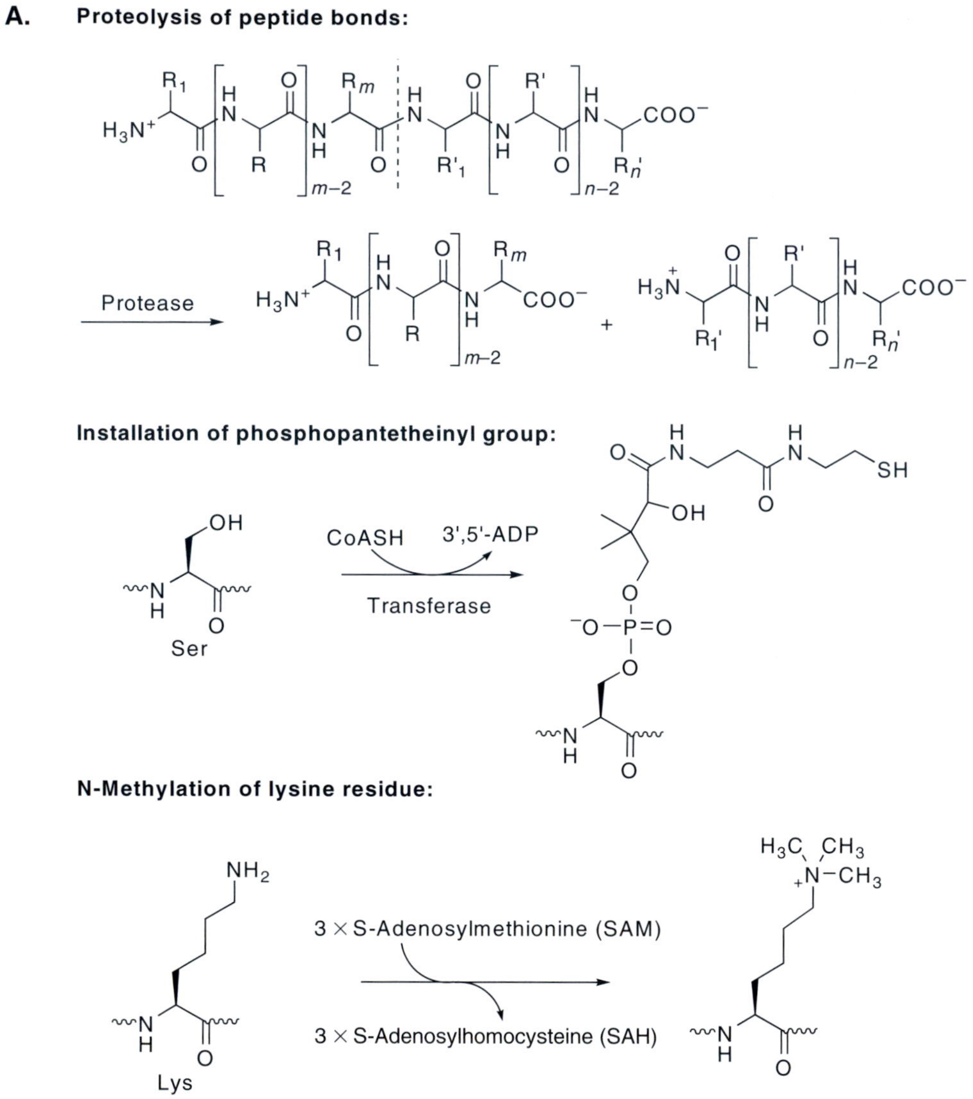

A.   Proteolysis of peptide bonds:

Installation of phosphopantetheinyl group:

N-Methylation of lysine residue:

Figure 1.12 Irreversible versus reversible posttranslational modifications. **A.** Irreversible changes include the proteolysis of peptide bonds, installation of phosphopantetheinyl groups in fatty acid synthase, and N-methylation of lysine residues in histones.

Some posttranslational modifications are freely reversible *in vivo* although stable *in vitro*, consistent with the observations that there is dedicated enzymatic machinery for covalent group removal as there is for covalent group addition. The subsets of posttranslational modifications generally thought to be dynamically reversible are protein phosphorylations and acylations (Figure 1.12B). The biological rationale for reversibility is often that such protein modifications are involved in signaling cascades. If phosphorylated and acylated forms of certain proteins are carrying signal

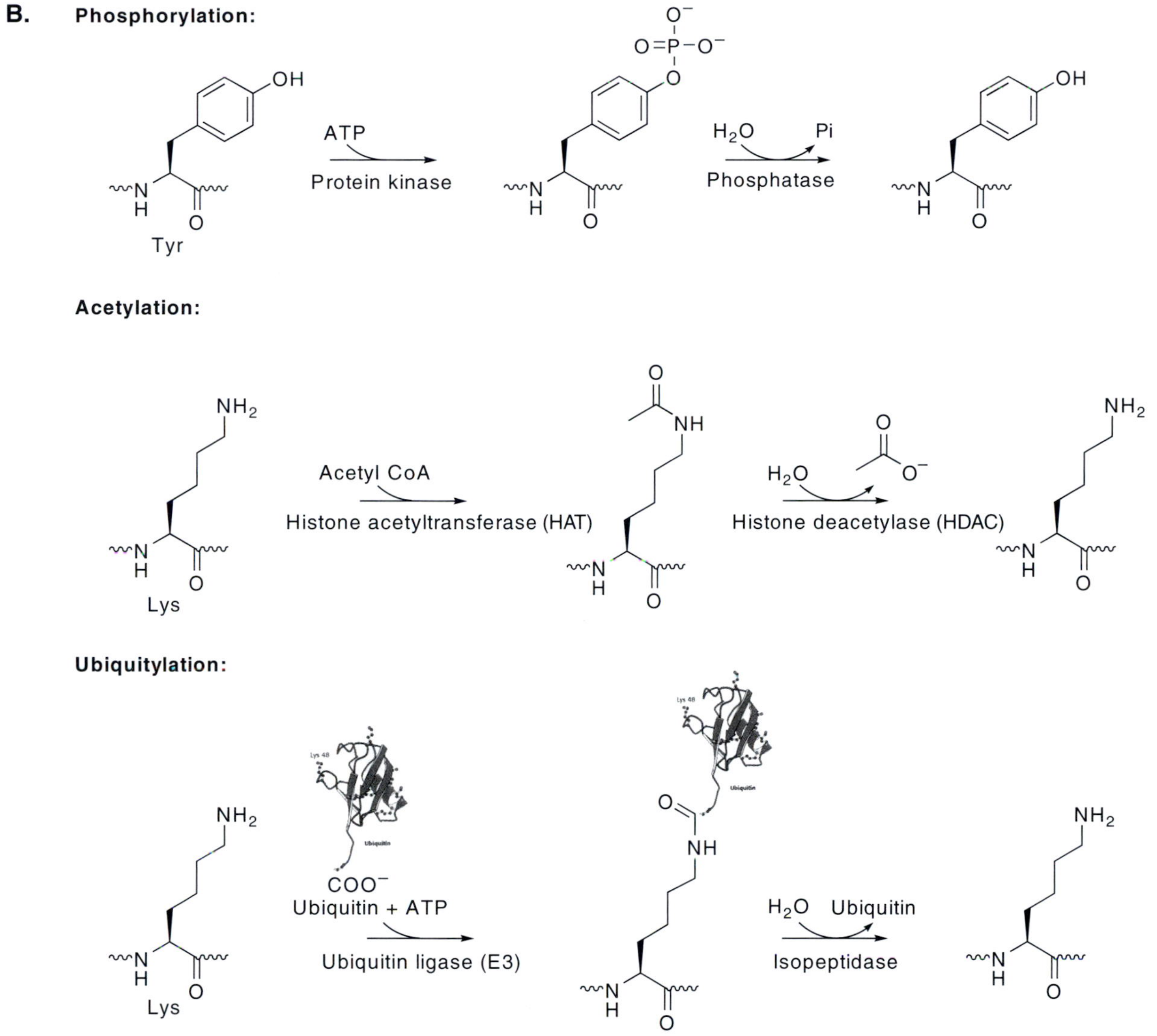

B.   Phosphorylation:

Acetylation:

Ubiquitylation:

Figure 1.12 **B.** Reversible covalent modifications include protein phosphorylations (due to the balance of kinases and phosphatases), histone acetylations (due to the balance of histone acetyltransferases and histone deacetylases), and protein ubiquitylations (due to the balance of ligases and deubiquitylating hydrolases).

information, then for signaling to be of finite duration one route to the termination of the signal is to undo the posttranslationally introduced covalent modification. Thus, protein kinase action is balanced by phosphoprotein phosphatase action and histone acetyltransferases are countered by histone deacetylases. The activity and micro distribution of the kinases/phosphatases and the acetylases/deacetylases (Chapter 6) controls the lifetime and mole fraction of protein isoforms phosphorylated or acetylated, respectively, and thereby is one index of the strength and duration of signaling. Any discussion of the logic and organization of protein kinases and protein acetylases needs to include the countervailing phosphatases and deacetylases to provide biologic coherence to the strategies. While there are 500 protein kinases in the human proteome (the human kinome), as estimated from bioinformatics predictions, there are about 150 protein phosphatases, suggesting that there does not need to be an exact one-to-one pairing of opposing posttranslational catalysts. Additionally, there may be more promiscuity of protein phosphatases and/or the likelihood that some phosphoproteins are not readily dephosphorylated. The acylation of proteins with ubiquitin markers by large families of ubiquitin ligases to create isopeptide linkage to lysine side chains is also reversible, balanced by a multitude of deubiquitylating hydrolases, as noted in Chapter 9.

Methods for Detection of Protein Posttranslational Modifications

The classical route to discovery of posttranslational modifications came from studies on individual proteins by investigators dedicated to the elucidation of the function and/or structure of a particular protein. In the early days of the characterization of amino acid constituents of proteins, it was noted that certain proteins had covalently associated phosphate groups, ultimately leading to the identification of the phosphate moieties as phosphoserine and phosphothreonine residues. In the study of the coenzyme forms of the vitamins lipoate and biotin, work on carboxylase and dehydrogenase enzyme complexes of primary metabolism revealed covalent attachment of biotin and lipoate through amide linkages to specific lysine side chains and these attachments were essential for enzymatic activity. In amino acid metabolism the coenzyme pyridoxal phosphate (PLP) enables many enzymatic transformations of amino acid groups in substrates. In the deamination of histidine to the α,β-unsaturated enoate urocanate, PLP was not present or required. For some years it was thought that the enzyme histidine ammonia lyase contained a covalent pyruvoyl moiety as the electron sink required for deamination catalysis. X-ray structural studies of the enzyme revealed that the active site instead contains a 4-methylidene-5-imidazole-5-one (MIO) that functions as an electrophilic coenzyme (Schwede et al., 1999). We will take up MIO in Chapter 12, dealing with the autocatalytic modifications of proteins, as MIO arises from $Ala_{142}Ser_{143}Gly_{144}$ in the precursor form of the enzyme. We noted above that x-ray analysis of a methanobacterial protein turned up pyrrolysine as the 22nd proteinogenic amino acid. It is likely that other posttranslational modifications will continue to be revealed in structural proteomics programs.

  Much of the contemporary interest in determining which proteins have modifications at specific sets of amino acid residues is in a proteomics context, driven by the genomics databases. For example, it is possible to predict phosphorylation sites in all of the proteins of the proteome based on knowledge of the consensus sites for families of protein kinases.

It is also possible to predict membrane association sites in proteins and within that group to predict subsets that will be subject to modification with glycosyl phosphatidylinositol (GPI) anchors. To establish which members of such GPI-anchored proteins are actually expressed at cell surfaces in given tissues and what fraction actually bear the GPI anchor requires techniques that can deal not just with one purified protein at a time for posttranslational characterization but rather that identify the GPI-anchored proteome. Analogously, one may wish to identify all proteins bearing one ubiquitin tag or multiple ubiquitin tags in a given cell type under particular conditions, over a specified time course. With His_6-tagged ubiquitin gene transfection into such cells, one can purify all the ubiquitylated proteins by His_6 affinity columns, setting up an enriched fraction for which ubiquitin content can be evaluated and located by mass spectrometry (MS) (Peng et al., 2003).

A variety of bioinformatics efforts are collected in the June 2004 issue of *Proteomics* (edited by R. Appel and A. Bairoch) and include papers on the Mod-Prot database (Farriol-Mathis et al., 2004) that predicts common types of posttranslational modifications of proteins and also the Unimod database (Creasy and Cottrell, 2004) for use in mass spectrometric applications.

Enormous attention in recent years has been devoted to methods to characterize the phosphoproteome (Ficarro et al., 2002), given estimates that one-third of eukaryotic proteins may be posttranslationally modified with a phosphate group at some point in their life cycle. Historically, phosphoproteins could be detected by the incorporation of radioactivity from $[^{32}P]PO_4^{3-}$ into γ-$[^{32}P]ATP$ and subsequently into protein phospho forms that could be distinguished from nonphosphorylated forms on two-dimensional gel electrophoresis due to the charge distinctions imparted by the phosphate group(s). Much attention in recent times has been focused on mass spectrometric approaches to give improved resolution and sensitivity (Liebler, 2002).

Phosphoproteins can be prepurified from nonphosphoproteins by binding to metal affinity columns, fragmented to phosphopeptides by protease digestion, and subjected to mass spectrometry (MALDI or nanoelectrospray), to identify the specific sequences containing the phosphate groups. The mass increase of the $-PO_3$ group is +80 Da, easily detectable over the predicted sequence of peptides. Phosphotyrosyl (pY) linkages are stable in the mass spectrometer and are readily detected. Phosphoseryl (pS) and phosphothreonyl (pT) linkages are more labile to β-elimination and have presented more problems for mass detection but several protocols have been reported for measuring the elimination product peptides or ones where alkyl thiols have been added to derivatize the dehydropeptides.

Table 1.2 Mass Changes on Some Common Posttranslational Modifications of Protein Residues

Type of Modification	Mass Changes	Example
Acetylation	42	Histone Lys N-acetylation
Amidation	−1	C-Terminal amide
Biotinylation	226	Pyruvate carboxylase
Carbamoylation	43	β-Lactamase (D type)
γ-Carboxylation	44	Glu residues in prothrombin
C-Mannosylation	162	Erythropoietin receptor
Cys sulfenic acid	16	NADH peroxidase
Deamidation	1	Rho (Gln$_{63}$)
N-Terminal acyl diacylglyceride (tripalmitate)	789	Bacterial lipoproteins
Dimethylation	28	Histone Arg dimethylation
Farnesylation	204	Ras GTPase
Formylglycine	−17	Maturation of sulfatases
Geranylgeranylation	272	Rab GTPases
O-GlcNAc	203	Sp1 transcription factor
O-Glucosylation	162	Notch
N-Glycoproteins	>2000	Variable: Initial tetradecasaccharyl Glc$_3$Man$_9$GlcNAc$_2$ chain trimmed to many variants
GPI lipid anchors	>1000	Variable, depending on lipid and ethanolamine chains
Hydroxylation	16	HIF-1α
Lipoylation	188	Lipoyl transacetylase component of pyruvate dehydrogenase
Methylation	14	Histone Lys monomethylation
Myristoylation	210	Src kinase
Nitrosylation	29	Tubulin
Palmitoylation	238	Ras GTPase
Phosphorylation	80	Protein kinase A
Phosphopantetheinylation	339	Fatty acid synthase
Sulfation	80	CCR5 receptor
Transamidation	−17	Transglutaminase products
Trimethylation	42	Histone Lys N-trimethylation

Mass spectrometry is the preferred method for detection of the other common groups introduced by posttranslational modification enzymes (Table 1.2), including acetyl groups (+42 Da), methyl groups (+14 Da), and long-chain acyl groups such as myristoyl (+210 Da) and farnesyl (+204 Da). Taking advantage of unit mass resolution, it has been possible to detect Gln to Glu deamidation in an active site peptide from the Rho protein by mass spectrometry with a change of only +1 Da (Buetow et al., 2001). Likewise, one can detect formation of a disulfide link from two cysteine SH residues (−2 Da). The GPI anchor addition (> +1000 Da), N-glycosylations with branched oligosaccharides (a heptameric oligosaccharide contributes more than 1000 Da in increased mass), and ubiquitylations (> +8000 Da per ubiquitin) are readily detected by mass change. We noted above that phosphoseryl and phosphothreonyl groups can be labile to β-elimination by mass spectrometry, so instead of a +80 Da change, one may actually see a –18 Da change from the dehydroalanine and dehydrobutyrine residues left after loss of inorganic phosphate from the peptide. Sulfated tyrosyl residues (+80 Da) are also labile, as are glucosylations (+162 Da), and may be difficult to localize by tandem mass spectrometry but should be amenable to the tandem MS method called electron capture dissociation. In some cases it is possible to get sequence information on the complex glycan chains in N-linkage in glycoproteins by mass spectrometry (Chapter 10).

The overarching analytical problems in determining the range and scope of posttranslational modifications in proteomes are created by the dynamic and substoichiometric nature of posttranslational modifications. In any cell at any time point only a subset of proteins may be expressed. As a simple example, in yeast one may see 2000 of the 6000 encoded proteins expressed in a given growth condition. Of the proteins being expressed, suppose 2000 were undergoing sets of posttranslational modifications: controlled proteolytic processings, phosphorylations, ubiquitylation and sumoylation, N- and O-glycosylation, acetylations, myristoylations, methylations, lipoylations, phosphopantetheinylations, biotinylations, etc. There will be heterogeneity both in space and time for the fraction of a given protein that is modified. For example, if there are 5000 copies per cell of a transmembrane receptor or a protein kinase, only a fraction of each protein population may be modified (e.g., the receptor monoubiquitylated for internalization, or a MAP kinase phosphorylated). That MAP kinase will require double phosphorylation, generating a pTXpY sequence in the activation loop for gain of full activity. The monophosphorylated pT or pY forms may have less than 1% of the catalytic activity of the pTXpY form but may represent substantial fractions of the kinase population. The abundance range for proteins undergoing phosphorylations may differ by 100–1000 fold. While it may be easy to detect the high-abundance proteins, the low-abundance members of the phosphoproteome, perhaps of great functional importance given that they act catalytically, may be very difficult to record. This analytical challenge for evaluating posttranslational modifications in proteomes will be an active area of protein research for some time to come.

Organization and Framework of This Book

This book places posttranslational modifications into a small number of categories, by reaction catalyzed, to examine the chemical logic of the transformation. The approach focuses on the nucleophilic protein side chain that reacts and the electrophilic donor cosubstrate being appropriated and redirected from small molecule primary metabolism. Chapters are organized according to the type of covalent posttranslational modification introduced. A number of cases where poorly nucleophilic amides, from Asn side chains in N-glycosylation or backbone amides in autocatalytic MIO or GFP cofactor formation, are activated are also analyzed in detail for chemical logic.

The properties of the posttranslational modification enzymes are also evaluated and codified but their mechanisms of action are not elaborated in detail. That is the subject of other volumes (Kyte, 1995; Silverman, 2000; Walsh, 1979). The functional consequences of the covalent modifications to the proteins are summarized in order to elucidate some unifying rules and molecular principles. Examples of tandem actions of multiple types of posttranslational modifications are presented, such as the farnesylation, the limited proteolysis and C-terminal methylation that localizes Ras proteins to the cell membrane, or the complex acetylations, phosphorylations, and ubiquitylations that control the functions and lifetime of the mammalian transcription factor p53.

Analysis of the types of covalent modifications that proteins undergo posttranslationally provides the foundation for asking questions of dynamics, stoichiometry, and signal integration. Deciphering the functional attributes of such modified proteins will enable better resolution of how the chemical diversity introduced on protein side chains by posttranslational modification expands the repertoire of proteomes for the enhanced scope of reactivity and for the control of information flow.

The coverage is not meant to be encyclopedic, either for every type of protein covalent modification known, or for all facets of a given class of posttranslational modification, or for all known biological consequences, especially in signaling pathways. This selectivity is necessary in a single volume that covers protein kinases, proteases, ubiquitylations of proteins, glycosylations, histone modifications, and decorations of proteins with various types of lipid anchors. The literature citations are biased purposely towards recent review articles to provide the reader entry to the literature and/or to recent papers, rather than historical coverage that is available in earlier works.

Finally, although many enzymes generate covalent adducts between a side chain of an amino acid residue in the active site and a substrate fragment at some stage in catalysis, these *transient* covalent adducts are not taken up in this volume. Only covalent protein modifications that are not transient are covered. In that regard there are enzymes, at rest between catalytic cycles, that are posttranslationally modified, such as the phosphoserine form of phosphoglucomutase and the acetyl form of the bacterial citrate lyase. These phosphoryl and acetyl groups participate in catalysis in each catalytic cycle and are replaced, by phosphoryl and acetyl groups derived from substrates, at the end of each catalytic cycle (Regni et al., 2004; Walsh, 1979).

Given that the book is organized by the chemical frameworks of posttranslational modification reactions of proteins, the order of specific topic presentation is somewhat arbitrary. Because protein phosphorylation is so pervasive in eukaryotic metabolism, the subject of protein kinases and phosphatases are taken up first, in Chapter 2. Then the cognate sulfuryl group transfers are addressed in Chapter 3, where the similar chemical logic of introducing a charged group at a previously neutral site in a protein substrate is discussed. SO_3^- group transfers are much more restricted than PO_3^{2-} enzymatic transfers to proteins.

Chapter 4 deals with one of the many facets of cysteine thiolate reactivity—namely, redox transformations, including disulfide bond formation and reductive reversal. These reactions are important to understand, given the prevalence of the S–S bond as a covalent modification in so many proteins in nonreducing compartments.

Chapters 5 and 6 take up two posttranslational modifications prevalent in the writing and reading of the histone code: methylation and acetylation. Both the one-carbon alkylation (methyl) and the two-carbon acylation (acetyl) occur mostly on lysine side chains although arginines are also modified. The N-acetylations are readily reversed hydrolytically but the N-methylations are not, creating distinct constraints on rewriting the histone code.

Chapter 7 examines the several enzymatic strategies for posttranslational modification of proteins to send them to membrane microenvironments, including acylation at the N-terminus and at cysteine thiolate side chains. Comparable to the alkylation/acylation modification dichotomy noted for histone, there is parallel acylation versus alkylation logic for lipid anchors, with stability consequences. The alkyl groups for lipids are the C_{15} farnesyl and C_{20} geranylgeranyl isoprene-based anchors, added at or near the C-termini of many members of the small GTPase families. The glycosyl phosphatidylinositol (GPI) lipid anchors are hybrids of glycerol-based phospholipids and oligosaccharide modifications.

Chapter 8 examines the posttranslational versions of proteolysis that happen in the life cycle of all proteins in organisms. The chapter starts with the classification of the thousands of proteases into four mechanistic types: serine, cysteine, aspartyl, and zinc proteases. Then the major types of proteases found in different locales in prokaryotic cells are discussed as a prelude for those acting in euakaryotic cells, including the chambered proteases of bacteria and the proteasomes of eukaryotes. Quality control and inventory functions are delineated and then the set of proteolytic cleavages that occur in transit of proteins through eukaryotic secretory pathways are noted. This includes signal peptidases, proprotein convertases, plasma membrane shedding proteases, caspases in programmed cell death cascades, and the RIP proteases that cleave protein substrates at intramembrane sites.

Chapter 9 takes up ubiquitylation and ubiquitin-like protein tags, as information-rich posttranslational markings. The diversity of the E3 ligases that recognize specific partner proteins for the polyubiquitylation that sends them to proteasomes for unthreading and digestion to limit peptides are examined, including rules for substrate selectivity and the timing of ubiquitylation. The monoubiquitylation of membrane proteins and their internalization and sorting in the trans Golgi vesicular network are taken up. The growing list of ubiquitin-recognizing domains in proteins that recognize, direct, and sort both mono- and polyubiquitylated proteins are described. Of the ten or more ubiquitin-like small protein tags, the most information is known about SUMO and SUMOylations of dozens of partner proteins.

Chapter 10 takes up protein glycosylation, in an overview mode for both N- and O-glycosylation of protein side chains. It starts with the pre-assembly of the tetradecasaccharyl-PP-dolichol oligosaccharyl donor for the initiation of N-glycan chain transfer and then examines the common steps in N-glycan trimming, ER protein quality control, and the maturation and branching by glycosyltransferases in the Golgi complex. O-glycosylation of Notch protein, for example, is under intensive study and this topic is summarized.

Chapter 11 focuses on the use of NAD as a cosubstrate for posttranslational modification, donating the ADP-ribosyl moiety in a net glycosyl transfer, largely under the catalytic direction of a variety of bacterial toxins and exoenzymes. A variety of GTPase superfamily members in vertebrate hosts are targets for modification and functional disruption.

Chapter 12 deals with the posttranslational hydroxylation of protein side chains, starting with the classical studies on proline hydroxylation in collagen maturation. Much current interest is on two types of posttranslational hydroxylation of the oxygen sensing subunit HIF-1α of the hypoxia inducible transcription factor. The non-heme iron hydroxylases that act on one Asn and two Pro side chains in HIF-1α are discussed along with the observations that the proline hydroxylations enable the recognition of HIF-1α for ubiquitylation and proteolysis.

Chapter 13 takes up the covalent modifications of proteins that are autocatalyzed. These can be protein kinase autophosphorylations or poly ADP-ribose polymerase decorating itself with hundreds of ADP-ribosyl groups in DNA damage response. The central themes of this chapter, though, are three-fold. One is the oxidative conversion of tyrosyl and tryptophanyl active site residues to quinonoid side chains in the active catalytic forms of amine oxidases. The second is the conversion of tripeptide moieties in loop regions of specific proteins into cyclic structures by intramolecular attack of amide nitrogens on their peptide carbonyl neighbors. This occurs in MIO formation in histidine ammonia lyase and in the cyclization/autoxidation reactions that generate the chromophores in green fluorescent proteins (GFP) and the related DsRed chromoproteins.

The third theme is the autocatalytic cleavage of a specific peptide bond in a precursor form of a protein via intramolecular ester/thioester intermediates. These intermediates can decompose with protein cleavage, as happens to generate the N-terminal nucleophile of the β subunit of proteasomal proteases, or to generate the N-terminal pyruvoyl group required for several amino acid decarboxylase catalysts. Most intriguingly this is the pathway in the hundred or more protein self-splicing ligations that go on in bacterial, archaebacterial, and yeast proteins.

Chapter 14 takes up posttranslational modifications that introduce three coenzymes—lipoate, biotin, and phosphopantetheine—covalently tethered to their apo proteins, with the business end of the coenzymes at the end of a swinging arm tether. These three classes of enzymes play crucial roles in enzymes of primary metabolism, and act to ferry acyl groups (lipoate and phosphopantetheine) or CO_2 between different active sites of multimodular enzyme systems for differentiated catalytic transformations.

Chapter 15 concludes the survey of major types of protein posttranslational modifications with two topics. One carries on the theme of the covalent tethering of CO_2, in this case the vitamin K-dependent fixation of CO_2 in up to a dozen closely-spaced glutamate side chains, creating the γ-carboxyGlu (Gla) residues that become high-affinity bivalent sites for binding clusters of calcium ions. The second topic is posttranslational amide bond formation and encompasses the transglutaminase-mediated cross-linking of proteins through the conversion of glutamine side chains to protein–γ-glutamyl cross-links. The oxidative cleavage of C-terminal glycine residues occurs by enzymatic hydroxylation and then decomposition of the hemiaminal products to leave the NH_2 of the glycine moiety behind as a C-terminal amide. Finally, protein circularization is known in plants, producing the protein cyclic form by net amide synthesis.

Chapter 16 summarizes the topics that have been covered in the previous chapters and notes efforts underway to expand the genetic code beyond 22 amino acids by evolving orthogonal pairs of tRNAs and tRNA synthetases that will activate nonproteinogenic amino acids and insert them at a given site in a protein by suppression of a stop codon in the particular mRNA. Over a dozen nonproteinogenic amino acids have been incorporated into target proteins *in vivo* by *E. coli*.

References

Black, D. "Mechanisms of alternate pre-messengerRNA splicing," *Annu. Rev. Biochem.* **72**:291–316 (2003).

Buetow, L., G. Flatau, K. Chiu, P. Boquet, and P. Ghosh. "Structure of the Rho-activating domain of *Escherichia coli* cytotoxic necrotizing factor 1," *Nat. Struct. Biol.* **8**: 584–588 (2001).

Chen, H., M. G. Thomas, S. E. O'Connor, B. K. Hubbard, M. D. Burkart, and C. T. Walsh. "Aminoacyl-S-enzyme intermediates in beta-hydroxylations and alpha,beta-desaturations of amino acids in peptide antibiotics," *Biochemistry* **40**:11651–11659 (2001).

Creasy, D. M., and J. S. Cottrell. "Unimod: Protein modifications for mass spectrometry," *Proteomics* **4**:1534–1536 (2004).

Farriol-Mathis, N., J. S. Garavelli, B. Boeckmann, S. Duvaud, E. Gasteiger, A. Gateau, A. L. Veuthey, and A. Bairoch. "Annotation of post-translational modifications in the Swiss–Prot knowledge base," *Proteomics* **4**:1537–1550 (2004).

Ficarro, S. B., M. L. McCleland, P. T. Stukenberg, D. J. Burke, M. M. Ross, J. Shabanowitz, D. F. Hunt, and F. M. White. "Phosphoproteome analysis by mass spectrometry and its application to *Saccharomyces cerevisiae*," *Nat. Biotechnol.* **20**:301–305 (2002).

Hantschel, O., and G. Superti-Furga. "Regulation of the c-Abl and Bcr-Abl tyrosine kinases," *Nat. Rev. Mol. Cell Biol.* **5**:33–44 (2004).

Hao, B., W. Gong, T. K. Ferguson, C. M. James, J. A. Krzycki, and M. K. Chan. "A new UAG-encoded residue in the structure of a methanogen methyltransferase," *Science* **296**:1462–1466 (2002).

Johnson, L. N., and R. J. Lewis. "Structural basis for control by phosphorylation," *Chem. Rev.* **101**:2209–2242 (2001).

Krishna, R. G., and F. Wold. "Post-translational modification of proteins," *Adv. Enzymol. Relat. Areas Mol. Biol.* **67**:265–298 (1993).

Kyte, J. *Mechanism in Protein Chemistry*. Garland: New York (1995).

Liebler, D. C. *Introduction to Proteomics*. Humana Press: Totowa, NJ (2002).

Maniatis, T., and B. Tasic. "Alternative pre-mRNA splicing and proteome expansion in metazoans," *Nature* **418**:236–243 (2002).

Paulus, H. "Protein splicing and related forms of protein autoprocessing," *Annu. Rev. Biochem.* **69**:447–496 (2000).

Peng, J., D. Schwartz, J. E. Elias, C. C. Thoreen, D. Cheng, G. Marsischky, J. Roelofs, D. Finley, and S. P. Gygi. "A proteomics approach to understanding protein ubiquitination," *Nat. Biotechnol.* **21**:921–926 (2003).

Petsko, G., and D. Ringe. *Protein Structure and Function*. New Science Press: Sunderland, MA (2003).

Polycarpo, C., A. Ambrogelly, A. Berube, S. M. Winbush, J. A. McCloskey, P. F. Crain, J. L. Wood, and D. Soll. "An aminoacyl-tRNA synthetase that specifically activates pyrrolysine," *Proc. Natl. Acad. Sci. U.S.A.* **101**:12450–12454 (2004).

Regni, C., L. Naught, P. A. Tipton, and L. J. Beamer. "Structural basis of diverse substrate recognition by the enzyme PMM/PGM from *P. aeruginosa*," *Structure (Cambridge)* **12**:55–63 (2004).

Schwede, T. F., J. Retey, and G. E. Schulz. "Crystal structure of histidine ammonia-lyase revealing a novel polypeptide modification as the catalytic electrophile," *Biochemistry* **38:**5355–5361 (1999).

Shabb, J. B. "Physiological substrates of cAMP-dependent protein kinase," *Chem. Rev.* **101:**2381–2411 (2001).

Silverman, R. B. *The Organic Chemistry of Enzyme-Catalyzed Reactions*. Academic Press: San Diego (2000).

Srinivasan, G., C. M. James, and J. A. Krzycki. "Pyrrolysine encoded by UAG in archaea: Charging of a UAG-decoding specialized tRNA," *Science* **296:**1459–1462 (2002).

Steiner, D. F. "The proprotein convertases," *Curr. Opin. Chem. Biol.* **2:**31–39 (1998).

Tsien, R. Y. "The green fluorescent protein," *Annu. Rev. Biochem.* **67:**509–44 (1998).

Walsh, C. T. *Enzymatic Reaction Mechanisms*. Freeman: San Francisco (1979).

Protein Phosphorylation by Protein Kinases

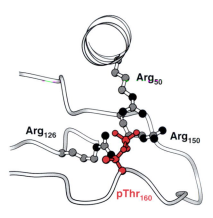

Charge pairing of an anionic phosphate group of pThr with cationic Arg
side chains can alter protein conformation.

The posttranslational modification of proteins by phosphorylation of protein side
chains is highly evolved and widely distributed as a mechanism for creating diversity
in the proteomes of eukaryotes. Protein phosphorylation is also known in prokaryotes
but is much less pervasive in bacterial protein metabolism and regulation, perhaps
reflecting the increase in signaling complexity in nucleated single cells and multicellular organisms (Lee and Goodbourn, 2001).

   The side chains that get phosphorylated in proteins are most commonly serine,
threonine, and tyrosine, reflecting the nucleophilic behavior of the –OH side chains
(Figure 2.1A). The cosubstrate is the universal phosphoryl donor adenosine triphosphate (ATP), with transfer of the γ-PO_3^{2-} to the attacking alcoholate side chain.
Mg^{2+} ions form bidentate chelates to two of the anionic phosphate oxygens in ATP,
lowering the barrier for regioselective phosphoryl transfer to an incoming electron-

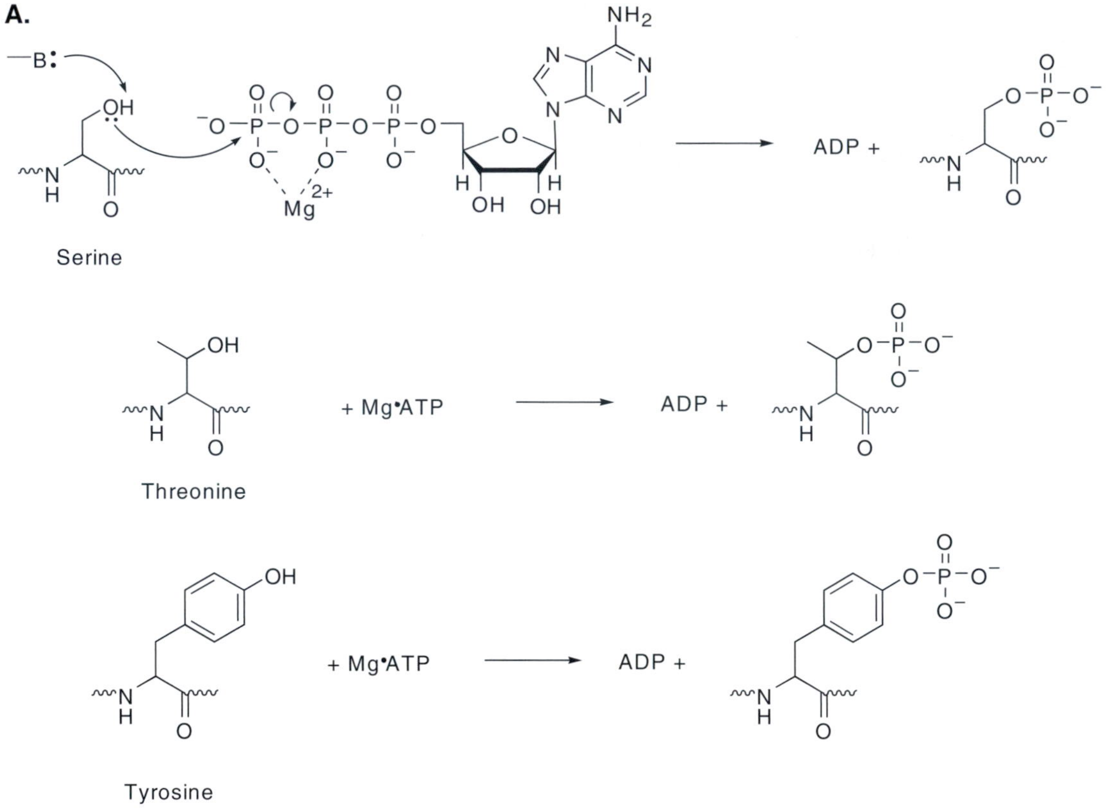

A.

Figure 2.1 Side chains in proteins that get phosphorylated by protein kinases. **A.** The OH groups of serine and threonine attack the γ-PO_3^{2-} of ATP to form phosphoSer (pSer = pS) or phosphoThr (pThr = pT) side chains. Tyrosine kinases transfer the γ-phosphoryl group to yield phosphotyrosine (pTyr = pY).

rich nucleophile. An imidazole ring nitrogen of histidine side chains in proteins, mostly in prokaryotic histidine kinase autophosphorylations, also can become phosphorylated (Figure 2.1B). Phosphoryl transfer then proceeds from the His–NPO_3 in the active sites of transmembrane sensor kinases to the carboxylate side chains of an aspartate of a partner protein that acts as response regulator. As schematized, the phosphoAsp form of the response regulator may be the active form of the transcriptional regulator. The transferring –PO_3^{2-} covalently transmits the chemical signal in bacterial and fungal two-component signaling systems (Hoch, 1995).

About one-third of the potential 30,000 proteins in the human proteome are estimated to be substrates for phosphorylation (Cohen, 2000) at some stage in their life cycle in eukaryotic cells. ATP-dependent phosphorylation enzyme catalysts are termed kinases (Walsh, 1979) and the subset working on protein substrates are termed protein kinases (PKs). There are just over 500 predicted kinases in the human proteome. This has been termed the human kinome (Manning et al., 2002) (Figure 2.2) and is about four times the size of the yeast kinome and double the size of the fly

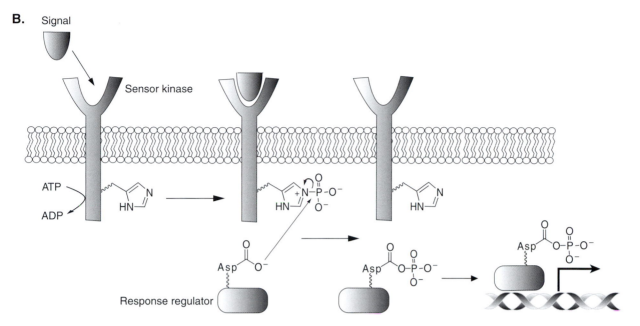

Figure 2.1 B. Two-component bacterial regulatory systems initiate signaling with autophosphorylation on the imidazole ring of a His residue by the histidine kinase sensor component. The PO_3^{2-} is then transferred to the β-carboxylate of an Asp residue in the second protein component, the response regulator.

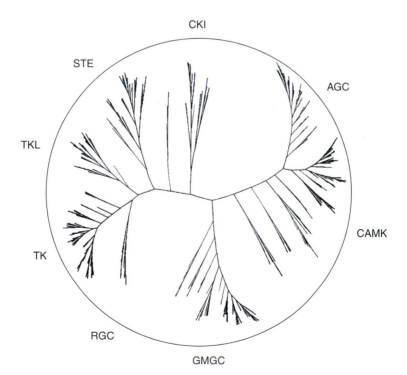

Figure 2.2 Schematic dendrogram of the human kinome (reprinted with permission from Manning et al., 2002).

kinome. The enormous number of protein kinases indicates both pervasiveness of this chemical class of protein modification and the more or less specific pairing of protein kinases and particular protein substrates, indicating some mechanisms of extended sequence recognition beyond the Ser, Thr, or Tyr side chain in a target protein.

Some PKs phosphorylate only one partner protein substrate while others such as protein kinase A (Shabb, 2001) or casein kinases (Pinna, 2003) can phosphorylate serines and threonines in more than 100 and more than 300 proteins, respectively. Some proteins get phosphorylated at multiple sites. This includes protein substrates such as a pair of serines, Ser_{32} and Ser_{36}, in the negative regulatory subunit of IκB that target the doubly modified protein for ubiquitin tagging and subsequent proteolytic destruction (Figure 2.3A) (Chapter 9). The protein kinases themselves undergo multisite phosphorylation. For example, the MAP kinase cascade component enzyme

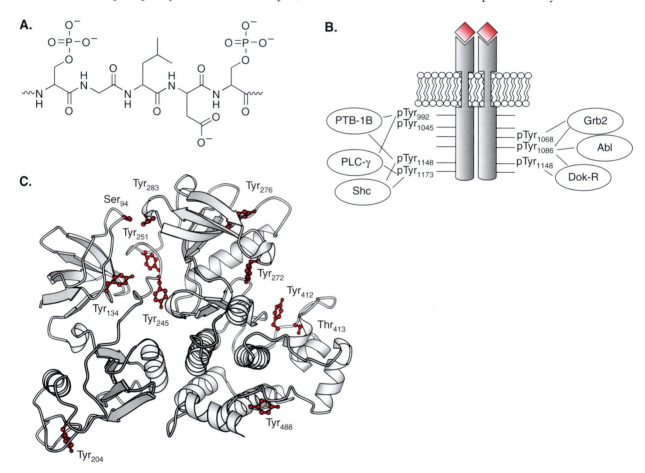

Figure 2.3 Multiple phosphorylations at different side chains within a particular protein substrate for one or more protein kinases. **A.** Enzymatic phosphorylation of two nearby serine side chains in the IκB subunit of the IκB-NFκB complex that anchors the NFκB transcription factor in the cytoplasm. **B.** Autophosphorylation *in trans* of multiple tyrosines in the activated dimer of the EGFR tyrosine kinase. **C.** Eleven phosphorylations (nine Tyr, one Thr, and one Ser) of Abl tyrosine kinase (figure made using PDB 1OPK).

MEK1 is phosphorylated on Ser_{217} and Ser_{221} (Cohen, 2002), while receptor tyrosine kinases autophosphorylate at multiple tyrosines in their cytoplasmic domains, as shown for the dimeric form of the EGF receptor. Distinct proteins are recruited to particular pY sites based on the consensus sequence around a given pY residue (Figure 2.3B) (Cohen, Pawson, and Nash, 2003).

A particular protein can be phosphorylated at serine and at tyrosine side chains for integration of different signals, as happens in the signaling enzyme MAP kinase (Chen et al., 2001). Up to 11 phosphorylations have been reported on the Abl tyrosine kinase, nine on tyrosine residues, one on Ser_{94}, and one on Thr_{413} (Hantschel and Superti-Furga, 2004) (Figure 2.3C). In protein-based signaling pathways, proteins that act in sequential steps can be phosphorylated by specific kinases, some of which act in cascades, where the catalytic action of each kinase to phosphorylate hundreds to thousands of protein substrates at each step can lead to dramatic amplifications of signals (Huang and Ferrell, 1996). Four parallel MAP kinase pathways report on and amplify signals from growth factors, integrins, oxidative stress, and the cytokine IL-1, respectively (Figure 2.4) (Hunter, 2000; Lee and Goodbourn, 2001).

Protein phosphorylation dynamics are crucial to the use of posttranslational phosphorylation in many eukaryotic signaling pathways. Initiation of signal may occur

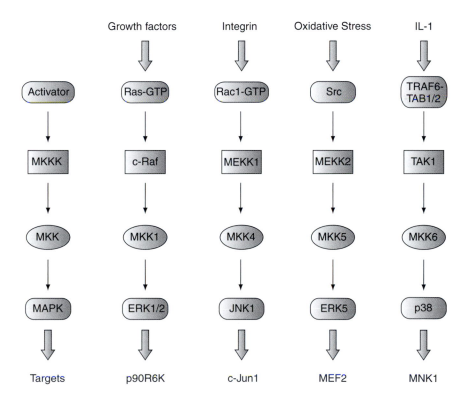

Figure 2.4 Amplification of signal in protein kinase cascades. Four parallel MAP kinase cascades relaying signals from growth factors, integrins, oxidative stress, or cytokines such as IL-1.

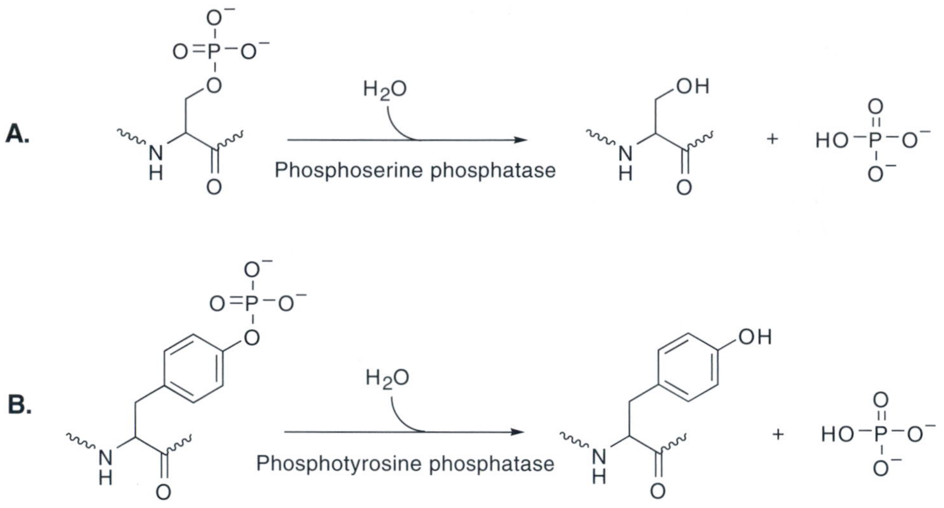

Figure 2.5 Removal of phosphoryl groups in proteins by action of phosphoprotein phosphatases. **A.** Action of phosphoserine/phosphothreonine phosphatases. **B.** Action of phosphotyrosine phosphatases.

when a protein side chain gets phosphorylated. Abrogation of signal can occur in general by two routes. One involves proteolytic destruction of the phosphorylated form of the protein. The second, and more common alternative, since it preserves the protein scaffold for another cycle of reversible phosphorylation, is hydrolytic removal of the $-PO_3^{2-}$ side chain (Figure 2.5). The phosphoserine (pSer, also abbreviated pS), phosphothreonine (pThr, also abbreviated pT) and phosphotyrosine (pTyr, also abbreviated pY) groups are chemically stable at physiological pH and temperature and require protein phosphatase action for hydrolysis, as will be examined later in this chapter. The balance of protein kinase/protein phosphatase towards each phosphoprotein determines the fractional stoichiometry of phosphorylation at each site. In contrast, phosphohistidine and aspartyl phosphate side chains in bacterial two-component regulatory systems are thermodynamically and kinetically labile to hydrolysis (Barrett and Hoch, 1998) and generally revert back to the dephospho state without separate protein phosphatase partners. Consonant with roles in signal transduction, most protein kinases, with casein kinase II an exception in being constitutively active (Pinna, 2003), are "off" in the basal state and their activity turned "on" by one or more signals.

The Special Utility of the Phosphate Group: Structure, Thermodynamics, and Mechanisms

In a classical review on the role of phosphates in biology, Westheimer (1987) noted several properties that made the phosphate group vital in biology. Inorganic phos-

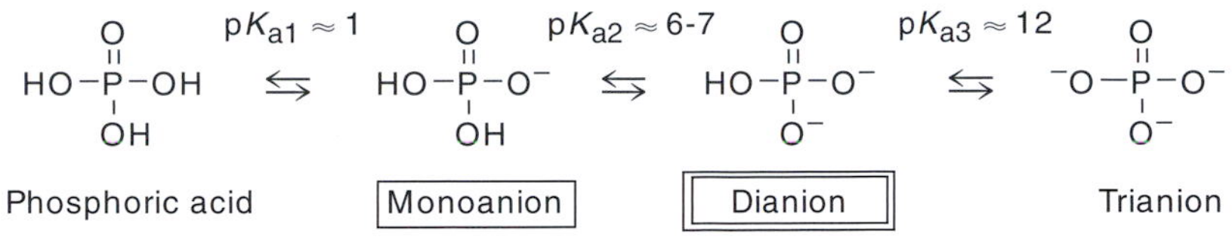

$$\underset{\text{Phosphoric acid}}{\text{HO}-\overset{\overset{\text{O}}{\|}}{\underset{\underset{\text{OH}}{|}}{\text{P}}}-\text{OH}} \;\overset{pK_{a1}\approx 1}{\rightleftharpoons}\; \underset{\boxed{\text{Monoanion}}}{\text{HO}-\overset{\overset{\text{O}}{\|}}{\underset{\underset{\text{OH}}{|}}{\text{P}}}-\text{O}^-} \;\overset{pK_{a2}\approx 6\text{-}7}{\rightleftharpoons}\; \underset{\boxed{\boxed{\text{Dianion}}}}{\text{HO}-\overset{\overset{\text{O}}{\|}}{\underset{\underset{\text{O}^-}{|}}{\text{P}}}-\text{O}^-} \;\overset{pK_{a3}\approx 12}{\rightleftharpoons}\; \underset{\text{Trianion}}{{}^-\text{O}-\overset{\overset{\text{O}}{\|}}{\underset{\underset{\text{O}^-}{|}}{\text{P}}}-\text{O}^-}$$

Figure 2.6 The three pK_a values of phosphoric acid: $pK_{a1} \approx 1$; $pK_{a2} \approx 6\text{-}7$; $pK_{a3} \approx 12$. Phoshoproteins at pH 7.5 will have >90% of each phosphoryl group as the dianion.

phate, at physiological pH, occurs predominantly in the dianion form, HPO_4^{2-}, of phosphoric acid ($pK_{a1} \approx 1$ and $pK_{a2} \approx 6$) (Figure 2.6). The charges provide aqueous solubility. The central phosphorus atom in the +5 oxidation state is electrophilic, while the peripheral oxygens, especially the negatively charged ones, are potential nucleophiles, generating double-headed reactivity to phosphate, to nucleophiles at phosphorus, and to electrophiles at oxygen (Walsh, 1979). Phosphate groups are stable at physiological pH but can undergo a reversible addition/elimination reaction catalyzed by enzymes that lower the relevant energy barriers to reaction. Inorganic pyrophosphate, $P_2O_7^{4-}$, is correspondingly the tetraanion of pyrophosphoric acid at physiological pH, in which one of the oxygen anions of inorganic phosphate has attacked the central phosphate of a second molecule, eliminating water as the phosphoric anhydride linkage is formed. The anhydride linkage in pyrophosphate is more destabilized than the ester linkage in the monophosphate ester (e.g., of glucose-6-PO_3^{2-}), and so is chemically more reactive in phosphoryl transfer. Phosphodiester linkages are found in the central β-phosphate of triphosphate chains, such as in ATP (Figure 2.7), and in the internucleotide linkages of RNA and DNA, the key covalent connecting bond of nucleic acids.

ATP, the phosphoryl donor cosubstrate in essentially all protein kinase reactions (GTP substitutes in a small subset), has the three side chain phosphates in the triphosphate chain in phosphoric anhydride linkages. This is the structural basis of its thermodynamic activation for transfer of either the α-, β-, or γ-phosphoryl moieties, for nucleotidyl, pyrophosphoryl, or phosphoryl transfer to some cosubstrate nucleophile respectively (Walsh, 1979) (Figure 2.7A). The K_{eq} for phosphoryl transfer to alcohol cosubstrates from ATP is about 10^2 in favor of phosphoryl transfer to serine side chains and about 10^1 towards Tyr side chains (Cole et al., 1999).

The mechanism of γ-PO_3^{2-} transfer to a protein Ser, Thr, or Tyr side chain involves regioselective attack of the side chain on the electrophilic γ-phosphorus of the Mg–ATP substrate due to specific orientation of the two substrates in the kinase active site and the geometry of the Mg chelate. This mechanism of phosphoryl transfer involves a dissociative transition state with early cleavage of the O-Pγ bond (Cole et al., 1999) (Figure 2.7B).

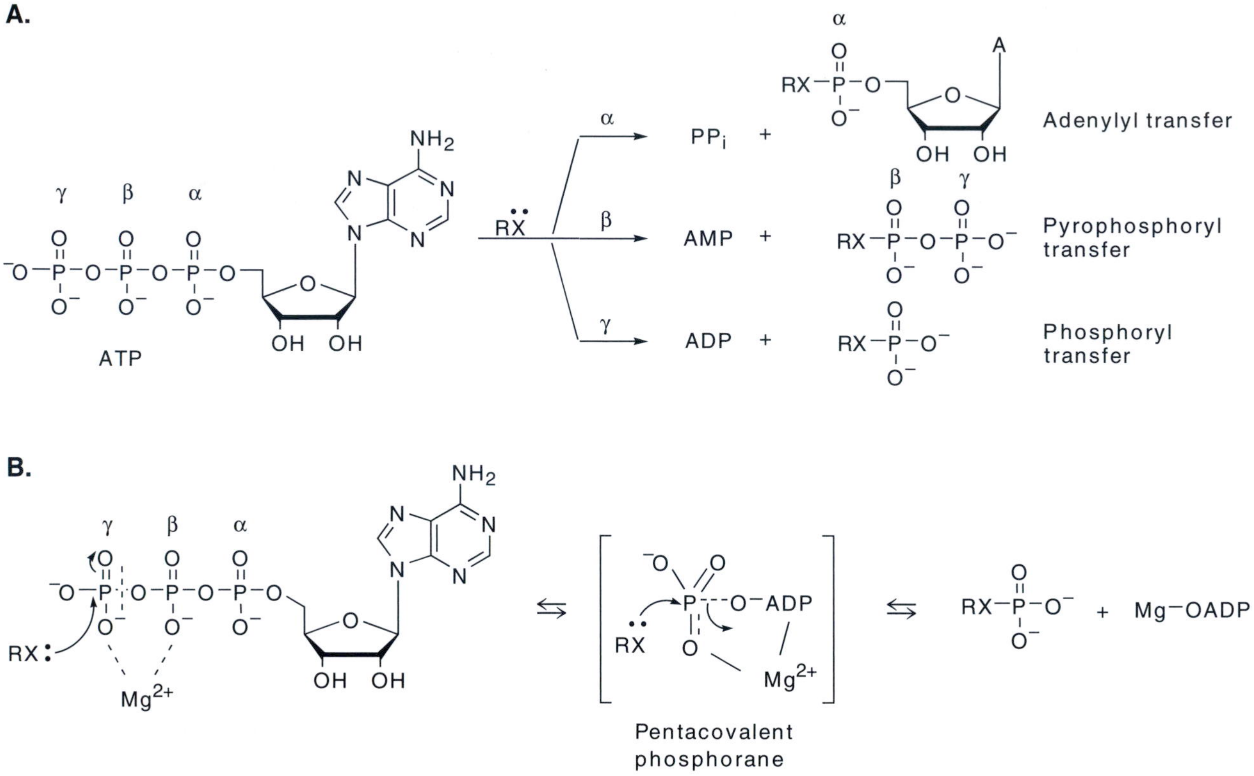

A.

B.

Pentacovalent
phosphorane

Figure 2.7 **A.** Attack of cosubstrate nucleophile on the electrophilic α-, β-, or γ-phosphorus atom of ATP results in adenylyl transfer, pyrophosphoryl transfer, or phosphoryl transfer, respectively. **B.** Phosphoryl transfer mechanisms involve initial formation of a pentacovalent phosphorane. The electron pair on the axial oxygen can reform the P=O double bond with expulsion of ADP to complete reaction in the forward direction.

The binding site for ATP is generally conserved in all the protein kinases, although there are small differences in orientation and identity of enzyme side chains that form the binding pocket that create differences in binding affinities. The binding sites for the protein moieties that become phosphorylated vary both in the major Ser/Thr versus Tyr side chain discrimination category (in general a deeper active site for the tyrosine side chain) and in the sequence around the Ser, Thr, or Tyr residue that is phosphorylated by different subsets of PKs. Peptide libraries are available in multiwell formats for screening the sequence specificity of newly purified kinases to establish their selectivity. Several other experimental methods have been devised to identify protein kinase substrates experimentally (Berwick and Tavare, 2004) and a number of informatics tools, such as PhosphoSite (Hornbeck, 2004), are available.

ATP is admirably designed and balanced as cellular energy currency (Walsh, 1979). Kinetically the tetraanion of ATP is stable in aqueous solution, yet thermodynamically the pyrophosphate linkages are sufficiently activated to drive phosphorylation of cosubstrate nucleophilic side chains far in the direction of γ-PO_3 transfer.

Structural Basis for Control of Protein Conformation by Phosphorylation

The introduction of the tetrahedral phosphate dianion ($pK_{a2} \approx 6$) to one or more side chains of a folded protein creates the opportunity for local charge pairing and hydrogen bonding that can create subtle to dramatic reorganization of the local protein microenvironment that drives conformer reorganizations (Johnson, 2001). The electrostatic interactions and the hydrogen bond possibilities to any of the four oxygens in the modified protein–OPO_3^{2-} group probably account for the pervasive use of posttranslational phosphoryl transfers to drive much of protein-based signaling in eukaryotic cells. To evaluate the changes enabled by protein side chain phosphorylation requires determination of the protein structure in both the dephospho and phospho forms. Where this has been accomplished (Johnson, 2001) one sees two types of strategies by which side chain phosphate dianions act as organizing/reorganizing centers for local protein architecture. One is interaction of the –PO_3^{2-} group with main chain nitrogens at the N-termini of α helices, the dianion interacting electrostatically with the δ^+ charge of the helix dipole (Figure 2.8A), as illustrated in isocitrate dehydrogenase and the KIX/phosphoKID complex.

A second common interaction motif is charge pairing of the phosphate dianion with one or more arginine side chains through the planar, bidentate cationic guanidinium group. This interaction is well illustrated in glycogen phosphorylase where $pSer_{14}$ makes hydrogen bonds to two Arg side chains, Arg_{69} in the same subunit and Arg'_{43} from the other subunit in the enzyme dimer (Figure 2.8B). The gain of the intra-subunit contact in the phospho form of the protein drives an allosteric conformational change in both subunits to activate the enzyme for its catalytic function of cleaving glycogen to glucose-1-P (Barford et al., 1991; Johnson, 2001). In contrast to the allosteric activating effect of serine side chain phosphorylation in glycogen phosphorylase is the direct inhibition of the active site of the citrate cycle enzyme isocitrate dehydrogenase (Dean and Koshland, 1990). Posttranslational phosphorylation of Ser_{113} at the start of an α helix in the enzyme blocks substrate access to the active site by electrostatic and steric modes, causing direct inhibition. We shall note below that activation of protein kinases involves phosphorylation of a Thr–OH (pS/pT kinases) or Tyr–OH (pY kinases) in an activation segment that then reorganizes and creates an enzyme conformer able to bind ATP and protein cosubstrate in a productive orientation for γ-PO_3^{2-} transfer.

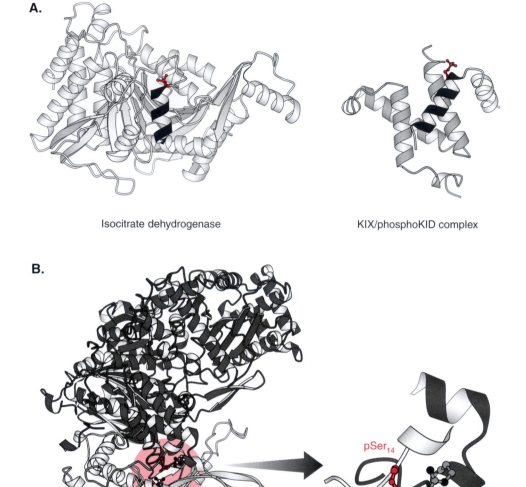

A.

Isocitrate dehydrogenase   KIX/phosphoKID complex

B.

pSer$_{14}$

Arg'$_{43}$

Arg$_{69}$

Figure 2.8 Two modes of interaction of phosphoSer/phosphoThr dianionic side chains with motifs in proteins. **A.** Electrostatic interaction of the phosphorylated side chain with the positive helix dipole at the N-terminus of an α helix, exemplified by ICDH and the KIX/phosphoKID pair (figure made using PDBs 4ICD for isocitrate dehydrogenase and 1KDX for KIX). **B.** Electrostatic interaction of the phosphate dianion with bidentate cationic guanidinium termini of arginine side chains. Exemplified is the orientation of pSer$_{14}$ with Arg$_{69}$ and Arg'$_{43}$ in glycogen phosphorylase (figure made using PDB 1GPA).

Domains Dedicated to Recognition of Phosphoproteins

In addition to the local conformational changes that protein side chain phosphoryla-tion induces by interation with cationic arginine side chains *in cis* or in heteroligo-meric subunit interfaces *in trans* noted above, there are several types of independently folding protein domains and modules that recognize the phosphorylated forms of proteins. These phosphoprotein recognition domains are key modular elements in phosphorelay-based signaling cascades in cells. Pawson and Nash (2003) recently noted that five domains have been described that show affinity for pS or pT side chains while two are selective for pY sequences (Kuriyan and Cowburn, 1997; Yaffe and Elia, 2001). The MH2 domain in the signal transducer SMAD proteins that act downstream of the transmembrane serine kinase that is the receptor for the growth factor TGFβ recognizes pS and/or pSXpS side chains. Forkhead-associated (FHA) domains (Figure 2.9) (Durocher and Jackson, 2002) seem selective for pT side chains. Additionally, pS motifs are recognized by 30-kDa 14-3-3 proteins (Yaffe, 2002a) or by the WW domain in the peptidyl prolyl isomerase Pin1, or by tandem repeats of WD40 domains. There are seven human isoforms of 14-3-3 proteins (Yaffe, 2002a). The pS side chain of the phosphorylated form of the dual specificity (pT and pY hydrolysis) protein phosphatase Cdc25C is complexed by three conserved cationic side chains, K_{49}, R_{56}, and R_{127}, in the 14-3-3 partners. A sixth pS recognition domain, the BRCT domain, has since been detected in the 100-residue C-terminal domain of the BRCA1 protein, associated with high penetrance of breast cancer (Manke et al., 2003; Xu et al., 2003), suggesting the full complement of phosphopep-tide recognition domains may not yet be catalogued. X-ray analysis of the BRCT domain with a bound pSer peptide corresponding to $pSer_{990}$ of a DNA helicase part-ner protein to the BRCT domain shows high recognition selectivity (Clapperton et al., 2004).

The canonical pY recognition module is the 100-residue Src homology 2 (SH2) (Figure 2.9) domain first found in the Src tyrosine kinase, where interaction of a con-served Arg side chain in the SH2 domain interacts with the pY residue being recog-nized. In the founding SH2 member Src, the C-terminal pY_{527} binds to an Arg side chain in the SH2 domain to create an autoinhibitory conformation in the enzyme at rest (Schindler et al., 1999; Xu et al., 1999). A second domain for binding to pY domains is the phosphotyrosine binding (PTB) domain (Figure 2.9), which recog-nizes pY side chains, often in NPXpY sequences, by a distinct motif from that used by SH2 domains (Yaffe, 2002b). Both PTB and SH2 domains function in signaling by the insulin receptor, whose cytoplasmic juxtamembrane domain has an NPEY sequence upstream of the tyrosine kinase domain. The *in trans* autophosphorylation of the NPEY to NPEpY sequence occurs when insulin is ligated to the extramembrane domain. The NPEpY motif recruits an adaptor protein known as IRS (IRS1 and IRS2 forms exist) via its PTB motif. IRS1 has 18 tyrosine residues, and is observed to be multiply phosphorylated on tyrosines after recruitment to the acti-

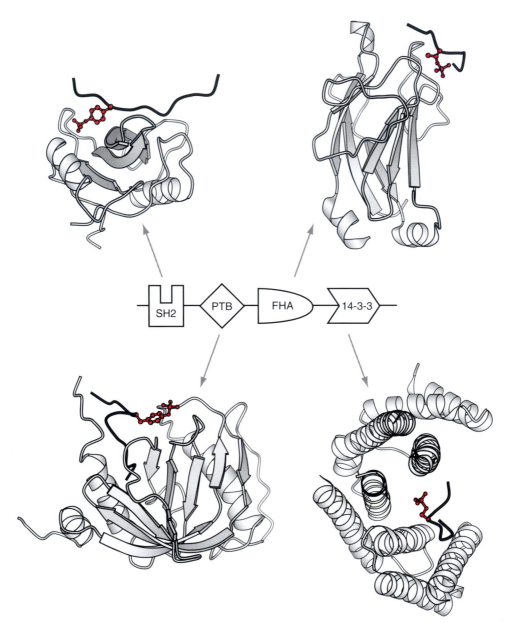

Figure 2.9 Protein domains that recognize and bind to phosphorylated side chains of proteins. Forkhead-associated (FHA) domains and 14-3-3 domains are selective for pS and/or pT residues. SH2 domains and PTB domains are selective for pY residue recognition. Phosphopeptide interactions with each domain are represented (figure made using PDBs 1LCJ for SH2, 1SHC for PTB, 1G6G for FHA, and 1QJA for 14-3-3).

vated insulin receptor kinase. Now the pY_n form of docked IRS1 can recruit many proteins that read the distinct sequences around the pY residues via their SH2 domains. Thus, SH2-domain-containing proteins that dock on pY_n IRS1 include phosphatidyl inositol-3 kinase (PI-3K), the adaptor Grb2, and the SH2-containing protein tyrosine phosphatase Shp2.

The existence of eight phosphoprotein interaction domains discovered to date shows the multiple ways a dianionic pS, pT, or pY side chain can signal to and reorganize protein conformations as a result of protein kinase action (Pawson and Nash, 2003).

Detection of Protein Phosphorylation: The Phosphoproteome

Historically, protein phosphorylation was detected by the transfer of radioactive ^{32}P from γ-[^{32}P]ATP to protein cosubstrate (Cohen, 2002). The covalent attachment of radiolabel to the protein could be detected in multiple ways, most readily by precipitation of the protein in aqueous trichloroacetic acid, washing with buffer, and scintillation counting. This detects ^{32}P-Ser, ^{32}P-Thr, and ^{32}P-Tyr linkages that are acid stable, but not P-His, P-Asp, or P-Glu linkages that are labile in acid. Since the latter are almost never used in eukaryotic protein phosphorylations, this limitation is not substantial. The ^{32}P radioactivity assay establishes irreversible covalent attachment to the protein. Acid hydrolysis of the protein followed by chromatography and autoradiography of the amino acid hydrolysate allows separation of ^{32}P-Ser, ^{32}P-Thr, and ^{32}P-Tyr, to establish the identity of the type of side chain phosphorylated. The localization of phosphorylation at a specific residue in the sequence of a protein and the determination of fractional occupancy at one or more side chains is more challenging, and these days is almost always accomplished not with radioactivity but by mass spectrometry (Kalume et al., 2003; Mann et al., 2002; Zhou et al., 2001), as noted below.

Another rapid probe for protein phosphorylation has been the development of phosphoprotein-specific antibodies. Antibodies to phosphotyrosine have been raised that are of high affinity and low sequence selectivity such that many pY proteins are detected by Western blotting procedures after gel electrophoresis (Figure 2.10). Comparably robust antibodies that detect pS and pT side chains in proteins have been slow to develop. There are sequence-selective pS/pT antibodies (e.g., to RXRXXpT or to pSXR sequences; Zhang et al., 2002) that are useful for evaluating subsets of Ser/Thr protein kinases, but these are not of general screening utility. Machida et al. (2003) have also proposed using libraries of phosphotyrosine-specific SH2 domains (discussed later in the chapter) as probes for pY-protein profiles.

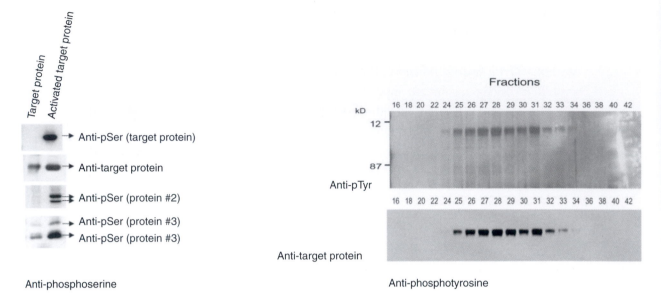

Figure 2.10 Detection of some pS- and pY-containing proteins by anti-pY- or anti-pS-antibodies (reproduced with permission from Zhang et al., 2002, for anti-pS and Izaguirre et al., 1999 for anti-pY).

For a proteomics or systems biology perspective one wants to determine the phosphorylation state, not just of one protein or a small set of proteins, but the phosphorylation status of many or all proteins in a cell in response to a particular stimulus. Mass spectrometry (MS) is by far the leading technology for evaluation of the phosphoproteome. Thousands of potential phosphorylation sites exist in proteins. For the yeast genome with 114 predicted protein kinases (Manning et al., 2002) and 6000 proteins, about 2000 phosphorylation sites might be expected. The human phosphoproteome might be more than one order of magnitude larger. A variety of protocols and approaches to catalog phosphoproteomes by mass spectrometry have been reported (Ficarro et al., 2002; Liebler, 2002; Manning et al., 2002).

Proteins to be evaluated for phosphorylation (e.g., in mixtures such as yeast cytosolic crude extract), are typically digested proteolytically, by proteases such as trypsin, to generate limit peptides. The peptides may then be subjected to liquid chromatography–mass spectrometry (LC–MS) to determine m/z ratios. Since the full complement of all the tryptic peptides in the yeast genome is known, the observed mass ratios can be compared to predicted values for the peptides to identify their protein of origin. Each phosphorylation adds 80 Da (PO_3H) (Table 1.2) to the molecular weight of a Ser, Thr, or Tyr residue in a peptide. To determine which residue of a tryptic peptide is phosphorylated, in principle one can perform tandem MS–MS to fragment a given m/z peptide in the second MS experiment and obtain the sequence. The

MS–MS spectrum for the peptide VPQLEIVPNpSAEER shown in Figure 2.11A allows determination of the presence of a labile pS residue and its placement in the sequence from the fragmentation pattern. For identification and mapping of pY sites in proteins, a protein band detected on an SDS–PAGE gel can be eluted, digested to peptide fragments, and analyzed by MS (Figure 2.11B). In this way the pY_{1162} of the SH_2-containing PTPase SHIP-2 was identified (Liebler, 2002).

Complications tend to arise in two ways. If phosphopeptides are present in low abundance, including low fractional abundance compared to the matched unphosphorylated peptide (e.g., suppose only 5% of a given protein where in phospho form it would be outnumbered 20:1 by the dephosphopeptide and the signal would be low), then their ionization may be suppressed and not detected. Second, the phosphoryl group in pS and pT but not pY linkages is subject to facile α,β-elimination in the sequencing mode to yield $HOPO_3$ (loss of 98 mass units) and the corresponding dehydroalanyl and dehydroaminobutyryl peptides (Figure 2.11C). The loss of the 98 mass units from pS and pT can be used to advantage to identify pS and pT sites. The

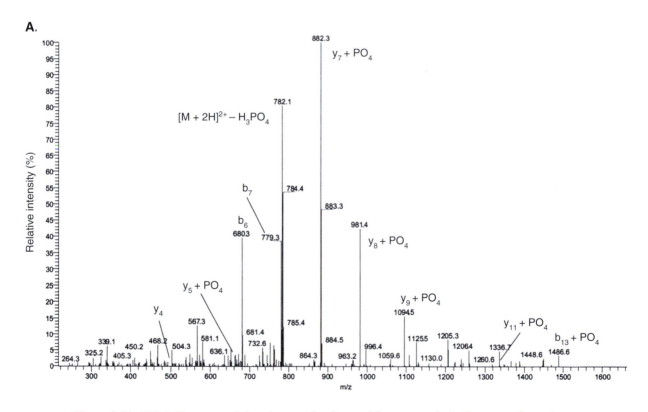

Figure 2.11 MS–MS spectra of phosphopeptides detected from proteolytic cleavages of proteins. **A.** MS–MS of the 2+ ion of the peptide VPQLEIVPNpSAEER from bovine casein (reproduced with permission from Liebler, 2002).

phosphorylated phenolic linkage of pY is stable and can be detected in positive ion mode as an iminium ion at *m/z* 216 for conclusive identification (Steen et al., 2001, 2002). Using this pY-specific iminium ion scanning on a high-resolution quadropole time-of-flight tandem mass spectrometer, Steen et al. (2003) were able to map nine pY sites in the 185-kDa oncogenic fusion protein Bcr-Abl. Six of these phosphotyro-

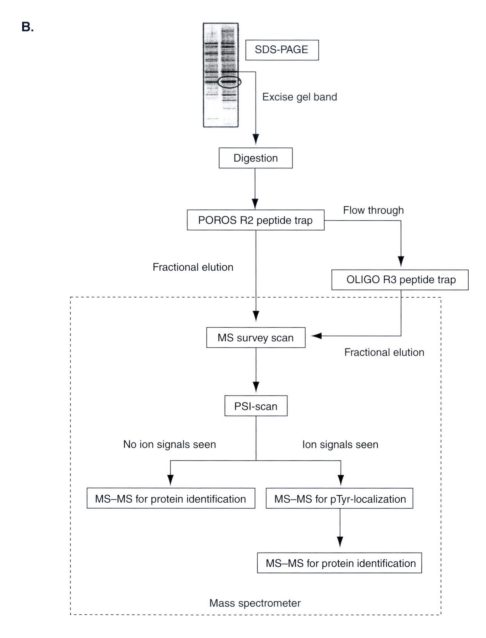

Figure 2.11 **B.** MS–MS localization of pY_{1162} of the PTPase SHIP-2 (reproduced with permission from Steen et al., 2002).

sine sites were novel and not predicted with high confidence by three of the common phosphorylation motif prediction programs.

To solve the low abundance problem Hunt and colleagues (Ficarro et al., 2002) have taken tryptic peptides, methylated their C-termini to cover up the negative charge such that only side chain phosphates in pS, pT, and pY residues are anionic,

B.

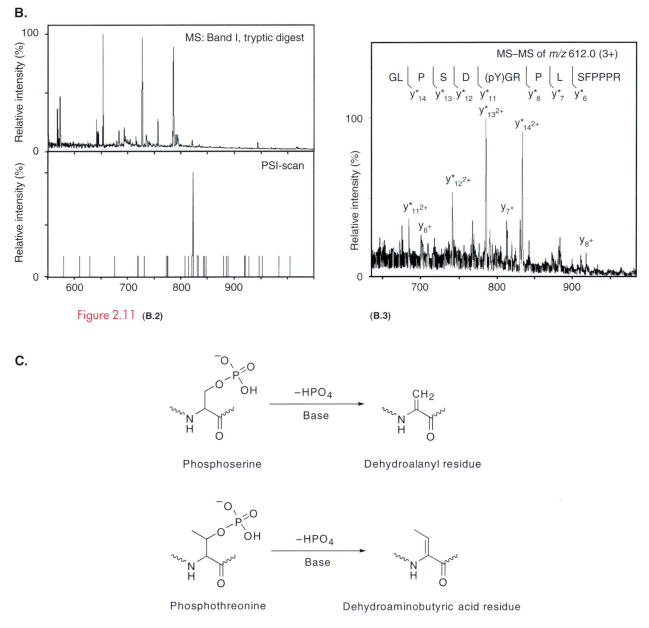

Figure 2.11 (B.2)

(B.3)

C.

Figure 2.11 **c.** Loss of HOPO₃ from pS and pT peptides by β-elimination to generate dehydroalanine and dehydroaminobutyrine residues at the sites of elimination.

and then concentrated the phosphopeptides by immobilized metal affinity columns (IMAC), where the immobilized metal cations bind the anionic phosphopeptides but not neutral peptides. The phosphopeptides are then eluted and subjected to LC–ESI mass spectrometry. From crude yeast protein extract, 216 phosphopeptides at 5 femtomole sensitivity were identified, representing 383 sites of phosphorylation on 171 different proteins. Of these, 60 were monophosphorylated peptide fragments, 145 were diphosphopeptides and 11 were triphosphopeptides, validating the multiple-site phosphorylation of specific proteins and locating all the phosphorylation sites at specific residues in particular proteins. The IMAC step probably selectively enriches for the doubly and triply phosphorylated peptides so these are doubtless overrepresented in terms of abundance in the cell. This approach has promise for inventorying cellular phosphoproteomes in different cell types under different conditions. Ficarro et al. (2002) were able to detect not only abundant phosphoproteins (e.g., heat shock proteins) but also several rare phosphoproteins at low abundance. One can do pairwise comparisons of such phosphoproteomes by using CH_3OH for esterification in the control set and CD_3OH for esterification of the perturbed set to evaluate concentration changes of particular phosphopeptides in response to different stimuli.

One would like to know the timing of protein phosphorylations in cells in response to some stimulus. A prototypic analysis is of the phosphoproteome driven by external epidermal growth factor binding to its cell surface receptor, EGFR, a tyrosine kinase (Blagoev et al., 2004). Using different forms of mass-labeled arginine in cell culture as markers for different time points, Blagoev could show that subsequent to a 40-fold increase in autophosphorylation of the EGFR within one minute, a total of 81 pY proteins could be detected via a two-stage analysis: anti-pY protein immunoprecipitation followed by proteolysis and MS analysis of pY peptides. A time series from 0–20 minutes revealed the kinetics of this phosphoproteome cascade, essential information for ordering the protein components in this growth factor signaling cascade. Of the 81 pY proteins detected, 31 were novel. This approach can be utilized with any phosphoprotein cascade, although the pS and pT reagents are still suboptimal.

Serine/Threonine Kinase Action

Substrate Specificity

Estimates of the ratio of protein phosphorylation in mammalian cells on serine and threonine residues versus tyrosine residues varies from >90% to >99.9% on serine and threonine. The relative abundance of pS to pT to pY residues has been listed as 1800:200:1 (Mann et al., 2002). The serine and threonine side chain phosphorylations are carried out by the approximately 400 pS/pT protein kinases that in turn comprise 80% of the mammalian kinome. The particular sequences in which pS and pT residues are found derive from the composite specificity of all the pS/pT kinases in operation in the particular cell type under the specific spatio-temporal conditions of a given microenvironment. Over the three decades of protein kinase research, a few dozen protein kinases have been characterized for catalytic efficiency and consensus specificity for PO_3^{2-} transfer from ATP. Some pS/pT kinases, such as PKA and PKC, act at serine residues with a nearby basic residue (Shabb, 2001). Others, such as casein kinase II (CKII), act at S/TXXD/E sequences (Pinna, 2003), where the acidic side chain is a main determinant. A third group, represented by cell division kinases (Cdks) are proline-directed serine kinases, recognizing some SP sequences for serine phosphorylation.

The range of protein substrates posttranslationally phosphorylated by particular protein kinases can also vary dramatically. For the cyclic-AMP-dependent PKA, over 100 protein substrates have been identified to date (Shabb, 2001). At the other extreme the kinases that operate in MAP kinase phosphorylation cascades appear to have narrow specificity for phosphorylation of the activation loop in their designated partner downstream kinase.

Consistent with its role of responding to concentrations of the master regulatory molecule cAMP that induce dissociation of the inhibitory regulatory subunits and allow the free catalytic subunit to function (Figure 2.12), PKA has effects on many layers of cellular protein machinery. The 100 known protein substrates are involved in modulation of signaling pathways mediated by G-protein-coupled receptors, by MAP kinase cascades, and by multiple transcription factors from CREB to NFκB to NFAT. Histone H3 phosphorylation at Ser_{10} by PKA is part of writing the histone code (Chapter 6). PKA phosphorylation of glycogen synthase kinase and pyruvate kinase regulates carbohydrate metabolic flux while phosphorylation of proteins in K^+, Na^+, and Cl^- channels regulate ion conductance (Shabb, 2001). The preferred consensus site for PKA in protein cosubstrates is RRXS/T with serine phosphorylated about 12 times more often than threonine. Lysine can replace the second arginine in many PKA sites. Tandem double phosphorylation of serines is observed with PKA, for example at $Ser_{345}Ser_{346}$ of the β_2-adrenergic receptor and $Ser_{28}Ser_{29}$ of the thyroid nuclear hormone receptor T3Rα (Shabb, 2001). Phosphorylation of the target protein transcription factor CREB (cAMP response element binding protein) at Ser_{133} by PKA enables CREB to assemble with the coactivator protein CBP and bind with high affinity at the cAMP response elements (CRE) in the promoter regions of genes activated by cAMP elevation. The CREB–Ser–OPO_3^{2-}–CBP complex can recruit RNA polymerase to the CRE promoters and activate target gene transcription.

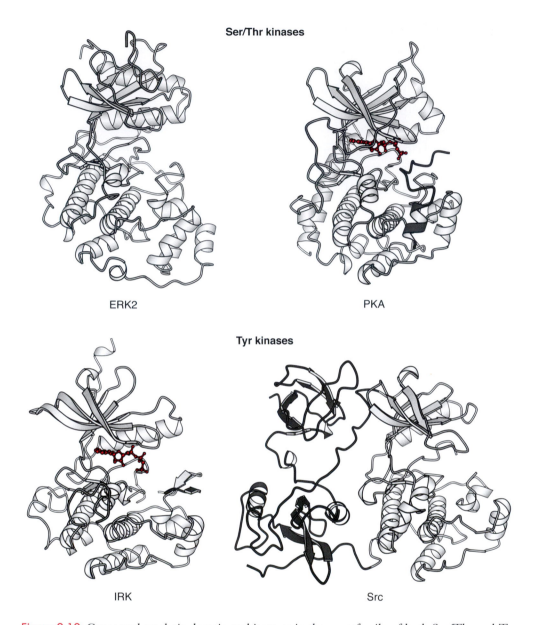

Ser/Thr kinases

ERK2

PKA

Tyr kinases

IRK

Src

Figure 2.12 Conserved catalytic domain architecture in the superfamily of both Ser/Thr and Tyr kinases: ERK2 and PKA as representative Ser/Thr kinases, and IRK and Src as prototypic Tyr kinase domains (figure made using PDBs 2ERK for ERK2, 1CDK for PKA, 1IR3 for IRK, and 1FMK for Src).

Kinase Subfamilies and Regulatory Strategies for Catalytic Domain Activation

The pS/pT kinome family has been divided into subfamilies of homologous kinases, including a PKA/PKG/PKC family where PKG is activated by cGMP and PKC by calcium ions and lipids, and these number 63 members in the human kinome (Manning et al., 2002). Analogously there are 74 protein kinases activated by calcium and the Ca^{2+} binding protein calmodulin (CaM Kinases) and there are 47 MAP kinase family members and 12 cell division kinases.

All the protein kinases are thought to have a common architecture in the catalytic domain, based on the determination of more than 30 catalytic domain x-ray structures (Johnson, 2001). Figure 2.12 shows the fold of two Ser/Thr kinases, ERK2 and PKA, and two Tyr kinases [the insulin receptor kinase (IRK) and Src]. In accord with the general rule that kinases are in the "off" mode in the absence of some signaling stimulus to turn them "on" as phosphoryl transfer catalysts, there are multiple modes for keeping the protein kinases' domains in the inactive state at rest. One mechanism is to have regulatory domains, either *in cis* or *in trans* as separate subunits. Over 100 modular protein domains have been detected, many in proteins in signaling pathways (Hunter, 2000). These represent combinatorial platforms on which protein–protein interactions, such as kinases with substrates and regulators, both small molecules (Newton, 2001) and proteins, can be assembled.

The regulatory domains or subunits can be autoinhibitory or they can be activating (Kobe and Kemp, 1999). The PKA regulatory logic is to have separate regulatory (R) subunits that bind tightly to the catalytic (C) subunits in an inactive R_2C_2 complex at rest. When cAMP levels rise as a consequence of adenylyl cyclase action, cAMP binds to the R subunits and causes dissociation (Figure 2.13A). The free C subunits are now active PO_3 transfer catalysts and can reach catalytic turnover numbers of about 30 transfers per second (Johnson et al., 2001). Autoinhibitory domains can instead be intramolecular as in myosin light chain kinase where the inhibitory domain folds back into the active site and fills it as a pseudosubstrate inhibitor, blocking access of substrate proteins (Kobe and Kemp, 1999) (Figure 2.14).

The regulatory subunits can instead be *activating* if they permit productive orientation of ATP and protein substrate in the active site of the catalytic subunit. This is the logic used by the two subunit cyclin dependent kinases (Brown et al., 1999), where the cyclin subunits are positive regulators and increase k_{cat} over basal activity 300-fold (Johnson, 2001; Lowe et al., 2002) while protein K_m is lowered 100-fold for an overall gain in catalytic efficiency, k_{cat}/K_m, of $>10^5$.

An additional general mode of controlling activity of the two-lobed protein kinase catalyst domains (Johnson, 2001; Johnson et al., 2001), a set of PK catalytic domains, is phosphorylation of a threonine side chain in an activation segment or loop near the active site. Phosphorylation serves as an organizing element as noted above, by hydrogen bonding and electrostatic interactions with conserved arginine side chains, to control conformation and location, driving reorganization to convert inactive catalytic domain conformers into active species where orientation of ATP and/or the protein cosubstrate is productive for rapid phosphoryl transfer.

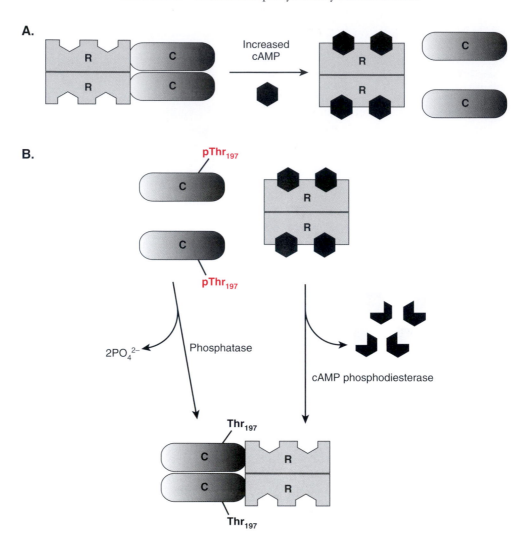

Figure 2.13 **A.** Schematic for dissociation of regulatory subunits from protein kinase A as cyclic AMP concentration rises. **B.** Subsequent activation of the catalytic C subunit by phosphorylation in the activation loop generates fully active kinase. Dephosphorylation by phosphatases and hydrolysis of cAMP by cAMP phosphodiesterase allow regeneration of the inactive R_2C_2 resting form of protein kinase A.

As a prototype, PKA requires conversion of T_{197} to pT_{197} in the activation segment (Figure 2.15A) for the liberated catalytic domain to be substantially active (Figure 2.13B). Analogously, human Cdk2 needs to be phosphorylated on T_{160} (Johnson, 2001) by a partner kinase to achieve the active state. In Cdk2, as in PKA, the activation is two-step: binding (Cdk2) or release (PKA) of the regulatory subunit, allowing access of the threonyl-OH side chain in the activation loop for subsequent phosphorylation by a partner kinase. In PKA the pT_{197} now hydrogen bonds and charge pairs with an R guanidinium ion side chain to reorganize the active site. In Cdk2 the

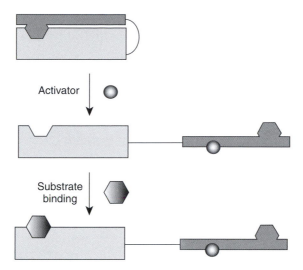

Figure 2.14 Occupancy of the active site of myosin light chain kinase by an intramolecular inhibitory domain. Activator binding to inhibitory domain competes with the catalytic domain and frees up the catalytic domain for substrate binding.

phosphoryl group in the pT_{160} form binds to three arginine residues (R_{50}, R_{126}, and R_{150}) to drive the activation segment (Figure 2.15B) and adjacent active site reorganization to recognize protein substrate for productive binding. The cyclin A subunit binding optimizes orientation of bound ATP while the pT_{160} network of interactions sets up protein cosubstrate orientation, two steps for the two substrates to be optimally positioned (Harper and Adams, 2001). The multiple-step logic for PK activation controls the "off/on" switching in time and space. The crucial phosphorylation of the threonyl side chain in the activation loop of PKA and also PKB and PKC can be carried out by the phosphoinositide dependent protein kinase, PDK1, linking phospholipids-phosphorylation-based signals and protein-kinase-based signaling.

MAP kinases, noted below, require tandem phosphorylation of two side chains in the activation loops before significant pS/pT kinase activity is exhibited by the catalytic domain, another double fail-safe mechanism. The ERK2 x-ray structure shows that this MAP kinase family member, once doubly phosphorylated in its activation segment at both T_{183} and Y_{185}, can use the pT_{183}/pY_{185} pair of phosphates to make contact with five arginines (R_{68}, R_{146}, and R_{170} from pT_{183} and R_{189} and R_{192} from pY_{185}) to drive refolding of the activation segment and part of the active site (Figure 2.15C). This sets up the active site to recognize both a serine side chain in a protein partner and the following proline residue, explaining the –SP-recognition of this proline-directed pS kinase family (Johnson, 2001). The conformation changes also drive the dimerization K_D from 20 μM down to 7.5 nM, enabling dimerization and subsequent translocation of activated ERK2 to the nucleus.

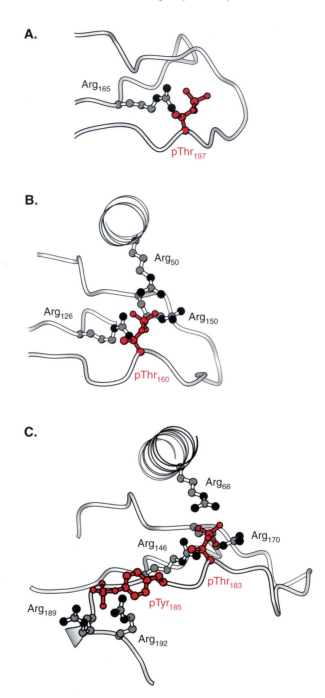

Figure 2.15 **A.** Phosphorylation of PKA on its activation segment converts Thr_{197} to $pThr_{197}$, allowing electrostatic interactions with Arg_{165} that drive consequent reorganization of that region of the active site (figure made using PDB 1CDK). **B.** Analogous conversion of Thr_{160} to $pThr_{160}$ during activation of the catalytic subunit of Cdk2 generates electrostatic interactions with Arg_{50}, Arg_{126}, and Arg_{150} (figure made using PDB 1JST). **C.** Interactions of the doubly phosphorylated activation segment $pT_{183}XpY_{185}$ in MAP kinase with five arginines (Arg_{68}, Arg_{146}, and Arg_{170} to pT_{183} and Arg_{189} and Arg_{192} to pY_{185}) (figure made using PDB 2ERK).

Parallel Cascades of Kinases

The mitogen activated protein kinases (MAPKs), such as ERK2 noted in the preceding paragraph, are famous for carrying protein-based signaling information from plasma membrane to nucleus via parallel kinase cascades of three sequentially acting kinases (Chen et al., 2001; Johnson and Lapadat, 2002), from yeast to man. The tandemly acting kinases are termed MAPKKK (also designated MKKK), MAPKK, and MAPK, in that order (Figure 2.4), with the MAP kinase kinase kinase the most upstream member of the cascade and MAPK the most downstream. The MAPKKK component phosphorylates the activation segment of MAPKK on a serine side chain, by the logic noted above. The pS form of MAPKK is now activated and phosphorylates the activation loop of MAPK to activate it. As noted above, double phosphorylation of the MAPK loop is required for substantial activity and a separate pY kinase is also required to make pY_{185}. The trio of kinases leads to cooperative activation of the signaling pathway and a steep signal amplification response (Huang and Ferrell, 1996).

MAP kinases have been implicated in many cellular responses, including gene expression, mitosis, cell movement, cell death pathways, responses to stress, and responses to growth factors (mitogens). There are separate well-defined MAPK cascades (Johnson and Lapadat, 2002) that can act in parallel with particular kinase isoforms selecting their upstream and downstream partner kinases to mediate responses to different stimuli arriving at the outside of the cell (Figure 2.4). Growth factors activate one pathway, via the Ras GTPase as proximal membrane/cytosolic activator. The MKKK is c-Raf. It phosphorylates MKK1 and then activated MKK1 activates ERK1/2. One of the targets of activated MAPK is the p90R6K (ribosome small subunit 6 kinase) on the way to regulation of mitosis and meiosis.

A second MAPK cascade can be activated by integrin binding to its membrane receptors, activating the Rac1 GTPase and then turning on the cascade of MEKK1, MKK4, and JNK1 with phosphorylation of the transcription factor c-Jun1 as a downstream output from JNK1. A third cascade starts with oxidative stress as input, activates the tyrosine kinase Src and then the cascade of MEKK2, MKK5, and ERK5 to phosphorylate myocyte enhancing factor-2 (MEF2). A fourth example in immune cells to turn on cytokine gene expression involves IL-1 as stimulus, TRAF6/TAB1/2 as proximal activators, and the cascade of TAK1, MKK6, and p38 to lead ultimately to phosphorylation of a fourth kinase—the MNK1 enzyme.

The activation principles of all of these parallel MAPK cascades follow the molecular logic of activation loop phosphorylation to drive inactive kinase domains to active conformations by $-PO_3^{2-}...Arg^+$ interactions, where specific recognition of partner proteins, on their activation loops, leads to controlled phosphoryl transfers.

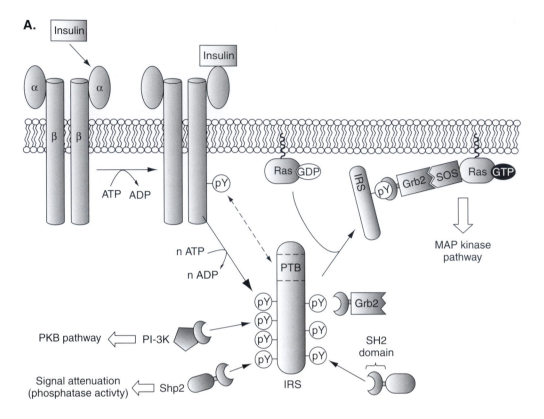

Figure 2.16 Signaling pathways proceed via cascades of protein phosphorylations and dephosphorylations. **A.** The insulin receptor tyrosine kinase (IRK) recruits IRS1 to the autophosphorylated insulin receptor by the PTB domain of IRS1. After IRS1 docks onto pY-IR, up to 18 Y residues can then be phosphorylated. The pY residues of IRS1 can independently recruit partner proteins with SH2 and/or PTB domains, each of which can nucleate or carry a separate arm of a downstream signaling pathway.

Protein Anchors and Adaptors in Phosphoprotein Signaling

One of the devices used to channel PKA to so many diverse protein substrates in eukaryotic cells involves A kinase anchoring proteins (AKAPs), a family of proteins with PKA binding domains and additional targeting domains for particular subcellular locales (Pawson and Scott, 1997). The AKAPs can act as scaffolds for assembly of both PKA and sets of protein substrates in particular regions of cells. For example, the N-myristoylated form (see Chapter 7) of AKAP-18 uses the N-terminal fatty acyl moiety to localize at the plasma membrane and there binds both the PKA catalyst and the L-type Ca^{2+} channel subunit as phosphorylation substrate (Hunter, 2000). Motifs in the anchoring proteins interact with sequences in the regulatory domains of PKA.

There are comparable scaffolds for other protein kinases, including PKC for localization at membrane interfaces, thus facilitating activation by lipids (Newton, 2001). In the MAP kinase cascades, the multiple kinases acting tandemly can be assembled on a common scaffold protein to amplify the flux of sequential kinase activation.

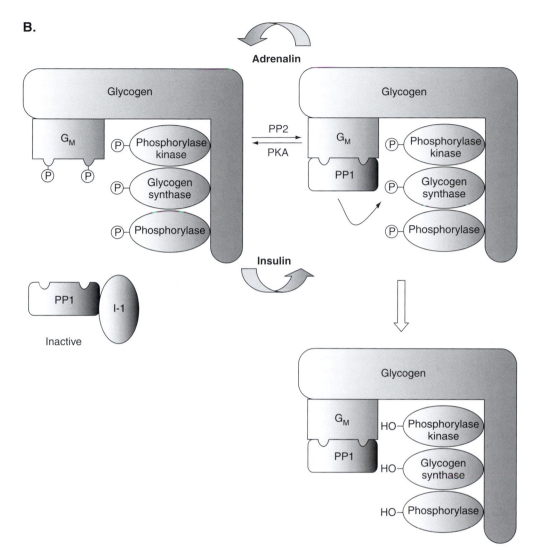

Figure 2.16 **B.** The mobilization of glycogen to glucose-1-P molecules is controlled by the phosphorylation states of phosphorylase kinase, glycogen synthase, and phosphorylase associated with glycogen particles. The pS/pT protein phosphatase PPI is recruited to the G_M subunit and then can dephosphorylate the neighboring phosphorylated proteins on the particle.

The scaffold protein Ste5 was first detected in the yeast mating factor MAP kinase cascade (Elion, 2001) and has led to subsequent appreciation of mammalian counterparts.

We noted above that signaling by the insulin receptor tyrosine kinase involves recruitment of the adaptor insulin receptor substrate (IRS) proteins by the NPEpY sequence in the autophosphorylated receptor and then the IRS proteins get phosphorylated on multiple tyrosine residues. Each pY site thus generated can recruit a specific SH2-containing partner protein to assemble sets of protein complexes. This is a two-stage assembly using pY residues as the phosphorelay signal at each of the two stages, with anchor and adaptor proteins, respectively (Figure 2.16A). This anchor logic also extends to the protein phosphatases discussed subsequently in this chapter where catalytic phosphatase subunits are directed to particular subcellular locations by regulatory subunits, which act also as anchoring elements for particular target structures. This is illustrated for the insulin-initiated recruitment of PP1 to glycogen particles by the targeting subunit G_M, bringing the protein phosphatase to the neighborhood of phosphorylase kinase, glycogen synthase, and phosphorylase molecules affiliated with and working on the glycogen substrates (Figure 2.16B). PP1 can then convert the phospho forms of these enzymes to dephospho forms and alter the balance of glycogen synthesis versus breakdown. The balance of enzymatic phosphoryl group addition and hydrolytic removal is modulated not only by the intrinsic activity of kinases and phosphatases, but their relative concentrations are modulated by scaffold, anchor, and adaptor proteins.

Transmembrane Serine Kinases

The human kinome contains 12 protein serine kinases that are predicted to be transmembrane proteins. The prototype is the kinase that acts as receptor for the cytokine tumor growth factor β (TGFβ). Engagement of TGFβ as protein ligand to the extramembrane domain is thought to dimerize the receptor serine kinase (RSK) and lead to transphosphorylation of the two catalytic domains in the cytoplasm (Figure 2.17), activating the kinase domains. The activated kinase domains then phosphorylate specific partner proteins, SMADs, which can move to the nucleus as dimers and act as transcription factors to up-regulate the TGFβ-responsive genes. These receptor serine kinases are homologs to the large class of receptor tyrosine kinases discussed in the next section.

Tyrosine Kinase Action

There are 91 protein tyrosine kinases (PTKs) predicted in the human kinome (Manning et al., 2002). This 20% of the 500 total protein kinases carries about 10% of the total phosphoprotein flux in cells. Of the 91 PTKs, 59 are transmembrane receptor tyrosine kinases (RTKs) analogous to the RSKs discussed previously, while 32 are nonreceptor tyrosine kinases.

The catalytic domains of the PTKs, both receptor and nonreceptor types, conform to the general architecture seen with the pS/pT kinases (Figure 2.12). Just as the serine/threonine kinases typically require phosphorylation of a threonine side chain in the activation segment to organize an active conformer, so the tyrosine kinases in general require conversion of one or more tyrosines to a pY residue in the comparable activation loop. The catalytic domain of the FGF receptor kinase requires

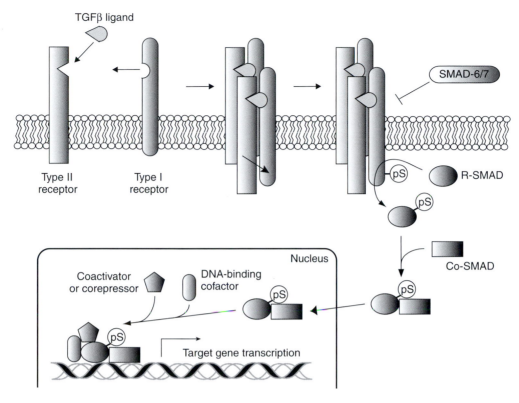

Figure 2.17 The TGFβ receptor is a transmembrane serine kinase. Engagement of the growth factor protein TGFβ on the extramembrane side of the receptor leads to dimerization and the autophosphorylation *in trans* of the cytoplasmic domains, one kinase domain phosphorylating serine(s) on the other chain of the dimer. The pS forms of the receptor recruit partner proteins (SMADs) that oligomerize and transmit signals to the nucleus for selective gene activation.

$pY_{653}pY_{654}$ in the loop for full activity while the catalytic domain of the insulin receptor has $pY_{1152}pY_{1162}pY_{1163}$. The nonreceptor kinases of the Src family have a comparable tyrosine (e.g., pY_{416}) that allows gain of catalytic activity (Kuriyan and Cowburn, 1997; Schindler et al., 1999; Xu et al., 1999). Typically there are additional noncatalytic domains that serve regulatory functions, both for recognition of activating ligands and/or for autoinhibition to keep the tyrosine kinase activity off in the nonsignaling, basal states. Failure to achieve tight regulation leads to loss of growth control and oncogenesis in both RTKs and nonreceptor PTKs.

In the RTKs, the extracellular domains recognize protein ligands such as insulin, nerve growth factor (NGF), epidermal growth factor (EGF) (Burgess et al., 2003), platelet derived growth factor (PDGF), fibroblast growth factor (FGF), and vascular endothelial growth factor (VEGF). These RTKs thus serve as the cognate receptors for these hormones and growth factors: insulin receptor (IR), NGFR, EGFR, PDGFR, FGFR, and VEGFR. The EGFR and its ligand EGF are schematized in

Figure 2.18 (Jones and Kazlauskas, 2001), which also notes that the kinase domain of the EGFR fits the paradigm of protein kinase folds. The external domains of each RTK have multiple domains, some of which function in ligand recognition. Each RTK has a single transmembrane domain. The cytoplasmic portions of RTKs have a

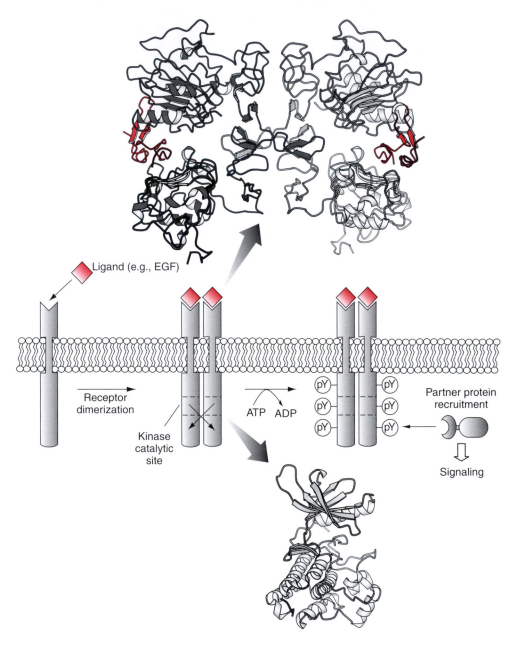

Figure 2.18 Prototypic organization of a transmembrane receptor tyrosine kinase, the EGF receptor (EGFR), and recognition of its ligand EGF. Binding of EGF to the external receptor domain has an *in trans* activating effect on the cytoplasmic tyrosine kinase domain [figure made using PDBs 1IVO (extracellular) and 1M14 (intracellular)].

juxtamembrane domain of about 40 residues, a catalytic domain of about 250 residues, and then a C-terminal domain of variable length. Ligand binding to the external domains typically induces RTK dimerization and the transphosphorylation of tyrosine residues in the cytoplasmic domains, at the juxtamembrane regions, in C-terminal domains, and sometimes in insert regions in the catalytic kinase domains (Hubbard and Till, 2000). Multiple phosphorylations create a tandem array of pY domains that recruit partner proteins that contain specialized pY-binding domains. These pY sites are in addition to the activation loop tyrosine phosphorylations noted in the previous paragraph.

Two such pY-binding domains are SH2 (Src homology 2) and PTB (phosphotyrosine binding) domains (Figure 2.9). The 100-residue SH2 domains (first found in the Src family of PTKs) recognize the dianionic character of the phosphate and the longer reach of pY over pS/pT side chains to enable specificity for phosphotyrosine. Particular SH2 domains can discriminate specific pY sequences by extended recognition of side chains at –3 to +3 sites and so enable selective binding to subsets of pY sequences in PTKs and in other pY proteins to recruit distinct partners (Hunter, 2000; Kuriyan and Cowburn, 1997; Pawson and Nash, 2003). Once recruited, the partner proteins can be phosphorylated on tyrosine and serve to initiate cytoplasmic signaling pathways. For example, the insulin receptor autophosphorylates itself in its cytoplasmic domain and through those pY side chains recruits scaffold proteins known as insulin receptor substrates (IRS). Up to a dozen or more tyrosine residues in IRS proteins docked on the activated IR can be converted to pY residues (White, 1994). These serve in turn as sequence-selective pY sites to recruit the next shell of partner proteins for building protein complexes to carry out insulin-initiated signaling (Figure 2.16A).

Analogously, the nerve growth factor receptor TrkA, once engaged by NGF externally, autophosphorylates in its juxtamembrane region, thereby using the NPXpY sequence to recruit the adaptor protein Shc via its PTB domain, phosphorylating Shc on a tyrosine residue. That pY residue then recruits the adaptor proteins that bring the Ras GTPase to the complex and start a MAP kinase cascade. The FGF receptor instead autophosphorylates itself at a C-terminal tyrosine residue, which mediates recruitment of the enzyme phospholipase Cγ (PLC) through its SH2 domain. PLC cleaves phospholipids at the membrane interface to release diacylglycerol, an activating ligand for protein kinase C (Hubbard and Till, 2000; Newton, 2001).

The Src subfamily of about 10 nonreceptor PTKs, including Lyn, Fyn, Hck, and Lck, contain canonical Src homology 1, 2, and 3 domains. SH1 is the catalytic tyrosine kinase domain, SH2 is the pY binding domain, and SH3 is a 60-residue domain that binds polyproline type peptide regions. The x-ray structures of Src and Lck revealed the structural basis for how Src type PTKs keep themselves in the "off" state at rest (Schindler et al., 1999; Xu et al., 1999). Inactive Src, at rest, is phosphorylated by the Src homolog C-terminal Src kinase (Csk) on a C-terminal tyrosine at residue 527. The SH2 domain binds this pY_{527} intramolecularly while the SH3 domain also binds an intramolecular peptide linker region (Figure 2.19). The two conformation constraints thus imposed allosterically keep the catalytic domain in an inactive con-

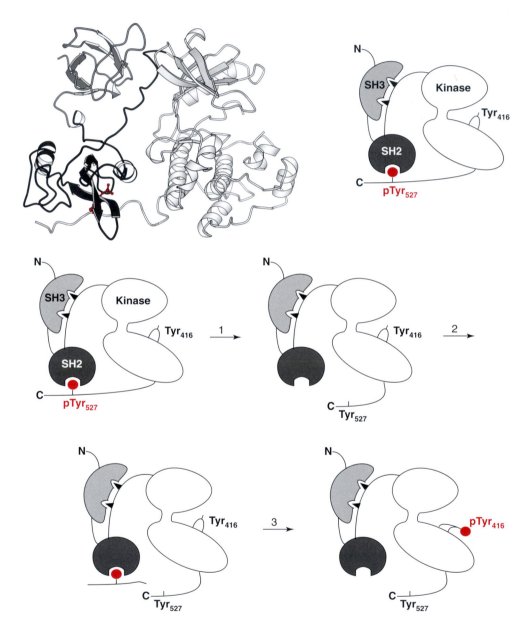

Figure 2.19 Multidomain structure of Src and homologous nonreceptor tyrosine kinases. Auto-inhibition by *in cis* complexation of the SH2 and SH3 domains. Activation by dephosphorylation of pY_{527}, competition for SH2 by pY protein ligands, and phosphorylation of Y_{416} in the activation segment (figure made using PDB 1FMK).

formation. Enzymatic dephosphorylation of pY_{527}, followed by autophosphorylation of Y_{416} in the activation loop, disengages the SH2 domain and organizes the activation loop and the active site for multistep gain of activity. The Src PTKs are N-myristoylated (see Chapter 7) and this lipid anchor helps with membrane association and localization in the vicinity of substrate proteins.

The Janus kinases (JAKs) function as dissociable partner subunits for the transmembrane single subunit cytokine receptors, which lack a covalently embedded tyrosine kinase domain, such as the interferon and interleukin receptors (Hunter, 2000). The JAKs behave as tyrosine kinase subunits on cytokine ligation to the extracellular domains of their receptors. Autophosphorylation of JAKs recruits the SH2 domain-containing STAT proteins (signal transducers and activators of transcription). Once in the vicinity of the activated JAKs, the STAT proteins are phosphorylated on tyrosine, and the pY side chain electrostatically and via hydrogen bonding drives dimerization of the pY STATs. Now the dimeric STATs migrate to the nucleus and act as transcription factors, directly binding to DNA promoter elements of genes responsive to the cytokines for target cell proliferation or differentiation.

Histidine Kinases and Phosphoaspartate Response Regulators

Protein histidine kinases are found most extensively in bacteria, partnered with protein substrates for phosphorylation, on a conserved aspartate residue, that act as response regulators. These two-component systems are utilized to sense external signals and transmit that information for selective gene activation in response. Typically, the sensor histidine kinase is a transmembrane protein, functionally analogous to eukaryotic receptor tyrosine kinases, with extracellular domains that sense the specific external signal, such as osmolarity or high Mg^{2+} levels, peptide pheromones, or antibiotics such as vancomycin (Hoch, 1995; Walsh, 2003).

Each histidine kinase has a transmembrane domain and then cytoplasmic domains that include dimerization domains and the catalytic kinase domain. The CheA histidine kinase (Figure 2.20) involved in bacterial chemotaxis lacks the transmembrane domain but has the dimerization and His kinase domains. On activation by external ligand and presumed dimerization, the kinase domains act *in trans* to autophosphorylate a specific histidine in the partner chain via ATP as donor substrate for the $-PO_3^{2-}$ group. The histidine converted to N-phosphohistidine at one of the imidazole ring nitrogens (Figure 2.1B) is often in the region of the dimerization domain. Again, this logic parallels the covalent autophosphorylation of tyrosyl residues by RTKs noted previously. Now the mechanisms by which RTK and His kinase autophosphorylations are read by cells diverge. Whereas pY side chains are stable, thermodynamically and kinetically, and get read by adaptor domains, such as the SH2 and PTB domains, the $N–PO_3$ linkage in phosphohistidine retains thermo-

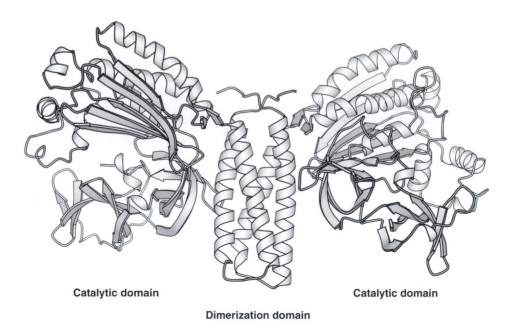

Catalytic domain   **Catalytic domain**

Dimerization domain

Figure 2.20 Histidine kinases: catalytic and dimerization domains of the CheA kinase involved in bacterial chemotaxis. Transphosphorylation occurs in one active site in which ATP is used to phosphorylate a His residue in the other subunit of the dimer (figure made using PDB 1B3Q).

dynamic activation as a phosphoryl donor under physiological conditions. Signaling proceeds by direct transfer of this activated PO_3^{2-} from the His residue of the sensor kinase to an aspartate side chain in the N-terminal domain of the partner response regulator protein.

Response regulators are generally, but not always, two domain proteins (the CheA chemotactic protein is a variant form without the extramembrane or transmembrane domains). An N-terminal domain, with a nucleophilic Asp, serves as receiver, and a C-terminal domain serves as the output domain, most commonly a DNA binding domain that can serve as a transcriptional activator or repressor by sequence-specific DNA binding to promoters of target genes (Hoch, 1995; West and Stock, 2001). The attack of the Asp–COO^- side chain of the receiver domain on the His–N–PO_3^{2-} is catalyzed by the receiver domain after specific complexation of the two proteins in the bacterial cytoplasm. Introduction of the tetrahedral phosphate in the receiver domain of the response regulator protein allows reorganization of the ligation sphere to a bound Mg^{2+} and propagation of a conformational change to the C-terminal domain, activating it for productive binding to DNA sequences (Figure 2.1). The aspartyl–$COOPO_3^{2-}$ group is a mixed carboxylic–phosphoric anhydride linkage and intrinsically much more labile to uncatalyzed hydrolysis than pSer, pThr, pTyr, and pHis residues.

The autodecomposition of the phosphoAsp forms of response regulators back to dephospho forms and P_i can occur in minutes to hours and can represent a built-in clock mechanism for limiting the duration of the response. Indeed, there are very few examples of protein phosphoaspartyl phosphatases known, presumably because this short lifetime obviates most of the need for such catalysis. There is such a bona fide phosphatase, SpoE, in the *Bacillus subtilis* sporulation pathway noted below (Fabret et al., 1999). It may be that the short lifetimes of pAsp side chains in bacterial response regulator proteins are perfectly adapted for the short time frames over which bacteria receive and integrate external stimuli. While the molecular logic of the transmembrane histidine kinases parallel that of the eukaryotic RTKs, the structure of the histidine kinase catalytic domain is not that of a typical pS or pY kinase. Rather, the ATP-binding domain is analogous to the ATP-cleaving active sites in the heat shock protein Hsp90 and the GyrB subunit of DNA Gyrase (Ban and Yang, 1998), suggesting evolution in a distinct protein fold lineage.

There are 14 pairs of sensor kinases and response regulators detected in the *Streptococcus pneumoniae* genome, and 63 or 64 in the opportunistic pathogen *Pseudomonas aeruginosa*, reflecting its extensive capacity to respond to many microenvironmental inputs and changes (Walsh, 2003). Two-component sensor kinase/response regulators are involved in virulence in several bacteria, including *Staphylococcus aureus* and the vancomycin-resistant *Enterococcus faecalis*.

Three of the most extensively characterized bacterial two-component systems are the EnvZ/OmpR osmosensing pair, the CheA/CheY chemotaxis system, and the Spo0 sporulation pathway in the Gram-positive bacterium *B. subtilis* (Fabret et al., 1999; West and Stock, 2001). The CheA protein is atypical in that it has the cytoplasmic His kinase domain but not the transmembrane or extramembrane domains. It uses transmembrane chemotaxis receptors (Chapter 5) for signal inputs. The Spo0 system is intriguing as an elaboration of the two-component phosphorelay system into a four-component phosphorelay system, with alternating pHis–pAsp/pHis–pAsp intermediates in the KinA/B→Spo0F→Spo0B→Spo0A (Figure 2.21) cascade. This

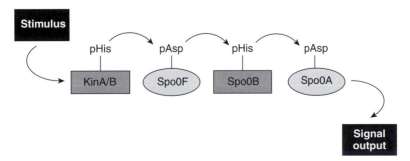

Figure 2.21 A four-step phosphorelay (His → Asp → His → Asp) in the Spo0 pathway for sporulation in *Bacillus subtilis*.

four-step phosphorelay, essentially a duplication of the pHis, pAsp basic two-component relay, approaches the lengths of eukaryotic MAPKKK, MAPKK, and MAP kinase cascades. The multiple steps presumably allow integration of information at each of the phosphoprotein intermediates. The pAsp in Spo0A is the substrate for the SpoE phosphatase activity as a modulator of sporulation signal duration and intensity.

In yeast, there is one histidine kinase (Saito, 2001) termed Sln1 and two receiver domain proteins, Ssk1 and Skn7, reflecting two parallel two-component pathways. The Sln1–Ssk1 pair is involved in osmoregulatory responses in *S. cerevisiae*, while the Sln1–Skn7 pair is a His–Asp phosphorelay involved in oxidative stress response. Each of these signals feeds into more traditional MAP kinase pathways in yeast, tieing together histidine kinase logic and pS/pY kinase logic in one pathway. In other lower eukaryotes and plants, His–Asp protein phosophorelay logic is used to sense external molecules such as oxygen, ethylene, cAMP, and cytokinins (Saito, 2001).

Turning "Off" Protein Kinase Signaling

With hundreds of protein kinases, pS/pT kinases, and pY kinases potentially operating in eukaryotic cells, there need to be strategies for controlling the duration and intensity of protein kinase signaling. Three possibilities immediately come to mind. One is to remove the activating stimulus for particular protein kinases or sets of protein kinases. When the ligands are small molecules, their concentration could be lowered to abrogate the initiating signal. Thus, cAMP can be cleaved hydrolytically to AMP by cAMP phosphodiesterases. As [cAMP] falls, it will dissociate from the PKA regulatory subunits and they can reassociate with PKA catalytic subunits to return to the inactive R_2C_2 heterotetramer. (There is, however, the troublesome matter that pT_{197} needs to be returned to the dephospho-T_{197} state to get the affinity of the catalytic subunit for the regulatory subunit back to the starting point). Analogously, the elevated Ca^{2+} concentration that activates the calmodulin-dependent PKs can be pumped out of cells or back into the endoplasmic reticulum to reverse the activating signal and enable dissociation of Ca^{2+} from calmodulin.

A second route to reversal of protein kinase activation can be proteolysis. This is famously the strategy in the cyclin-dependent cell cycle kinases where ubiquitylation (see Chapter 9) of the cyclin subunits targets them for rapid destruction by the proteasome. Absent the activating cyclin subunits, the Cdk catalytic subunits lose orders of magnitude in catalytic efficiency.

In the case of Wnt signaling in animal development, cytoplasmic β-catenin is continuously phosphorylated and thereby signals for its polyubiquitylation and proteasome destruction by a phosphorylation-directed ubiquitin ligase (Amit et al., 2002).

The phosphorylation-directed proteolytic destruction signal requires the tandem action of two protein kinases: casein kinase I (CKI), then glycogen synthase kinase-3β (GSK-3β) (Ali et al., 2001). CKI recognizes Ser_{45} of β-catenin and the subsequent pS form of catenin serves as a priming signal for GSK-3β to act at three nearby upstream catenin side chains—T_{41}, S_{37}, and S_{33}. In the absence of the CKI phosphorylation, β-catenin is a poor substrate for GSK-3β. GSK-3β finds β-catenin in a complex presented by a pair of scaffold proteins, Axin and APC, which GSK-3β also phosphorylates. When Wnt3A protein is secreted by neighboring cells it comes to the outside of the target cell, binds to the transmembrane receptor frizzled, and activates the cytoplasmic protein disheveled in a phosphorylation process. Phospho-disheveled then prevents the action of GSK-3β on β-catenin. Now the monophospho-$pSer_{45}$ form of catenin is not recognized by the ubiquitin ligase, so it is not destroyed continuously by the proteasomes. As β-catenin concentration builds up in the cytoplasm it can transit to the nucleus and act as transcriptional coactivator of TCF/LEF-mediated gene transcription.

The third form of control of PK action is the reversal of the covalent posttranslational phosphoryl modification on phosphoproteins, including the hydrolytic removal of the activating pT and pY loop residues in the many protein kinases. This is the strategy that utilizes protein phosphatase catalytic action.

The Aurora A Ser/Thr kinase acts during the cell cycle in the nucleus on proteins associated with the centromere and appears to move to different subnuclear structures during the cell cycle (Carmena and Earnshaw, 2003). It gets phosphorylated at three sites—Ser_{59}, Thr_{295}, and Ser_{349} (Littlepage et al., 2002). The $pThr_{295}$ is in the activation loop and is required for catalysis. Dephosphorylation of this pThr residue by protein phosphatase 1 (described below) leads to deactivation. The $pSer_{59}$ may direct the kinase for recognition by a phosphoprotein-recognizing subunit of the polyubiquitylating ligase that marks the kinase for proteolytic destruction (see Chapters 8 and 9).

Phosphoprotein Phosphatases

The most straightforward route to control the myriad effects of protein phosphorylations is to reverse the action of protein kinases—that is, to hydrolyze the phosphoryl moieties from pS, pT, and pY sites in proteins (Figure 2.5). Just as there are serine/threonine and separate tyrosine kinase families of catalysts, there are phosphoserine/phosphothreonine-specific and phosphotyrosine-specific families of phosphoprotein phosphatases (Kennelly, 2001), with about 150 protein phosphatases total in the human genome (Jackson and Denu, 2001), some 107 of them as pY-specific phosphatases (Alonso et al., 2004). In yeast the genome encodes 37 phosphoprotein phosphatases (Virshup, 2000). While the net hydrolysis of pS and pT versus pY side chains is highly related, the catalytic mechanisms for the phosphoryl transfer catalysts are quite distinct. The pS/pT phosphatases are typically metalloenzymes, using Fe^{3+} and Zn^{2+} in a bimetallic center to activate water for the hydrolytic attack on the phosphorus of the PO_3 group (Figure 2.22A). By contrast, phosphotyrosine phosphatases (PTPases) use covalent catalysis, adding the nucleophilic thiolate side chain of the active site cysteine into the phosphoryl group, thus releasing the tyrosyl leaving group and creating a covalent $S–PO_3$ thiophosphoryl enzyme intermediate (Figure 2.22B). This covalent intermediate is decomposed hydrolytically in a second half-reaction in which water attacks the electrophilic phosphorus atom. In Chapter 4 we note the susceptibility of the reactive thiolate in PTPase active sites to oxidation by oxidants such as the HOOH generated by growth factors during signaling.

Phosphatase specificity can be evaluated against candidate proteins and also against arrays of phosphopeptides spotted on plates (Espanel et al., 2002). The balance of kinase/phosphatase concentration and catalytic efficiency in a given microen-

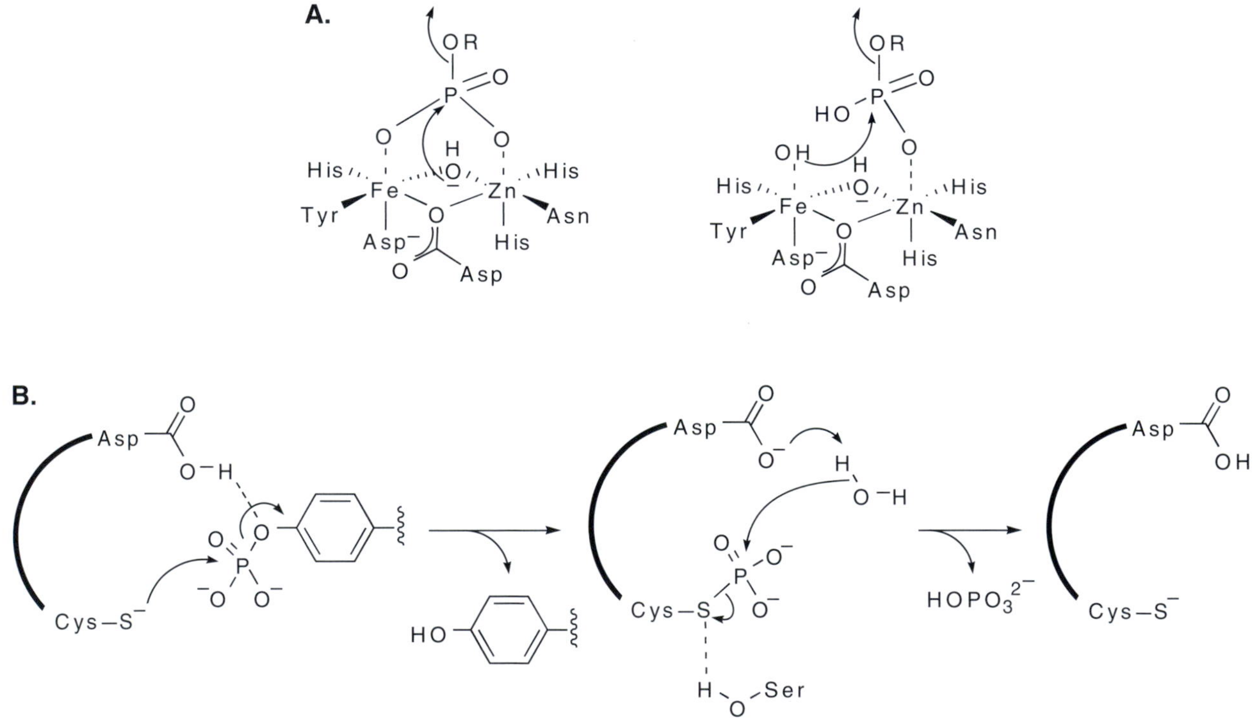

Figure 2.22 Phosphoprotein phosphatases. **A.** Direct hydrolysis mechanism by bimetallic center at the active sites of pS/pT protein phosphatases such as PP1. **B.** Nucleophilic catalysis in PTPase catalysis, generating a covalent thiophosphoryl covalent enzyme intermediate in a two-step enzyme phosphorylation-dephosphorylation reaction sequence.

vironment, towards a given phosphoprotein site, will determine the net amount of the phosphoproteome at any point in time and space within cells.

Protein Serine Phosphatases

Two major families of eukaryotic phosphoprotein phosphatases are encompassed by PP1 and PP2A members. The catalytic subunit of PP1 enzymes are active for hydrolytic removal of the $-PO_3^{2-}$ group from pS and pT side chains in a wide range of sequence contexts. Selectivity is imposed by at least 30 distinct variants of regulatory subunits that interact with a common sequence motif (–R/KR/KV/IXF/Y–) on the back side of the catalytic subunit to direct the PP1 holo enzyme to different target sites within particular cell types (Figure 2.17) (Aggen et al., 2000). In liver a 124-kDa subunit targets PP1 to glycogen particles where it can control the phosphorylation status of glycogen synthase, glycogen phosphorylase, and glycogen phosphorylase kinase, modulating glycogen accumulation and breakdown and opposing the kinases that phosphorylate those three enzymes. Analogously in striated muscle, an M110 subunit targets that PP1 holo enzyme to myosin and can balance the kinase activity of such enzymes as myosin light chain kinase. Nuclear targeting subunits direct PP1 activity towards nuclear phosphoproteins. Several protein inhibitors of PP1 isoforms are known to provide another layer of regulation in the dynamics of phosphoprotein-level control. An additional layer of regulation in PP1-G_M complexes is the phosphorylation of two serine side chains in the G_M subunit (Ser_{46} and Ser_{65}), which can cause G_M dissociation and release of the catalytic phosphatase subunit from glycogen particles and/or activation of the catalytic subunit towards phospho forms of glycogen synthase (Cohen, 2000).

PP2A protein phosphatases also are complexes of catalytic and regulatory subunits, one of which is a scaffolding element, the A subunit with 15 repeats of a 39-residue domain (Virshup, 2000). PP2A trimers can be found in association with the GSK-3β/β-catenin complexes that are central to Wnt signaling noted previously. Since the phosphorylation status of catenin determines its lifetime and efficacy as a transcription factor, the PP2A activity is a key component.

Two protein pSer phosphatases are specific for dephosphorylating the C-terminal domain (CTD) of RNA polymerase II (PolII) (Kamenski et al., 2004). The CTD can consist of up to 52 repeats of the heptad YSPTSPS, with Ser_2 and Ser_5 sites undergoing reversible phosphorylation and dephosphorylation. The dephosphoCTD form is required for PolII initiation of mRNA translation, while hyperphosphorylated CTD forms are required for mRNA elongation where the anionic phosphate-rich CTD recruits mRNA processing factors. Then, for recycling of PolII to start another transcription cycle, the CTD must be dephosphorylated at the dozens of pSer sites in the

7×52 residues of the CTD. The x-ray structure of the Scp1 CTD phosphatase (Kamenski et al., 2004) suggests this catalyst may transfer the PO_3^{2-} group from pSer residues on the CTD substrate to an active site Asp, generating a transient pAsp intermediate in each of the dozens of catalytic dephosphorylation cycles required to regenerate the form of PolII competent for mRNA initiation.

Dual Specificity Phosphoprotein Phosphatases

There is a family of phosphoprotein phosphatases that are transitional between pS/pT phosphatases and pY phosphatases by virtue of the fact that they can hydrolyze phosphoryl groups on all three kinds of side chains. They are dual-specificity phosphatases (DSPases) (Keyse, 2000), and while they can accelerate hydrolysis of both types of phosphorpoteins, they use the covalent thiophosphoryl enzyme mechanistic route typical of PTPases detailed below. There are 12 DSPases in the yeast proteome and 10 identified in humans. The preferred substrates for DSPases that act as MAP kinase phosphatases are thought to be various isoforms of MAP kinases that contain the pTXpY dyad required for kinase activity (Jackson and Denu, 2001; Zhan et al., 2001). The DSPases typically have a noncatalytic N-terminal domain that recognizes the MAP kinase partner and then presents it in an intramolecular complex to the catalytic domain at the C-terminus of the DSPase. This prior binding of the ERK form of MAP kinase by the noncatalytic domain of the DSPase of MKP3 enhances catalytic efficiency for dephosphorylation of phospho forms of ERK by >4000-fold (Kim et al., 2003). The Cdc25A, B, and C isoforms of dual-specificity phosphatases show some selectivity for a given Cdk–cyclin pair.

Protein Tyrosine Phosphatases

Of the 90 PTPases in mammals, about 60 are transmembrane enzymes, termed receptor transmembrane phosphatases (RTPs) in analogy to RTKs. The RTKs are localized in the same two-dimensional plane of the cell membrane as the RTPs and these are the preferred substrates. The prototypic RTP is CD45 in lymphocyte membranes and also the leukocyte antigen related (LAR) PTPase (Majeti and Weiss, 2001). The model for regulated action of RTPases is ligand engagement on the extracytoplasmic domain, activation of the cytoplasmic catalytic PTPase domain, and action on pY residues on neighboring proteins, including cytoplasmic autophosphorylation sites of RTKs.

Of the many cytoplasmic PTPases, substantial attention has been paid to PTP1B, given evidence that its activity is relevant to controlling the duration and intensity of insulin signaling in some diabetics (Elchebly et al., 1999). There are two known

PTPases, Shp1 in lymphoid cells and Shp2 broadly expressed, with tandem SH2 domains at the N-terminus of the protein as internal regulatory domains (Neel et al., 2003). The two SH2 domains act somewhat differently. The most N-terminal SH2 acts as *in cis* ligand for a pY residue at the C-terminus of the phosphatase (Figure 2.23) and shuts down the PTPase domain, much the way internal ligation of the C-terminal pY by the Src SH2 domain keeps that PTK "off" in the basal state. When a pY-containing protein comes into the vicinity the C-SH2 is free to act as a recruiting domain. The N-SH2 domain releases the SH-PTP-pY tail, especially if the bound substrate protein has multiple pY residues, as in the pY_n forms of IRS1. Thus, double engagement of the tandem SH2 domains by pY-containing proteins frees up the PTPase catalytic domain with up to 100-fold gain in catalytic activity towards the recuited pY proteins as hydrolysis substrates.

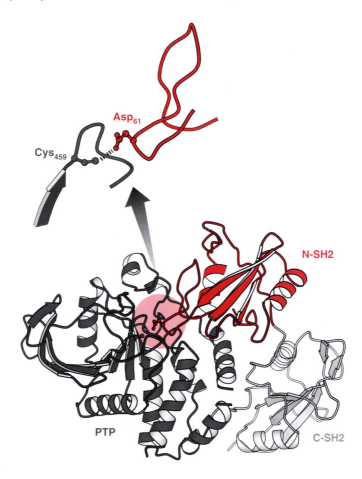

Figure 2.23 Structure of Shp2, a PTPase with tandem SH2 domains at the N-terminus for negative autoregulation in the basal state and for stimulation by pY protein substrates. One or more pY residues at the C-terminus act as *in cis* ligands for the N-SH2 domain. The active site Cys_{459} is coordinated to the side chain of Asp_{61} (figure made using PDB 2SHP).

A novel subfamily of PTPases has recently been reported by three groups (Li et al., 2003; Rayapureddi et al., 2003; Tootle et al., 2003) studying organ development in flies and mice. Proteins of the eyes absent (eya) family have a transcription-activating domain and a second eya domain with homology to phosphatases of the haloacid dehydrogenase superfamily. Assays of the eya domain showed robust PTPase activity towards pY peptide substrates and the pY forms of RNA polymerase II, and for auto-dephosphorylation of the pY form of the eya domain. This is a PTPase activity not utilizing a covalent $CysS–PO_3^{2-}$ intermediate. A conserved aspartate could play a catalytic role as general base or nucleophile. It is also the first example of a transcription factor with intrinsic phosphatase activity and can apparently convert a partner protein Dach1 from corepressor to coactivator by pY dephosphorylation. It will be of interest to see if additional members of this novel class of phosphotyrosyl protein phosphatases are discovered and what their range of substrate activities will be.

References

Aggen, J. B., A. C. Nairn, and R. Chamberlin. "Regulation of protein phosphatase-1," *Chem. Biol.* **7**:R13–23 (2000).

Ali, A., K. P. Hoeflich, and J. R. Woodgett. "Glycogen synthase kinase-3: Properties, functions, and regulation," *Chem. Rev.* **101**:2527–2540 (2001).

Alonso, A., J. Sasin, N. Bottini, I. Friedberg, A. Osterman, A. Godzik, T. Hunter, J. Dixon, and T. Mustelin. "Protein tyrosine phosphatases in the human genome," *Cell* **117**:699–711 (2004).

Amit, S., A. Hatzubai, Y. Birman, J. S. Andersen, E. Ben-Shushan, M. Mann, Y. Ben-Neriah, and I. Alkalay. "Axin-mediated CKI phosphorylation of beta-catenin at Ser 45: A molecular switch for the Wnt pathway," *Genes Dev.* **16**:1066–1076 (2002).

Ban, C., and W. Yang. "Crystal structure and ATPase activity of MutL: Implications for DNA repair and mutagenesis," *Cell* **95**:541–552 (1998).

Barford, D., S. H. Hu, and L. N. Johnson. "Structural mechanism for glycogen phosphorylase control by phosphorylation and AMP," *J. Mol. Biol.* **218**:233–260 (1991).

Barrett, J. F., and J. A. Hoch. "Two-component signal transduction as a target for microbial anti-infective therapy," *Antimicrob. Agents Chemother.* **42**:1529–1536 (1998).

Berwick, D. C., and J. M. Tavare. "Identifying protein kinase substrates: Hunting for the organ-grinder's monkeys," *Trends Biochem. Sci.* **29**:227–232 (2004).

Blagoev, B., S. E. Ong, I. Kratchmarova, and M. Mann. "Temporal analysis of phosphotyrosine-dependent signaling networks by quantitative proteomics," *Nat. Biotechnol.* **22**:1139–1145 (2004).

Brown, N. R., M. E. Noble, J. A. Endicott, and L. N. Johnson. "The structural basis for specificity of substrate and recruitment peptides for cyclin-dependent kinases," *Nat. Cell Biol.* **1**:438–443 (1999).

Burgess, A. W., H. S. Cho, C. Eigenbrot, K. M. Ferguson, T. P. Garrett, D. J. Leahy, M. A. Lemmon, M. X. Sliwkowski, C. W. Ward, and S. Yokoyama. "An open-and-shut case? Recent insights into the activation of EGF/ErbB receptors," *Mol. Cell* **12:**541–552 (2003).

Carmena, M., and W. C. Earnshaw. "The cellular geography of aurora kinases," *Nat. Rev. Mol. Cell Biol.* **4:**842–854 (2003).

Chen, Z., T. B. Gibson, F. Robinson, L. Silvestro, G. Pearson, B. Xu, A. Wright, C. Vander-bilt, and M. H. Cobb. "MAP kinases," *Chem. Rev.* **101:**2449–2476 (2001).

Clapperton, J. A., I. A. Manke, D. M. Lowery, T. Ho, L. F. Haire, M. B. Yaffe, and S. J. Smer-don. "Structure and mechanism of BRCA1 BRCT domain recognition of phosphory-lated BACH1 with implications for cancer," *Nat. Struct. Mol. Biol.* **11:**512–518 (2004).

Cohen, P. "The origins of protein phosphorylation," *Nat. Cell Biol.* **4:**E127–130 (2002).

Cohen, P. "The regulation of protein function by multisite phosphorylation," *Trends Biochem. Sci.* **25:**596–601 (2000).

Cole, P. A., D. Sondhi, and K. Kim. "Chemical approaches to the study of protein tyrosine kinases and their implications for mechanism and inhibitor design," *Pharmacol. Ther.* **82:**219–229 (1999).

Dean, A. M., and D. E. Koshland, Jr. "Electrostatic and steric contributions to regulation at the active site of isocitrate dehydrogenase," *Science* **249:**1044–1046 (1990).

Durocher, D., and S. P. Jackson. "The FHA domain," *FEBS Lett.* **513:**58–66 (2002).

Elchebly, M., P. Payette, E. Michaliszyn, W. Cromlish, S. Collins, A. L. Loy, D. Normandin, A. Cheng, J. Himms-Hagen, C. C. Chan, C. Ramachandran, M. J. Gresser, M. L. Tremblay, and B. P. Kennedy. "Increased insulin sensitivity and obesity resistance in mice lacking the protein tyrosine phosphatase-1B gene," *Science* **283:**1544–1548 (1999).

Elion, E. A. "The Ste5p scaffold," *J. Cell Sci.* **114:**3967–3978 (2001).

Espanel, X., M. Huguenin-Reggiani, and R. H. Van Huijsduijnen. "The SPOT technique as a tool for studying protein tyrosine phosphatase substrate specificities," *Protein Sci.* **11:**2326–2334 (2002).

Fabret, C., V. A. Feher, and J. A. Hoch. "Two-component signal transduction in *Bacillus subtilis*: How one organism sees its world," *J. Bacteriol.* **181:**1975–1983 (1999).

Ficarro, S. B., M. L. McCleland, P. T. Stukenberg, D. J. Burke, M. M. Ross, J. Shabanowitz, D. F. Hunt, and F. M. White. "Phosphoproteome analysis by mass spectrometry and its application to *Saccharomyces cerevisiae*," *Nat. Biotechnol.* **20:**301–305 (2002).

Hantschel, O., and G. Superti-Furga. "Regulation of the c-Abl and Bcr-Abl tyrosine kinases," *Nat. Rev. Mol. Cell Biol.* **5:**33–44 (2004).

Harper, J. W., and P. D. Adams. "Cyclin-dependent kinases," *Chem. Rev.* **101:**2511–2526 (2001).

Hoch, J. A., and T. J. Silhavy. *Two Component Signal Transduction*. ASM Press: Washington, DC (1995).

Hornbeck, P., I. Chabra, J. M. Kornhauser, E. Skrzypek, and B. Zhang. "PhosphoSite: A bioin-formatics resource dedicated to physiological protein phosphorylation," *Proteomics* **4:**1551–1561 (2004).

Huang, C. Y., and J. E. Ferrell, Jr. "Ultrasensitivity in the mitogen-activated protein kinase cascade," *Proc. Natl. Acad. Sci. U.S.A.* **93:**10078–10083 (1996).

Hubbard, S. R., and J. H. Till. "Protein tyrosine kinase structure and function," *Annu. Rev. Biochem.* **69:**373–398 (2000).

Hunter, T. "Signaling—2000 and beyond," *Cell* **100:**113–127 (2000).

Izaguirre, G. L., P. Aguirre, B. Ju, B. Aneskievich, and J. Haimovich. *J. Biol. Chem.* **274:**37012–37020 (1999).

Jackson, M. D., and J. M. Denu. "Molecular reactions of protein phosphatases—insights from structure and chemistry," *Chem. Rev.* **101:**2313–2340 (2001).

Johnson, D. A., P. Akamine, E. Radzio-Andzelm, M. Madhusudan, and S. S. Taylor. "Dynamics of cAMP-dependent protein kinase," *Chem. Rev.* **101:**2243–2270 (2001).

Johnson, G. L., and R. Lapadat. "Mitogen-activated protein kinase pathways mediated by ERK, JNK, and p38 protein kinases," *Science* **298:**1911–1912 (2002).

Johnson, L. N., and R. J. Lewis. "Structural basis for control by phosphorylation," *Chem. Rev.* **101:**2209–2242 (2001).

Jones, S. M., and A. Kazlauskas. "Growth factor-dependent signaling and cell cycle progression," *Chem. Rev.* **101:**2413–2423 (2001).

Kalume, D. E., H. Molina, and A. Pandey. "Tackling the phosphoproteome: Tools and strategies," *Curr. Opin. Chem. Biol.* **7:**64–69 (2003).

Kamenski, T., S. Heilmeier, A. Meinhart, and P. Cramer. "Structure and mechanism of RNA polymerase II CTD phosphatases," *Mol. Cell* **15:**399–407 (2004).

Kennelly, P. J. "Protein phosphatases—a phylogenetic perspective," *Chem. Rev.* **101:**2291–2312 (2001).

Keyse, S. M. "Protein phosphatases and the regulation of mitogen-activated protein kinase signalling," *Curr. Opin. Cell Biol.* **12:**186–192 (2000).

Kim, Y., A. E. Rice, and J. M. Denu. "Intramolecular dephosphorylation of ERK by MKP3," *Biochemistry* **42:**15197–15207 (2003).

Kobe, B., and B. E. Kemp. "Active site-directed protein regulation," *Nature* **402:**373–376 (1999).

Kuriyan, J., and D. Cowburn. "Modular peptide recognition domains in eukaryotic signaling," *Annu. Rev. Biophys. Biomol. Struct.* **26:**259–288 (1997).

Lee, M., and S. Goodbourn. "Signalling from the cell surface to the nucleus," *Essays Biochem.* **37:**71–85 (2001).

Li, X., K. A. Oghi, J. Zhang, A. Krones, K. T. Bush, C. K. Glass, S. K. Nigam, A. K. Aggarwal, R. Maas, D. W. Rose, and M. G. Rosenfeld. "Eya protein phosphatase activity regulates Six1-Dach-Eya transcriptional effects in mammalian organogenesis," *Nature* **426:**247–254 (2003).

Liebler, D. C. *Introduction to Proteomics.* Humana Press: Totowa, NJ (2002).

Littlepage, L. E., H. Wu, T. Andresson, J. K. Deanehan, L. T. Amundadottir, and J. V. Ruderman. "Identification of phosphorylated residues that affect the activity of the mitotic kinase Aurora-A," *Proc. Natl. Acad. Sci. U.S.A.* **99:**15440–15445 (2002).

Lowe, E. D., I. Tews, K. Y. Cheng, N. R. Brown, S. Gul, M. E. Noble, S. J. Gamblin, and L. N. Johnson. "Specificity determinants of recruitment peptides bound to phospho-CDK2/cyclin A," *Biochemistry* **41:**15625–15634 (2002).

Machida, K., B. J. Mayer, and P. Nollau. "Profiling the global tyrosine phosphorylation state," *Mol. Cell. Proteomics* **2:**215–233 (2003).

Majeti, R., and A. Weiss. "Regulatory mechanisms for receptor protein tyrosine phosphatases," *Chem. Rev.* **101:**2441–2448 (2001).

Manke, I. A., D. M. Lowery, A. Nguyen, and M. B. Yaffe. "BRCT repeats as phosphopeptide-binding modules involved in protein targeting," *Science* **302:**636–639 (2003).

Mann, M., S. E. Ong, M. Gronborg, H. Steen, O. N. Jensen, and A. Pandey. "Analysis of protein phosphorylation using mass spectrometry: Deciphering the phosphoproteome," *Trends Biotechnol.* **20:**261–268 (2002).

Manning, G., D. B. Whyte, R. Martinez, T. Hunter, and S. Sudarsanam. "The protein kinase complement of the human genome," *Science* **298:**1912–1934 (2002).

Neel, B. G., H. Gu, and L. Pao. "The 'Shp'ing news: SH2 domain-containing tyrosine phosphatases in cell signaling," *Trends Biochem. Sci.* **28:**284–293 (2003).

Newton, A. C. "Protein kinase C: Structural and spatial regulation by phosphorylation, cofactors, and macromolecular interactions," *Chem. Rev.* **101:**2353–2364 (2001).

Pawson, T., and P. Nash. "Assembly of cell regulatory systems through protein interaction domains," *Science* **300:**445–452 (2003).

Pawson, T., and J. D. Scott. "Signaling through scaffold, anchoring, and adaptor proteins," *Science* **278:**2075–2080 (1997).

Pinna, L. A. "The raison d'être of constitutively active protein kinases: The lesson of CK2," *Acc. Chem. Res.* **36:**378–384 (2003).

Rayapureddi, J. P., C. Kattamuri, B. D. Steinmetz, B. J. Frankfort, E. J. Ostrin, G. Mardon, and R. S. Hegde. "Eyes absent represents a class of protein tyrosine phosphatases," *Nature* **426:**295–298 (2003).

Saito, H. "Histidine phosphorylation and two-component signaling in eukaryotic cells," *Chem. Rev.* **101:**2497–2509 (2001).

Schindler, T., F. Sicheri, A. Pico, A. Gazit, A. Levitzki, and J. Kuriyan. "Crystal structure of Hck in complex with a Src family-selective tyrosine kinase inhibitor," *Mol. Cell* **3:**639–648 (1999).

Shabb, J. B. "Physiological substrates of cAMP-dependent protein kinase," *Chem. Rev.* **101:**2381–2411 (2001).

Steen, H., M. Fernandez, S. Ghaffari, A. Pandey, and M. Mann. "Phosphotyrosine mapping in Bcr/Abl oncoprotein using phosphotyrosine-specific immonium ion scanning," *Mol. Cell. Proteomics* **2:**138–145 (2003).

Steen, H., B. Kuster, M. Fernandez, A. Pandey, and M. Mann. "Detection of tyrosine phosphorylated peptides by precursor ion scanning quadrupole TOF mass spectrometry in positive ion mode," *Anal. Chem.* **73:**1440–1448 (2001).

Steen, H., B. Kuster, M. Fernandez, A. Pandey, and M. Mann. "Tyrosine phosphorylation mapping of the epidermal growth factor receptor signaling pathway," *J. Biol. Chem.* **277:**1031–1039 (2002).

Tootle, T. L., S. J. Silver, E. L. Davies, V. Newman, R. R. Latek, I. A. Mills, J. D. Selengut, B. E. Parlikar, and I. Rebay. "The transcription factor Eyes absent is a protein tyrosine phosphatase," *Nature* **426:**299–302 (2003).

Virshup, D. M. "Protein phosphatase 2A: A panoply of enzymes," *Curr. Opin. Cell Biol.* **12:**180–185 (2000).

Walsh, C. T. *Enzymatic Reaction Mechanisms.* Freeman: San Francisco (1979).

Walsh, C. T. *Antibiotics: Actions, Origins, Resistance.* ASM Press: Washington, DC (2003).

West, A. H., and A. M. Stock. "Histidine kinases and response regulator proteins in two-component signaling systems," *Trends Biochem. Sci.* **26:**369–376 (2001).

Westheimer, F. H. "Why nature chose phosphates," *Science* **235:**1173–1178 (1987).

White, M. F., and C. R. Kahn. "The insulin signaling system," *J. Biol. Chem.* **269:**1–4 (1994).

Xu, L., Y. Wei, J. Reboul, P. Vaglio, T. H. Shin, M. Vidal, S. J. Elledge, and J. W. Harper. "BTB proteins are substrate-specific adaptors in an SCF-like modular ubiquitin ligase containing CUL-3," *Nature* **425:**316–321 (2003).

Xu, W., A. Doshi, M. Lei, M. J. Eck, and S. C. Harrison. "Crystal structures of c-Src reveal features of its autoinhibitory mechanism," *Mol. Cell* **3:**629–638 (1999).

Yaffe, M. "How do 14-3-3 proteins work? Gatekeeper phosphorylation and the molecular anvil hypothesis," *FEBS Lett.* **513:**53–57 (2002a).

Yaffe, M. B. "Phosphotyrosine-binding domains in signal transduction," *Nat. Rev. Mol. Cell Biol.* **3:**177–186 (2002b).

Yaffe, M. B., and A. E. Elia. "Phosphoserine/threonine-binding domains," *Curr. Opin. Cell Biol.* **13:**131–138 (2001).

Zhan, X. L., M. J. Wishart, and K. L. Guan. "Nonreceptor tyrosine phosphatases in cellular signaling: Regulation of mitogen-activated protein kinases," *Chem. Rev.* **101:**2477–2496 (2001).

Zhang, H., X. Zha, Y. Tan, P. V. Hornbeck, A. J. Mastrangelo, D. R. Alessi, R. D. Polakiewicz, and M. J. Comb. "Phosphoprotein analysis using antibodies broadly reactive against phosphorylated motifs," *J. Biol. Chem.* **277:**39379–39387 (2002).

Zhou, H., J. D. Watts, and R. Aebersold. "A systematic approach to the analysis of protein phosphorylation," *Nat. Biotechnol.* **19:**375–378 (2001).

CHAPTER 3

Sulfuryl Transfers: Action of Protein Sulfotransferases and Aryl Sulfatases

Sulfation of two adjacent tyrosines ($Tyr_{14,15}$) in the chemokine receptor CCR5.

The sulfuryl group, $-SO_3^-$, is an anionic tetrahedral group quite similar to the phosphoryl group. The $-SO_3^-$ group is monoanionic at physiological pH while the $-PO_3$ group can can carry one or two negative charges ($pK_{a2} \approx 6$) at physiological pH. Like the PO_3^{2-} dianion discussed in Chapter 2, the SO_3^- moiety has an electropositive second-row atom (sulfur) at its core that can be captured in nucleophilic attack, proceeding via pentacovalent intermediates (Figure 3.1A). While the $-PO_3$ group is used much more extensively in cellular energy metabolism than $-SO_3^-$, there are analogies in the way that these two tetrahedral anionic groups are utilized for protein posttranslational modifications. Covalently sulfated proteins are formed by protein sulfotransferase action in analogy to protein phosphotransferase (kinase) action.

SO_3^- group transfer can also be reversed with hydrolytic sulfatases opposing the action of sulfotransferases. Covalent sulfated enzyme intermediates are generated

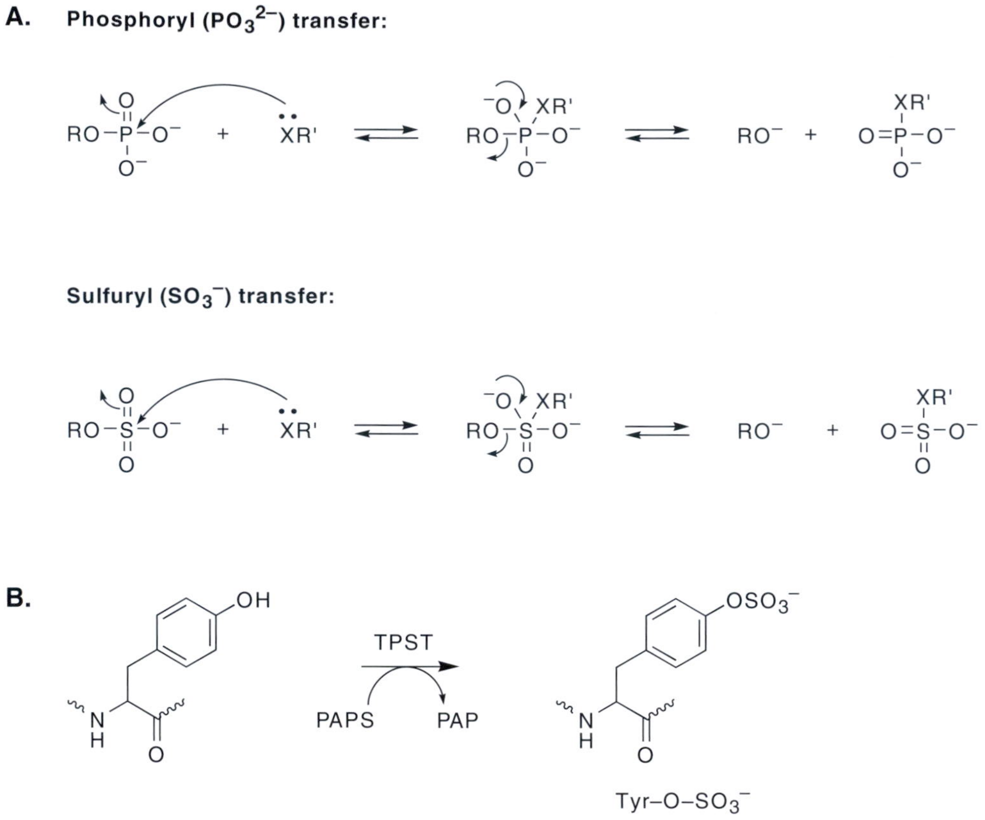

A.   **Phosphoryl (PO$_3{}^{2-}$) transfer:**

Sulfuryl (SO$_3{}^-$) transfer:

B.

Tyr–O–SO$_3{}^-$

Figure 3.1 Sulfuryl group transfers in protein posttranslational modification. **A.** Attack at the electrophilic sulfur of sulfate esters by cosubstrate nucleophiles. **B.** Sulfotransferase action yields sulfated tyrosyl residues in proteins.

during sulfatase action in analogy to the phosphoenzyme intermediates in PTPase action described in Chapter 2. Both the protein–SO$_3{}^-$ and the protein–OPO$_2$H$^-$ forms have an increase of 80 Da and are diagnostic molecular weight increases for mass spectrometric detection of these posttranslational modifications. As noted in Chapter 1, the protein/peptide sulfate esters can be labile in MS and thus do not give quantitative data readily.

Tyrosyl Protein Sulfotransferases (TPSTases)

As just discussed in the previous chapter, protein kinases are classified as serine/threonine kinases or tyrosine kinases based on the attacking nucleophilic side chain in protein cosubstrates. The analogous protein sulfotransferases have been believed to be specific for tyrosine modification (Figure 3.1B). However, a recent proteomic approach (Medzihradszky et al., 2004) has turned up O-sulfoserine and O-sulfothreonine residues in proteins by mass spectroscopic analysis of proteolyzed peptide fragments. How extensive these Thr and Ser O-sulfations will be in proteomes remains to be determined in subsequent studies.

Two highly related tyrosyl protein sulfotransferase genes are found in humans, encoding two constitutively active TPSTases. Both TPST1 and TPST2 are found in the Golgi network of cells and act on proteins passing through the secretory network. TPSTases have a short N-terminal region followed by a 17-residue membrane-spanning region, a 40-residue stem region, and a 310–320-residue catalytic domain that is oriented towards the lumenal surface of the Golgi compartment (Moore, 2003). Thus, the transmembrane TPSTases recognize proteins that are transiting through the Golgi lumen. This localization explains why membrane proteins and secreted proteins can bear sulfo-tyrosyl posttranslational modifications but cytosolic and nuclear proteins do not.

There does not appear to be a strict sequence that determines if a tyrosyl side chain of a protein undergoing transit through the secretory compartments will get sulfated, but there is a strong correlation with negatively charged glutamate side chains in the vicinity, up to three Glu/Asp residues ± 5 residues of Y residues that get sulfated. The correlations of subsets of acidic sequences at sites of known tyrosine sulfation have led to the Sulfinator prediction program (www.expasy.org/tools/sulfinator/) for protein regions that will incorporate $-SO_3^-$ groups. Such predictions suggest one out of 20 proteins secreted by HepG2 cells will bear one or more $-SO_3^-$ groups and up to one in three proteins secreted by fibroblasts. Other predictions suggest up to 2100 proteins in mice may be sulfated, suggesting this is a common posttranslational modification (Moore, 2003). Another estimate is that 1% of all tyrosyl residues in the eukaryotic proteome are sulfated (Baeuerle and Huttner, 1987). To date only a small fraction of this number of proteins have been experimentally verified to undergo sulfation.

In contrast to the pervasive use of protein phosphorylation, protein sulfurylation does not seem to be a common strategy in signal transduction pathways, but sulfation of the terminus of the thyroid stimulating hormone (TSH) receptor N-terminus is required for interaction of receptor with the TSH ligand. It appears protein–SO_3^- moieties are not readily reversible by biological catalysts. There are no established cases of hydrolytic removal of the sulfate from tyrosyl groups in proteins (i.e., there are no protein sulfatases). There *are* sulfatases for removal of SO_3^- groups in small molecules as we shall note in a subsequent section of this chapter. Tyrosine sulfate, presumably from extracellular proteolysis, is excreted in the urine (Moore, 2003).

The sulfuryl donor to protein tyrosyl residues is an anhydride in analogy to the thermodynamically activated anhydride side chain in ATP. In this case it is an $ADP–O–SO_3^-$ mixed phosphoric–sulfuric anhydride. The biological sulfuryl donor is generally not the simple ATP analog $ADP–O–SO_3^-$ (APS), but rather the 3′-phosphoAPS, known as phosphoAPS and abbreviated as PAPS (Figure 3.2A). PAPS

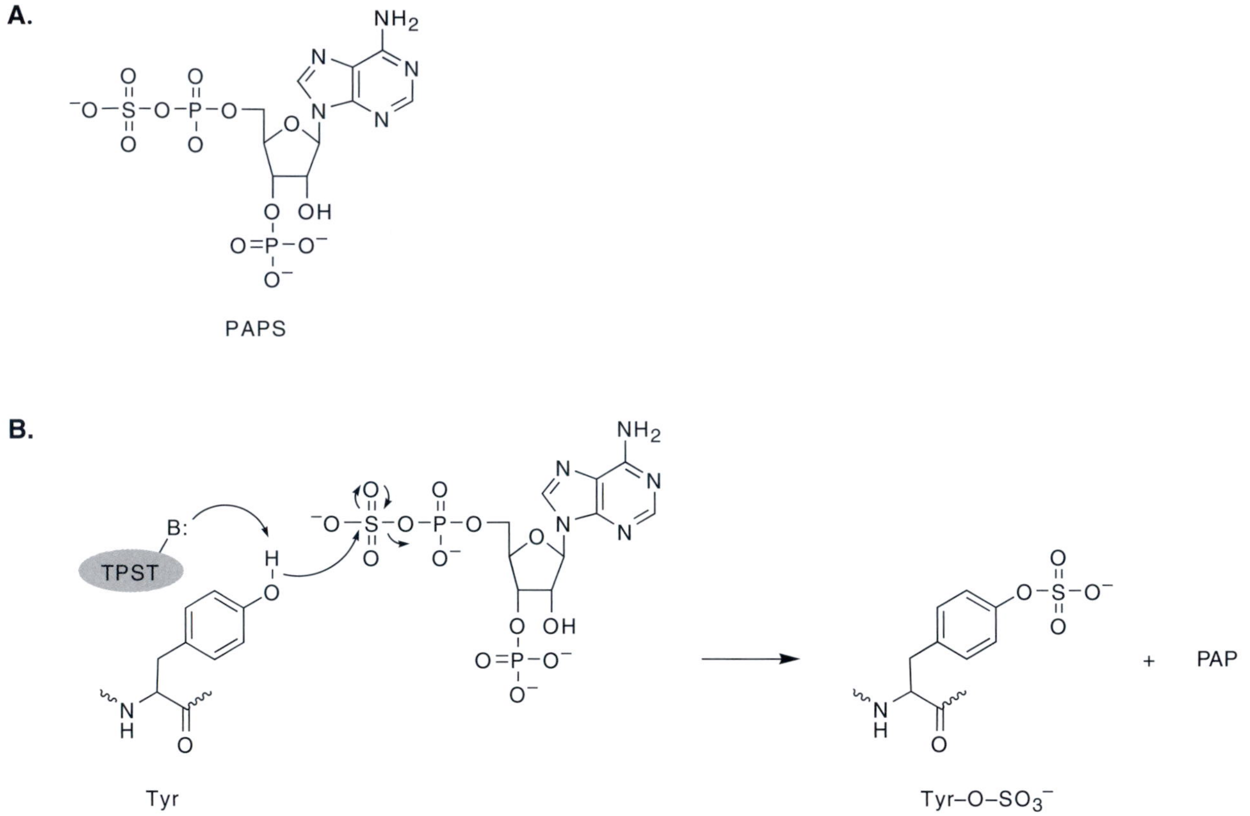

A.

PAPS

B.

Figure 3.2 **A.** PAPS, the biological sulfuryl donor, is thermodynamically activated for $-SO_3^-$ transfer in the sulfuric–phosphoric anhydride linkage. **B.** SO_3^- transfer from PAPS to the phenolate anion of tyrosyl residues.

has a very favorable thermodynamic potential for $-SO_3^-$ transfer and is invariably the biological sulfuryl donor (Figure 3.2B).

PAPS is required in the lumen of the Golgi compartment to be accessible to the active site of TPSTases as they bind the tyrosyl moieties of their protein substrates. There is a PAPS transporter for moving PAPS into the Golgi lumen from the cytoplasm (Figure 3.3). In turn, PAPS is assembled in the cytoplasm in two enzymatic steps from inorganic sulfate that enters cells via transporter proteins in the plasma membrane. Starting with inorganic sulfate (SO_4^{2-}) and ATP, ATP sulfurylase uses one of the sulfate oxygens as a nucleophile to attack the β-P atom in ATP, generating APS and PP_i. Then APS kinase catalyzes the reaction of APS with a second molecule of ATP to phosphorylate the 3′-OH of APS and yield PAPS. Thus, PAPS has been produced via fragmentation of two ATP molecules, a measure of its high group transfer potential as a sulfurylating agent.

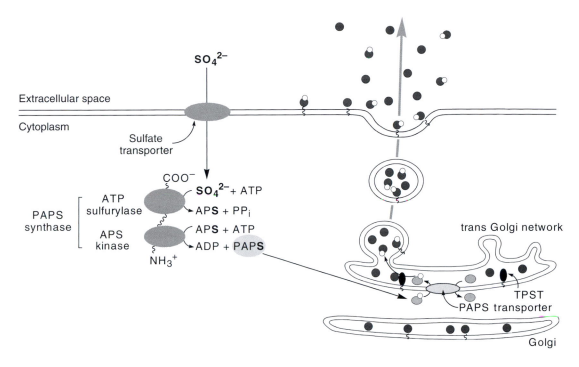

Figure 3.3 Schematic of PAPS transporter moving PAPS into the Golgi lumen for access to the active sites of TPTases (redrawn from Moore, 2003).

Examples of secreted proteins that are sulfated on tyrosines relevant for function are the blood clotting proteins factor V and factor VIII, where the sulfation of one Tyr is important for the high-affinity interaction with its partner protein, von Willebrand's factor. Type III and type V collagens are sulfated (Moore, 2003). Also, the P-selectin glycoprotein ligand (PSGL-1) needs to be sulfated for selective recognition of P-selectin (Kehoe and Bertozzi, 2000).

The chemokine receptor CCR5 on cell surfaces of T lymphocytes has been of interest since natural variants provide resistance to HIV infection, the first indication that it was a coreceptor with CD4 for HIV entry (Cormier et al., 2000; Farzan et al. 1999; Farzan et al. 2000). The interaction of CCR5 with its chemokine ligands—MIP-1α, MIP-1β, and RANTES—has been shown to depend on sulfation of up to four tyrosine residues in the N-terminal region of CCR5 (Figure 3.4). Likewise, HIV uptake into macrophages is reduced by mutations converting these tyrosines to phenylalanines (Farzan et al., 1999). This has intensified interest in characterizing the kinetics and stoichiometries of each of the four tyrosines sulfated in CCR5 by TPST1 and TPST2 (Seibert et al., 2002), and in searching for TPST inhibitors. Other chemokine receptors, such as CXCR4 and CCR2b, are also sulfated on N-terminal tyrosines, leading to speculation that this may be a general posttranslational modification of chemokine receptors and account for 5–100-fold effects on chemokine ligand affinities.

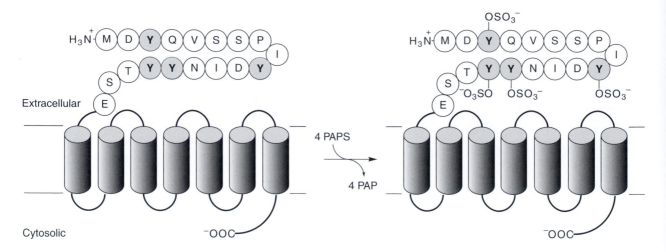

Figure 3.4 Enzymatic sulfation of four tyrosine residues in the N-terminus of the CCR5 receptor. Model for orientation of the seven transmembrane receptor CCR5 in plasma membranes.

As schematized in Figure 3.4, CCR5 has the seven transmembrane helical secondary structure characteristic of G-protein-coupled receptors (Seibert et al., 2002). The N-terminal 18 residues are predicted to be at the extracellular surface of the plasma membrane and contain tyrosine (Y) residues at 3, 10, 14, and 15, with two aspartate residues at 2 and 11 and a glutamate at 18. Consistent with the neighboring anionic side chains, these tyrosines are predicted and observed to be sulfated. Mutagenesis studies and competition with sulfated versions of the N-terminal peptide in chemokine ligand binding assays indicated the importance of sulfation at Y_{10} and Y_{14} over Y_3 and Y_{15}. Direct evaluation of the action of recombinant TSPT1 and TSPT2 on the CCR5 2–18 N-terminal peptide was carried out by a kinetic analysis that monitored peptide sulfation by HPLC, with several chromatographically resolvable sulfated peaks (Figure 3.5) (Seibert et al., 2002). The presence of the $-SO_3^-$ on each of the four tyrosines was established by protease action and mass spectrometry where the SO_3 increment of +80 Da is a signature. Unlike the $-PO_3H$ (+80 Da) increment that is stable in the mass spectrometer, the sulfate linkage is labile and gives variable stoichiometry of sulfated and unsulfated tyrosines. Therefore, a combination of HPLC and MS analysis is required to determine the kinetics and stoichiometry of the multiple tyrosine sulfation in the N-terminal region of CCR5. By such analysis, Y_{14} and Y_{15} are independently sulfated first, then Y_{10} and finally Y_3 by both TPST1 and TPST2 (Seibert et al., 2002). In CCR5 molecules that are tetrasulfated, there will be a net negative charge of –11 on side chains 3–18, creating a strong negative surface region for interaction with partner proteins. It is highly likely that CCR5 N-termini are heterogeneous, with different fractional stoichiometries of sulfation, perhaps with

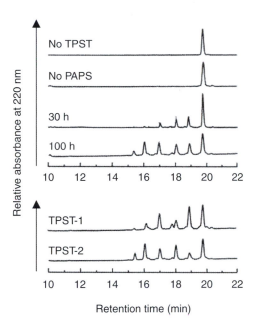

Figure 3.5 HPLC separation of sulfated peptides of the 2–18 N-terminal CCR5 peptide (reproduced with permission from Seibert et al., 2002).

Y_{14} and Y_{15} sulfation sites predominating. The functional consequences of this heterogeneity are unresolved but are perhaps analogous to heterogeneity of protein N-glycosylation, where the fractional stoichiometries are related to posttranslational modification enzyme specificity and catalytic efficiency for distinct sequences, but also on transit times through the secretory pathway.

Recently human antibodies from AIDS patients with good neutralizing activities against gp120 have been found to be sulfated on tyrosyl residues at the heavy chain CDR3 region, suggesting these antibodies mimic CCR5 recognition of gp120 (Choe et al., 2003). They are the first instance in which sulfation of antibodies was detected, reflecting that dedicated posttranslational modifications can occur during antibody transit through the secretory pathway to the cell surface.

Aryl Sulfatases

Although sulfotyrosyl linkages in proteins are not readily reversed due to the lack of protein sulfatases, there are varieties of $N–SO_3^-$ and $O–SO_3^-$ linkages that are cleaved hydrolytically by sulfatases. The physiological substrates are sulfated glycosaminoglycans, such as heparan sulfates, chondroitin sulfates, keratin sulfates, and dermatan sulfates with both $N–SO_3$ and $O–SO_3$ linkages of differing regiospecificity, sulfolipids such as the sugar-containing sphingolipid cerebroside-3-sulfate, sulfated peptides such as cholecystokinin, and steroid sulfates. Failure to degrade the sulfated polysaccharides in particular leads to lysosomal storage diseases. Eight such human diseases have been tied to monogenic defects in eight separate sulfatases (von Bulow et al., 2002). Thirteen sulfatases have been detected in humans, eight residing in lysosomes, the remainder in the endoplasmic reticulum and Golgi complex (von Figura et al., 1998). The best studied human sulfatases are "aryl" sulfatase A and "aryl" sulfatase B, named due to high levels of hydrolase activity on synthetic aryl sulfate esters. A principal substrate of aryl sulfatase A is probably cerebroside-3-sulfate, given its accumulation in those with a genetic deficiency for this enzyme. This sulfolipid is a constituent of myelin sheaths and lack of this sulfatase is involved in demyelinating disease (von Figura et al., 1998).

All the sulfatases, including a bacterial sulfatase from pseudomonads, require a posttranslational modification of an active site cysteine to a C_α-formylglycine (Fgly) before they attain catalytic activity (Figure 3.6). This was detected originally from clinical observations that a few patients had multiple sulfatase deficiencies. Purification of aryl sulfatase A and analysis of the active site tryptic peptide revealed that this region of the native enzyme had a molecular weight 18 Da less than expected. The peptide also reacted with carbonyl reagents, consistent with conversion of the active site cysteine into formylglycine. The *Klebsiella pneumoniae* sulfatase undergoes an analogous but distinct posttranslational modification of an active site serine to the aldehyde moiety of formylglycine. This modification appears to be a distinct mechanism to get to the same active enzyme (von Figura et al., 1998).

The x-ray structure of both aryl sulfatases A and B and the *Pseudomonas aeruginosa* sulfatase have been determined (Boltes et al., 2001; Lukatela et al., 1998), validating the formylglycine residue and showing remarkable homology to alkaline phosphatase. The sulfatases all crystallize with a divalent cation in the active site. Electron density in the 2.3-Å aryl sulfatase A map was interpreted in favor of the hydrate form of formylglycine and conclusively validated in the 1.3-Å map of the pseudomonal sulfatase. The aldehyde and the gem hydrate form of Fgly are in equilibrium, with the hydrate believed to be the active form of the enzyme, initiating catalysis by nucleophilic attack on a bound sulfate ester substrate (Figure 3.7A). This leads, through a pentacovalent

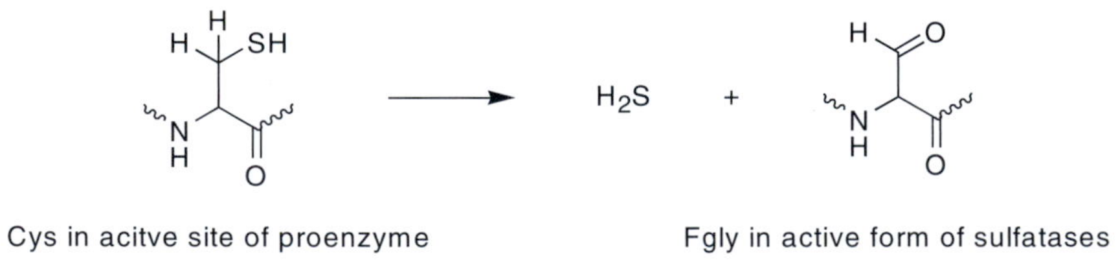

Cys in acitve site of proenzyme   Fgly in active form of sulfatases

Figure 3.6 Posttranslational modification of cysteine or serine to formylglycine to create the active form of aryl sulfatases.

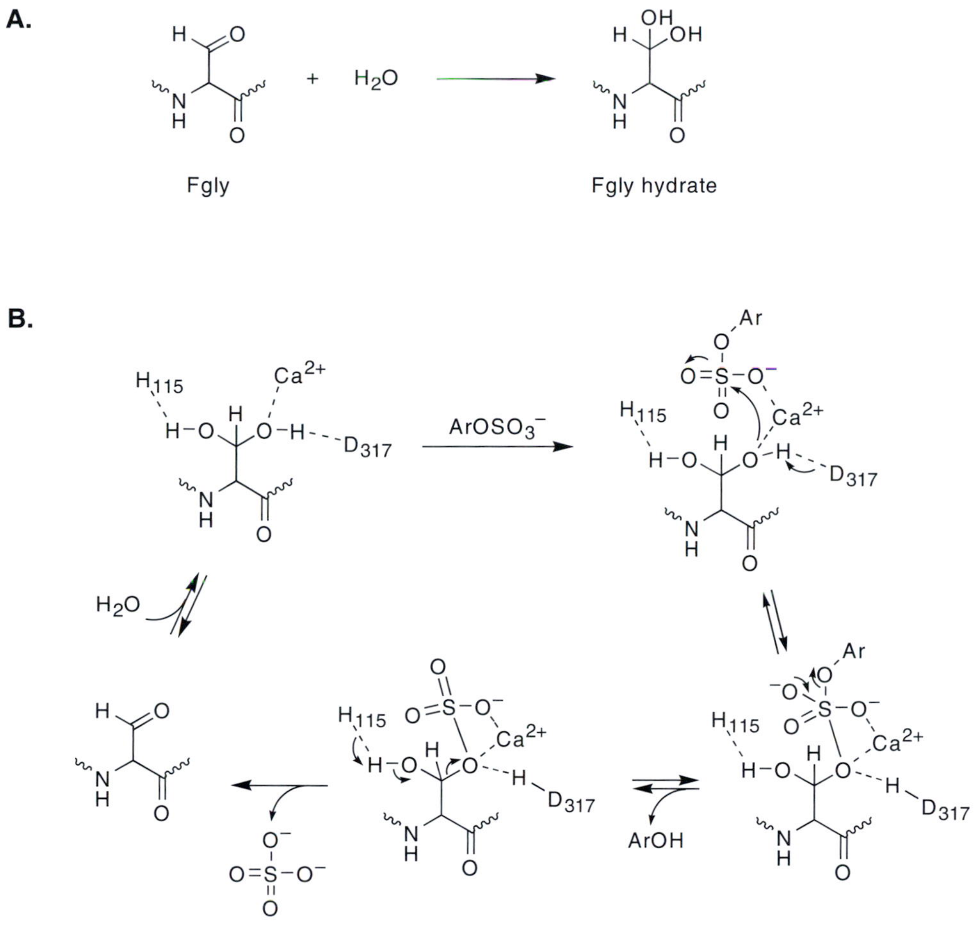

A.

Fgly   Fgly hydrate

B.

Figure 3.7 Catalytic mechanism for sulfatases. **A.** Equilibrium between the aldehyde form of Fgly and the geminal hydrate. **B.** The Fgly gem hydrate as nucleophile attacking the sulfur atom of sulfated substrates to generate a sulfoenzyme intermediate during substrate hydrolysis.

intermediate, to release of ROH and generation of a covalent sulfoenzyme (Figure 3.7B). The C–OSO$_3$ bond in that intermediate can now be cleaved by participation of the OH on the same carbon (the β-carbon of the Fgly) in expulsion of OSO$_3^{2-}$ as the Fgly reforms. To complete the catalytic cycle, the gem hydrate form of Fgly reforms by addition of water.

The sulfatases have an almost identical fold to the M^{2+} alkaline phosphatase that catalyzes a similar transformation, the net hydrolysis of phosphate monoesters (ROPO$_3^{2-}$) (Figure 3.8A). Indeed, the active site serine of alkaline phosphatase overlays with the position of the Fgly hydrate of aryl sulfatases. Each enzyme initiates its catalytic cycle by nucleophilic attack, the phosphatase capturing the P atom of the bound ROPO$_3$ substrate, on its way to pentacovalent phosphorane and then covalent

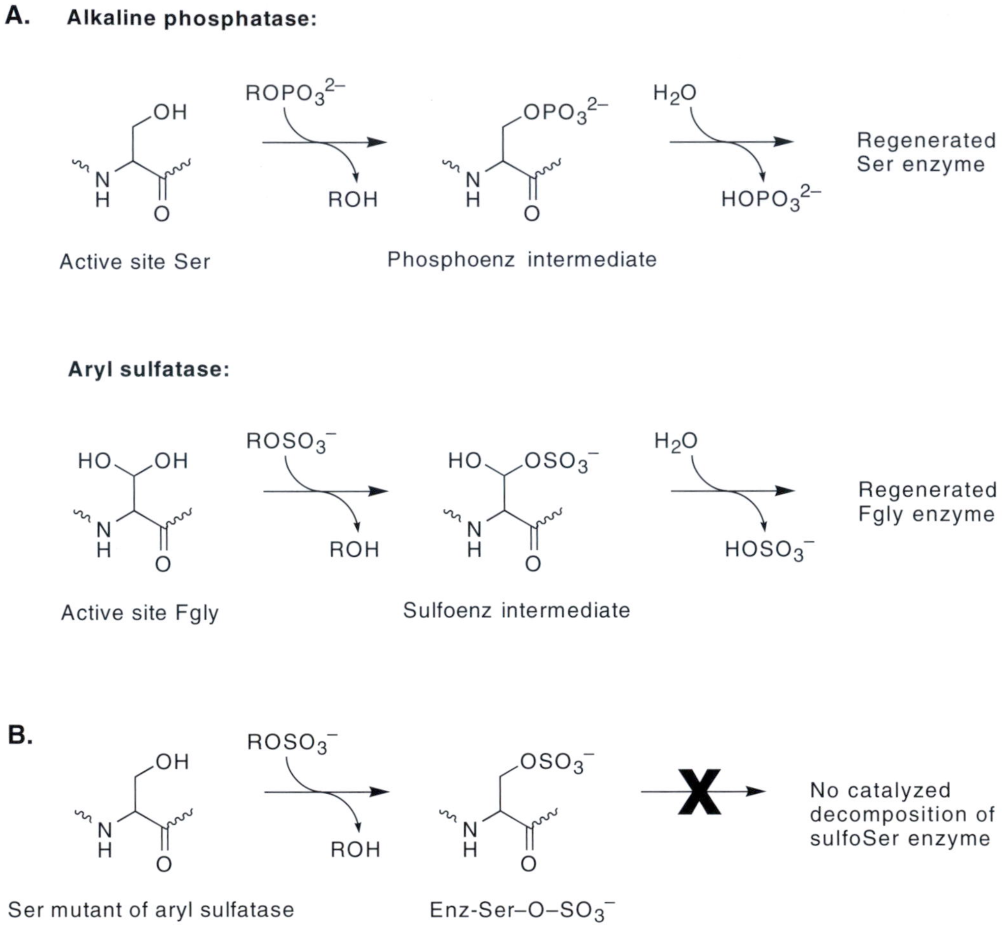

A.   **Alkaline phosphatase:**

Active site Ser   Phosphoenz intermediate

Aryl sulfatase:

Active site Fgly   Sulfoenz intermediate

B.

Ser mutant of aryl sulfatase   Enz-Ser–O–SO$_3^-$

Figure 3.8 **A.** Mechanism for alkaline phosphatase action on phosphate ester substrates, analogous to the sulfatase mechanism. **B.** The active site serine mutant of sulfatase gets stuck at the covalent sulfoenzyme intermediate stage.

phosphoenzyme intermediate. The first half-reaction of the sulfatase catalytic cycle mirrors the phosphatase but then diverges (Dierks et al., 1998). If a Cys-to-Ser mutant of a sulfatase is expressed, it is catalytically inactive. It can get through the first half of the reaction and make a sulfoseryl enzyme intermediate, but that persists as a stable species and does not turn over to hydrolytic products. Unlike the phosphatase phosphoseryl enzyme intermediate, which is sufficiently labile to be captured by hydrolysis in the second half-reaction, the sulfoseryl enzyme is too stable (Figure 3.8B). Now the rationale for the unusual posttranslational conversion of cysteinyl side chain to Fgly becomes clear (von Bulow et al., 2001). In the sulfatase Fgly hydrate the pK_a of the geminal OH is estimated to be five pK_a units more basic than the pK_a for a serine CH$_2$OH side chain. This increased basicity makes for a lower energy barrier for generating the alkoxide ion needed for intramolecular expulsion of HOSO$_3^-$ from the sulfoFgly covalent intermediate. It appears therefore that nature has engineered this posttranslational mutation to dial in the reactivity required to labilize the sulfoenzyme intermediate, a more difficult task than cognate phosphoenzyme hydrolysis. Since there are no proteinogenic amino acids with aldehyde (or ketone) side chains, the protein catalytic inventory is expanded through posttranslational modification. Then, rather than using the aldehyde form as an electron sink (compare with the pyruvoyl-enzyme prosthetic group usage in Chapter 12), the hydrate form is used with opposite polarity as the initiating nucleophile.

The enzyme, formylglycine-generating enzyme (FGE), responsible for the Cys-to-Fgly posttranslational modification in the activation of sulfatases occurs in the endoplasmic reticulum lumen (Dierks et al., 2003). Sulfatase proenzymes are translated with signal sequences that direct them into the endoplasmic reticulum on their way to later routing to the lysosomal vesicles (Moore, 2003). Aryl sulfatase A not only undergoes the Cys-to-Fgly modification, but also three N-glycosylations (Chapter 10) in transit through the secretory network. The Cys-to-Fgly modification may happen during refolding of the sulfatase protein as it is chaperoned across the endoplasmic reticular membrane. There is a consensus sequence CXPSRXXXL/MTGR/K/L, which when inserted into a protein transiting into the endoplasmic reticulum, is sufficient to convert the Cys to Fgly. Recent purifications of the FGE (Cosma et al., 2003; Dierks et al., 2003) have demonstrated an O_2 requirement. This is consistent with a mechanism in which the Cys residue in the proenzyme is monoxygenated at C_β to produce transiently the β-OH–β-SH side chain, a thiohemiacetal, which can rapidly

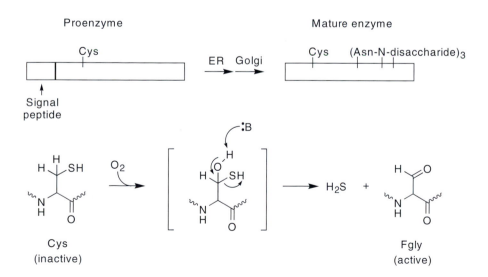

Figure 3.9 Proposal for the formylglycine-generating enzyme. Oxygenation of C_β of the cysteine in the pro form of sulfatases to create the thiohemiacetal form of Fgly. Loss of H_2S generates the active form of the sulfatase.

and spontaneously eliminate H^+ and SH^- to yield Fgly (Figure 3.9). It may be that FGE will be a nonheme oxygenase and thus fall into the category of the posttranslational modifying enzymes taken up in Chapter 11.

References

Baeuerle, P. A., and W. B. Huttner. "Tyrosine sulfation is a trans-Golgi-specific protein modification," *J. Cell Biol.* **105**:2655–2664 (1987).

Boltes, I., H. Czapinska, A. Kahnert, R. von Bulow, T. Dierks, B. Schmidt, K. von Figura, M. A. Kertesz, and I. Uson. "1.3 Å structure of arylsulfatase from *Pseudomonas aeruginosa* establishes the catalytic mechanism of sulfate ester cleavage in the sulfatase family," *Structure (Cambridge)* **9**:483–491 (2001).

Choe, H., W. Li, P. L. Wright, N. Vasilieva, M. Venturi, C. C. Huang, C. Grundner, T. Dorfman, M. B. Zwick, L. Wang, E. S. Rosenberg, P. D. Kwong, D. R. Burton, J. E. Robinson, J. G. Sodroski, and M. Farzan. "Tyrosine sulfation of human antibodies contributes to recognition of the CCR5 binding region of HIV-1 gp120," *Cell* **114**:161–170 (2003).

Cormier, E. G., M. Persuh, D. A. Thompson, S. W. Lin, T. P. Sakmar, W. C. Olson, and T. Dragic. "Specific interaction of CCR5 amino-terminal domain peptides containing sulfotyrosines with HIV-1 envelope glycoprotein gp120," *Proc. Natl. Acad. Sci. U.S.A.* **97**:5762–5767 (2000).

Cosma, M. P., S. Pepe, I. Annunziata, R. F. Newbold, M. Grompe, G. Parenti, and A. Ballabio. "The multiple sulfatase deficiency gene encodes an essential and limiting factor for the activity of sulfatases," *Cell* **113**:445–456 (2003).

Dierks, T., C. Miech, J. Hummerjohann, B. Schmidt, M. A. Kertesz, and K. von Figura. "Posttranslational formation of formylglycine in prokaryotic sulfatases by modification of either cysteine or serine," *J. Biol. Chem.* **273**:25560–25564 (1998).

Dierks, T., B. Schmidt, L. V. Borissenko, J. Peng, A. Preusser, M. Mariappan, and K. von Figura. "Multiple sulfatase deficiency is caused by mutations in the gene encoding the human C(alpha)-formylglycine generating enzyme," *Cell* **113**:435–444 (2003).

Farzan, M., T. Mirzabekov, P. Kolchinsky, R. Wyatt, M. Cayabyab, N. P. Gerard, C. Gerard, J. Sodroski, and H. Choe. "Tyrosine sulfation of the amino terminus of CCR5 facilitates HIV-1 entry," *Cell* **96**:667–676 (1999).

Farzan, M., N. Vasilieva, C. E. Schnitzler, S. Chung, J. Robinson, N. P. Gerard, C. Gerard, H. Choe, and J. Sodroski. "A tyrosine-sulfated peptide based on the N terminus of CCR5 interacts with a CD4-enhanced epitope of the HIV-1 gp120 envelope glycoprotein and inhibits HIV-1 entry," *J. Biol. Chem.* **275**:33516–33521 (2000).

Kehoe, J. W., and C. R. Bertozzi. "Tyrosine sulfation: A modulator of extracellular protein–protein interactions," *Chem. Biol.* **7**:R57–61 (2000).

Lukatela, G., N. Krauss, K. Theis, T. Selmer, V. Gieselmann, K. von Figura, and W. Saenger. "Crystal structure of human arylsulfatase A: The aldehyde function and the metal ion at the active site suggest a novel mechanism for sulfate ester hydrolysis," *Biochemistry* **37:**3654–3664 (1998).

Medzihradszky, K. F., Z. Darula, E. Perlson, M. Fainzilber, R. J. Chalkley, H. Ball, D. Greenbaum, M. Bogyo, D. R. Tyson, R. A. Bradshaw, and A. L. Burlingame. "O-Sulfonation of serine and threonine—mass spectrometric detection and characterization of a new posttranslational modification in diverse proteins throughout the eukaryotes," *Mol. Cell. Proteomics* (2004).

Moore, K. L. "The biology and enzymology of protein tyrosine-O-sulfation," *J. Biol. Chem.* **278:**24243–24246 (2003).

Seibert, C., M. Cadene, A. Sanfiz, B. T. Chait, and T. P. Sakmar. "Tyrosine sulfation of CCR5 N-terminal peptide by tyrosylprotein sulfotransferases 1 and 2 follows a discrete pattern and temporal sequence," *Proc. Natl. Acad. Sci. U.S.A.* **99:**11031–11036 (2002).

von Bulow, R., B. Schmidt, T. Dierks, N. Schwabauer, K. Schilling, E. Weber, I. Uson, and K. von Figura. "Defective oligomerization of arylsulfatase A as a cause of its instability in lysosomes and metachromatic leukodystrophy," *J. Biol. Chem.* **277:**9455–9461 (2002).

von Bulow, R., B. Schmidt, T. Dierks, K. von Figura, and I. Uson. "Crystal structure of an enzyme–substrate complex provides insight into the interaction between human arylsulfatase A and its substrates during catalysis," *J. Mol. Biol.* **305:**269–277 (2001).

von Figura, K., B. Schmidt, T. Selmer, and T. Dierks. "A novel protein modification generating an aldehyde group in sulfatases: Its role in catalysis and disease," *BioEssays* **20:**505–510 (1998).

Modifications of Cysteine and Methionine by Oxidation–Reduction

Reversible dithiol–disulfide interconversion at Cys–X–X–Cys sequences in thioredoxins.

The two sulfur-containing amino acids, cysteine and methionine, are subject to oxidation of the polarizable sulfur atom. Biological redox chemistry of the thiol group of cysteinyl residues is more extensive than on the thioether side chain of methionyl residues, but both have enzymes dedicated to reduction of oxidized forms of the amino acids (disulfides and sulfenic acids in the case of cysteine; sulfoxides in the case of methionine) (Jacob et al., 2003) (Figure 4.1).

In other contexts we note the extensive utilization of the thiol side chain of cysteinyl residues as a nucleophile in posttranslational protein modification, such as in S-palmitoylation for the attachment of the fatty acyl lipid anchor (Chapter 7), and in $S–PO_3^{2-}$ intermediate formation in protein tyrosine phosphatase action (Chapter 2). Both of these reactions take advantage of the accessible pK_a (around pH 8) for the dissociation of the thiol to thiolate: $Cys–SH \rightleftarrows H^+ + Cys–S^-$. The electron-rich thiolate, kinetically available near neutral pH, is the most powerful nucleophile in proteinogenic amino acid side chains.

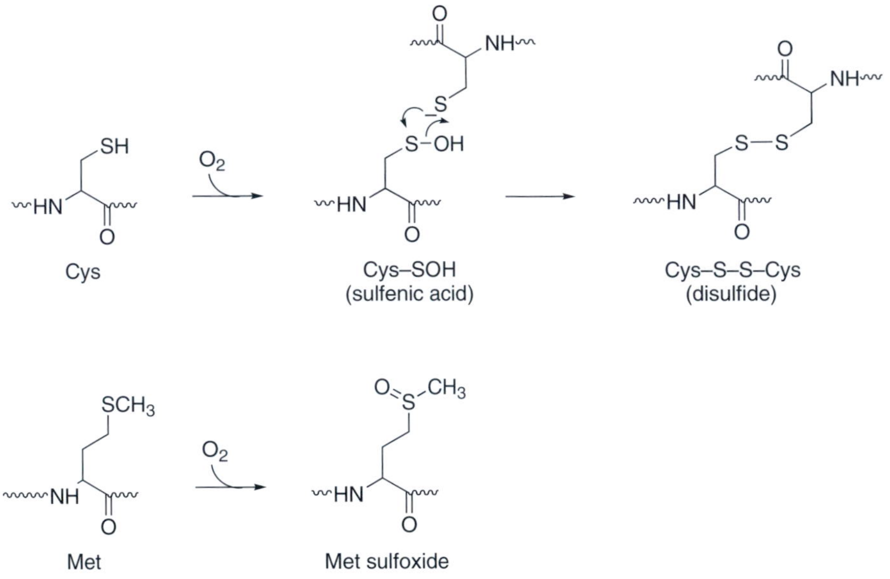

Figure 4.1 Reversible oxidation and reduction of cysteine and methionine side chains in proteins: thiol to sulfenic acid and disulfide; thioether to sulfoxide.

The electron-rich sulfur anion is also susceptible to oxidation (electron removal), either two at a time or one at a time. The one-electron route produces thiyl radicals (Figure 4.2). Two thiyl radicals can couple to form the disulfide (cystine). The two cysteine–SH to one cystine disulfide redox interconversion is the most common one that occurs to proteins in oxidizing microenvironments (Giles et al., 2003; Jacob et al., 2003).

The thiyl radical could alternatively be intercepted by OH•, the hydroxyl radical, to yield sulfenic acids, or by nitric oxide (NO•) to yield the S-nitroso species (Finkel, 2000; Jacob et al., 2003). Thiolate anions and thiyl radicals can also react with and decompose peroxides (peroxiredoxins) in the detoxification and termination of signaling mediated by reactive-oxygen-species-mediated signaling. In most proteins sulfenic acids are unstable, because the sulfur atom is sufficiently electrophilic in this oxidation state to be captured by neighboring cysteine thiolate chains, releasing water and leading to the disulfide (Figure 4.2). In this reaction manifold sulfenic acids are reaction intermediates, not final end products. There are some notable exceptions where the microenvironment of a particular protein isolates and stabilizes the sulfenic acid (Figure 4.3) sufficiently for characterization by x-ray diffraction or NMR spectroscopy, validating its existence, in such enzymes as NADH peroxidase and NADH oxidase (Claiborne et al., 1999).

Protein thiol oxidation occurs in reactions with reactive oxygen species (ROS). ROS arise by metabolic reduction of O_2. One-electron reduction of O_2 generates

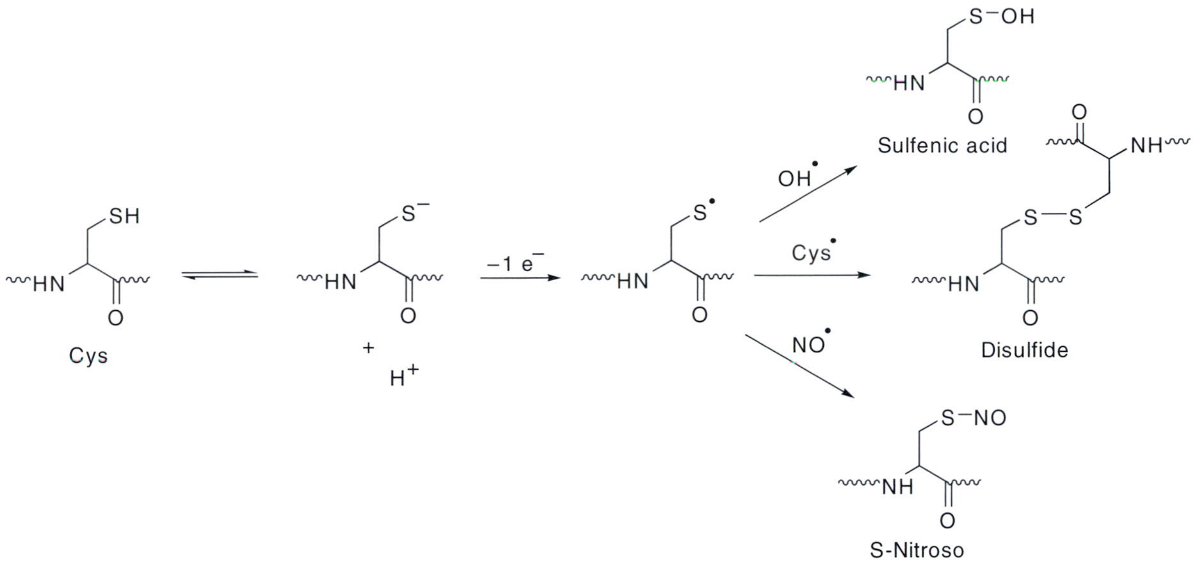

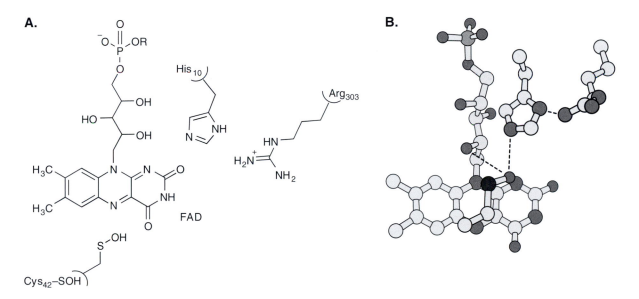

Figure 4.2 Oxidation of the electron-rich thiolate anion of cysteine residues. One-electron oxidation routes can lead to disulfides, sulfenic acids, or S-nitroso adducts.

A.

B.

Figure 4.3 Sulfenic acids as stable entities in proteins; stable Cys-SOH accumulating in the NADH peroxidase active site. **A.** Schematic representation of the active site. **B.** Structure of the active site. Carbons are pale gray, oxygens, phosphorus, and nitrogens are colored gray, and sulfur is colored black (figure made using PDB 1JOA).

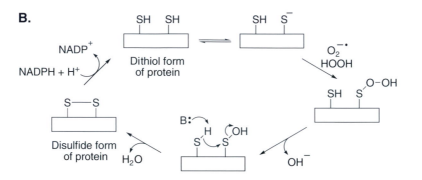

B.

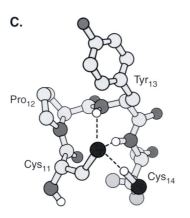

Figure 4.4 Reactive oxygen species (ROS). **A.** Superoxide, hydrogen peroxide, and hydroxyl radical. **B.** Scavenging of superoxide or peroxide by thiol pairs. **C.** Architecture of the CXXC disulfide pair in glutaredoxin-3. Carbons are colored pale gray, oxygens and nitrogens are colored gray, and sulfurs are colored black (figure made using PDB 1ILB).

superoxide ($O_2^{-}\bullet$), two-electron reduction yields hydrogen peroxide (HOOH), three-electron reduction yields hydroxyl radical (OH$\bullet$), and four-electron reduction takes oxygen all the way to water (Figure 4.4A). Superoxide and hydroxyl radical are short-lived metabolites while HOOH is longer-lived in biological milieus and is the predominant species scavenged by thiol oxidation to sulfenates and then to disulfides (Figure 4.4B). ROS are generated during electron passage down mitochondrial respiratory chains by the NADPH oxidase activity of white cells (respiratory burst) and by flavin- and iron-containing enzymes that generate peroxide and/or leak superoxide equivalents (Finkel, 2000; Giles, 2003; Giles et al., 2003). Among the proteins that

Table 4.1 Proteins Undergoing Reversible Ditihol–Disulfide Redox Conversion*

Protein	CXXC Motif	$E^{\circ\prime}$	Function
Thioredoxin (12 kDa)	CGPC	–270 mV	Reduction of disulfide bonds in proteins
Glutaredoxin (9 kDa)	CPYC	–233 mV	Reduction of ribonucleotide reductase
DsbA (21 kDa)	CGHC	–125 mV	Shuffling of disulfide bonds in proteins transiting to the ER
Protein disulfide isomerases (57 kDa)	CPHC	–127 mV	Reduction and isomerization of disulfide bonds in bacterial periplasm

*Adapted from Jacob, 2003.

participate in dithiol–disulfide redox interconversions in sulfur redox metabolism are small proteins such as glutaredoxins (Table 4.1 and Figure 4.4C).

Thiol to Disulfide: Oxidations, Reductions, and Disulfide Bond Isomerizations

The cytoplasm of prokaryotic cells and the cytoplasm and nucleus of eukaryotic cells are relatively reducing microenvironments [reduction potential ($E^{\circ\prime}$) = –0.27 V (Chivers, 1998)] exemplified by the ratio of reduced (GSH) to oxidized glutathione (GSSG), the γ-Glu-Cys-Gly tripeptide (Figure 4.5) that is the soluble thiol/disulfide buffer in most organisms. A ratio of 100/1:GSH/GSSG, at total concentrations of 1–10 mM, characterizes these reducing intracellular environments. In addition, the protein thioredoxin (Trx) (Holmgren, 1989) (Figure 4.6) is the protein-based equivalent (Table 4.1) of glutathione in cells, cycling between dithiol and disulfide forms in redox metabolism and accelerating the approach to the equilibrium of dithiol and disulfide forms of proteins. The high GSH/GSSG and Trx-SH/Trx-SS-Trx ratios are maintained by electron transfer from NADPH, the cellular reducing currency. Electron flow from NADPH to the disulfides of GSSG and the disulfide form of Trx are enabled by dedicated reductases, glutathione reductase and thioredoxin reductase, discussed later in this chapter. The favorable thermodynamic driving force of using electrons from NADPH ($E^{\circ\prime}$ = –320 mV) enables most proteins in those cellular compartments to have their cysteine side chains present as the reduced, nucleophilic thiols. There are special exceptions, such as the thioredoxins and thioredoxin reductase and glutathione reductase, which have disulfides of unusual kinetic stability.

In the periplasm of Gram-negative bacterial cells and the eukarotic secretory compartments starting with the endoplasmic reticulum (ER) [$E^{\circ\prime}$ = –0.17 to –0.18 V (Chivers, 1998)], the microenvironment is oxidizing, there is no effective dithiol buffering capacity, and proteins are found with cysteine thiols oxidized to disulfides. The

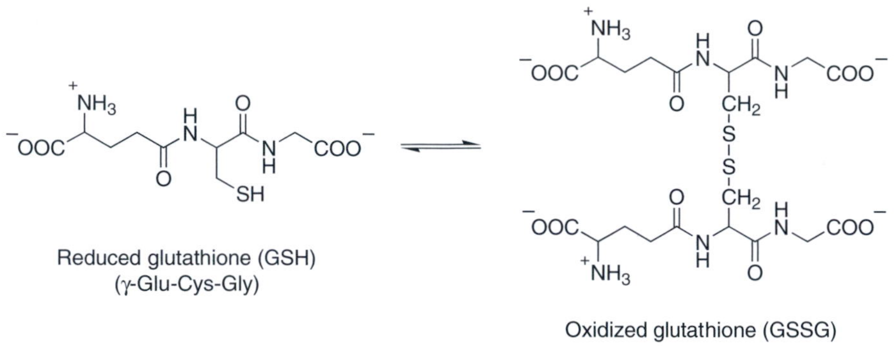

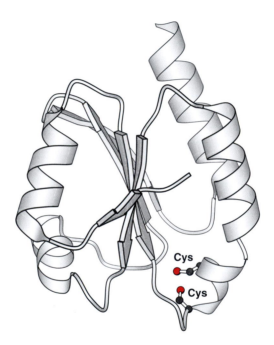

Reduced glutathione (GSH)
(γ-Glu-Cys-Gly)

Oxidized glutathione (GSSG)

Figure 4.5 Reduced and oxidized forms of the tripeptide glutathione, the small molecule thiol redox buffering agent in cells.

Cys

Cys

Figure 4.6 Structure of thioredoxin, a key protein in cellular thiol redox metabolism, with the active site cysteines highlighted. Carbons are colored dark gray and sulfurs are colored red (figure made using PDB 1R26).

ER, the primary compartment for the folding of proteins (Zapun et al., 1999) as they enter the secretory pathway of eukaryotic cells, is thought to be a transitional compartment with regard to thiol/disulfide potential—a compartment in which disulfide bonds form through thiol oxidations and in which they get reshuffled from non-native to native –S–S– connectivity by the agency of protein disulfide isomerases resident in the ER.

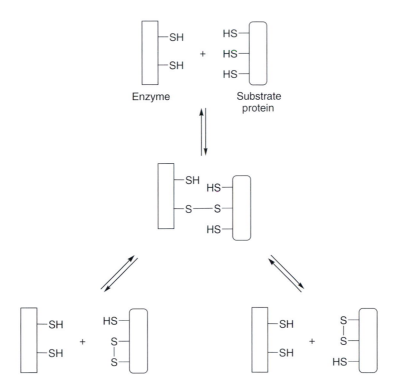

Figure 4.7 Thiol–disulfide interconversion and disulfide shuffling: the interplay of the isomerization and oxido-reduction of pairs of thiols on enzyme and substrate proteins.

As proteins transit from reducing to more oxidizing compartments there is a balance between dithiol to disulfide conversions and the reshuffling of disulfide bonds that may have been kinetically favored but traps regions of the proteins in non-native (nonfunctional) conformations (Figure 4.7). The reshuffling process operates by making use of the rapidly attained thiol–disulfide equilibrium via nucleophilic attack of one thiol on a disulfide to form a new disulfide and release one of the cysteine components of the original disulfide as free thiol. The final distribution of the equilibrium will be dictated by the energetics of the disulfide isomers, presumably reflecting the energy of the folded protein conformers. Disulfide bond formation is thought to proceed as early as the transit of unfolded proteins into the ER lumen, given reduction potentials for disulfides in unfolded proteins of –0.22 V (Chivers, 1998), halfway between the potential of the cytoplasm (reducing) and the ER (oxidizing). The approach to dithiol–disulfide equilibration is catalyzed by enzymes in the oxidizing compartments, such as the endoplasmic reticulum of eukaryotic cells and the periplasm of Gram-negative bacteria. One such enzyme is Ero1p in the ER lumen, a flavoprotein oxidizing protein dithiols to disulfides and passing the two electrons removed to O_2 to generate H_2O_2 (Gross et al., 2004).

The common motif for proteins that catalyze both dithiol–disulfide redox and dithiol–disulfide shuffling is typically a tetrapeptide sequence –CXXC– (Chivers et al., 1997) found in the small cytoplasmic protein thioredoxin (Table 4.1 and Figure 4.6) and in proteins with domains containing folds homologous to thioredoxin. The two cysteines in thioredoxin folds can shuttle between disulfide (oxidized) and dithiol (reduced) forms.

Thioredoxin has a dedicated electron transfer partner protein, thioredoxin reductase, which moves electrons from NADPH, the cellular reducing currency (Walsh, 1979), to the thioredoxin disulfide. Thioredoxin reductase has three redox sinks (Figure 4.8): 1. a noncovalently but tightly bound flavin coenzyme, FAD; 2. an active site $C_{59}XXC_{64}$ motif that accumulates as the disulfide in the enzyme at rest; 3. a selenosulfide at the C-terminus of the enzyme, involving Cys_{497}' and selenoCys$_{498}'$. Thioredoxin reductase is a homodimer, with two active sites that span the dimer interface. Each active site has the FAD and the Cys$_{59}$–S–S–Cys$_{64}$, but the Cys$_{497}'$–S–Se–Cys$_{498}'$ of that active site comes from the other dimer. Electrons flow from NADPH to FAD, then to the active site disulfide, generating the two-electron-reduced dithiol form. This reduced enzyme species can now reduce the selenosulfide. The active site geometry favors a charge transfer interaction of the Cys$_{64}$–SH with

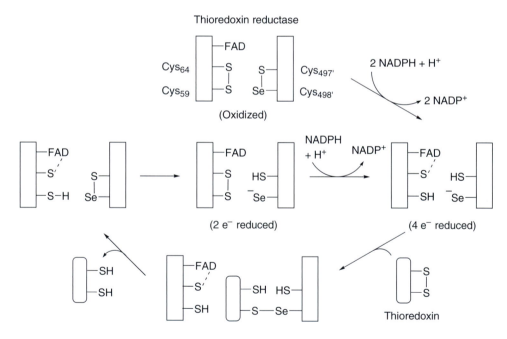

Figure 4.8 Thioredoxin reductase with three electron sinks that can participate in catalysis: FAD, the active site Cys$_{59}$–SS–Cys$_{64}$ disulfide, and the Cys$_{498}'$–S–Se–Cys$_{497}'$ selenosulfide from the other subunit in the homodimer. Reduction of the disulfide form of thioredoxin via a mixed seleno-sulfide intermediate.

the FAD, allowing the Cys_{59}–S^- to act as the nucleophile towards the selenosulifde. A second NADPH molecule can be oxidized to generate a four-electron-reduced form of the reductase with the $Cys_{59,64,497}'$ as the thiolates and $SeCys_{498}'$ as the selenate anion. The selenate anion is the stronger nucleophile and can attack a molecule of thioredoxin disulfide bound in the reductase active site to generate a transient mixed disulfide and then the thioredoxin dithiol as the enzyme selenosulfide linkage reforms. The net result is flow of electrons from NADPH to the thioredoxin disulfide, generating NADP and the thioredoxin dithiol form. The Trx–$(SH)_2$ is a potent reductant for other proteins in the cell, as noted in the next paragraph (Holmgren, 1989). The catalyst is the reductase cycling between two-electron- and four-electron-reduced forms.

Among the key partner proteins for thioredoxin is ribonucleotide reductase (RDR), converting all four ribonucleoside diphosphates to the four essential 2'-deoxyribonucleoside diphosphate precursors for DNA biosynthesis (Stubbe et al., 2003). In each catalytic cycle where RDR reduces a ribose to a 2'-deoxyribose on a substrate molecule, the active site dithiol of RDR receives those two electrons and gets oxidized to the cystine disulfide (Figure 4.9). To function catalytically the disulfide of RDR must be reduced back to the dithiol form. This is enabled by the dihydro form of thioredoxin. Ultimately the electrons in NADPH are utilized for each 2'-deoxyribonucleotide monomer produced by RDR action, with thioredoxin as the

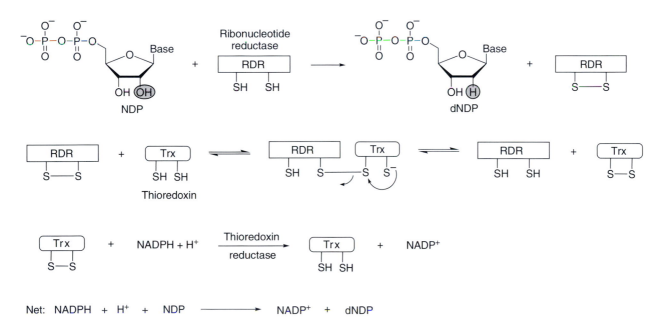

Figure 4.9 Thioredoxin in the dithiol form as a required cosubstrate for catalytic cycling of ribonucleotide reductase in NDP to dNDP conversions for DNA biosynthesis.

essential conduit. Thus, consideration of the roles of thioredoxin reductase and ribonucleotide reductase provides insight into the sulfur redox shuttle role for thioredoxin and related thioredoxin fold proteins (Table 4.1).

Oxidation of pairs of cysteine thiols to cystine disulfides, such as the three disulfides in each of 30 EGF domains (90 disulfides in all) of the cell surface receptor protein Notch, is favored thermodynamically during passage through the oxidizing secretory compartments to the cell surface. However, to obtain proteins with homogeneously ordered, regiospecific, native disulfide linkages requires kinetic accelerations by protein disulfide isomerases and dithiol/disulfide oxidase catalysts, as shown schematically in Figure 4.7. Insight into this disulfide forming and unscrambling process has been obtained by a combination of genetic and biochemical investigations of protein transit from cytoplasm to periplasm and then to cell envelope in *E. coli* (Kadokura et al., 2003).

The CXXC motif has been termed a redox rheostat (Chivers et al., 1997) in active sites of protein disulfide isomerases, where the identity of the XX residues can affect the reduction potential of the disulfide (Table 4.1). The disulfide reduction potentials of three proteins with CXXC motifs—thioredoxin ($E°' = -0.27$ V), protein disulfide isomerase ($E°' = -0.18$ V), and DsbA ($E°' = -0.10$ V) (Chivers, 1998)—are adapted to their cellular compartments of cytoplasm, endoplasmic reticulum, and bacterial periplasm, respectively. The tunable CXXC rheostat in turn affects the fraction of dithiol form present as mono- and bisthiolate and so indirectly controls thiolate nucleophilicity as well.

Four *E. coli* proteins, DsbA–DsbD (*d*isulfide *b*ond), are required for generation of the correct disulfide pairing in proteins sent outside the cytoplasmic compartment: flagella, pilins, periplasmic enzymes, and secreted protein toxins (Figure 4.10A). DsbD sits in the cytoplasmic membrane, with three pairs of cysteines in CXXC arrays on periplasmic domains. These can accept electrons from reduced thioredoxin (via thioredoxin reductase and ultimately NADPH). The dithiol/disulfide pairs in DsbD can interact with DsbC, which has protein disulfide isomerase (PDI) activity, resulting from two CXXC domains of its own (Figure 4.10B). DsbC acts to rearrange nonnative disulfides that may have formed kinetically to native, thermodynamically favored disulfides in client proteins in the periplasm or outer envelope. DsbA is the periplasmic dithiol oxidase, oxidizing pairs of cysteine residues to cystine disulfides in proteins secreted into the periplasm. DsbA is a thioredoxin family member and has the redox active CXXC motifs. Each time DsbA oxidizes a periplasmic protein, the DsbA active site disulfide is reduced to the dithiol (Kadokura et al., 2004). It has to dump those two electrons to catalyze another cycle. This occurs by dithiol–disulfide interchange with DsbB, a transmembrane protein. In turn, as DsbB accepts electrons, reoxidizing DsbA and preparing it for the next catalytic cycle, the reduced DsbD must now get rid of the newly acquired two electrons. This occurs by passage of electrons

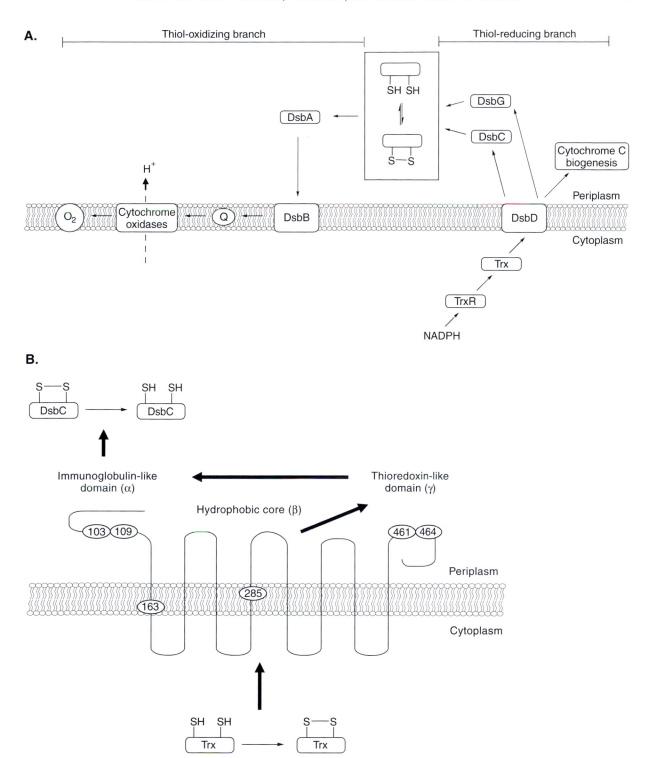

Figure 4.10 The cascade of Dsb protein action in *E. coli* to move electrons to the periplasm for thiol–disulfide oxido-reduction and disulfide bond scrambling. **A.** The pathway of electron flow from NADPH in the cytoplasm to the DsbA-D system in the membrane and periplasm. **B.** Membrane topology and electron flow in and out of DsbD. Numbered ovals represent the essential cysteines.

C.

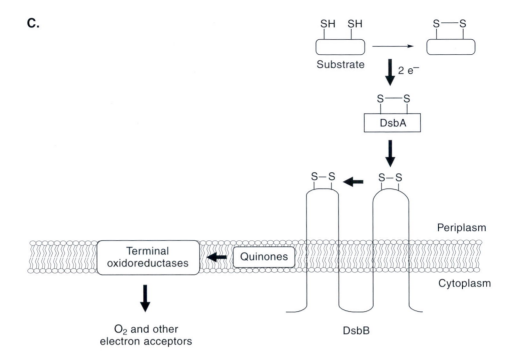

Figure 4.10 **c.** Oxidative pathway for disulfide bond formation in periplasmic proteins via action of the DsbA–DsbB couple.

from DsbB to ubiquinone in respiratory chains in the membrane (Figure 4.10C). Ultimately, the electrons removed from proteins that are oxidized in the periplasm and outer membrane of *E. coli* get sent through the respiratory chain and reduce molecular oxygen. Their route is through DsbA and DsbB. DsbC unscrambles non-native disulfide linkages in the process and it is kept in redox balance by DsbD in relay from cytoplasmic thioredoxin, as needed (Figures 4.10A and B).

Although eukaryotic systems have been much less extensively characterized, there are multiple protein disulfide isomerase (PDI) proteins in the ER compartment (Zapun et al., 1999), including isoforms (ER57p) that form complexes with calnexin and calreticulin and thus are recruited to act as PDIs for glycoproteins (Chapter 10) that are refolding in the ER. The PDIs are oxidized to disulfide forms for thiol interchange with Ero1p (noted earlier) and then the disulfide forms of PDIs in turn oxidize dithiols in proteins that have been translocated into the ER.

Thioredoxin Reductase Versus Glutathione Reductase: S–Se Versus S–S

In the above discussion we have noted that the small protein thioredoxin can function both as a dithiol/disulfide buffer and an equilibration catalyst for disulfides and dithiols in other proteins in cells and that the tripeptide glutathione plays a comparable role as a diffusible small molecule reductant in cells. As the oxidized thioredoxin disulfide (Trx–S–S–Trx) or the oxidized glutathione disulfide (GS–SG) forms accumulate, they are re-reduced using electrons from NADPH by the two enzymes thioredoxin reductase and glutathione reductase. These are two related enzymes with bound FAD and an active site CXXC disulfide in the active site as conduits for the electrons passing from NADPH to the Trx–S–S–Trx or GS–SG cosubstrates (Figures 4.5 and 4.6). The GSSG reductase does not have the third redox sink, the terminal selenosulfide that is present in thioredoxin reductase. Instead, the GSSG disulfide bound in the active site is reduced by direct thiol–disulfide interchange with Cys_{58}–S^-.

The mammalian thioredoxin reductase also has the capacity to reduce lipid hydroperoxides and hydrogen peroxide. This detoxification outcome is mediated by the enzyme's nucleophilic Cys-selenate anion. Reaction with ROOH yields the Cys-selenenic acid (Cys–Se–OH) that is then captured by the thiolate of Cys_{497}' to reform the selenylsulfide resting state of the enzyme (Figure 4.11A). This is a variant of the peroxiredoxin mechanisms discussed next.

Peroxiredoxins

When thioredoxin reductase acts, in a side reaction, to reduce hydrogen peroxide to two molecules of water, it is acting in a detoxification mode as a peroxiredoxin. Glutathione peroxidase and thioredoxin peroxidases are peroxiredoxins where the destruction of hydrogen peroxide and lipid peroxides is thought to be the primary physiological function of these enzymes as surveillance catalysts. Consistent with the catalytic redox strategy of Figure 4.11A, glutathione peroxidase also has selenocysteine incorporated at the active site and decomposes HOOH via the selenenic acid intermediate in each turnover. The thiol–selenol interchanges result in the net oxidation of two reduced glutathiones to the GSSG disulfide as HOOH/ROOH is reduced to HOH/ROH (Figure 4.11B).

Peroxiredoxins are constructed with thioredoxin-type structural folds, but are found in two subfamilies (Figure 4.11C)—namely, those with one thiol and those with a pair of thiols in the active site (Wood et al., 2003). In the one-thiol version, exemplified by hOrf6, the Cys–SOH intermediate must be captured by an external thiol, such as glutathione, to give a mixed protein-SSG disulfide that gets attacked by a second RSH to produce the starting hOrf6–SH and the RS–SG. In the two-thiol version of peroxiredoxins, the mechanism of Figure 4.11B occurs, for example in yeast thioredoxin peroxidase, HBP23, and the bacterial peroxiredoxin AhpC (Georgiou, 2002) with intramolecular breakdown of the Cys–Se–OH by a neighboring $CysS^-$ side chain.

A.   Thioredoxin reductase as peroxiredoxin

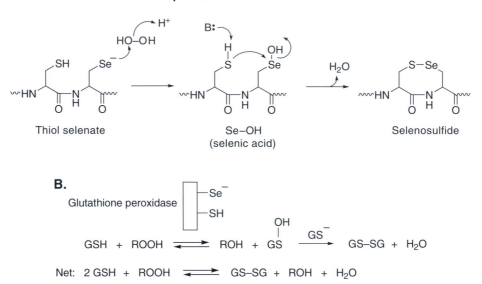

B.

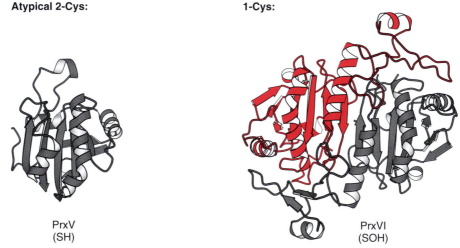

C.   Atypical 2-Cys: **1-Cys:**

PrxV   PrxVI
(SH)   (SOH)

Figure 4.11 Peroxiredoxin activity of **A.** thioredoxin reductase and of **B.** glutathione peroxidase. **c.** Peroxiredoxin structural classes (figure made using PDBs 1HD2 for PrxV, 1QQ2 for PrxI, 1PRX for PrxVI, 1QMV for PrxII, 1KYG for AhpC, and 1E2Y for TryP).

A variant of the single active site cysteine peroxidase is found in the *Enterococcus faecalis* NADH peroxidase (Claiborne et al., 1999). X-ray analysis has confirmed that the active site cysteine, Cys_{42}, cycles between the Cys–SH starting state and the Cys–SOH oxidized intermediate (Figure 4.3), with this sulfenic acid form detectable by a charge-transfer band with the FAD in the active site and thereby sufficiently stable to be crystallized and directly observed.

Typical 2-Cys:

Figure 4.11 C. (*continued*)

Peroxiredoxins have been suggested to be components of signaling pathways in which metabolically generated HOOH [e.g., from insulin (Mahadev et al., 2001), epidermal growth factor signaling (Lee, 1998), or cytokines (Wood, 2003)] is the proximal signal that acts to oxidize thiols in protein targets (to sulfenic acids and/or disulfides). The peroxiredoxins, by decomposing HOOH, will control the duration and intensity of HOOH signaling (and indirectly other reactive oxygen species). The stress-activated MAP kinase cascade can be turned on by bursts of H_2O_2 that overwhelm the buffering capacity of the peroxiredoxins (Veal et al., 2004).

Readouts of HOOH signaling include the increased phosphorylation status of the Src family tyrosine kinases in T cells after HOOH exposure, in substantial part due to the transient inhibition of protein tyrosine phosphatases. As noted in Chapter 2, PTPases have an active site cysteine thiolate that acts as a nucleophile to remove phosphoryl groups from phosphotyrosyl protein substrates via S-phosphocysteine covalent intermediates. The active site Cys–S$^-$ of PTPases is susceptible to oxidation and inactivation. The inactivation could be reversed by reduction back to the Cys–S$^-$ state. It had been assumed that the inactive form of PTPases was the Cys–SOH in analogy to the above cases, but recent structural studies have indicated that the oxidation state of the inactive PTPases is a sulfenylamide, arising by capture of the initial sulfenic acid by the isoamide form of the peptide bond at Ser$_{216}$, thus generating a sulfenylamide with altered geometry in the active site (Figure 4.12) (Salmeen et al., 2003; van Montfort et al., 2003). The sulfenylamide form appears to be a reversible protective pathway, cleaved back to starting enzyme by exogenous thiolate, thereby preventing further oxidation of Cys–SOH to Cys–SOOH (sulfinic acid), which is irreversible.

Over-oxidation of the reactive cysteine thiol in peroxiredoxins has also been detected, for example in the x-ray structure of thioredoxin peroxidase B (Schroder et al., 2000). The sulfenic acid reaction intermediate Cys$_{51}$–SOH, normally converted in each catalytic cycle to Cys$_{51}$–SS–Cys$_{172}$, might be captured instead by a second HOOH and oxidized irreversibly to the Cys$_{51}$–SOOH sulfinic acid, inactivating the peroxiredoxin. Recently a protein in yeast with the ability to reduce sulfinic acids has been detected (Biteau et al., 2003). It was further proposed that this might be related to a regulatory role of the peroxiredoxin, where normal levels of HOOH would be scavenged in cells, but a large burst would overwhelm the peroxiredoxin and allow hydrogen peroxide to oxidize other proteins and have signaling consequences. In this context thioredoxin peroxidase B is highly abundant in erythrocytes during cell maturation as a presumed surveillance agent against reactive oxygen species. Peroxiredoxins also reduce peroxynitrite, generated from the combination of superoxide and nitric oxide (Stamler et al., 2001).

OxyR and OhrR: Transcription Factors as Redox Sensors by Thiol Oxidation

Reactive oxygen species can turn on particular bacterial genes for expression of detoxification and repair systems (Fuangthong and Helmann, 2002). The induction of such oxidative stress stimulons (clusters of co-regulated genes) occurs by redox sensing through cysteine thiol and dithiol oxidation in transcription factors. The paradigm is the *E. coli* OxyR transcription factor, where formation of the Cys$_{199}$-

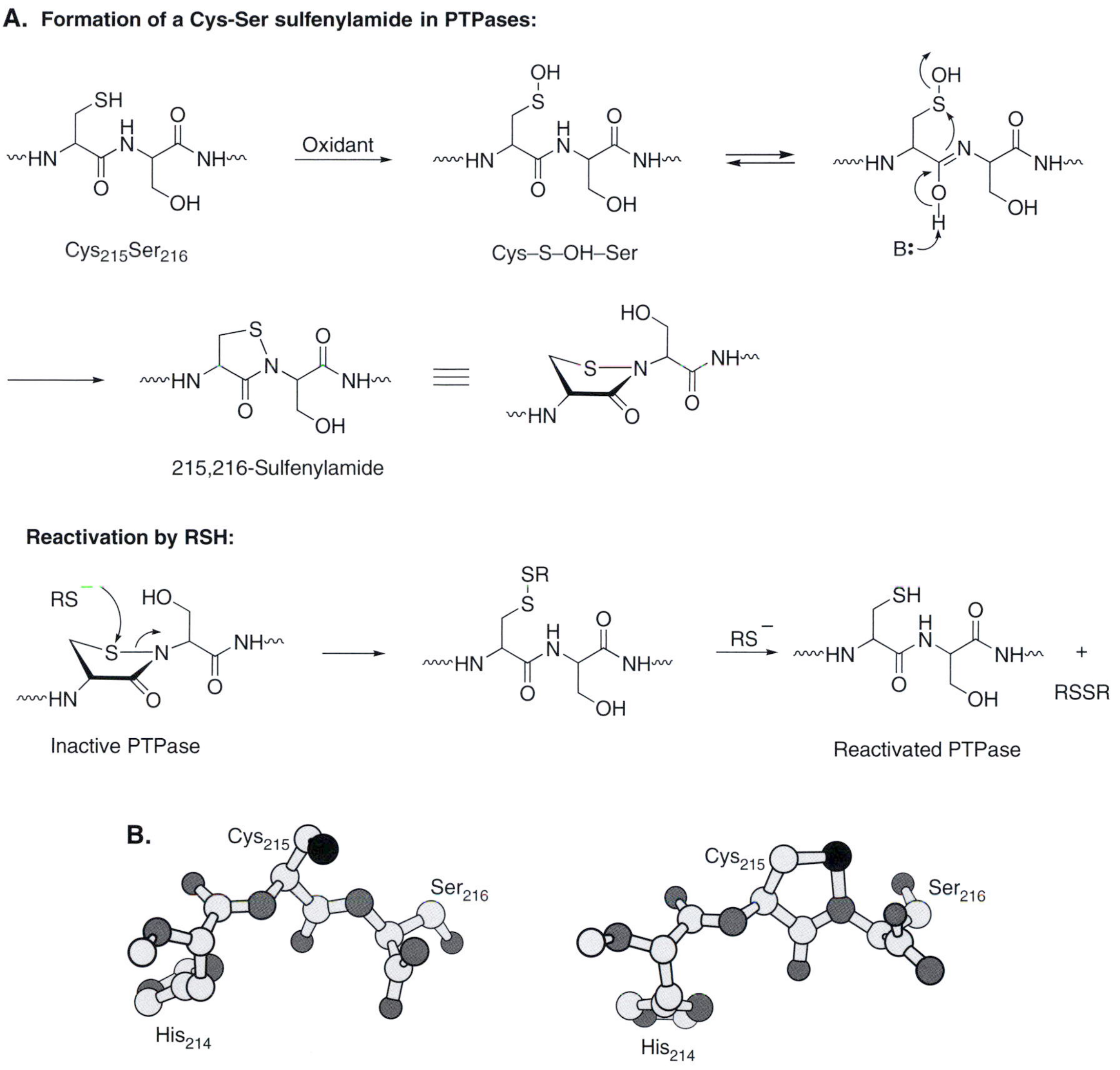

A. Formation of a Cys-Ser sulfenylamide in PTPases:

Cys$_{215}$Ser$_{216}$ Cys–S–OH–Ser   B:

215,216-Sulfenylamide

Reactivation by RSH:

Inactive PTPase   Reactivated PTPase

B.

Figure 4.12 Oxidation of the active site cysteine in PTPases to a sulfenylamide and reversible reduction by thiolates. **A.** Proposed mechanism. **B.** Structure of Cys$_{215}$Ser$_{216}$ in the PTP1B active site before oxidation (left) and after oxidation conversion to the sulfenylamide (figure made using PDBs 2HNP for the left panel and 1OES for the right panel).

Cys$_{208}$ disulfide leads to a conformational change in OxyR that activates transcription of target genes (Choi et al., 2001; Zheng et al., 1998). Disulfide formation in OxyR captures the rearrangement of β sheet into helix and generates a new β strand in the protein. Cys$_{199}$ is the sentinel thiol that has the thiolate anion oxygenated and then captured by Cys$_{208}$ (Figure 4.13). In *B. subtilis*, the OhrR transcription factor is a peroxide sensor by virtue of oxidation of a Cys–SH to a Cys–SOH (Fuangthong and Hel-

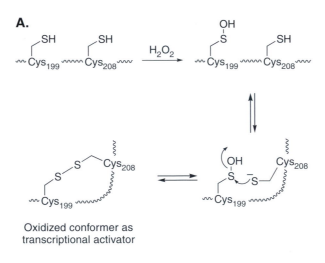

Oxidized conformer as
transcriptional activator

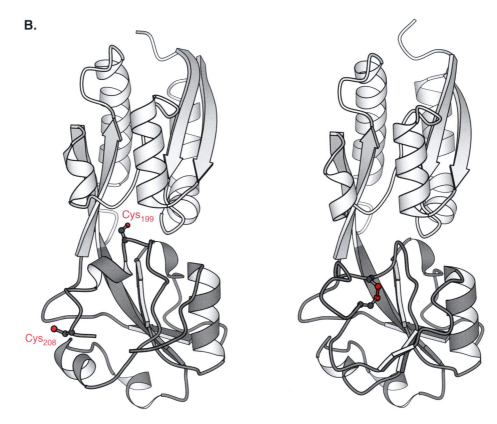

Figure 4.13 Sentinel function of Cys_{199} in the *E. coli* redox-sensing transcription factor OxyR. A. Conformational rearrangement driven by oxidant-mediated conversion of Cys_{199} and Cys_{208} to a disulfide. B. X-ray structure of the reduced (left panel) and oxidized (right panel) forms of OxyR. Carbons are colored dark gray and sulfurs are colored red (figure made using PDBs 1I69 for the left panel and 1I6A for the right panel).

mann, 2002). The thiol oxidation alters the OhrR conformation and lowers its affinity for inverted repeat sequences in the *ohrA* gene control region, derepressing this resistance gene. In mammalian systems the Fos/Jun heterodimeric transcription factor has oxidizable thiols that act as redox sensors (consult Claiborne et al., 1999, for a review).

S-Nitrosylation of Proteins

Thiyl radicals in proteins can react with the short-lived radical signal molecule nitric oxide to yield S-nitrosoCys side chains (Figure 4.2). Proteins in high abundance and/or with accessible and reactive thiolates, such as the cysteinyl proteases of the caspase family (Stamler et al., 2001), undergo inactivating S-nitrosylation. So, too, do membrane receptor proteins such as the ryanodine receptor and the NMDA receptor. It is not yet known whether there is protein-mediated specificity for nitrosylation or whether the pattern reflects some combination of protein abundance, particular cysteine thiolate reactivity, and propinquity to sources of NO. Two specific examples are S-nitrosylation of the IKKβ catalytic subunit required for NFκB activation and subsequent gene transcription (Reynaert et al., 2004), and S-nitrosylation of the ubiquitin ligase E3 protein parkin in the brain (see Chapter 9), thereby inhibiting its function and abrogating neuroprotective function (Chung et al., 2004).

Reversal of the S-nitrosyl modification can occur nonenzymatically via attack of thiols, such as glutathione (GSH) or the reduced form of thioredoxin, to yield the mixed disulfides as intermediates for subsequent thiol–disulfide interchanges.

O-Nitration of Tyrosyl Residues in Proteins

Nitric-oxide-derived reactive species can also be funneled to modify tyrosyl residues in proteins, generating the ortho-nitro derivative of tyrosine (3-nitroTyr) (Figure 4.14) (Radi, 2004). Tyrosine nitration is observed to increase during inflammatory responses (Schopfer et al., 2003) with modification of proteins at high abundance, including serum albumin, superoxide dismutase, histones, α-tubulin, and actin. The proximal nitrating species is thought to be nitrogen dioxide ($\bullet NO_2$), reacting at the diffusional limit with tyrosyl radicals at the ortho position. Nitrogen dioxide can arise by homolytic cleavage of peroxynitrous acid (ONOOH), in turn derived from nitric oxide ($\bullet NO$) and superoxide ($O_2^- \bullet$) (see Schopfer et al., 2003, for a review). The tyrosyl radicals in protein substrates, as well as the peroxynitrous acid homolysis, are thought to be generated by peroxidases. Ortho nitration lowers the pK_a of affected tyrosyl residues from 10.1 to 7.2, and thus raises the concentration of anionic species in the nitroTyr side chains. Proteomic analyses are in progress to evaluate the tissue

NO$^\bullet$ synthase: NADPH + Arginine + O$_2$ $\longrightarrow$ NADP$^+$ + Citrulline + H$_2$O + NO$^\bullet$

NO$^\bullet$ + O$_2^{-\bullet}$ + H$^+$ $\longrightarrow$ HOONO (Peroxynitrous acid)

HOONO $\longrightarrow$ NO$_2^\bullet$ + HO$^\bullet$

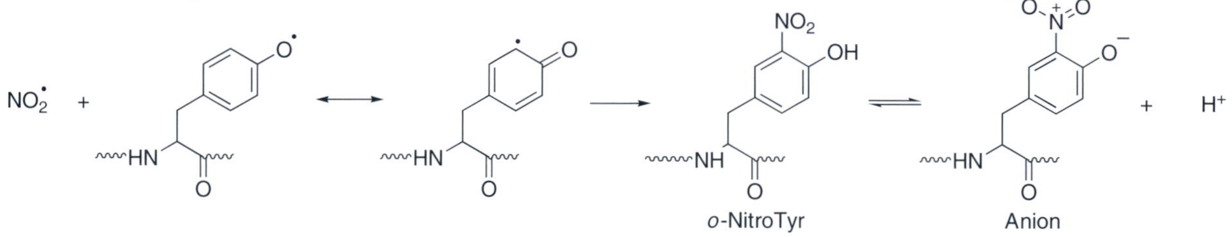

Figure 4.14 Ortho nitration of tyrosyl residues in proteins from $^\bullet$NO as initial oxidant.

localization of nitrated proteins, the fractional levels of posttranslational modification, and the regioselectivity of tyrosyl residues that get nitrated, as context for functional analysis. Specific tyrosines in albumin (Tyr$_{138,411}$), actin (Tyr$_{91,198,240}$), and tubulin (the C-terminal Tyr) are known to be nitrated (Schopfer et al., 2003).

Methionine Sulfoxide Formation and Reduction

Hydrogen peroxide also oxidizes protein methionine residues to methionine sulfoxide. A well-studied case is the oxidation of a methionine residue in the circulating α_1-proteinase inhibitor, which dramatically lowers its affinity for target serine proteases, such as elastase in lung tissue. Excess elastase activity is part of the pathology in emphysema. The methionine sulfoxidation reaction is mediated by superoxide and peroxide in the lungs.

The damage can be reversed when protein methionine sulfoxide residues are reduced enzymatically back to the thioether oxidation state by methionine sulfoxide reductases, which use the dithiol form of thioredoxin as the external reductant. The MsrA enzyme has been characterized and has an active site cysteine that serves as the nucleophile to attack the sulfoxide sulfur in a protein substrate (Boschi-Muller et al., 2000). The resultant adduct resolves with oxygen transfer to release the methionine residue in the protein product and the sulfenate form of methionine sulfoxide reductase. This enzyme CysS–OH is attacked by one of the thiolates in thioredoxin, followed by disulfide interchange to regenerate MsrA and yield oxidized thioredoxin (Figure 4.15). It can be seen that the thioredoxin-mediated reduction of methionine sulfoxide residues, using nucleophilic active site cysteines, sulfenic acid intermediates and disulfide final states, follows a strategy parallel to that for cysteine thiol redox chemistry.

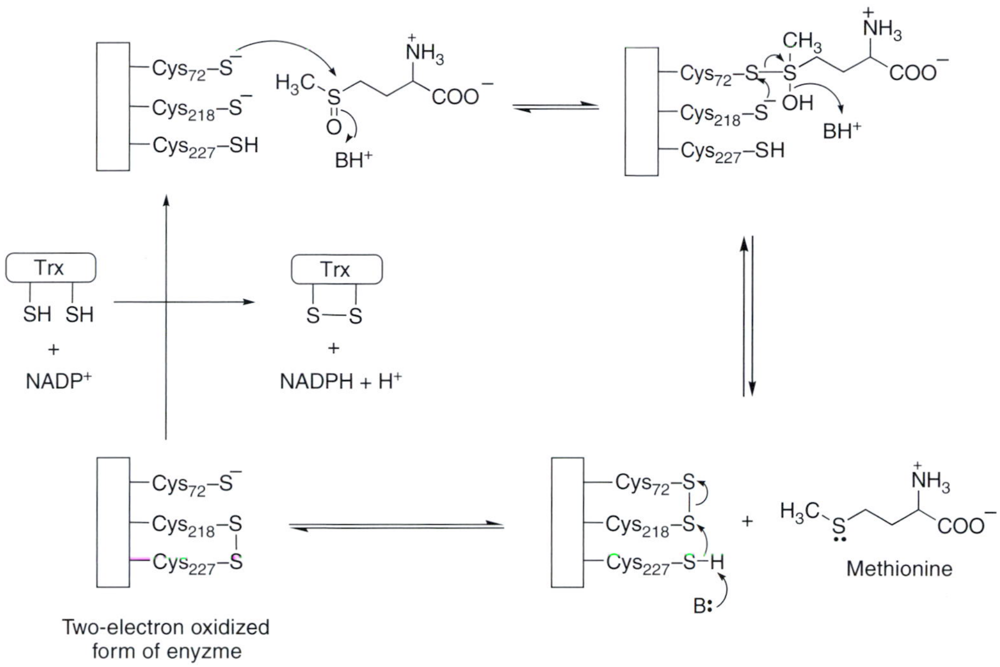

Figure 4.15 Catalytic cycle for human methionine sulfoxide reductase with the dithiol form of thioredoxin as cosubstrate.

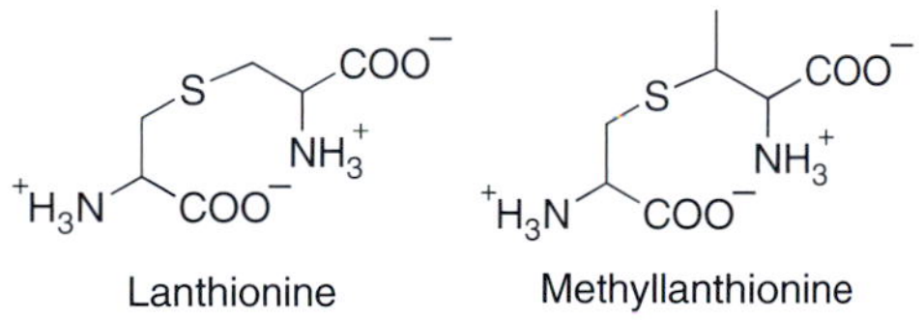

Figure 4.16 Thioether-containing bifunctional amino acids, lanthionine and methyllanthionine.

Cysteine Participation in Protein Thioether Formation

Two unusual sulfur-bridged double-headed amino acids, lanthionine and methyllanthionine (Figure 4.16) have been isolated from proteins of both prokaryotes and eukaryotes. The unusual thioether arises from the addition of a cysteine thiolate onto a dehydroalanine or dehydroaminobutryine residue in enzyme-mediated conjugate addition chemistry. The best studied systems are the bacterial lantibiotic (*lant*hionine-containing ant*ibiotics*) peptides and small proteins, such as the food preservative peptide nisin, where the thioether bridges provide conformational constraints that populate bioactive conformations for the peptides to exert antibacterial activity. They either target bacterial membranes to create transient pores or interact selectively with the peptidolipid inetermediate, lipid II, in bacterial cell wall biosynthesis (Walsh,

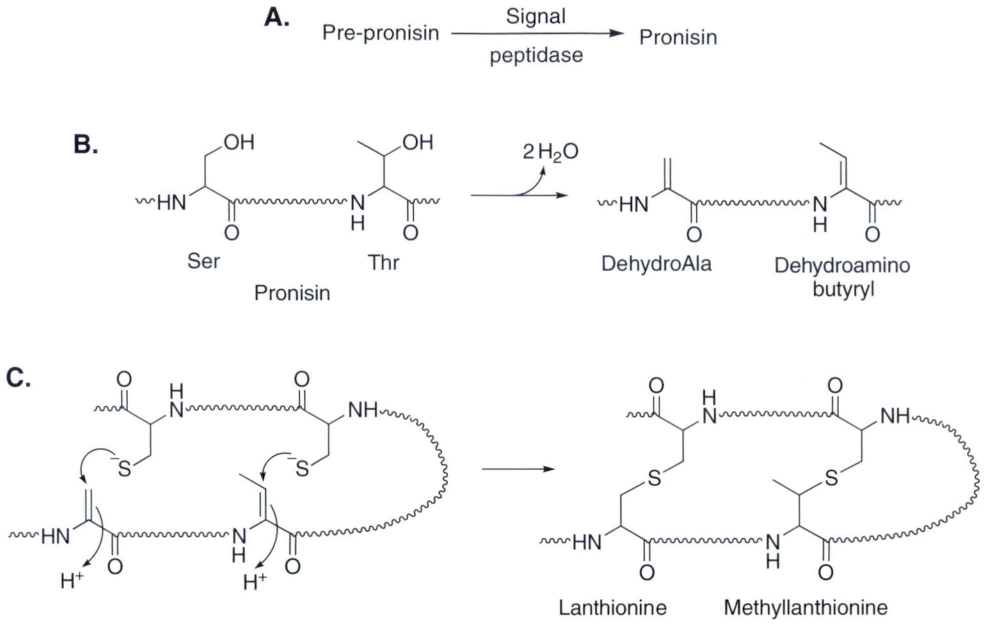

Figure 4.17 Maturation of the lantibiotic peptide nisin. **A.** Proteolytic removal of the signal peptide of pre-pronisin. **B.** Enzymatic dehydration of Ser and Thr residues in pronisin. **C.** The final steps in nisin maturation involve capture of dehydroAla and dehydroaminobutyryl side chains by thiolate side chains of Cys residues to create transannular thioether bridges in the formation of lanthionyl and methyllanthionyl residues.

2003). Nisin is encoded genetically in Gram-positive producer bacteria and the pre-pronisin gets cleaved by signal peptidase (see Chapter 8) to pronisin at the cell membrane (Figure 4.17A), where a series of posttranslational modifications occur. First, one or more serine or threonine residues are dehydrated by enzymes encoded in the lantibiotic biosynthetic cluster to yield the dehydroalanine or dehydroaminobutyrine side chains as acceptors (Figure 4.17B). Cysteine side chains that are at the right distance and orientation can then be activated by a dedicated thioether-forming enzyme to add into the terminus of the double bond of dehydroAla-dehydroaminobutyrine to yield the thio ethers of lanthionine and methyllanthionine residues (Figure 4.17C) (Xie et al., 2004). These are the rigidifying posttranslational steps that create the architecture of the active antibiotics. Subsequent proteolytic cleavage of the propeptide yields the mature lantibiotic that is exported from the producing cell to work on susceptible bacteria in the local environment. The net conversion of cysteine side chains is from thiolate to thioether, set up by the redox change at the β-carbon of the Ser and Thr side chains by the dehydrations. We shall examine analogous dehydrations of serine side chains to N-terminal dehydroalanines, which unravel to pyruvamide groups in the autocatalysis reactions of Chapter 13.

Thiopeptide Linkage in Methyl CoM Reductase

The last example of an unusual linkage of a sulfur atom in a protein does not involve a Cys or Met derivative but rather a divalent sulfur atom in place of an oxygen atom in a peptide bond. Instead of a normal peptide linkage at Gly_{465} in the enzyme methyl coenzyme M reductase (found in a methanogenic bacterium) (Grabarse et al., 2000), the oxygen of that peptide is replaced by sulfur. The purpose or origin of the sulfur atom is not known, although it is within 10 Å of the active site of this methane-generating enzyme. We note in Chapter 5 that this enzyme also contains unusual C-methylated amino acid residues.

References

Biteau, B., J. Labarre, and M. B. Toledano. "ATP-dependent reduction of cysteine-sulphinic acid by *S. cerevisiae sulphiredoxin*," *Nature* **425:**980–984 (2003).

Boschi-Muller, S., S. Azza, S. Sanglier-Cianferani, F. Talfournier, A. Van Dorsselear, and G. Branlant. "A sulfenic acid enzyme intermediate is involved in the catalytic mechanism of peptide methionine sulfoxide reductase from *Escherichia coli*," *J. Biol. Chem.* **275:**35908–35913 (2000).

Chivers, P. T., K. E. Prehoda, and R. T. Raines. "The CXXC motif: A rheostat in the active site," *Biochemistry* **36:**4061–4066 (1997).

Chivers, P. T., and R. Raines. *Protein Disulfide Isomerase: Cellular Enzymology of the CXXC Motif.* Pp. 487–505. Marcel Dekker: New York (1998).

Choi, H., S. Kim, P. Mukhopadhyay, S. Cho, J. Woo, G. Storz, and S. Ryu. "Structural basis of the redox switch in the OxyR transcription factor," *Cell* **105:**103–113 (2001).

Chung, K. K., B. Thomas, X. Li, O. Pletnikova, J. C. Troncoso, L. Marsh, V. L. Dawson, and T. M. Dawson. "S-Nitrosylation of parkin regulates ubiquitination and compromises parkin's protective function," *Science* **304:**1328–1331 (2004).

Claiborne, A., J. I. Yeh, T. C. Mallett, J. Luba, E. J. Crane III, V. Charrier, and D. Parsonage. "Protein-sulfenic acids: Diverse roles for an unlikely player in enzyme catalysis and redox regulation," *Biochemistry* **38:**15407–15416 (1999).

Finkel, T. "Redox-dependent signal transduction," *FEBS Lett.* **476:**52–54 (2000).

Fuangthong, M., and J. D. Helmann. "The OhrR repressor senses organic hydroperoxides by reversible formation of a cysteine-sulfenic acid derivative," *Proc. Natl. Acad. Sci. U.S.A.* **99:**6690–6695 (2002).

Georgiou, G. "How to flip the (redox) switch," *Cell* **111:**607–610 (2002).

Giles, N. M., G. I. Giles, and C. Jacob. "Multiple roles of cysteine in biocatalysis," *Biochem. Biophys. Res. Commun.* **300:**1–4 (2003).

Giles, N. M., A. B. Watts, G. I. Giles, F. H. Fry, J. A. Littlechild, and C. Jacob. "Metal and redox modulation of cysteine protein function," *Chem. Biol.* **10:**677–693 (2003).

Grabarse, W., F. Mahlert, S. Shima, R. K. Thauer, and U. Ermler. "Comparison of three methyl-coenzyme M reductases from phylogenetically distant organisms: Unusual amino acid modification, conservation and adaptation," *J. Mol. Biol.* **303:**329–344 (2000).

Gross, E., D. B. Kastner, C. A. Kaiser, and D. Fass. "Structure of Ero1p, source of disulfide bonds for oxidative protein folding in the cell," *Cell* **117:**601–610 (2004).

Holmgren, A. "Thioredoxin and glutaredoxin systems," *J. Biol. Chem.* **264:**13963–13966 (1989).

Jacob, C., G. I. Giles, N. M. Giles, and H. Sies. "Sulfur and selenium: The role of oxidation state in protein structure and function," *Angew. Chem., Int. Ed. Engl.* **42:**4742–4758 (2003).

Kadokura, H., F. Katzen, and J. Beckwith. "Protein disulfide bond formation in prokaryotes," *Annu. Rev. Biochem.* **72:**111–135 (2003).

Kadokura, H., H. Tian, T. Zander, J. C. Bardwell, and J. Beckwith. "Snapshots of DsbA in action: Detection of proteins in the process of oxidative folding," *Science* **303:**534–537 (2004).

Lee, S. R., K. Kwon, S. Kim, and S. G. Phee. "Reversible inactivation of PTP1B in A431 cells stimulated with epidermal growth factor," *J. Biol. Chem.* **273:**15366–15372 (1998).

Mahadev, K., X. Wu, A. Zilbering, L. Zhu, J. T. Lawrence, and B. J. Goldstein. "Hydrogen peroxide generated during cellular insulin stimulation is integral to activation of the distal insulin signaling cascade in 3T3-L1 adipocytes," *J. Biol. Chem.* **276:**48662–48669 (2001).

Radi, R. "Nitric oxide, oxidants, and protein tyrosine nitration," *Proc. Natl. Acad. Sci. U.S.A.* **101:**4003–4008 (2004).

Reynaert, N. L., K. Ckless, S. H. Korn, N. Vos, A. S. Guala, E. F. Wouters, A. van der Vliet, and Y. M. Janssen-Heininger. "Nitric oxide represses inhibitory kappaB kinase through S-nitrosylation," *Proc. Natl. Acad. Sci. U.S.A.* **101:**8945–8950 (2004).

Salmeen, A., J. N. Andersen, M. P. Myers, T. C. Meng, J. A. Hinks, N. K. Tonks, and D. Barford. "Redox regulation of protein tyrosine phosphatase 1B involves a sulphenyl-amide intermediate," *Nature* **423:**769–773 (2003).

Schopfer, F. J., P. R. Baker, and B. A. Freeman. "NO-dependent protein nitration: A cell signaling event or an oxidative inflammatory response?" *Trends Biochem. Sci.* **28:**646–654 (2003).

Schroder, E., J. A. Littlechild, A. A. Lebedev, N. Errington, A. A. Vagin, and M. N. Isupov. "Crystal structure of decameric 2-Cys peroxiredoxin from human erythrocytes at 1.7 Å resolution," *Struct. Fold. Des.* **8:**605–615 (2000).

Stamler, J. S., S. Lamas, and F. C. Fang. "Nitrosylation. The prototypic redox-based signaling mechanism," *Cell* **106:**675–683 (2001).

Stubbe, J., D. G. Nocera, C. S. Yee, and M. C. Chang. "Radical initiation in the class I ribonucleotide reductase: Long-range proton-coupled electron transfer?" *Chem. Rev.* **103:**2167–2201 (2003).

van Montfort, R. L., M. Congreve, D. Tisi, R. Carr, and H. Jhoti. "Oxidation state of the active-site cysteine in protein tyrosine phosphatase 1B," *Nature* **423:**773–777 (2003).

Veal, E. A., V. J. Findlay, A. M. Day, S. M. Bozonet, J. M. Evans, J. Quinn, and B. A. Morgan. "A 2-Cys peroxiredoxin regulates peroxide-induced oxidation and activation of a stress-activated MAP kinase," *Mol. Cell* **15:**129–139 (2004).

Walsh, C. T. *Enzymatic Reaction Mechanisms*. Freeman: San Francisco (1979).

Walsh, C. T. *Antibiotics: Actions, Origins, Resistance*. ASM Press: Washington, DC (2003).

Wood, Z., L. Poole, and P. A. Karplus. "Peroxiredoxin evolution and the regulation of hydrogen peroxide signalling," *Science* **300:**650–653 (2003).

Xie, L., L. M. Miller, C. Chatterjee, O. Averin, N. L. Kelleher, and W. A. van der Donk. "Lacticin 481: In vitro reconstitution of lantibiotic synthetase activity," *Science* **303:**679–681 (2004).

Zapun, A., C. A. Jakob, D. Y. Thomas, and J. J. Bergeron. "Protein folding in a specialized compartment: The endoplasmic reticulum," *Struct. Fold. Des.* **7:**R173–R182 (1999).

Zheng, M., F. Aslund, and G. Storz. "Activation of the OxyR transcription factor by reversible disulfide bond formation," *Science* **279:**1718–1721 (1998).

CHAPTER 5

Protein Methylation

Methylation of the ε-NH$_2$ of Lys side chains can proceed to trimethylLys.

Proteins can undergo posttranslational methylation (one-carbon transfer) at one or more nucleophilic side chains. Most common are methylations on nitrogen or oxygen atoms. Methylation on carboxylate side chains cover up a negative charge and add hydrophobicity. N-Methylation of lysines does not alter the cationic charge but does increase hydrophobicity. In particular, dimethylation and trimethylation of lysine side chains in proteins increase both hydrophobicity and steric bulk and can affect protein–protein interactions if they are in an interacting surface.

Protein residues methylated on nitrogen include the ε-amine of lysine, the imidazole ring of histidine, the guanidino moiety of arginine, and the side chain amide nitrogens of glutamine and asparagine (Figure 5.1A). These modifications are permanent and not readily reversible under physiological conditions. There are no known enzymes that reverse N-methylation, in contrast to N-acetylation (Chapter 6), where

A.   N-Methylations:

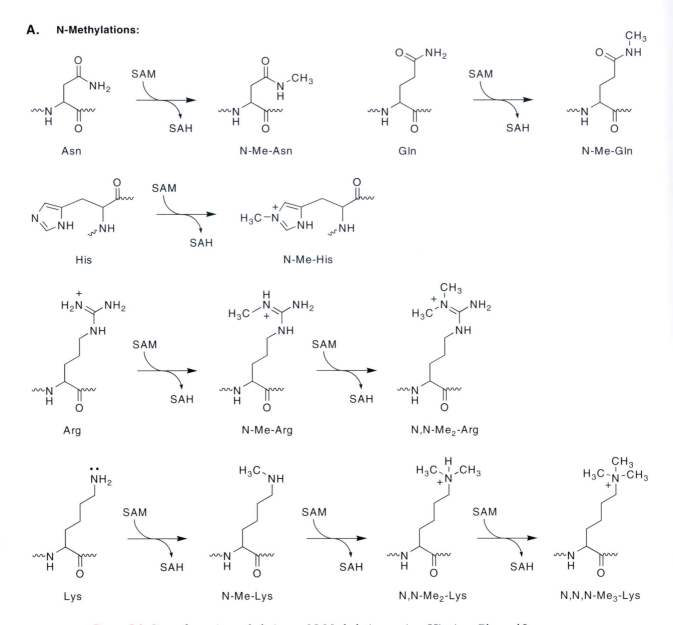

Figure 5.1 Sites of protein methylation. **A.** N-Methylation on Asn, His, Arg, Gln, and Lys.

enzyme-mediated hydrolysis or transfer to the ADP-ribose moiety of NAD occurs. N-Methyl groups on DNA bases can be oxygenated enzymatically, and deconvoluted to formaldehyde and the starting base (Sedgwick, 2004), and enzymes committed to oxidative demethylation of proteins have just recently been detected (Shi et al., 2005).

Proteins can also be methylated on oxygen atoms, such as the side chain carboxylates of glutamates and aspartates, to create methyl esters (Figure 5.1B). The lifetimes

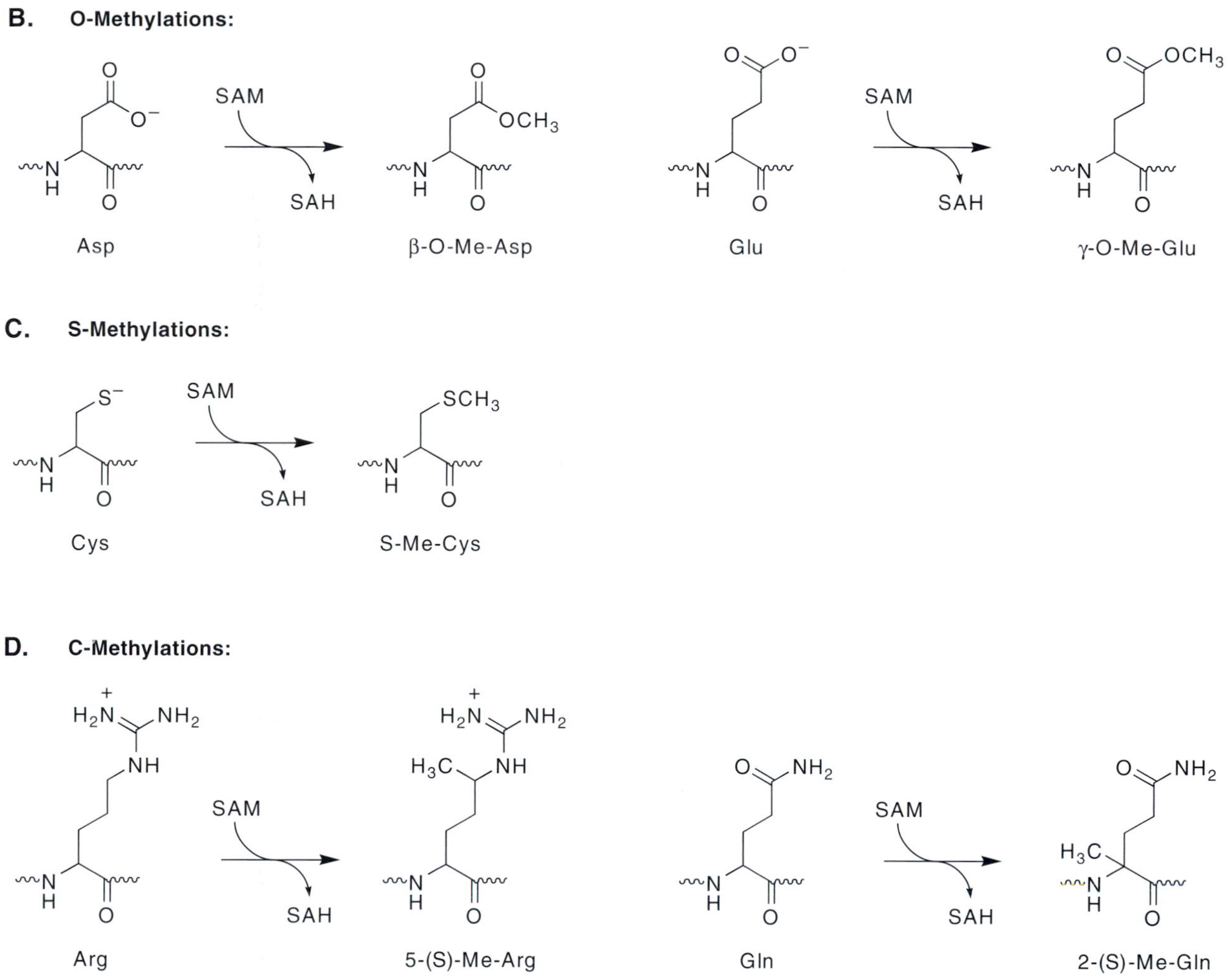

B.　O-Methylations:

Asp　　　　　β-O-Me-Asp　　　　　Glu　　　　　γ-O-Me-Glu

C.　S-Methylations:

Cys　　　　　S-Me-Cys

D.　C-Methylations:

Arg　　　　　5-(S)-Me-Arg　　　　　Gln　　　　　2-(S)-Me-Gln

Figure 5.1　**B.** O-Methylation on Glu and Asp side chain carboxylates. **C.** S-Methylation of cysteine thiolate. **D.** C-Methylation at methylene carbons of Arg and Gln residues.

of the protein methyl esters vary, with hydrolysis back to Glu and Asp residues as the outcome, and these lifetimes relate to function as we shall note.

Proteins can also be methylated on the thiolate side chain of cysteine (Figure 5.1C) and, most surprisingly, C-methylation of arginine and glutamine side chains has recently been detected by x-ray analysis of the structure of methyl coenzyme M reductase from methanogenic bacteria (Ermler et al., 1997) (Figure 5.1D).

The donor substrate for methyl group transfer is S-adenosylmethionine (Figure 5.2), abbreviated both as SAM and AdoMet in the biochemical literature. Transfer of the reduced one-carbon CH_3 fragment from SAM to both small molecules and

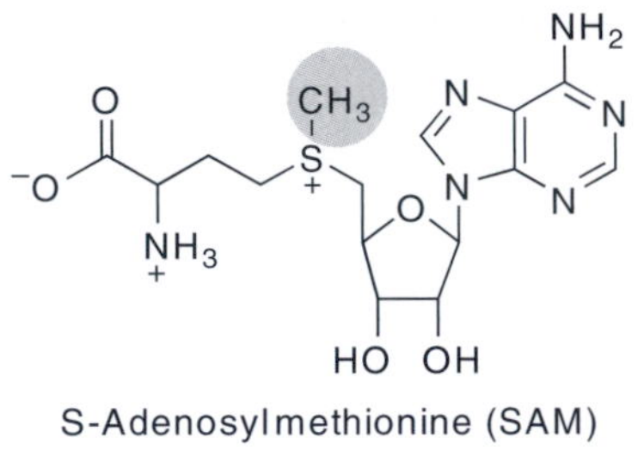

Figure 5.2 S-Adenosylmethionine (SAM), the one-carbon donor cosubstrate in protein methylations.

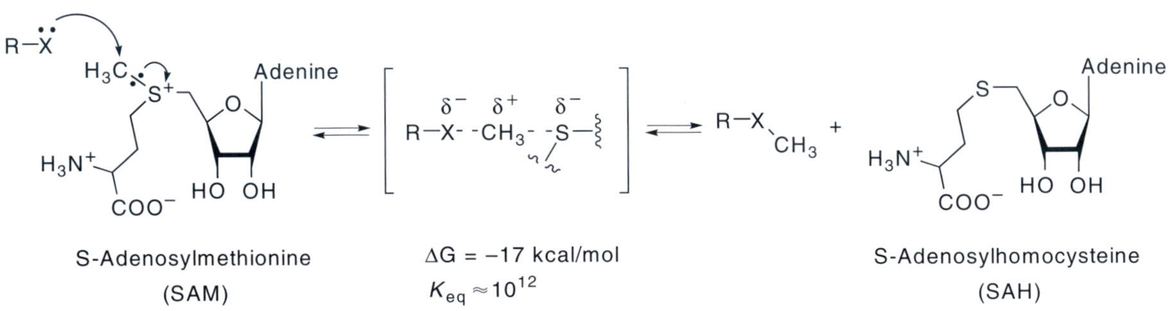

Figure 5.3 Polarity of methyl transfer from SAM: transfer of a CH_3^+ equivalent.

macromolecules (nucleic acids, proteins, and oligosaccharides) is a common chemical transformation in biology and SAM is claimed to be the second most frequently used substrate in enzymatic reactions after ATP. Cosubstrates get methylated on nitrogen, oxygen, and sulfur nucleophilic centers and also on some electron-rich carbon centers in small molecules.

The CH_3 fragment in SAM that undergoes transfer to nucleophiles is attached to a trivalent sulfur atom, in a sulfonium linkage with a formal positive charge on sulfur. This polarizes the CH_3 group for low-energy transfer as a "CH_3^+" equivalent that then undergoes transfer to an electron-rich attacking nucleophile (Figure 5.3). The electrons in the S^+–CH_3 bond remain on sulfur in the S-adenosylhomocysteine (SAH) product. This has the mechanistic and stereochemical characteristics of an S_N2 reaction. The thermodynamic activation of SAM as a methylating agent has been measured at –17 kcal/mol (Walsh, 1979). Compared to the –7 kcal/mol for phosphoryl transfer from ATP, there is enormous thermodynamic driving force for methyl transfer to the nucleophilic cosubstrate.

The catalysts for these posttranslational thermodynamically favored protein methylations, the protein methyltransferases, all have a related binding site for S-adenosylmethionine and clefts for binding the protein cosubstrate with the side chain to be methylated. From informatic and structural analyses, five families of protein methyltransferases have been proposed (Schubert et al., 2003a), suggesting five independent evolutionary paths to methyltransferases. Class I and class V appear to contain most of the *protein* methyltransferases, with the other classes dedicated to methylations of DNA, RNA, and small molecules.

Historically, protein methylation was detected by protein hydrolysis and amino acid analysis of liberated N-methyl amino acid. The N-methyl amino acids survive acid hydrolysis, but the O-carboxymethyl amino acids are labile to hydrolysis, and so are not detected. With the advent of radiolabeled $[^{14}C]CH_3$–SAM, it became possible to detect methylation stoichiometry and regiochemistry in peptide fragments. Most useful in contemporary contexts is mass spectrometry where introduction of each CH_3– moiety results in an increase of 14 mass units (+CH_3, –H) at the side chain undergoing methylation. Some arginine residues in histones can be dimethylated and some lysine residues trimethylated to cause diagnostic increases of 28 and 42 Da, respectively. Using very high resolution mass spectrometers, the 0.035 Da mass difference between trimethylation and acetylation can be distinguished (Pesavento, 2004).

Protein O-Methylation

Chemotaxis Receptor Protein Glutamate Carboxy Methyltransferase

Among the best studied protein methylation reactions are those that occur in bacterial chemotaxis on the cytoplasmic domains of the transmembrane receptors for aspartate (Tar), serine (Tsr), ribose and galactose (Tgr), and peptides (Tap). The Tar receptor is methylated on the γ-carboxylates of four conserved glutamate residues on two helices, α_6 and α_9, to yield the methyl esters (Figure 5.4A) (Bass and Falke, 1999). Methylation is catalyzed by the CheR enzyme and reversed by the CheB enzyme. The Tsr, Tgr, and Tap receptors are analogously methylated on glutamates and these four receptors have been known collectively as methyl-accepting chemotaxis proteins (MCPs) to reflect these modifications. The Tar receptor functions as a dimer, with two transmembrane helices per subunit making a four-helix bundle in the dimer. On binding of the aspartate ligand to the periplasmic domain, occupancy is transmitted through the membrane by a proposed swinging piston rotation of the transmembrane helices (Falke et al., 1997; Grebe and Stock, 1998), communicating ligand occupancy on the outside to the cytoplasmic domain (Figure 5.4B). This signal interfaces with

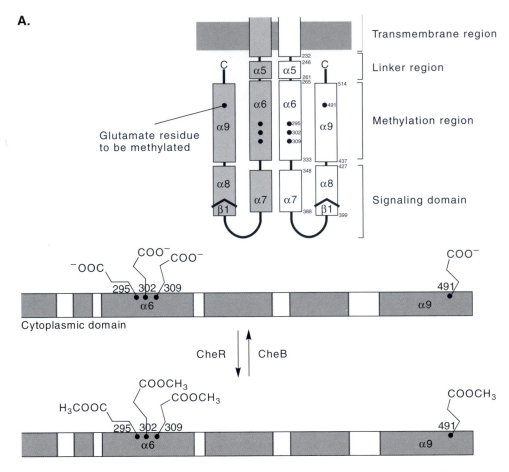

Figure 5.4 **A.** Methylation on the side chain glutamates of bacterial transmembrane methyl-accepting chemotaxis proteins.

the flagellar motors to control smooth swimming versus tumbling in the liganded versus unliganded states of the receptor (Bass and Falke, 1999).

The methylation state of the MCP glutamate side chains are controlled by a balance of the SAM-dependent methylase CheR (schematized in Figure 5.4B) and the opposing glutamate ester demethylase CheB (Figure 5.5). CheB has a catalytic domain and a regulatory domain. The regulatory domain can be phosphorylated by the histidine kinase CheA (Chapter 2), which activates the catalytic domain and leads to a lower steady state of MCP methylation. CheA activation in turn is influenced by the periplasmic ligand occupancy state of the MCPs. At high levels of aspartate and other MCP ligands, CheA is less active, CheB becomes dephosphorylated and less active, and the methylation state of the glutamate side chains in the MCPs rise, leading to increased frequency of tumbling (Figure 5.6).

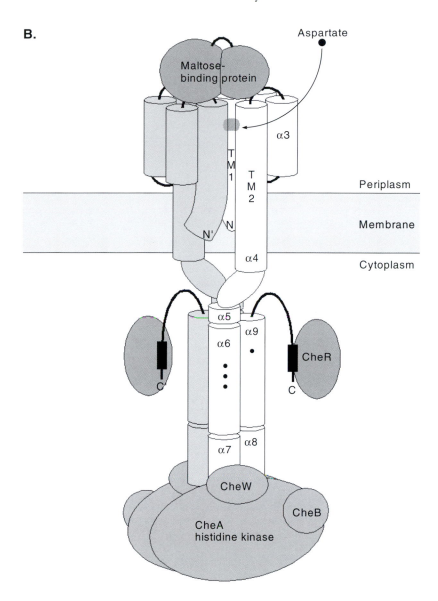

Figure 5.4 **B.** Effect of binding of aspartate ligand to the extracellular domain of the Tar protein: swinging piston rotation of the transmembrane helices.

Thus MCP glutamate side chain methylation is a central part of the signaling system in bacterial chemotaxis. The ability to reverse the signal by demethylation, in particular hydrolysis by CheB, requires that the MCP methylations be *reversible*. Only glutamate and aspartate side chain methylations have this chemical property. N- and S-methylations are not hydrolytically labile, providing a chemical rationale for why glutamate methylation has been selected in these signaling cascades.

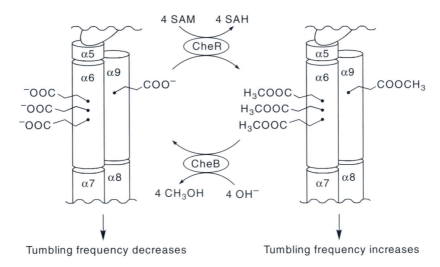

Figure 5.5 Control of methylation state of MCPs by opposing methyltransferase (CheR) and methylesterase activity (CheB).

Ras Carboxy Terminal Methyltransferase

As noted in Chapter 7, small GTPase (G) proteins are directed to associate with membrane microenvironments in eukaryotic cells by posttranslational modifications that add lipid substituents to the N-terminal and C-terminal regions of the G proteins. The lipid substituents can be either fatty acyl chains (e.g., myristoyl and palmitoyl) or isoprenoids of C_{15} (farnesyl) or C_{20} (geranylgeranyl) length. The C-terminal processing of the Ras, Rap, and Rab subfamilies of these GTPases is quite complex, as you will see in detail in Chapter 7. Typically, prenylation (C_{15} or C_{20}) occurs at one or two of the most downstream cysteine thiols, followed by proteolysis at the modified cysteine. The newly liberated carboxylate group is then methylated in the last step from SAM via a prenylated protein methyltransferase (Figure 5.7) (Clarke, 1992). The negative charge of the C-terminal COO^- is removed by methylester formation, presumably lowering the energy for insertion/association of the prenylated C-terminus into the membrane compartments of the cell. The prenylated protein methyltransferase of yeast is an integral membrane enzyme (Sapperstein et al., 1994).

Aspartate Carboxy Methyltransferase in Aging Protein Repair: Protein Isoaspartyl Methyltransferase (PIMT)

A third example of protein carboxylate O-methylation occurs when a protein repair methyltransferase acts to repair isoaspartate linkages that arise during the aging

Activation of CheB:

Aspartate binding

Autophosphorylated CheA:

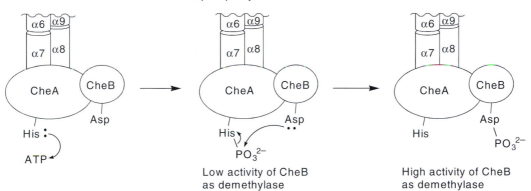

Figure 5.6 Integrated response to aspartate as chemo-attractant by Tar, CheA, CheR, and CheB by autophosphorylation and subsequent demethylation.

process (Clarke, 1985). These methyltransferases are found in both prokaryotes and eukaryotes and act to combat the spontaneous damage that can arise at asparagines and aspartate residues in proteins. In contrast to the chemotaxis proteins that get methylated to a stoichiometry of four residues (glutamates) per protein, the aspartate methylations involved in configurational reversal occur at very low fractional stoichiometries. Asparaginyl side chains are prone to slow hydrolytic nonenzymatic deamidation to yield aspartates and that process is irreversible. Asp side chains and Asn side chains can undergo intramolecular attack by the nitrogen in the amide of the downstream peptide bond in a strain-free five-member transition state to eliminate –OH (Asp) or –NH$_2$ (Asn) to create a disubstituted aminosuccinimide cyclic adduct.

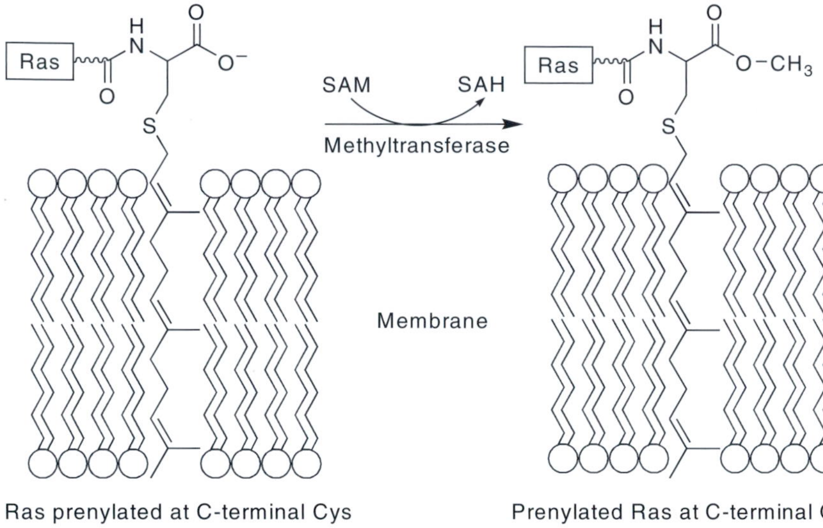

Figure 5.7 Action of Ras O-carboxy methyltransferase in a three-step pathway of Ras modification leading to localization at plasma membranes.

This is analogous to a late step in protein splicing (see Chapter 13). The aminosuccinimide adduct can reopen by attack of water at either carbonyl, as shown in Figure 5.8. One hydrolytic opening goes back to an L-Asp residue. The other route yields a β-Asp residue, also known as an L-isoaspartyl residue. This route leaves a β-amino acid embedded in the backbone of the protein and is a net insertion of a CH_2 group into the backbone. This can distort the protein conformation and affect function. Analysis of Asn_{67} in a loop of RNAse A indicated a half-life of 7.4 days at 37 °C and pH 8.2 for isoAsp formation (Skinner et al., 2000). Also, knockout of the bacterial or eukaryotic repair methyltransferase gene shortens life spans significantly, attesting to the importance of this repair process.

The question arises why and how a methyltransferase recognizing protein isoaspartyl residues (PIMT) could have a repair outcome. The strategy is that carboxy methylation, specifically of the isoaspartyl side chain carboxylate to create the methylester, provides enough thermodynamic activation that attack by the amido nitrogen is accelerated to regenerate the common succinimide intermediate (Figure 5.9), noted above as the adduct that partitions between hydrolysis to L-Asp and L-isoAsp. The PIMT thus drives repeated cycles of conversion of isoaspartyl residues to succinimide to Asp/isoAsp. In some peptide systems, the partition is about 1/3 for hydrolysis of succinimide to Asp/isoAsp, so 25% of the flux would go back to repaired protein backbone in each cycle (Griffith et al., 2001). This is an expensive backbone

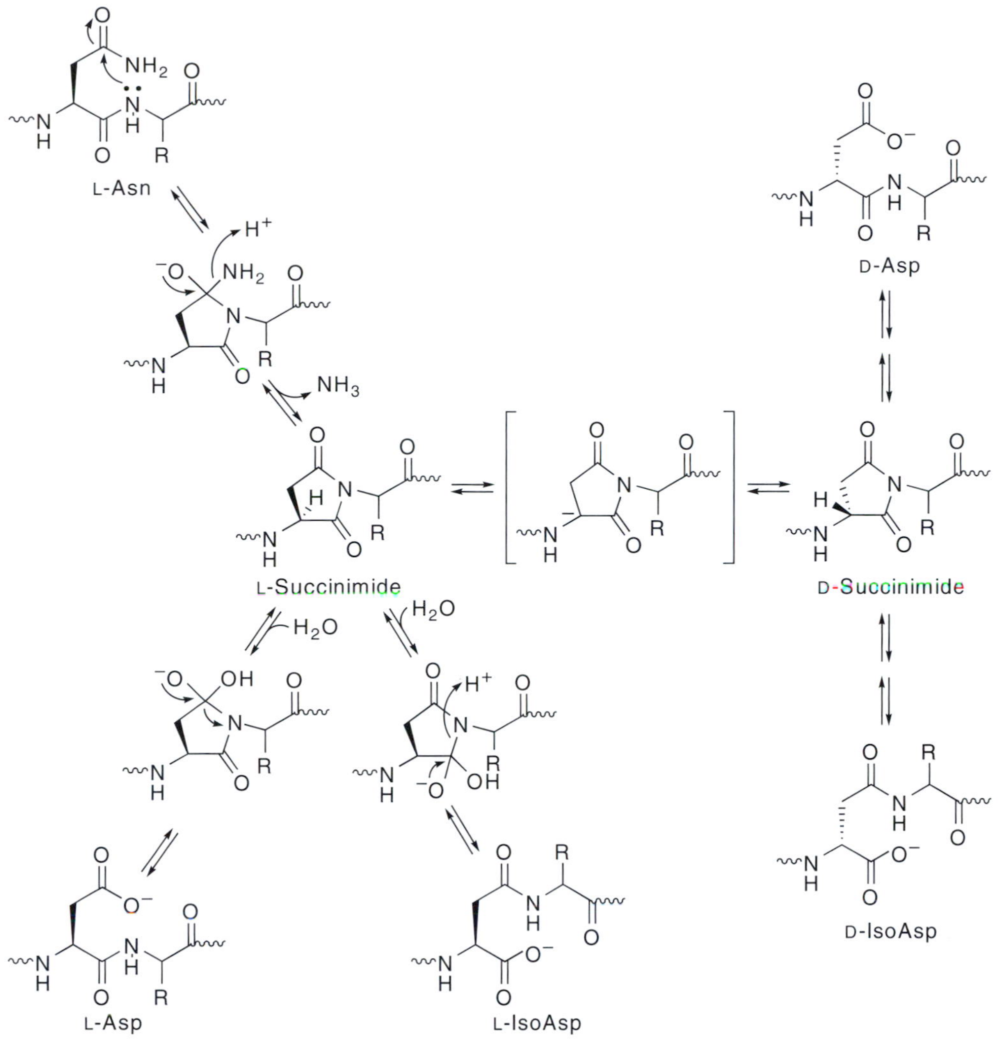

Figure 5.8 Nonenzymatic reactions of Asn or Asp via aminosuccinimide cyclic intermediates to yield Asp and IsoAsp residues.

repair in two senses. First, multiple cycles are needed to drive the β-aspartyl residue back to the α-aspartyl-containing protein backbone. Second, in each cycle a molecule of the highly activated SAM (ΔG = –17 kcal/mol) is consumed. Looked at another way, the favorable thermodynamics enable a remarkable rearrangement of the peptide backbone (β-aspartyl back to α-aspartyl) without breaking the peptide chain backbone connectivity. X-ray structural analysis of both the *Thermotoga maritima* and the *Pyrococcus furiosus* PIMTs reveal the orientation of SAM in the active site and an isoaspartyl tripeptide (Griffith et al., 2001; Skinner et al., 2000). The orientation of the

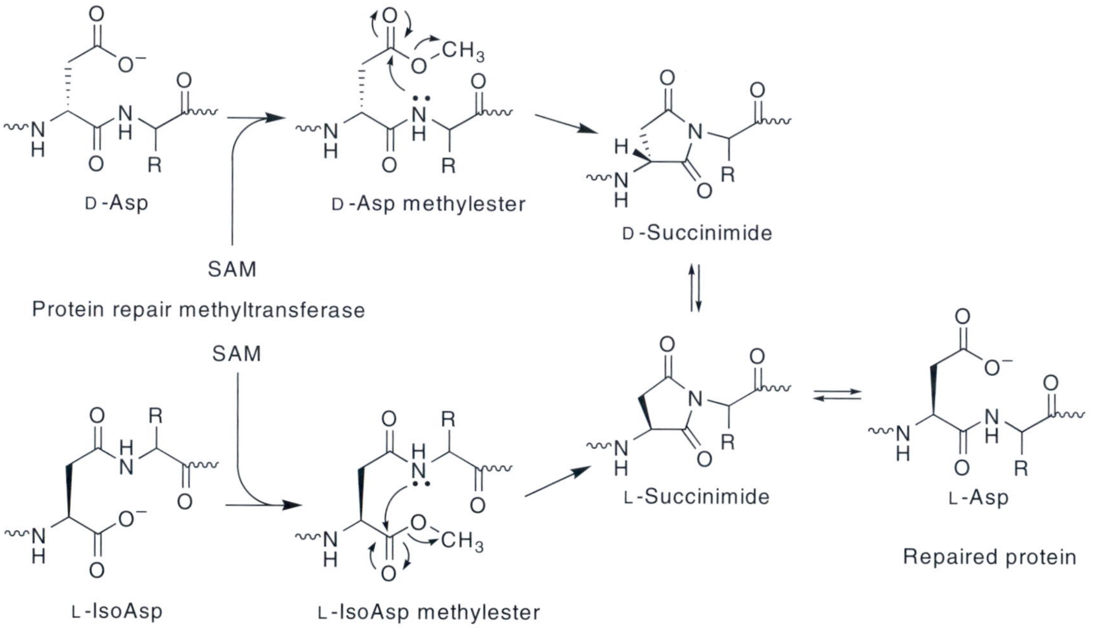

Figure 5.9 Role of protein isoaspartyl methyltransferase (PIMT) in repair of IsoAsp–β-peptide linkages in proteins. Cycling of the IsoAsp methylester through the common succinimide intermediate to drive β-aspartyl linkages back to α-aspartyl peptide bonds.

isoaspartyl peptide suggests that an L-aspartyl chain cannot be recognized, explaining methylase specificity for only the modified side chain.

In some cases the aminosuccinimide intermediate can epimerize by H_α exchange and then open to D-Asp side chains. The PIMT can also methylate D-Asp carboxylate side chains, at lower catalytic efficiency than L-isoAsp, and drive them back through the succinimide intermediate by multiple cycling to a fraction of L-Asp side chains.

Protein N-Methylation

Protein N-methylation, in contrast to O-methylation, is largely unidirectional and irreversible with no known enzymes that release the intact CH_3 group. Oxidative routes for CH_3–N bond cleavages in N-methylated proteins have received speculation, in analogy to oxidative N-demethylation of N-methylglycine. Recent studies, noted in the last section of the chapter, have now identified a regiospecific histone H3 methyllysine deaminase to counteract the action of histone methyltransferases at these Lys residues. A variety of nitrogen side chains are observed to function as competent nucleophilic cosubstrates for protein N-methyltransferases, including amino groups of lysines, the imidazole nitrogens of histidine, the guanidino groups of arginine, and even the carboxamido groups of glutamine.

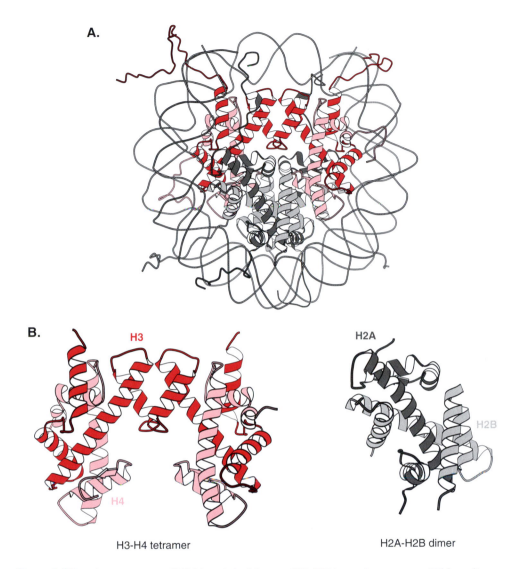

Figure 5.10 **A.** Arrangement of DNA and the histones H1–H4 in nucleosomes. **B.** Ribbon diagram of H3–H4 tetramers and H2A–H2B dimers (figure made using PDB 1EQZ).

Lysine N-Methylation

While a variety of proteins undergo posttranslational N-methylation of the ϵ-NH_2 of lysine residues, most of the recent interest has been on the multiple lysine and arginine methylations observed in the H3 and H4 subunits of histones. The placement of histones and DNA have been established within the core of nucleosome particles (Figure 5.10A) (Khorasanizadeh, 2004) and the structures of the H3–H4 tetramers and H2A–H2B dimers that make up the histone cores have also been established (Figure 5.10B). The tails of the H2A–H2B dimer and the H3–H4 tetramer are

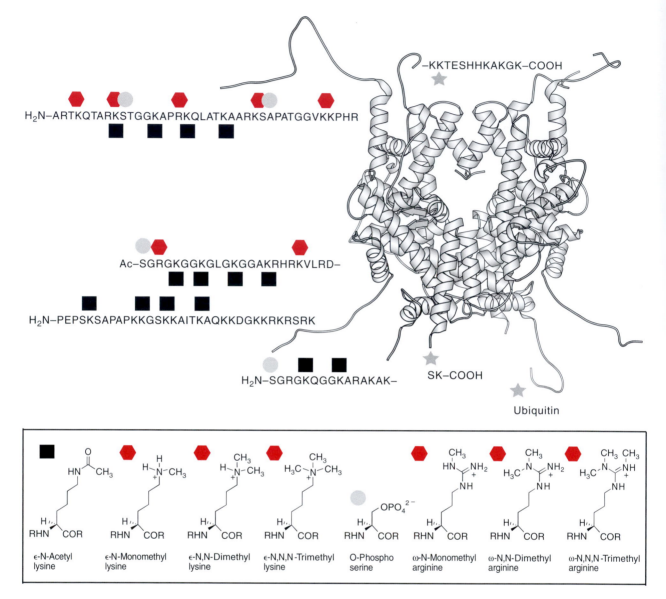

Figure 5.11 Schematic representation of the N-terminal tails of histones in a nucleosome core. The eight types of modifications on lysines, arginines, and serines are schematized. The N-methylations that are the subject of this chapter are shown in red (figure made using PDB 1EQZ).

schematized in Figure 5.11. This drawing demonstrates the accessibility of the histone N-terminal tails to proteins that will dock on the nucleosomes and either carry out posttranslational modifications (methylations, acetylations, phosphorylations, and ubiquitylation) or read the modifications that have been covalently placed there. The 28-residue N-terminal tail of histone H3 is the most highly modified

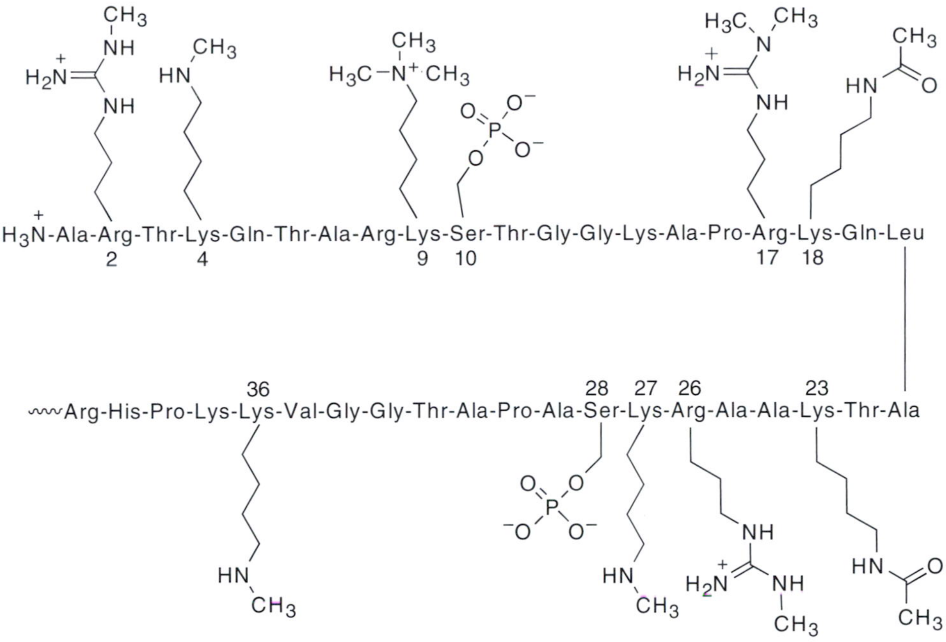

Figure 5.12 Eleven sites of posttranslational modifications in the first 36 residues of the N-terminal tail of histone H3 in nucleosomes. Nine of these are N-methylations of Lys or Arg.

posttranslationally. As noted in the next section, three arginines, at positions 2, 17, and 26, can be methylated. Four lysines, at positions 4, 9, 27, and 36, can also be methylated for a total of seven sites of methylation in a 36-residue stretch (Figure 5.12). Two other lysines, at positions 18 and 23, undergo acetylation (see Chapter 6) but not methylation. Lys_{79}, outside the N-terminal tail of H3, is also methylated (Ezhkova and Tansey, 2004). In the N-terminal tail of H4, Arg_3 can be methylated as can Lys_{20}. There are no lysine methylations seen in H2A or H2B, although there are lysine residues in the N-terminal tails and they do get acetylated (Marmorstein, 2001). Ser_{10} and Ser_{27} of H3 can be phosphorylated, for a total of 11 possible modifications in the first 36 residues of each of the two H3 chains in a nucleosome. The combinatorial possibilities for specific isoforms of histone H3 are astronomical ($2^4 \times 4^4 \times 3^3 = 110,592$ combinations possible in principle), even with the strong likelihood that only some of the sites are used in any one of the $\sim 10^7$ nucleosomes that package the human genome, with only a few patterns likely to predominate.

In addition to the multiplicity of sites for regiochemical lysine methylation, as in H3, there is also the possibility of mono-, di-, or tri-N-methylation at each lysine residue, increasing the combinatorial possibilities and forming one of the pillars of the "histone code" hypothesis, elaborated more fully in Chapter 6 (Turner, 2002). As

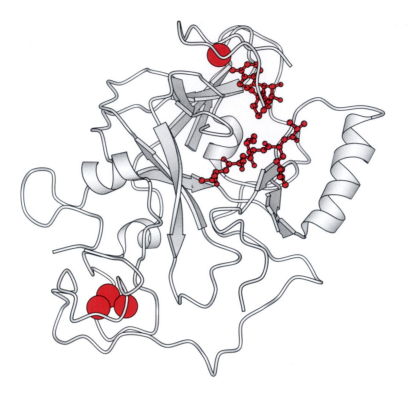

Figure 5.13 The SET domain in the trimethyllysine-forming DIM5 methyltransferase with a bound trimethylLys peptide and SAH cofactor in red. The red spheres are zinc atoms (figure made using PDB 1PEG).

noted in Chapter 6, histone acetylation is readily reversed by hydrolytic histone deacetylases, but there is no comparable reversing hydrolase mechanism for Lys–N–CH$_3$ linkages so N-methylation is presumably an irreversible component to the writing and reading of the histone code.

Protein methyltransferases selective for each of the four lysines (K$_4$, K$_9$, K$_{27}$, and K$_{36}$) have been detected by a combination of genetic and biochemical assays. All but one of these methyltransferases [an H3-K$_{79}$ MTase (Min et al., 2002)] has a conserved 130-amino-acid domain termed a SET domain (from supressor-enhancer of zeste, trithorax) that forms β sheet catalytic cores (Lachner et al., 2003). The SET domain and surrounding sequence constitute the SAM-binding site and a narrow buried pocket for binding and presentation of the histone lysine side chains to be methylated (Figure 5.13) in a substrate β strand that completes an enzyme-substrate β sheet. Mammals have an estimated 70 SET genes in 7–10 gene families (Lachner and Jenuwein, 2002), while *S. cerevisiae* has seven SET genes, allowing genetic evaluation of each of the seven encoded lysine methyltransferases. If all the SET-domain-containing proteins do indeed have protein N-methyltransferase activity, there will be many routes to specific combinations of regioselectively methylated Lys residues.

Structure–function studies on the DIM5 (decrease in DNA methylation), an H3-K_9 methyltransferase from *N. crassa* (Figure 5.13), have shown that it makes predominantly the trimethylLys$_9$ product with only a small amount of Me$_1$ and Me$_2$ intermediates (Zhang et al., 2003), consistent with *in vivo* patterns of H3 methylation at this lysine residue. By contrast an H3-K_4-specific methyltransferase known as SET7/9 makes the monomethyl-ϵ-Lys$_4$ product, again consistent with *in vivo* patterns. Zhang et al. (2003) posited from the x-ray structure that volume control in the lysine binding channels was a prime determinant of the monomethylation versus trimethylation state of lysine side chain products, with the SET7/9 channel unable to accommodate a dimethylated or trimethylated Lys side chain. This was validated by mutations (F281Y in DIM5, and Y305F in SET7/9) that contracted volume in DIM5 and turned it into a monomethylating enzyme and expanded active site volume in SET7/9 and increased flux to a dimethylLys peptide product. These results raise the prospect of re-engineering SET domain lysine methyltransferases to alter mono-, di-, and trimethyllysine product distributions and evaluate the effects on the histone code.

The difference in size and hydrophobicity of momomethyl- versus trimethylLys side chains is utilized for specific recruitment of partner proteins via their chromodomains. This is emphasized in the structure of the chromodomain of the protein HP1 in complex with an H3-derived peptide bearing trimethylLys$_9$ (Figure 5.14A) (Khorasanizadeh, 2004). Six residues of the H3 peptide form a β strand that completes a β sandwich in the chromodomain and the hydrophobic trimethylLys$_9$ side chain packs into an aromatic cage comprised of three aromatic side chains in the chromodomain (Figure 5.14B). Since the N-methylations alter hydrophobicity and enable recruitment of coactivators via methyllysine recognition via chromodomains one might expect large effects on gene activation profiles from changes in mono-, di-, and trimethyllysine fractional distributions at the five lysines in H3 and H4.

The conversion of H3-K_4 to a trimethylated product is very likely due to kinetic processivity, with the second and third methylations occurring before release of the product from the previous catalytic cycle. In this connection various lysine methyltransferases have low turnover numbers and high pH optima, as though the histone substrate forms that bind are those with free base ($-NH_2$) forms of lysine side chains as the competent species for binding. Structural analyses confirm the lack of an obvious active site base and suggest enforced propinquity of bound SAM and Lys–NH_2 side chains may be the major step for catalysis.

The role of specific SET-domain-containing histone methyltransferases in genomic silencing is under intensive investigation (Lachner and Jenuwein, 2002).

A.   **B.**

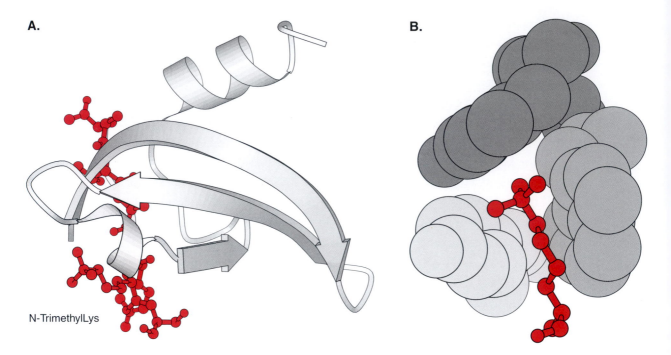

N-TrimethylLys

Figure 5.14 **A.** Structure of the chromodomain of HP1 complexed with a trimethylLys peptide.
B. Hydrophobic cavity of the chromodomain for recognition of the hydrophobic Me_3NH moiety of
the trimethylLys residue (figure made using PDB 1KNE).

The balance of methylation at $H3-K_4$ versus $H3-K_9$, for example, seems to correlate
with gene silencing, genomic instability, and active versus inactive chromatin. Methylation status at these lysines can differentially enhance or repress binding of transcription factor complexes and coregulators, and can interact with other posttranslational
modifications, such as acetylation, phosphorylation, and ubiquitylation, to enable
combinatorial titrations of gene activity (Turner, 2002). It is known that ubiquitylation of the K_{123} side chain of H2B by the ubiquitylating E3 ligase RAD-6 (see Chapter 9) is a prerequisite for the subsequent trimethylation of K_9 on H3 (Sun, 2002). It is
likely that the prior ubiquitylation of the H2B tail recruits two of the ATPase subunits
of the proteasome cap, Rpt4 and Rpt6 (see Chapter 9) (Ezhkova and Tansey, 2004),
perhaps by ubiquitin-recognizing motifs in the ATPases. These ATPases may then
travel with RNA polymerase II, unwinding the chromatin and enabling access of the
Set1 and Dot1 methylases, respectively, to get to K_4 and K_9 of H3 (Ezhkova and
Tansey, 2004).

The SET9 methyltransferase works poorly on histones in nucleosomes but has
recently been shown to N-methylate a specific lysine side chain in a nonhistone protein, TAF10, one of the TBP-associated factors, enhancing its interaction with RNA
polymerase (Kouskouti et al., 2004).

This observation indicates expanded roles for posttranslational methylation of distinct protein targets in control of gene transcription, including the transcription factor p53. As is discussed in Chapters 6 and 9, the C-terminal region of p53 is a regulatory region, analogous to the N-terminal tails of histones that can be multiply modified by acetylation and ubiquitylation at a series of Lys residues. The SET9 methyltransferase will N-methylate Lys_{372} in the p53 C-terminal domain, increasing its lifetime, perhaps by blocking adjacent ubiquitylation (Chapter 9) or by recruiting a partner protein via the N-Me-Lys residue that prevents further modification in the C-terminal tail (Chulkov, 2004).

Arginine N-Methylation

Arginine residues are known to be N-methylated in several proteins, including fibroblast growth factor and nucleolin. Specific Arg N-methyltransferases are known, some of which make the monomethyl and some the dimethylarginine products (schematized in Figure 5.12), as one or two SAM donor substrates are consumed. Two such arginyl N-methyltransferases that have effects on gene transcription have been studied recently. One is the PRMT1 (Mowen et al., 2001) that methylates the conserved Arg_{31} side chain in the N-terminal domain of the dimeric transcription factor STAT1. STAT1 and the six related mammalian STATs (signal transducers and activators of transcription) induce transcription of genes in response to interferons α and β and other cytokines. Most of the focus on posttranslational modifications of SH2 domain-containing STATs has been on the phosphorylation of Tyr_{701} by Janus kinases (JAKs), which is required to induce STAT dimerization via SH2 domains across the dimers, and subsequent translocation to the nucleus. Arg_{31} methylation adds a second type of modification to STAT control.

PRMT1 had been observed to associate with the cytoplasmic region of one of the subunits of interferon receptors, perhaps allowing presentation to STATs when they bind to the IFN receptors. The N-methylation of Arg_{31} in the N-terminus of STAT1 does not affect tyrosyl (or $seryl_{727}$ by MAP kinases) phosphorylation or STAT translocation, but does block the binding of a partner protein to nuclear STATs. This protein, protein inhibitor of activated STATs (PIAS1), when complexed with STATs prevents binding to the target gene enhancer regions and so abrogates transcriptional activation. Arg_{31} N-methylation introduces a hydrophobic tail on the arginyl side chain and may thereby disrupt the STAT–PIAS1 recognition and allow STAT to dock on target DNA. PRMT1 also methylates Arg residues in histone H4 and thereby is classifiable as a histone N-methyltransferase (Khorasanizadeh, 2004).

The arginyl N-methyltransferase CARM1 (cofactor associated R methylase) is observed to have multiple effects from methylation of arginine residues in distinct protein substrates. It had been known that CARM1 is an enzyme that methylates multiple Arg residues in the N-terminal tail of histone H3 (Arg_2, Arg_{17}, and Arg_{26}). CARM1 is recruited by protein–protein interactions with coreceptors for transcription factors such as the p160 class (Xu, 2001), which have two recruitment domains at the C-terminus, AD1 and AD2, which recruit p300 and CARM1, respectively. The p300 protein has histone acetyltransferase activity (see Chapter 6) and is thought to act first on H3, followed by CARM1, to create a particular pattern of lysine acetylation and arginine methylation (Figure 5.15). In addition to the histone

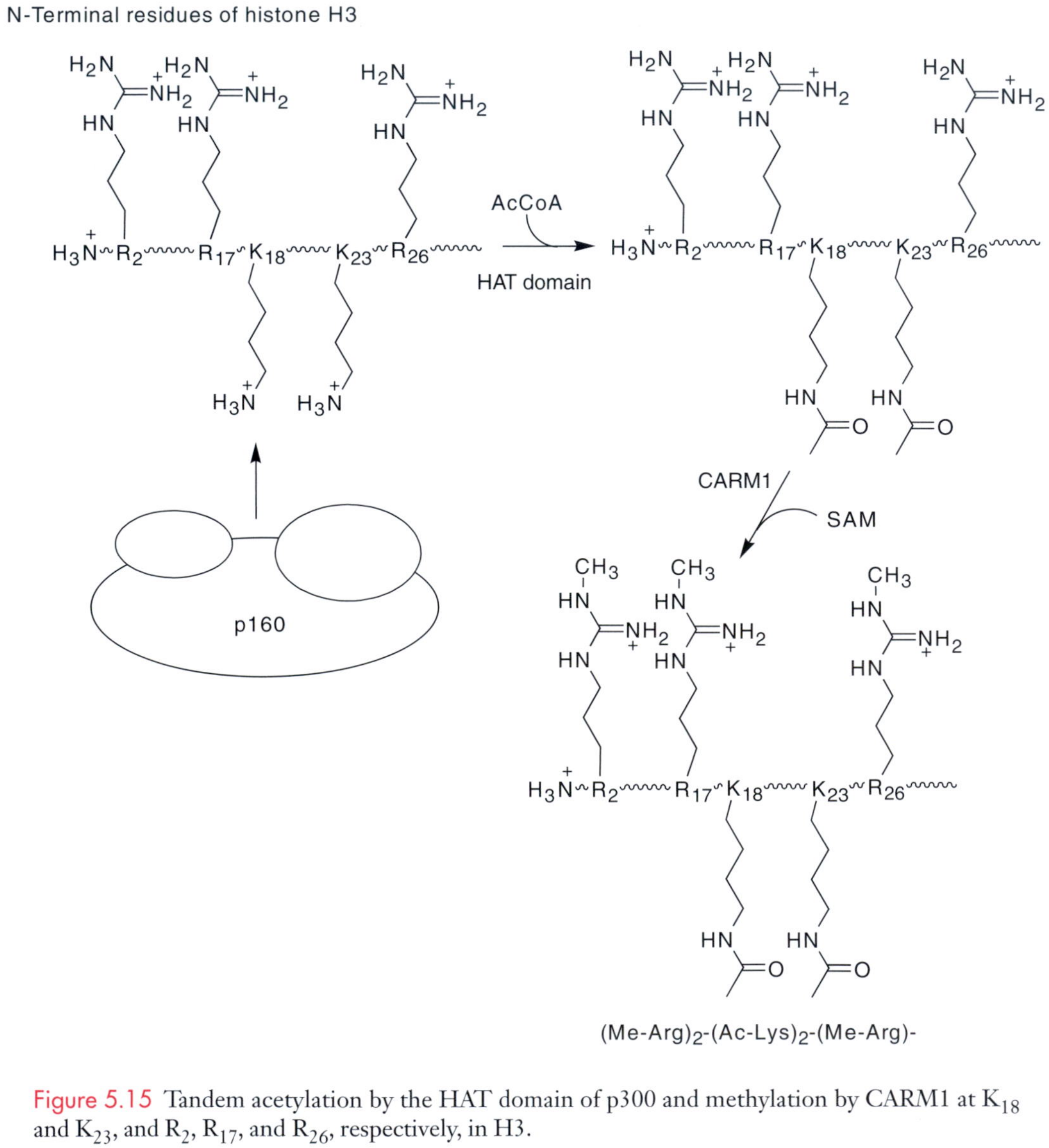

N-Terminal residues of histone H3

Figure 5.15 Tandem acetylation by the HAT domain of p300 and methylation by CARM1 at K_{18} and K_{23}, and R_2, R_{17}, and R_{26}, respectively, in H3.

H3 methylation activity, CARM1 has been observed to methylate its partner p300 on Arg_{580}. R_{580} is in the KIX domain of p300 and methylation disrupts a hydrogen bond to Tyr_{640}, destabilizing a helix in KIX. In turn, this disrupts the interaction of KIX with the KID (kinase inducing domain) of CREB, a key transcription factor for cAMP-mediated gene transcription. Thus CARM1, *by its catalytic activity*, acts as a *corepresso*r for cAMP signaling. At the same time CARM1 can still function as a *coactivator* of nuclear-hormone-receptor-mediated gene activation pathways because the R_{580} methylation state is not utilized in coreceptor recruitments.

Histidine N-Methylation

Histidine N-methylation is occasionally found in proteins, including the methanobacterial methyl-CoM reductase, functioning in the last stage of methane biogenesis. The function of this rare modification in the methane-generating enzyme is unkown. It is noted in Chapter 11 that a single His residue in the eukaryotic protein synthesis elongation factor protein EF2 is methylated three times in the process of conversion to diphthamide. Again, the precise function of this His trimethylation is obscure, although it does target that residue for ADP-ribosylation by diphtheria toxin (see Chapter 11).

Glutamine and Asparagine N-Methylation

N-Methylation of asparagines and glutamine carboxamido nitrogens has been observed in proteins. Such amide nitrogens are certainly not traditional nucleophiles, but biological systems have overcome the lowered reactivity of Asn carboxamido nitrogens in all the posttranslational N-glycosylations of proteins (see Chapter 10). Both orientation and electronic effects are likely to decrease the resonance stabilization in the amide and enhance nitrogen nucleophilicity. γ-N-Methylasparagine (Figure 5.16A) has been found at one specific residue in the tetrapyrrole-containing proteins of phycobilisomes involved in light harvesting in cyanobacteria and red algae (Klotz and Glazer, 1987), but the function of this posttranslational modification is unclear.

A comparable amide N-methylation, a δ-N-Me-glutamine residue (Figure 5.16A), is found in a GGQ motif in ribosomal release factors involved in termination of protein synthesis. Release factors are proteins that recognize stop codons in mRNA and then hydrolyze the full length peptidyl-tRNA (Figure 5.16B) to effect the covalent disconnection of nascent proteins from tRNA. Methylation of the Q in the GGQ motif is required for efficient release in each of the two *E. coli* release factors, RF1 and RF2.

A.

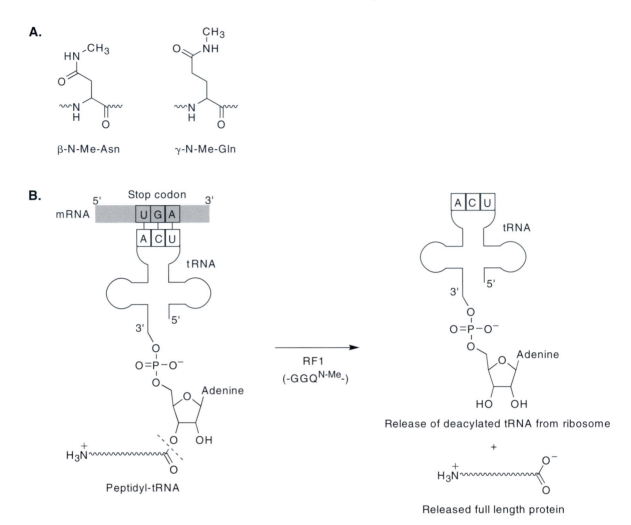

β-N-Me-Asn   γ-N-Me-Gln

B.

Figure 5.16 **A.** Enzymatic methylation of the carboxamido nitrogens of Asn and Gln residues in proteins. **B.** N-Methylation of the GGQ motif of ribosome release factors creates the active form of the enzyme for peptidyl-tRNA deacylation at the ribosome.

The N-methyl-Gln residue was detected and identified by mass spectrometry. The GGQ^{Me} motif is hypothesized to help position the attacking water molecule in the peptidyl-tRNA hydrolysis.

The enzyme responsible for the GGQ motif methylation was first identified as HemK in *E. coli* but has been renamed PrmC (protein release factor methylase). A comparable Prm enzyme has been detected in humans. X-ray analysis of the *Thermotoga maratima* PrmC methylase (Schubert et al., 2003b) shows architecture comparable to purine N-methyltransferases and a hydrogen bond network that polarizes the

Q side chain in the GGQ tripeptide moiety for the S_N2 attack on the transferring methyl group of bound cosubstrate SAM. In the free PrmC the GGQ loop is on the surface but is thought to reorganize on binding of the release factors to the ribosomal stop codons with bound peptidyl-tRNA.

Protein C-Methylation

Methyl CoM Reductase: S-, N-, and C-Methylation

The enzyme methyl coenzyme M reductase, found in methanogenic bacteria, catalyzes the last step in biogenesis of the million tons of CH_4 produced biologically on the planet each year. The reaction involves net protonolysis of a CH_3–SR thioether bond via the intermediacy of a methyl-nickel organometallic intermediate, formed by intermediate methyl transfer to the nickel porphinoid coenzyme F430 at the active site of the enzyme (Figure 5.17A). In addition to the catalytic methyl transfers, there are posttranslational methylations of one of the subunits of the enzyme.

Determination of the x-ray structure of the $\alpha_2\beta_2\gamma_2$ structure of methyl CoM reductase (Ermler et al., 1997) revealed four methylated amino acid residues in each α subunit: 1N-Me-His$_{257}$, S-Me-Cys$_{452}$, 2-Me-Gln$_{400}$, and 5-Me-Arg$_{271}$. The methylations were confirmed by mass spectrometry of chymotryptic peptide fragments. All four methyl groups derive from SAM (Selmer et al., 2000) and are probably added posttranslationally to the α subunits before association with the β and γ subunits since the α subunits are buried in the holo enzyme. His$_{257}$ is in the active site and the other methylated side chains are proximal, suggesting these are functional modifications, arising from methyl–enzyme intermediates on the reaction pathway to methane.

The His N-methylation and Cys S-methylation have precedents in other protein posttranslational modifications, but the sterically demanding C-methylation of the Gln and Arg side chains are to date unique. The paucity of C-methylations of proteins probably reflects the insufficient nucleophilicity of side chain carbon centers as nucleophiles. Selmer et al. (2000) noted that the methylations of Gln$_{400}$ and Arg$_{271}$ were both regiospecific (C_2 and C_5, respectively) and stereospecific (2S and 5S). They also suggested possible mechanisms in which the C_2 of Gln or the C_5 of Arg were deprotonated—namely, via C–H cleavage to give a carbanion that could be stabilized by the adjacent amide NH, participating as an ammonium ylid (adjacent + and – charges) (Figure 5.17B). There is as yet no evidence on these mechanisms, nor is it known if the strictly anaerobic milieu of the methanogens is consequential for these protein C-methylations.

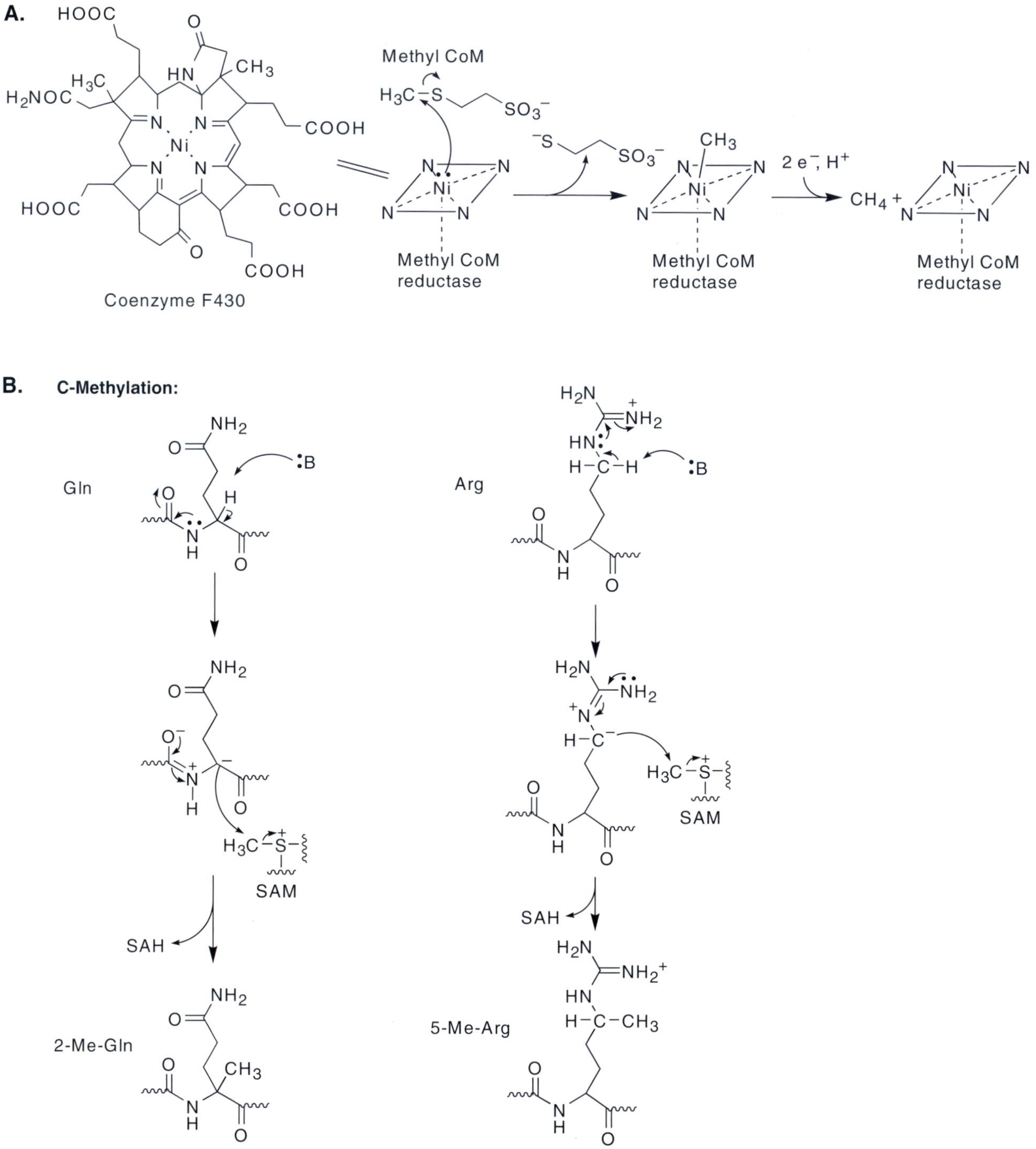

A.

Coenzyme F430

B.   **C-Methylation:**

Gln

Arg

2-Me-Gln

5-Me-Arg

Figure 5.17 **A.** Reaction catalyzed by the methane-forming enzyme methyl coenzyme M reductase in methanogenic archaebacteria. **B.** Mechanistic proposal for auto-C-methylation by ammonium ylid formation to produce 2-Me-Gln and 5-Me-Arg residues.

Removal of N-Methyl Groups in Proteins by Deimination of MethylArg or Deamination of MethylLys Residues

Protein Arginine Deimination: Competition with Arg N-Methylation

It was noted earlier in this chapter that protein Asp and Glu methylesters are labile to hydrolysis and their lifetimes (minutes to hours) may be modulated by specific protein microenvironments. By contrast, N-methylated Lys and Arg residues are indefinitely stable to hydrolysis. Given the absence of known protein N-demethylases, it has been an open question if and how N-methylation, for example of histone H3 and H4 tails, are reversed.

There are enzymes known that hydrolyze the guanidine side chain of Arg residues in proteins, a net deimination (Figure 5.18) to a citrulline (Cit) residue. Four human variants of peptidyl arginine deiminase (PADI) are known (Vossenaar et al., 2003). Of these, PADI4 is nuclear in location. PADI4 has been shown to deiminate $Arg_{2,8,17,26}$ in histone H3, the same four Arg residues that are mono- and dimethylated by the CARM1 methyltransferase (Cuthbert et al., 2004). PADI4 action results in transcriptional repression and so opposes CARM1 action for gene transcriptional activation (Figure 5.18). Enzyme assays with PADI4 indicate deimination of monomethylArg side chains but not of dimethylArg chains. Thus, PADI4 and CARM1 compete for posttranslational modification of Arg side chains in H3 (Cuthbert et al., 2004).

It is unclear how the hydrolytic deimination of Arg to Cit residues in H3 and other proteins is reversed. There are citrulline transaminases known that work on the free nonproteinogenic amino acid but as yet none that work as peptidyl citrulline transaminases. It has also been proposed that Cit-containing H3, as well as N-methylLys and N-methylArg forms of histones, are replaced by cycles of partial or complete nucleosome dissociation and reassociation with unmodified histones.

A Specific Deaminase for N-MethylLys$_4$ in Histone H3

The N-monomethylation and N,N-dimethylation of Lys_4 in histone H3 that is associated with transcriptional repression can be reversed by an FAD-dependent amine oxidase, termed LSD1 (lysine specific deaminase) that will take off one or both N-methyl groups by oxidation of the $CH_3–NH–$ moiety to $CH_2=N–$, an imine product. The imine linkage is subject to spontaneous addition of water and release of the one-carbon fragment as formaldehyde. As shown in Figure 5.19, N, N–Me_2–Lys_4 of H3 can be converted successively to the monomethyl and then the free Lys_4 side chain by two rounds of action by LSD1, reversing the initial action of the histone methylase. In contrast, LSD1 would not oxidiatively deaminate the trimethylLys$_4$,

Histone H3 tail

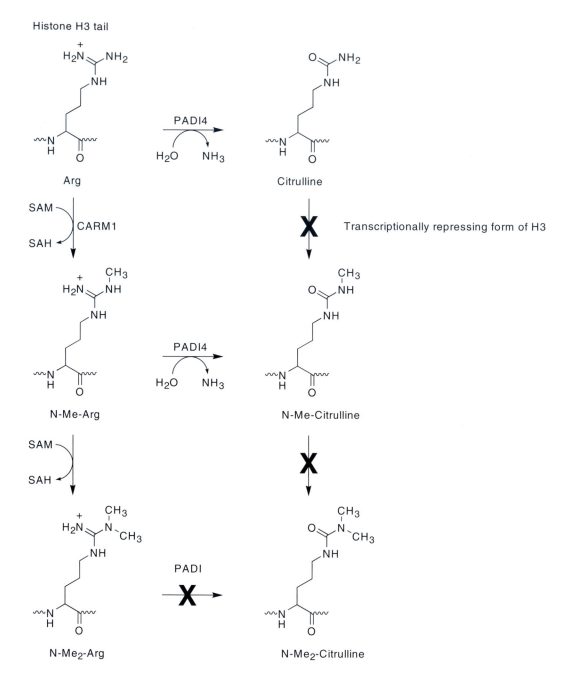

Transcriptionally repressing form of H3

Transcriptionally activating form of H3

Figure 5.18 Peptidyl arginine deiminase isoform 1 (PAD1) by hydrolytic deimination of the guanidine groups in Arg residues (e.g., in histone N-terminal tails) can antagonize the transcriptional activation induced by CARM1-mediated N-methylations.

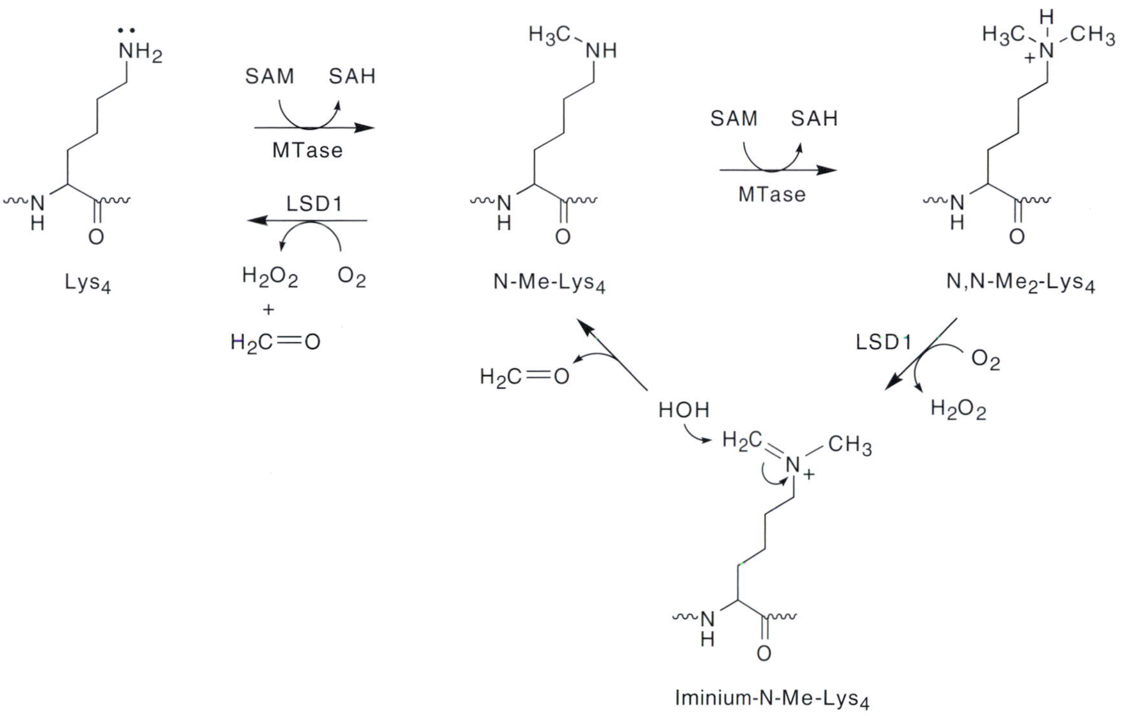

Figure 5.19 The ε-NH$_2$ of Lys$_4$ in histone H3 can be sequentially monomethylated, then dimethylated via two SAM molecules and catalytic action of a methyltransferase (MTase). The N-CH$_3$ groups are oxidatively removed as formaldehyde (CH$_2$=O) by action of a regioselective lysine specific deaminase (LSD1) completing a cycle with LSD1 opposing the MTase for gene activation/repression.

consistent with the existence of an alternate enzyme or mechanism for return of this derivative to unmodified Lys$_4$. Furthermore, LSD1 is selective for demethylation of Lys$_4$ and does not act on methylLys$_{9,36,79}$ in H3, suggesting this may be the first in a family of N-methylhistone deaminases (Shi et al., 2004).

References

Bass, R. B., and J. J. Falke. "The aspartate receptor cytoplasmic domain: In situ chemical analysis of structure, mechanism and dynamics," *Struct. Fold. Des.* **7**:829–840 (1999).

Chulkov, S., J. K. Kurash, J. R. Wilson, B. Xiao, N. Justin, G. S. Ivanov, K. McKinney, P. Tempst, C. Prives, S. J. Gamblin, N. A. Barlev, and D. Reinberg. "Regulation of p53 activity through lysine methylation," *Nature* **432**:353–360 (2004).

Clarke, S. "Protein carboxyl methyltransferases: Two distinct classes of enzymes," *Annu. Rev. Biochem.* **54**:479–506 (1985).

Clarke, S. "Protein isoprenylation and methylation at carboxyl-terminal cysteine residues," *Annu. Rev. Biochem.* **61**:355–386 (1992).

Cuthbert, G. L., S. Daujat, A. W. Snowden, H. Erdjument–Bromage, T. Hagiwara, M. Yamada, R. Schneider, P. D. Gregory, P. Tempst, A. J. Bannister, and T. Kouzarides. "Histone deimination antagonizes arginine methylation," *Cell* **118**:545–553 (2004).

Ermler, U., W. Grabarse, S. Shima, M. Goubeaud, and R. K. Thauer. "Crystal structure of methyl-coenzyme M reductase: The key enzyme of biological methane formation," *Science* **278**:1457–1462 (1997).

Ezhkova, E., and W. P. Tansey. "Proteasomal ATPases link ubiquitylation of histone H2B to methylation of histone H3," *Mol. Cell* **13**:435–442 (2004).

Falke, J. J., R. B. Bass, S. L. Butler, S. A. Chervitz, and M. A. Danielson. "The two–component signaling pathway of bacterial chemotaxis: A molecular view of signal transduction by receptors, kinases, and adaptation enzymes," *Annu. Rev. Cell Dev. Biol.* **13**:457–512 (1997).

Grebe, T. W., and J. Stock. "Bacterial chemotaxis: The five sensors of a bacterium," *Curr. Biol.* **8**:R154–R157 (1998).

Griffith, S. C., M. R. Sawaya, D. R. Boutz, N. Thapar, J. E. Katz, S. Clarke, and T. O. Yeates. "Crystal structure of a protein repair methyltransferase from *Pyrococcus furiosus* with its L-isoaspartyl peptide substrate," *J. Mol. Biol.* **313**:1103–1116 (2001).

Khorasanizadeh, S. "The nucleosome: From genomic organization to genomic regulation," *Cell* **116**:259–272 (2004).

Klotz, A. V., and A. N. Glazer. "Gamma-N-methylasparagine in phycobiliproteins. Occurrence, location, and biosynthesis, *J. Biol. Chem.* **262**:17350–17355 (1987).

Kouskouti, A., E. Scheer, A. Staub, L. Tora, and I. Talianidis. "Gene-specific modulation of TAF10 function by SET9-mediated methylation," *Mol. Cell* **14**:175–182 (2004).

Lachner, M., and T. Jenuwein. "The many faces of histone lysine methylation," *Curr. Opin. Cell Biol.* **14**:286–298 (2002).

Lachner, M., R. J. O'Sullivan, and T. Jenuwein. "An epigenetic road map for histone lysine methylation," *J. Cell Sci.* **116**:2117–2124 (2003).

Marmorstein, R. "Protein modules that manipulate histone tails for chromatin regulation," *Nat. Rev. Mol. Cell Biol.* **2**:422–432 (2001).

Min, J., X. Zhang, X. Cheng, S. I. Grewal, and R. M. Xu. "Structure of the SET domain histone lysine methyltransferase Clr4," *Nat. Struct. Biol.* **9**:828–832 (2002).

Mowen, K. A., J. Tang, W. Zhu, B. T. Schurter, K. Shuai, H. R. Herschman, and M. David. "Arginine methylation of STAT1 modulates IFNalpha/beta-induced transcription," *Cell* **104**:731–741 (2001).

Pesavento, J. J., Y.-B. Kim, G. K. Taylor, and N. L. Kelleher. "Prescriptive annotation of histone modifications: A new approach for streamlined characterization of proteins by top down mass spectrometry," *J. Am. Chem. Soc.* **126**:3386–3387 (2004).

Sapperstein, S., C. Berkower, and S. Michaelis. "Nucleotide sequence of the yeast STE14 gene, which encodes farnesylcysteine carboxyl methyltransferase, and demonstration of its essential role in α-factor export," *Mol. Cell Biol.* **14**:1438–1449 (1994).

Schubert, H. L., R. M. Blumenthal, and X. Cheng. "Many paths to methyltransfer: A chronicle of convergence," *Trends Biochem. Sci.* **28:**329–335 (2003a).

Schubert, H. L., J. D. Phillips, and C. P. Hill. "Structures along the catalytic pathway of PrmC/HemK, an N5-glutamine AdoMet-dependent methyltransferase," *Biochemistry* **42:**5592–5599 (2003b).

Sedgwick, B. "Repairing DNA-methylation damage," *Nat. Rev. Mol. Cell Biol.* **5:**148–157 (2004).

Selmer, T., J. Kahnt, M. Goubeaud, S. Shima, W. Grabarse, U. Ermler, and R. K. Thauer. "The biosynthesis of methylated amino acids in the active site region of methyl–coenzyme M reductase," *J. Biol. Chem.* **275:**3755–3760 (2000).

Shi, Y., F. Lan, C. Matson, P. Mulligan, J. R. Whetstine, P. A. Cole, R. A. Casero, and Y. Shi. "Histone demethylation mediated by the nuclear amine oxidase homolog LSD1," *Cell* **119:**941–953 (2004).

Skinner, M. M., J. M. Puvathingal, R. L. Walter, and A. M. Friedman. "Crystal structure of protein isoaspartyl methyltransferase: A catalyst for protein repair," *Struct. Fold. Des.* **8:**1189–1201 (2000).

Sun, Z. W., and C. D. Allis. "Ubiquitination of histone H2B regulates H3 methylation and gene silencing in yeast," *Nature* **418:**104–108 (2002).

Turner, B. M. "Cellular memory and the histone code," *Cell* **111:**285–291 (2002).

Vossenaar, E. R., A. J. Zendman, W. J. van Venrooij, and G. J. Pruijn. "PAD, a growing family of citrullinating enzymes: Genes, features and involvement in disease," *BioEssays* **25:**1106–1118 (2003).

Walsh, C. T. *Enzymatic Reaction Mechanisms.* Freeman: San Francisco (1979).

Xu, W., H. Chen, K. Du, H. Asahara, M. Tini, B. M. Emerson, M. Montminy, and R. M. Evans. "A transcriptional switch mediated by cofactor methylation," *Science* **294:**2507–2511 (2001).

Zhang, X., Z. Yang, S. I. Khan, J. R. Horton, H. Tamaru, E. U. Selker, and X. Cheng. "Structural basis for the product specificity of histone lysine methyltransferases," **12:**177–185 (2003).

CHAPTER 6

Protein N-Acetylation

NAD-dependent deacetylation of ε-acetyl-Lys residues in histones generates
2'-O-ADP-ribose as coproduct

Protein acylations occur with short-chain and long-chain acyl CoAs as donor substrates to nucleophilic side chains in proteins. The long-chain myristoyl CoA and palmitoyl CoA donors are used to attach the C_{14} and C_{16} acyl chains as lipid anchors to direct the modified proteins to cellular membranes, as detailed in Chapter 7. The short-chain acetyl CoA, discussed in this chapter, is utilized in protein acetylation for two distinct biological purposes. One is the acetylation of the N-termini of a large fraction of eukaryotic proteins, an irreversible modification occurring cotranslationally. The other is the acetylation of proteins, most famously histones and transcription factors that affect selective gene transcription and chromatin structure. These are regulatory events and are reversed by N-deacetylase enzymes. The two N-acetylations are depicted (Figures 6.1A and B), for N-terminal acetylation and lysine ε-NH_2 acetylation, respectively.

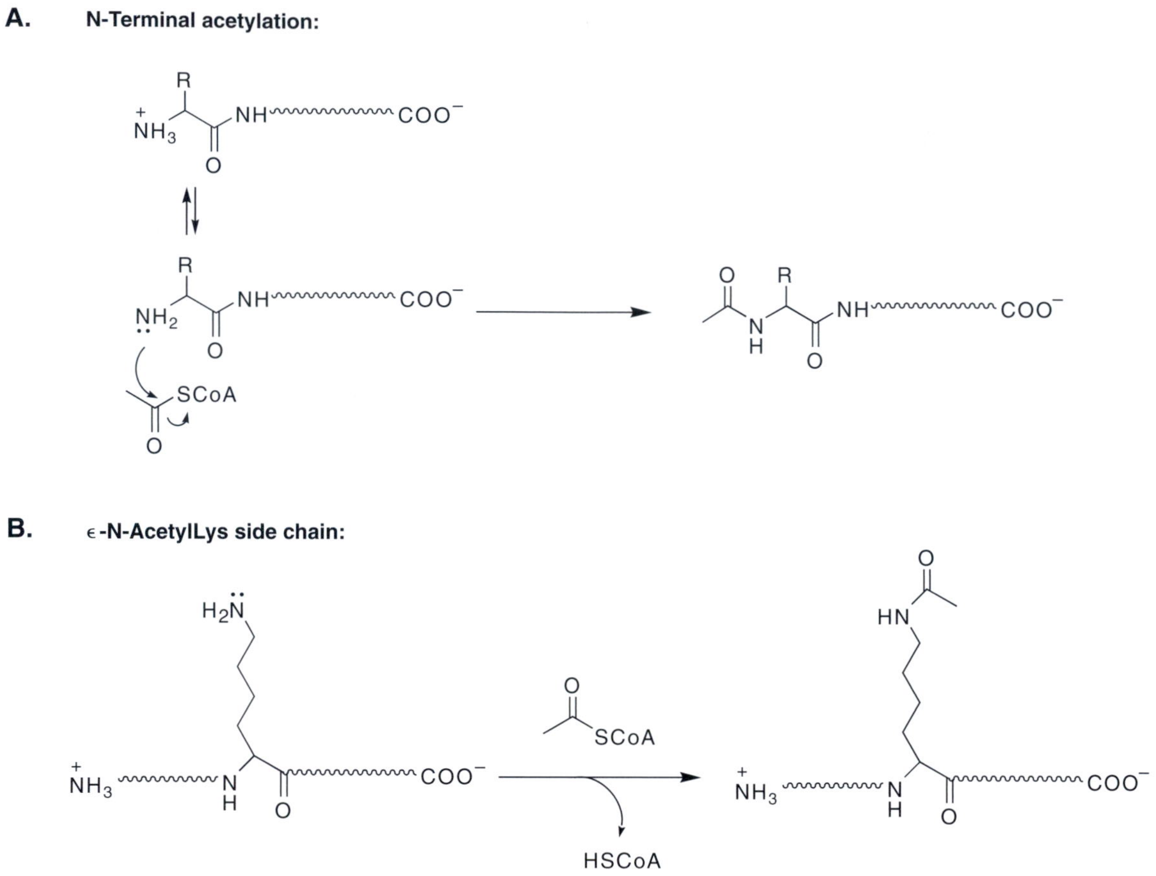

A.   N-Terminal acetylation:

B.   ε-N-AcetylLys side chain:

Figure 6.1 Acetyl CoA as acetyl donor for protein N-acetylation. **A.** Acetylation of the N-terminal amino groups of proteins. **B.** Acetylation of the ε-NH$_2$ of lysine residues.

N-Terminal Acetylation of Eukaryotic Proteins

Enzymatic acetylation of the N-terminus of proteins occurs in about 50% of yeast proteins, and up to 80–90% of higher eukaryotic proteins. In contrast, it is very rare in prokaryotes, where 807 of 810 *E. coli* proteins examined were not acetylated at the N-terminus (Polevoda and Sherman, 2003). The three *E. coli* proteins acetylated were ribosomal subunits S5 (N-Ac-Ala-Arg-), S18 (N-Ac-Ala-His-), and L12 (N-Ac-Ser-Ile-). Also, the protein synthesis elongation factor EFTu is N-acetylated. Since *E. coli* protein synthesis uses N-formylmethionine as the initiating residue, the fMet on the three ribosomal proteins must be deformylated, then cleaved to reveal Ala and/or Ser as the new N-termini in S8, S15, and L12, which then becomes acetylated (Figure 6.2A).

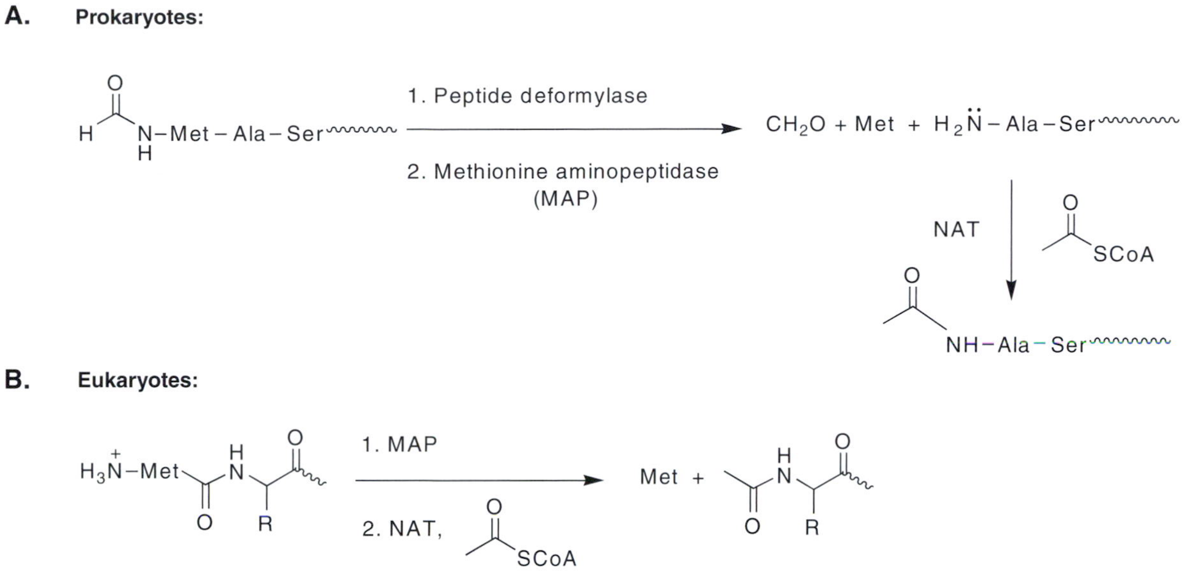

A.   Prokaryotes:

B.   Eukaryotes:

Figure 6.2 **A.** N-Terminal acetylation of a few *E. coli* proteins occurs after deformylation and hydrolytic removal of the N-terminal methionine. **B.** Action of methionine aminopeptidase (MAP) and N-acetyltransferase (NAT) at the amino termini of eukaryotic proteins for N-terminal acetylation.

The highly prevalent eukaryotic acetylations of protein N-termini can occur on intact N-terminal methionine amino groups or can occur after the cotranslational hydrolytic clipping of the N-terminal methionine by one of two methionine aminopeptidases (MAPs) (Polevoda and Sherman, 2000 and 2003) (Figure 6.2B). Both the methionine aminopeptidase and the N-acetyltransferase (NAT) enzymes act cotranslationally, when the emerging protein chain has grown to 20–50 residues and is still elongating on the ribosome. The specificity determinants for MAP action determine whether it is the N-terminal methionine or the amino group of what had been residue 2 in the nascent protein chain that is available for the NAT to N-acetylate. MAP cleaves methionyl residues rapidly when the second residue on the growing protein is small, such as Gly, Ala, Ser, Cys, Thr, Pro, and Val, so those residues are highly represented in eukaryotic proteins as N-termini. In turn, these uncovered Ser, Ala, Gly, Thr, and the persisting Met_1 are the major residues that get N-acetylated.

Yeast genetics has provided insights into the existence of three NATs (NATA, B, and C in fungi), with orthologs in higher eukaryotes, and individual NAT knockouts have allowed for NAT isozyme specificity evaluation (summarized in Polevoda and Sherman, 2003). NATB and NATC act as acetyltransferases for yeast proteins with the original Met_1 still in place. NATB recognizes Met-Glu, Met-Asp, Met-Gln, and Met-Met sequences, while NATC N-acetylates at Met-Ile, Met-Leu, Met-Trp, and Met-Phe, reflecting preferences for acidic or hydrophobic side chains at residue 2.

NATA catalyzes acetyl transfer to the NH_2 group of Ser, Ala, Gly, and Thr termini, the predominant side chains liberated from MAP action.

The biological significance of N-terminal acetylation of the bulk of eukaryotic proteins is unclear. While yeast genetics reveals phenotypic defects, the subset of essential proteins for N-acetylation has not been fully deconvoluted and less is known for higher eukaryotes. Eubacterial and archaeal proteins are rarely acetylated at the N-terminus. Doubtless there will be a continuum of functions detected on a case-by-case basis. For example, the gag protein of the yeast L-A dsRNA virus must be acetylated at the N-terminus for effective virion assembly, and tropomyosin has to be N-acetylated for interaction with actin cables (Polevoda and Sherman, 2003). Some actins undergo a variant of N-terminal acetylation, in which the N-terminal Met-Glu sequence is N-acetylated, the N-acetyl-Met-Glu cleaved by a specific exopeptidase, to yield the H_2N–Glu terminus that is then acetylated to give the mature actin N-terminus of N-acetyl-Glu.

N-Acetylation of Lysine-ε-NH_2 Side Chains

Most of the contemporary interest in protein N-acetylation is not on N-terminal acetylation but rather on the regiospecific modification of particular lysine side chains in proteins, starting with the N-terminal tails of the four histones—H2A, H2B, H3, and H4—that form the octameric histone cores of nucleosomes, $(H2A)_2(H2B)_2,(H3)_2(H4)_2$, around which 145–147 base pairs of DNA is wrapped in chromatin structures. As noted also in Chapter 5 in the section on histone tail N-methylations, the N-termini of the four histones are not involved in the core structure and so are flexible, unstructured, and available for posttranslational modification, including phosphorylation, methylation, ubiquitylation, and acetylation (Figure 6.3) (Marmorstein, 2001; Turner, 2002). These covalent markings are proposed to be signals, the histone code (Strahl and Allis, 2000) that gets read by proteins of the gene transcriptional machinery for selective activation or repression of genes in particular regions of chromatin. In analysis of such a histone code, one will need to understand how the code is written (by the posttranslational modification catalysts in response to integrated signaling inputs) and also how the code is read (by transcriptional machinery and the coactivator and corepressor components).

The recent widespread efforts to characterize histone lysine acetyltransferases stemmed from the discovery that some transcriptional coactivators turned out to be histone acetyltransferases (HATs) (Berger, 1999; Kouzarides, 2000). In keeping with the posttranslational modification involved in signaling, the acetyl groups on histone lysine side chains can be removed reversibly by histone deacetylases (HDACs). Some transcriptional corepressor proteins turned out to have HDAC activity, providing further

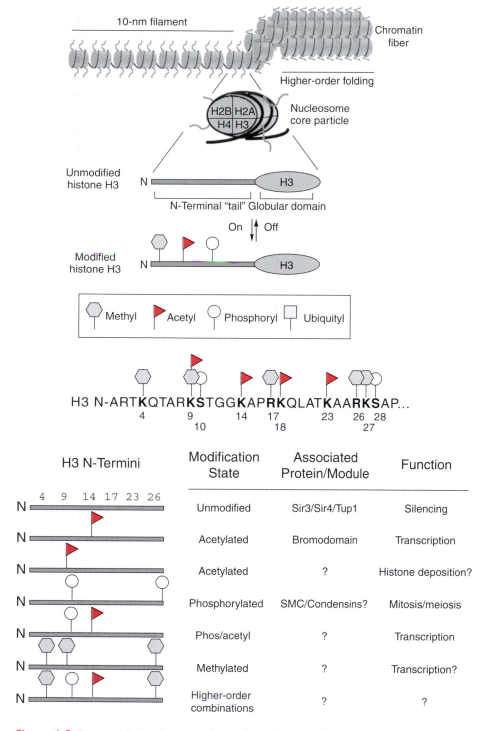

impetus for the characterization of the balance of HAT and HDAC activity (Figures 6.4A and B) to understand the dynamic integration of the histone code over time.

The N-acetylation of histones can be detected by mass spectrometry, with an increase of 42 mass units for each acetyl group introduced, or by use of radioactive *acetyl CoA as cosubstrate to monitor protein covalent radioactivity, or by using N-acetyllysine-specific antibodies in chromatin immunoprecipitation (ChIP) assays for qualitative detection of acetylated histone levels. With these methodologies it could be established that there are two lysines on H2A (K_5 and K_9), four on H2B (K_5, K_{12}, K_{15}, and K_{20}), four on H3 (K_9, K_{14}, K_{18}, and K_{23}), and four on H4 (K_5, K_8, K_{12}, and K_{16}) that can undergo HAT-mediated acetylations, for a total of 14 possible acetylation sites (Figure 6.4B). Two copies of each histone mean 28 potential acetylation sites per octamer in each nucleosome. It has been reported that yeast histones can show up to 13 acetylated residues per octamer core (Kouzarides, 2000; Turner, 2002), reflecting almost 50% posttranslational utilization under the measured conditions. The combinatorial possibilities are potentially immense for distinct isoforms, but there appear to be preferential patterns of modifications as noted below.

Since lysine side chains are cationic at physiological pH, N-acetylation will quench the positive charges. If three or four of the four lysine side chains in H3 or H4 were acetylated, as can happen in nucleosomes where promoters are actively transcribed, then the charge quenching and the consequent electrostatic weakening of interaction of histone tails with negatively charged DNA could contribute to opening up of the chromatin in that microenvironment (Berger, 1999; Kouzarides, 2000).

Alternatively, or perhaps additionally, the acetyl groups provide additional bulk and acetylated lysine side chains can be specifically recognized, for example by bromodomains in partner proteins (Turner, 2002), and recruit them to those nucleosomes. Histone hyperacetylation is correlated with transcriptional activation and histone hypoacetylation (HDAC activity exceeding HAT activity) is correlated with chromatin regions of gene silencing. The reversibility of histone acetylation and deacetylation increases the likelihood of heterogeneity of acetyl content at any nucleosome at any time point in a replicating cell.

Recent mass spectrometric analysis (Zhang et al., 2002) indicates mixtures of a tetraacetyllysine form (at $K_{5,8,12,16}$) of histone H4, one triacetyl regioisomer (at $K_{8,12,16}$), one diacetyl regiosiomer (at $K_{12,16}$), and a single monoacetyl regiosiomer (at K_{16}). This pattern suggests directionality (a "zipper" model) to the kinetics of HAT and HDAC action at the N-tail of histone H4 in nucleosomes, where the lysine side chains nearest the N-terminus are acetylated most slowly and/or deacetylated most rapidly, a constraint on models of chromatin remodeling dynamics. Another MS-centered analysis of H4 modifications, with algorithm matching to deal with the

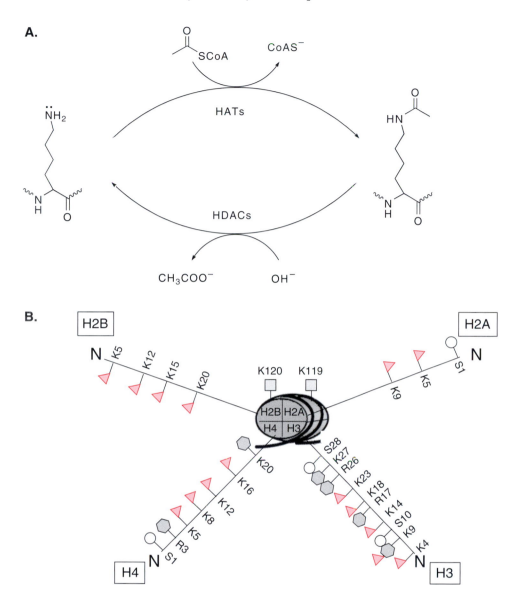

Figure 6.4 **A.** The acetylation levels on histone tails are balanced by the opposing activities of histone acetyltransferases (HATs) and histone deacetylases (HDACs). **B.** Posttranslational modifications of 26 residues in the tails of histones 2A, 2B, 3, and 4. The lysine acetylations are represented by triangles, lysine methylations by octagons, and serine phosphorylations by circles (adapted from Felsenfeld and Groudine, 2003).

large number of combinatorial predictions, utilized butyrate-treated Hela cells to stimulate transcription and gave an H4 fragment with +238 Da, matching N-terminal acetylation, acetylation at all four of K_5, K_8, K_{12}, and K_{16}, and dimethylation at K_{20} (Pesavento, 2004) (Figure 6.5).

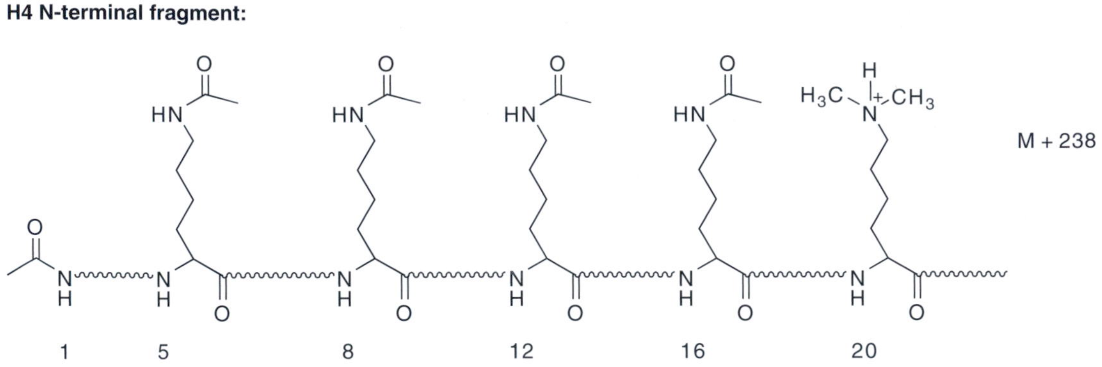

H4 N-terminal fragment:

M + 238

Figure 6.5 Detection of a hexa-modified isoform of histone H4 by MS analysis and algorithm matching. The increase in mass of 238 Da over the unmodified H3 tail peptide could be matched to the specific isoform shown.

Histone Acetyltransferases: Families of HATs

The prototypic transcriptional coactivator found to contain HAT activity is the yeast protein Gcn5 and it has become the founding member of the GNAT (Gcn5-related N-acetyltransferase) superfamily, which includes human PCAF (*p*300 and *CBP a*sso-ciated *f*actor) (Cheung et al., 2000; Roth et al., 2001). Other coactivators found to contain embedded HAT domains include the p300 and its functional twin the CREB binding protein (CBP), the steroid receptor coactivator (SRC1), and the activated thyroid and retinoid receptor (ACTR). Additionally, TATA box-associated proteins (TAFs) such as TAF$_{II}$250 have HAT domains among other domains (Brown et al., 2000). X-ray analysis of several of the NAT (GNAT) catalytic domains display bind-ing sites for acetyl CoA in close apposition to sites for the ϵ-NH$_2$ of lysine side chains in histone-derived peptides that were co-crystallized with CoASH (Marmorstein, 2001; Roth et al., 2001; Turner, 2002), suggesting direct N-acetylation reaction mechanisms (Figure 6.6). However, the molecular basis of regioselective acetylation of particular lysines in protein substrates is not yet deciphered and fully predictable.

Histone acetyltransferase domains are often parts of multidomain proteins, con-sistent with the multifunctional roles for transcriptional coactivators in protein–protein recognition and assembly of protein complexes. For example, both Gcn5 and PCAF have HAT domains upstream of 110-residue domains termed bromodomains (Figures 6.7A and B). Structural and functional analysis (Marmorstein, 2001) has established that the bromodomains bind N-acetyllysine peptides, suggesting that these are domains involved in intra- and interprotein recognition of the acetylated histones after HAT domains have completed their modifications (Khorasanizadeh,

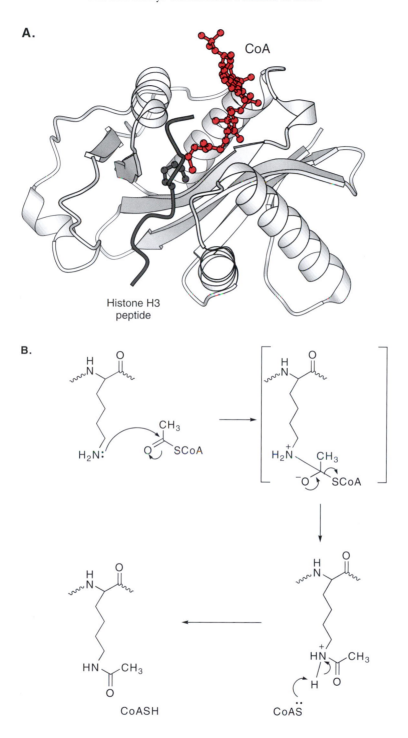

Figure 6.6 **A.** X-ray structure of an N-acetyltransferase domain co-crystallized with CoA and histone H3 substrate peptide (figure made using PDB 1QSN). **B.** Direct acetyl transfer from acetyl CoA to Lys-ϵ-NH$_2$.

2004). The CBP/p300 pair have five domains, including bromo- and HAT domains (Figure 6.7A), consistent with roles as global transcriptional activators (Marmorstein, 2001). $TAF_{II}250$ has a kinase domain, followed by a HAT domain, and then a tandem pair of bromodomains. The pair of bromodomains is proposed to have synergistic recognition for pairs of acetyllysine side chains, such as at K_{14} and K_{18} on H3 or K_{12} and K_{16} on H4. This would be part of the combinatorial machinery that could read the histone code and distinguish the positional loading and occupancy of acetyl groups on histone tails. The structures of bromodomains from Gcn5 (Figure 6.7C) and also from CBP complexed with ϵ-acetylLys peptides have been solved (Figure 6.7 B) (Khorisanizadeh, 2004) to provide insights into how these protein domains are recruited to the modified lysine side chains.

In turn, these multidomain coactivators, with their embedded HAT domains, are parts of large complexes such as the SAGA complex in yeast with at least 14 proteins and a particle molecular weight of about 2 Megadaltons. The SAGA complex contains Gcn5 and shows acetyltransferase activity selective for the lysines on histone H3 (Berger, 1999). In humans, comparable mega complexes of transcription factors, coactivators with HAT and bromodomains, scaffold proteins, regulatory subunits, and TBP interacting proteins, can also show H3 NAT selectivity. Complexes containing the MYST N-acetyltransferase family modify H4 lysines more selectively, while complexes with p300/CBP tend to acetylate lysine side chains on H3 and H4 equivalently and also on H2A and H2B (Berger, 1999), consistent with global coactivator roles. Writing and reading the histone code probably occurs in the context of these multiprotein machines, where the acetylation state of particular histones in specific nucleosomes in part controls chromatin condensation and gene promoter accessibility.

Histone acetylation is only part of the posttranslationally introduced code. Phosphorylation of serine side chains, notably Ser_{10} (and Ser_{28}) of H3 may precede and set the local charge context for acetylation of adjacent K_9 or K_{14} within the H3 chain or predispose a mega complex to dock and acetylate K_{16} on one of the two H4 tails in the histone octamer, perhaps by the MOF HAT (Roth et al., 2001). The combination of $P\text{-}Ser_{10}$ on H3 and $N\text{-}Ac\text{-}K_{16}$ on H4 appears to be a marker for X chromosome transcription (Roth et al., 2001; Strahl and Allis, 2000). The pathways for preferential or ordered regiospecific multiple modifications (acetylations, methylations, and phosphorylations) of histone tails will continue to be deciphered and functional consequences evaluated. The multiple-site modifications are reminiscent of the multiple-site phosphorylations of proteins (Kouzarides, 2000), singly and in complex, such as the 11 sites of phosphorylation of the Abl tyrosine kinase noted in Chapter 2.

Multiple automodifications of the HAT domain of p300 have been reported (Thompson et al., 2004) on expression in *E. coli*. Thirteen lysine side chains of the HAT domain become acetylated, 10 of them in an 18-residue hinge region. Hyper-

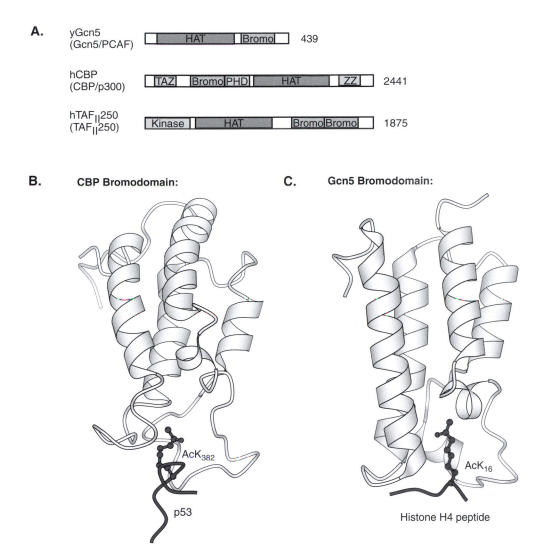

Figure 6.7 **A.** Bromodomains in coactivator proteins serve as recognition domains for N-acetylated lysine-containing proteins such as histones. **B.** CBP bromodomain structure in complex with a C-terminal p53 peptide containing N-acetylLys$_{382}$ (figure made using PDB 1JSP). **C.** X-ray structure of Gcn5 bromodomain complexed with histone H4 peptide containing ϵ-N-acetylLys$_{16}$ (figure made using PDB 1E6I).

acetylation of the hinge region activates the p300 HAT by up to 10-fold for the catalytic efficiency of the acetylation of external peptide substrates (and presumably external proteins). This raises the prospect of autoinhibitory loops in other HATs at rest, in analogy to protein kinases (Chapter 2), and suggests that autoacetylation of lysine side chains in such loops could autoactivate HAT domains—that is, HATs could be acetyl-gated acetylases (Pugh, 2004).

HATs and FATs

In addition to side chains of lysine residues in histones as substrates, the HAT domains, embedded in the multimodular coactivators that are parts of the megadalton transcriptional machinery that assemble at promoter regions of genes, can acetylate lysine side chains in transcription factors, presumably also recruited to such complexes. When the catalytic acetyltransferase domains work on transcription factors, they have often been termed FATs (factor acetyltransferases) rather than HATS. Sometimes this can affect DNA binding. The transcription factors acetylated include p53, E2F1, EKLF, and the hemopoietic protein GATA1 (Kouzarides, 2000). The lysines that are acetylated are in regions adjacent to the DNA binding domains of these transcription factors. The net result of such acetylation appears to be enhancement of DNA binding and greater activity in transcription. On the other hand, the scaffold protein HMG1, found in enhanceosomes, is acetylated within the DNA binding region and this acetylation blocks DNA binding. Transcription factor acetylation by HATs can also alter protein–protein interactions, as occurs when the binding of *Drosophila* TCF to its coactivator armadillo is interrupted, and analogously when the binding of nuclear hormone receptor to ACTR is abrogated on lysine acetylation. The stability of transcription factor proteins can also be affected by acetylation, as E2F1 has a longer half-life after lysine modification (Kouzarides, 2000).

Just as distinct sequential posttranslational modifications can be part of pathways in Ras activation (Chapter 7), phosphorylation of p53 (on serine residues) can stimulate subsequent acetylation of p53. For example, recruitment of p300/CBP is enabled by phosphorylation and, once recruited, p300/CBP can then acetylate the C-terminus of p53. In fact, there are five lysines clustered in the C-terminal regulatory domain of p53 (K_{370}, K_{371}, K_{372}, K_{381}, and K_{382}) that can be acetylated *in vitro* and *in vivo* by p300/CBP acting as a FAT (Li et al., 2002) (Figure 6.8). Exposure of cells to stress increases CBP acetylation of p53. Acetylation in this region can greatly increase DNA binding, perhaps driven by a conformational change as the charge and side chains of the lysine residues are altered. It turns out that these lysines are also the sites of Mdm2-mediated ubiquitylation of p53 to send it to the proteasome for degradation (Chapter 9). Thus, acetylation opposes ubiquitylation at the same side chains and the acetylation/ubiquitylation balance can control the lifetime of p53 molecules and the transcription of p53-responsive genes.

Another transcription factor where multiple acetylations can occur is the RelA isoform of NFκB, where up to five lysines in different subdomains can be acetylated by p300, CBP, or PCAF. Different effects can ensue from regiospecific acetylation. At K_{122} and K_{123}, for example, acetylation leads to decreased transcriptional activation,

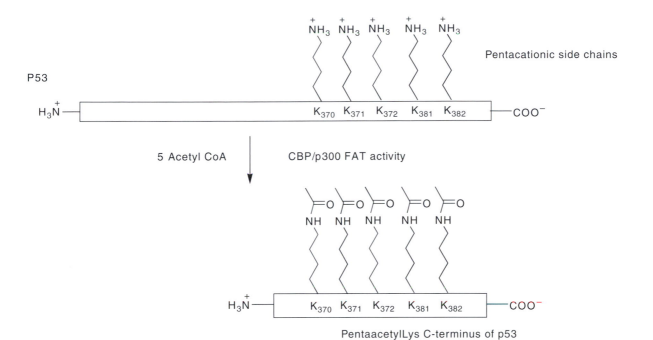

Figure 6.8 Multiple acetylations of lysine residues at the C-terminus of p53 by phosphorylated p300/CBP acting progressively as a FAT.

while at K_{310} acetylation leads to net transcriptional activation (Chen and Greene, 2004). The pattern of RelA acetylation is in turn controllable by prior posttranslational phosphorylation at Ser_{276} and Ser_{311}, emphasizing a temporal hierarchy to waves of such posttranslational modifications. In addition to phosphorylation of substrates, phosphorylation of the proteins containing the HAT domains can alter the HAT activity (Kouzarides, 2000). This is known for CBP, which is phosphorylated by cyclin/CDKs and for Gcn5 phosphorylation by DNA–protein kinase.

Writing the code clearly must also involve co-coordination of lysine methylation and acetylation. Recall from Chapter 5 that the H3 tail has three arginines (R_2, R_{17}, and R_{26}) and four lysines (K_4, K_9, K_{27}, and K_{36}) that can get methylated by one or more histone methyltransferases. Three additional lysines in H3 get acetylated (K_{14}, K_{18}, and K_{23}) (Figure 6.4B). The combinatorial possibilities at these 10 basic side chains can create 77,760 distinct molecular forms of histone H3. One lysine, K_9, is known to be targeted by both methyltransferases and acetyltransferases in mutually exclusive competition. Perhaps the methylation versus acetylation state at K_9 can be read to send the subsequently accumulating posttranslational acetylations or methylations in different directions. Phosphorylation and ubiquitylation of histones can also play in higher-order combinations. For example, as noted above, Ser_{10} and Ser_{28} can be phosphorylated during meiosis/mitosis (Strahl and Allis, 2000), and phosphorylation and acetylation combinations are known. In yeast, the ubiquitylation of H2A at K_{120} directs methylation on H3 at K_4 (Strahl and Allis, 2000) (Figure 6.4B). This may reflect part of the dynamics of continual updating of the code such that "sets of markers of the local state of genetic material" (Strahl and Allis, 2000) produce distinct patterns that can be read and integrated for distinct downstream readouts. Communication is also built in through noncatalytic domains, bromodomains that recognize the acetyllysine side chains, and chromodomains (Chapter 5) that recognize N-methyllysine side chains for recruitment and integration of the hybrid states at the protein–protein level in transcription factor mega complexes. NMR analysis has revealed how the bromodomain of the transcriptional coactivator CBP interacts with the N-acetylLys$_{382}$ side chain of p53 (Mujtaba et al., 2004) (Figure 6.7B). One of the analytical problems in analyzing the code is that the substoichiometric modification of a given histone molecule, H3, at any of the sites of acetylation, phosphorylation, methylation, or ubiquitylation can generate high heterogeneity and distributions of many posttranslationally modified forms. A recent report on the chemoenzymatic synthesis of histones allows purposeful stoichiometric modification at any residue in any of the histone tails—just the set of reagents that will be useful to look at N-methylase, HAT, and HDAC specificity preferences (He et al., 2003).

Histone Deacetylases (HDACs) and Sirtuins

HDACs

Although there are no deacetylases known that remove the acetyl groups at the N-termini of eukaryotic proteins, there are deacetylases that reverse histone acetylation at the ϵ-NH$_2$ of lysine residues. Whereas HATs are generally coactivating for transcription, histone deacetylases (HDACs) work in opposition as corepressors (Ng and Bird, 2000; Roth et al., 2001; Turner, 2002). Maintaining the lysine side chains in histone N-terminal tails in hypoacetylated states is associated with chromosome condensation, silenced promoters, and the absence of transcription in those regions of chromatin (Verdin, 2003). Remarkably, there are 16 human genes thought to encode histone deacetylases, reflecting their proposed importance in writing and rewriting the histone code for genome transcription. The 16 encoded deacetylases fall into two distinct families (Grozinger and Schreiber, 2002). There are nine HDACs that fit the pattern of simple hydrolases that release the acetyl moiety as acetate (Figure 6.4A). The catalytic domains are prototypic for zinc-dependent hydrolases, activating an amide bond for water attack. The sirtuins, members of the second family, are not simple

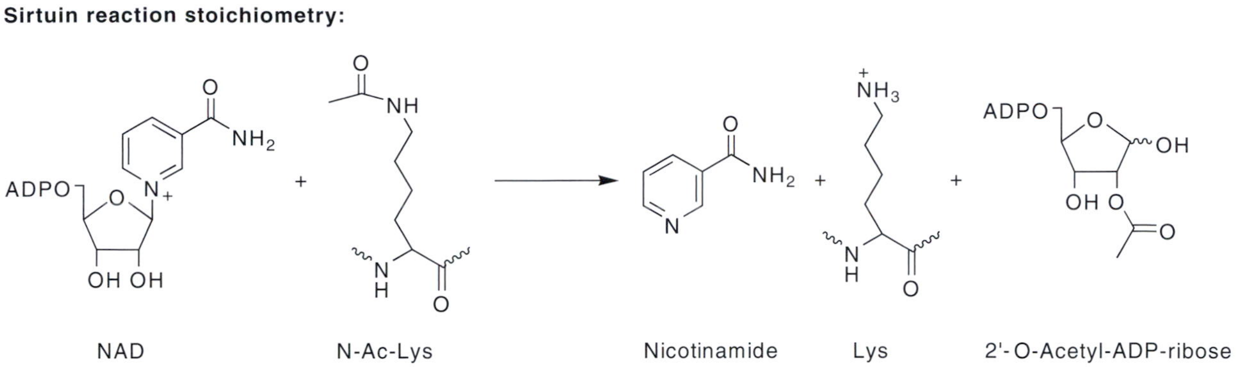

Sirtuin reaction stoichiometry:

Figure 6.9 The reaction stoichiometry for deacetylation by NAD-dependent sirtuins.

hydrolases. They cleave an NAD cosubstrate molecule in every turnover, and generate nicotinamide and 2′-O-acetyl-ADP-ribose as coproducts (Figure 6.9). We take up the HDACs first.

In some analogy with HATs, HDACs are often found as components in multiple-protein complexes that are carrying out surveillance of the histone tails in nucleosomes. HDAC1 and HDAC2 are found in SIN, NuRD, and CoREST complexes that do nuclear remodeling and silencing (Grozinger and Schreiber, 2002; Ng and Bird, 2000). HDACs 3–7 are distributed into different complexes, including the NcoR and SMRT complexes. Again, in analogy to HAT-containing proteins, HDACs themselves are thought not to bind directly to DNA or even to histones, but to be recruited to nucleosomes by transcription factors or nuclear hormone receptor proteins. Competition can occur between HDACs and HATs for the same partner proteins. HDAC4 and p300, for example, compete for MEF2. When the HAT is bound to MEF2, the complex acts as a gene-activating switch. When HDAC4 is bound, it switches to a silencing complex. HDACs 4, 5, and 7 can be phosphorylated by protein kinases, attract a 14-3-3 phosphoserine-binding protein partner (Chapter 2), and can be shuttled to the cytoplasm where they will be kept away from the chromatin and so become nonfunctional. HDAC isoform selectivity may be intrinsic for particular acetylLys sequences or may be imposed by where partner proteins recruit the HDACs.

Sirtuins

The sirtuin class of gene-silencing proteins was named for the prototype yeast Sir2 protein, where "Sir" stands for "silent information regulator." The five yeast and seven human sirtuin protein isoforms are all thought to function catalytically as

NAD-dependent histone deacetylases. The oxidized form of the NAD coenzyme is cosubstrate and is consumed stoichiometrically for every lysine side chain deacetylated. This connects histone deacetylation to the energy charge of cells and phenotypically Sir genes are related to long life and aging in yeast (Guarente and Kenyon, 2000). The coproducts are the deacetylated lysine side chain in the histone, nicotinamide, and an acetylated derivative of ADP-ribose, 2′-O-ADP-ribose. The initially formed 2′-O-acetyl ester of ADP-ribose equilibrates to the 3′-O-ester with a half-life of under an hour, but it is the 2′-O-acetyl-ADP-ribose that is the sirtuin product (Denu, 2003; Jackson and Denu, 2002; Sauve and Schramm, 2003). The Sir2 enzyme fragments NAD at an early stage of the reaction to generate the ribooxacarbenium ion, an electrophile strong enough to capture weak nucleophiles. The N-acetyllysine side chain of a histone is bound in the Sir active site (Avalos et al., 2002; Min et al., 2001) and the weakly nucleophilic acetyl oxygen captures the carbenium ion at C1 to generate a peptidyl-O-alkyl imidate (Figure 6.10A). Neighboring group participation of the 2-OH of the ribose on the imidate will lead to elimination of the H_2N–lysine–histone product and yield a cyclic 1,2-acyloxonium ion. Its capture by water and directed decomposition yields the 2′-O-acetyl-ADP-ribose. The x-ray structure of a yeast sirtuin yHst2 has been solved in complex with a histone peptide and the product 2′-O-acetyl-ADP-ribose (Zhao et al., 2003), and complexes of Sir2 with acetylated peptides and NAD (Figure 6.10B) suggest substrate binding tunnels and coupled conformation changes that provide structural clues to this redox-coupled protein deacetylation process (Avalos et al. 2004).

Why do sirtuins and HDACs both exist? Given that acetyllysine deacetylation is a thermodynamically favored reaction mediated by the zinc-dependent HDACs in a straightforward amide bond hydrolysis, why spend metabolically valuable NAD to also do this job by sirtuins? Sirtuins are important genetically for establishing and maintaining silencing via hypoacetylation of H3 and H4 tails (Imai et al., 2000). They have roles in the life-span and maintenance of genomic stability (Brachmann et al., 1995; Lin et al., 2000). Sirtuins are recruited into complexes, two major ones in yeast, which lead to the silencing of gene expression at telomeres and also at the ribosomal DNA locus and specifically keep histones hypoacetylated in this region of chromatin (Grozinger and Schreiber, 2002). This may require specific partners available to sirtuins but not HDACs. It is also possible that NAD, as a monitor of energy balance, is coupled mechanistically in some as yet unknown manner. Finally, there has been speculation (but no evidence) that the nicotinamide or 2′-O-acetyl-ADP-ribose fragments may have some special signaling role to play.

Sirtuins also deacetylate acetyllysine side chains in tubulins (North et al., 2003) and in bacterial acetyl CoA synthetase (Starai et al., 2003). Intracellular NAD levels, or the nicotinamide metabolite, could act to regulate Sir2 and homologs (Sauve and

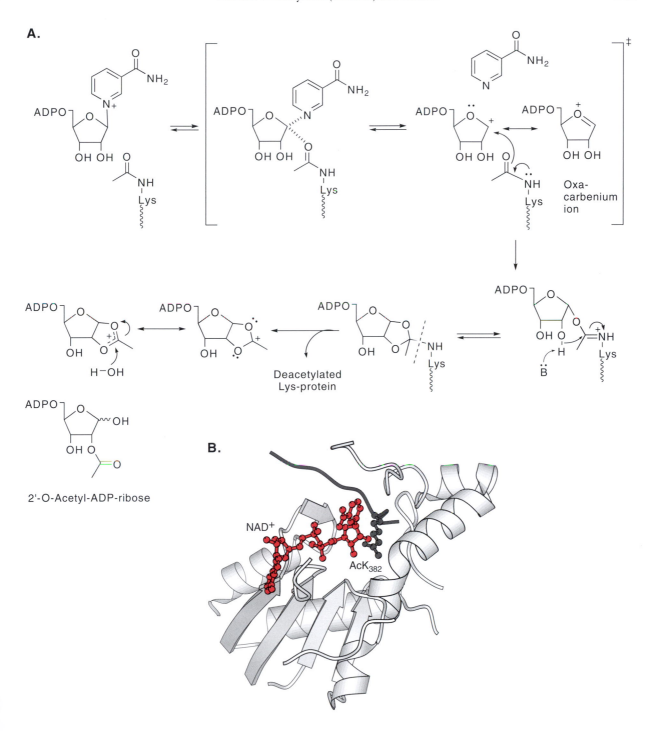

Figure 6.10 **A.** Fragmentation of NAD to the ribooxacarbenium ion and generation of a peptidyl-O-alkyl imidate that decomposes to free lysine side chain and 2′-O-acetyl-ADP-ribose. **B.** Structure of yeast sirtuin yHst2 in complex with the p53 C-terminal peptide containing N-acetyl-K_{382} and NAD (figure made using PDB 1MA3).

Schramm, 2003). Since bacterial genomes have Sir2 homologs but not histones, their functions may have been ancient and evolved from other protein deacetylation functions.

References

Avalos, J. L., J. D. Boeke, and C. Wolberger. "Structural basis for the mechanism and regulation of Sir2 enzymes," *Mol. Cell* **13**:639–648 (2004).

Avalos, J. L., I. Celic, S. Muhammad, M. S. Cosgrove, J. D. Boeke, and C. Wolberger. "Structure of a Sir2 enzyme bound to an acetylated p53 peptide," *Mol. Cell* **10**:523–535 (2002).

Berger, S. L. "Gene activation by histone and factor acetyltransferases," *Curr. Opin. Cell Biol.* **11**:336–341 (1999).

Brachmann, C. B., J. M. Sherman, S. E. Devine, E. E. Cameron, L. Pillus, and J. D. Boeke. "The SIR2 gene family, conserved from bacteria to humans, functions in silencing, cell cycle progression, and chromosome stability," *Genes Dev.* **9**:2888–2902 (1995).

Brown, C. E., T. Lechner, L. Howe, and J. L. Workman. "The many HATs of transcription coactivators," *Trends Biochem. Sci.* **25**:15–19 (2000).

Chen, L. F., and W. C. Greene. "Shaping the nuclear action of NF-kappaB," *Nat. Rev. Mol. Cell Biol.* **5**:392–401 (2004).

Cheung, W. L., S. D. Briggs, and C. D. Allis. "Acetylation and chromosomal functions," *Curr. Opin. Cell Biol.* **12**:326–333 (2000).

Denu, J. M. "Linking chromatin function with metabolic networks: Sir2 family of NAD(+)-dependent deacetylases," *Trends Biochem. Sci.* **28**:41–48 (2003).

Grozinger, C. M., and S. L. Schreiber. "Deacetylase enzymes: Biological functions and the use of small-molecule inhibitors," *Chem. Biol.* **9**:3–16 (2002).

Guarente, L., and C. Kenyon. "Genetic pathways that regulate aging in model organisms," *Nature* **408**:255–262 (2000).

He, S., D. Bauman, J. S. Davis, A. Loyola, K. Nishioka, J. L. Gronlund, D. Reinberg, F. Meng, N. Kelleher, and D. G. McCafferty. "Facile synthesis of site-specifically acetylated and methylated histone proteins: Reagents for evaluation of the histone code hypothesis," *Proc. Natl. Acad. Sci. U.S.A.* **100**:12033–12038 (2003).

Imai, S., C. M. Armstrong, M. Kaeberlein, and L. Guarente. "Transcriptional silencing and longevity protein Sir2 is an NAD-dependent histone deacetylase," *Nature* **403**:795–800 (2000).

Jackson, M. D., and J. M. Denu. "Structural identification of 2'- and 3'-O-acetyl-ADP-ribose as novel metabolites derived from the Sir2 family of beta-NAD$^+$-dependent histone/protein deacetylases," *J. Biol. Chem.* **277**:18535–18544 (2002).

Khorasanizadeh, S. "The nucleosome: From genomic organization to genomic regulation," *Cell* **116**:259–272 (2004).

Kouzarides, T. "Acetylation: A regulatory modification to rival phosphorylation?" *EMBO. J.* **19:**1176–1179 (2000).

Li, M., J. Luo, C. L. Brooks, and W. Gu. "Acetylation of p53 inhibits its ubiquitination by Mdm2," *J. Biol. Chem.* **277:**50607–50611 (2002).

Lin, S. J., P. A. Defossez, and L. Guarente. "Requirement of NAD and SIR2 for life-span extension by calorie restriction in *Saccharomyces cerevisiae*," *Science* **289:**2126–2128 (2000).

Marmorstein, R. "Protein modules that manipulate histone tails for chromatin regulation," *Nat. Rev. Mol. Cell Biol.* **2:**422–432 (2001).

Min, J., J. Landry, R. Sternglanz, and R. M. Xu. "Crystal structure of a SIR2 homolog-NAD complex," *Cell* **105:**269–279 (2001).

Mujtaba, S., Y. He, L. Zeng, S. Yan, O. Plotnikova, Sachchidanand, R. Sanchez, N. J. Zeleznik-Le, Z. Ronai, and M. M. Zhou. "Structural mechanism of the bromodomain of the coactivator CBP in p53 transcriptional activation," *Mol. Cell* **13:**251–263 (2004).

Ng, H. H., and A. Bird. "Histone deacetylases: Silencers for hire," *Trends Biochem. Sci.* **25:**121–126 (2000).

North, B. J., B. L. Marshall, M. T. Borra, J. M. Denu, and E. Verdin. "The human Sir2 ortholog, SIRT2, is an NAD$^+$-dependent tubulin deacetylase," *Mol. Cell* **11:**437–444 (2003).

Pesavento, J. J., Y.-B. Kim, G. K. Taylor, and N. L. Kelleher. "Prescriptive annotation of histone modifications: A new approach for streamlined characterization of proteins by top down mass spectrometry," *J. Am. Chem. Soc.* **126:**3386–3387 (2004).

Polevoda, B., and F. Sherman. "N-alpha-terminal acetylation of eukaryotic proteins," *J. Biol. Chem.* **275:**36479–36482 (2000).

Polevoda, B., and F. Sherman. "N-terminal acetyltransferases and sequence requirements for N-terminal acetylation of eukaryotic proteins," *J. Mol. Biol.* **325:**595–622 (2003).

Pugh, B. F. "Is acetylation the key to opening locked gates?" *Nat. Struct. Mol. Biol.* **11:**298–300 (2004).

Roth, S. Y., J. M. Denu, and C. D. Allis. "Histone acetyltransferases," *Annu. Rev. Biochem.* **70:**81–120 (2001).

Sauve, A. A., I. Celic, J. Avalos, H. Deng, J. D. Boeke, and V. L. Schramm. "Chemistry of gene silencing: The mechanism of NAD$^+$-dependent deacetylation reactions," *Biochemistry* **40:**15456–15463 (2001).

Sauve, A. A., and V. L. Schramm. "Sir2 regulation by nicotinamide results from switching between base exchange and deacetylation chemistry," *Biochemistry* **42:**9249–9256 (2003).

Starai, V. J., H. Takahashi, J. D. Boeke, and J. C. Escalante-Semerena. "Short-chain fatty acid activation by acyl-coenzyme A synthetases requires SIR2 protein function in *Salmonella enterica* and *Saccharomyces cerevisiae*," *Genetics* **163:**545–555 (2003).

Strahl, B. D., and C. D. Allis. "The language of covalent histone modifications," *Nature* **403:**41–45 (2000).

Thompson, P. R., D. Wang, L. Wang, M. Fulco, N. Pediconi, D. Zhang, W. An, Q. Ge, R. G. Roeder, J. Wong, M. Levrero, V. Sartorelli, R. J. Cotter, and P. A. Cole. "Regulation of the p300 HAT domain via a novel activation loop," *Nat. Struct. Mol. Biol.* **11:**308–315 (2004).

Turner, B. M. "Cellular memory and the histone code," *Cell* **111:**285–291 (2002).

Verdin, E., F. Dequiedt, and H. G. Kasler. "Class II histone deacetylases: Versatile regulators," *Trends Genet.* **19:**286–294 (2003).

Zhang, K., K. E. Williams, L. Huang, P. Yau, J. S. Siino, E. M. Bradbury, P. R. Jones, M. J. Minch, and A. L. Burlingame. "Histone acetylation and deacetylation: Identification of acetylation and methylation sites of HeLa histone H4 by mass spectrometry," *Mol. Cell Proteomics* **1:**500–508 (2002).

Zhao, K., X. Chai, and R. Marmorstein. "Structure of the yeast Hst2 protein deacetylase in ternary complex with 2'-O-acetyl ADP ribose and histone peptide," *Structure (Cambridge)* **11:**1403–1411 (2003).

Protein Lipidation

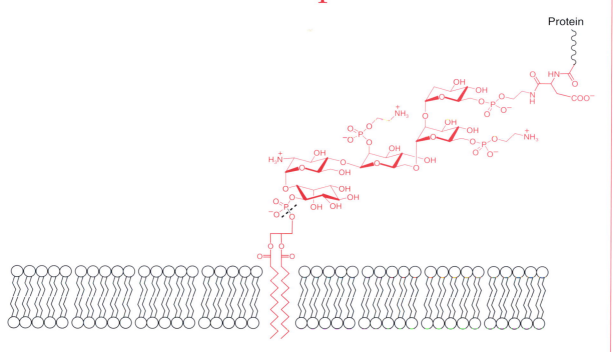

The GPI anchor attaches the C-terminus of a protein to membranes via an oligosaccharide–phospholipid tether.

A large number of proteins undergo covalent modifications with one or more lipid anchors that help to target the modified proteins to particular membranes (e.g., the endoplasmic reticulum, Golgi complex, lysosome, axon, dendrite, or plasma membrane) in eukaryotic cells. There are four major types of lipid anchors appended enzymatically to proteins by distinct modification strategies: one at the N-terminus (N-myristoylation), one at the C-terminus [glycosyl phosphatidylinositol (GPI) anchor], and two at cysteine thiolates, proximal to membrane surfaces (S-acylation and S-prenylation) (Bijlmakers and Marsh, 2003; Glomset et al., 1990). S-Acylations can occur at cysteines at any point in the protein, while prenylations are found at cysteines typically within four residues of the carboxyl terminus.

In the myristoylation and palmitoylation of proteins, the donors of the acyl groups are myristoyl CoA (C_{14}) and palmitoyl CoA (C_{16}), respectively. They are captured by nucleophiles on the protein undergoing modification, the N-terminal amine of Gly_1 in myristoylation, and the thiolate of one or more cysteinyl residues for S-palmitoylation (Figures 7.1A and B). (There are some rare cases of transfers of long chain acyl groups to Ser and Thr hydroxyl side chains.) Cysteine thiolate side chains are also the

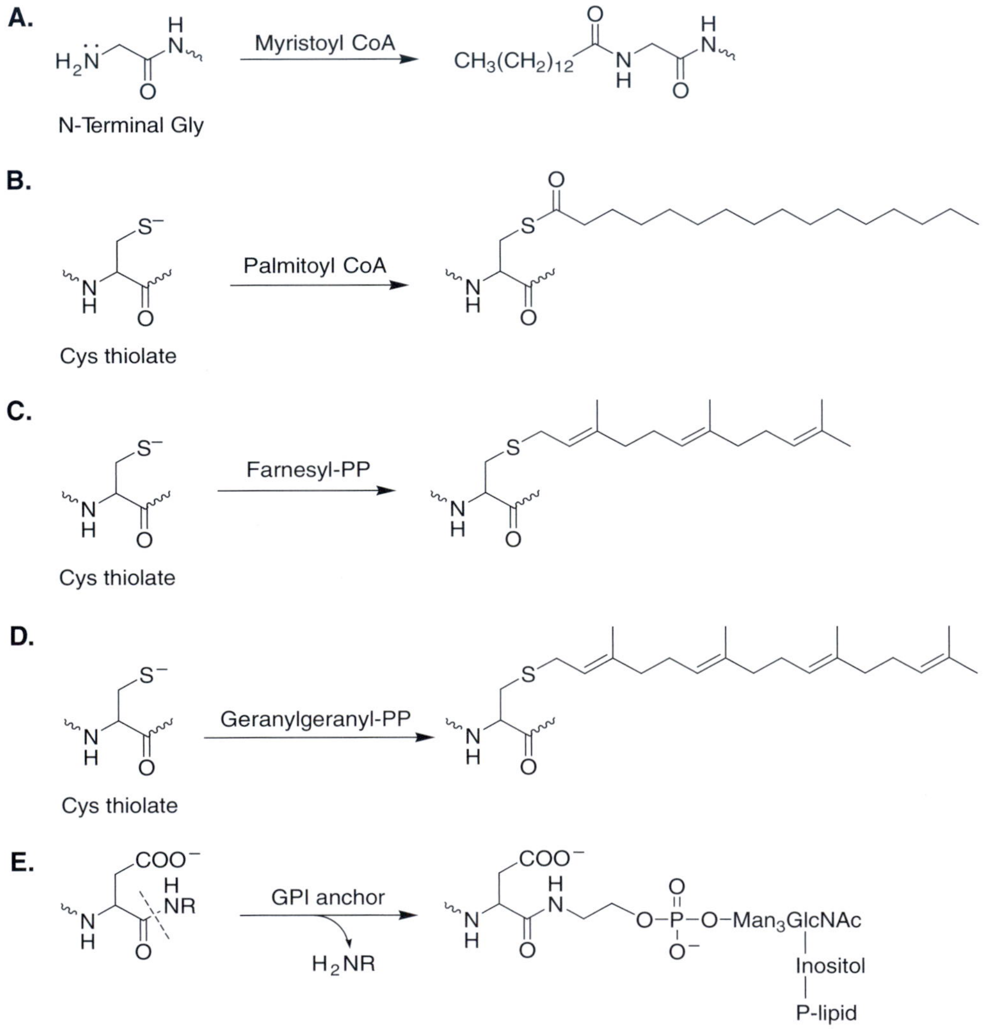

Figure 7.1 Lipid groups added posttranslationally to proteins. **A.** Myristoylation of the N-terminal glycine residue. **B.** Thiol ester formation by palmitoylation of cysteine thiolate side chains. **C.** Formation of C_{15} farnesyl thioether. **D.** Formation of C_{20} geranylgeranyl thioether to Cys residues. **E.** Attachment of a GPI anchor to the C-terminal acyl group of proteins by the amino group of the ethanolamine portion of the GPI anchor.

nucleophiles in prenyl transfers of C_{15} (farnesyl) and C_{20} (geranylgeranyl) chains from the corresponding farnesyl-PP and geranylgeranyl-PP lipids (Figures 7.1C and D), with displacement of the PP_i moiety and attachment of carbon-1 of the farnesyl and geranylgeranyl chains to the attacking thiolate to form thioether linkages. The GPI anchor comprises a phospholipid [phosphatidylinositol (PI)] coupled to a tetrasaccharide chain ending in a phosphoethanolamine. The terminal NH_2 of the ethanolamine moiety is the attacking nucleophile in the posttranslational modification, a reverse of polarity from the above three examples, where the incoming lipid is the electrophilic partner in the modification reaction (Figure 7.1E). The protein receiving the GPI anchor thus functions as the acyl donor, in a net **transamidation** sequence: breaking an amide (peptide) bond in the protein near the C-terminus as the ethanolamine-glycolipid anchor forms the new C-terminal amide.

A distinct protein lipidation that is a hybrid of the reactions in Table 7.1 occurs at N-terminal cysteine thiol groups of bacterial membrane proteins. The N-terminal cysteine in the modified protein is an internal cysteine in the preprotein precursor that is liberated to become the new N-terminus by hydrolytic cleavage of the upstream peptide bond during membrane transit by bacterial signal peptidase action (see Chapter 8). The N-terminal signal peptide is cleaved by signal peptidase II (Paetzel et al., 2002) on the periplasmic side of the cytoplasmic membrane after enzymatic transfer of a diacylglyceride (from phosphatidylglycerol) to the Cys at what will be the downstream side of the cleavage site (Figure 7.2). The attack of the cysteine thiolate displaces the phosphoglycerol head from the phospholipid donor and generates a thioether linkage to the residual diacylglycerol moiety (Sankaran and Wu, 1994). A subsequent N-acylation (analogous to N-myristoylation) of the N-terminal NH_2 of the DAG–S–Cys_1 residue then ensues, creating a lipoprotein with three acyl chains on the first Cys residue. Bioinformatics analysis of the signal peptidase II sites predicts hundreds of such lipoproteins in Gram-negative eubacteria (Bairoch and Apweiler, 2000; Juncker et al., 2003): 101 for *E. coli*, 186 for *Pseudomonas aeruginosa*, and 103 in the Gram-positive bacterium *Bacillus subtilis*. Site-specific posttranslational proteoly-

Table 7.1 Four Types of Lipid Modification Strategies

Strategies	Outcome
C-Terminal addition of GPI anchors (P-lipid pentasaccharylamine) via **transamidation** (nucleophile on anchor)	Amide link
N-Terminal **addition of myristoyl** electrophile from myristoyl CoA to N-terminal Gly	Amide link
S-Acylation, largely by palmitoyl CoA on Cys thiolates close to membrane interfaces	Thioesters are labile
S-Prenylation of cysteines, usually two to four residues from C-terminus	Thioether linkage stable

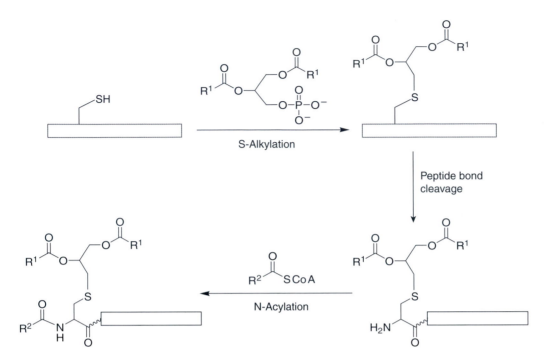

Figure 7.2 Bacterial S,N-lipoprotein formation. Attachment of diacylglycerol to cysteine thiolate followed by proteolysis and N-acylation to create Cys_1 with three acyl groups attached via N- and S-substituents.

sis and covalent attachment of a carbohydrate group also occurs at the other end, the C-terminus, of surface proteins in the Gram-positive *Staphylococcus aureus*. Enzymes called sortases (Pallen et al., 2001) cleave surface preproteins at an LPXT sequence near the C-terminus and mediate a net transamidation to the Gly_5 cross-bridge of the peptidoglycan layer of the bacterial cell wall (Figure 7.3A).

An additional variant of a C-terminal modification of a protein during proteolytic cleavage, in analogy to GPI-anchor attachment, where the protein chain is the acyl donor and the lipid is the nucleophile that becomes attached to the C-terminus, is in the formation of the protein–cholesterol ester arising on autocleavage of Hedgehog proteins discussed in Chapter 13 (Figure 7.3B) (Porter et al., 1996). The purpose of the lipidation, in this case with a sterol, is to tether the protein product for two-dimensional diffusion in the plane of the plasma membrane. There may be more examples to discover. The Wnt-1 protein, known previously to be S-palmitoylated, has also been proposed to be covalently modified by ganglioside, although the structure is yet to be determined (Willert et al., 2003).

Three of the four lipid modifications in Table 7.1 are posttranslational. The N-myristoylation reactions are cotranslational, while the nascent peptide chain is still attached to the ribosomes and before folding has progressed. All four of the lipida-

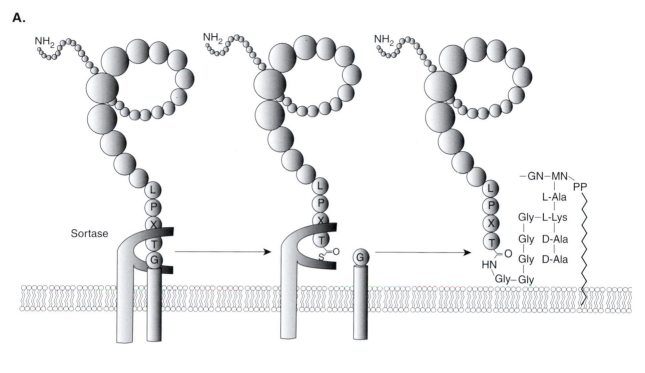

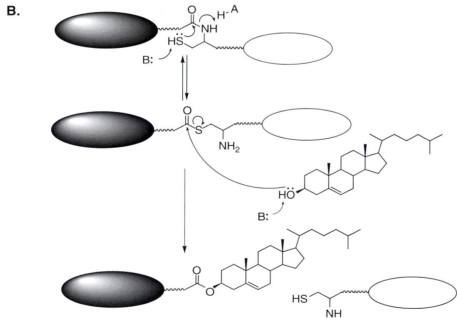

Figure 7.3 C-Terminal covalent lipidation of proteins. A. Sortase-mediated transamidation and transfer of outer membrane bacterial proteins to the glycine bridge of peptidoglycan molecules. B. Hedgehog-mediated formation of a protein–cholesterol ester via cleavage of the single-chain Hedgehog and transfer of the amino terminal fragment to the 3-OH of cholesterol.

tions increase the hydrophobicity of the modified protein and increase affinity for the membranes. There are multiple examples with two or more types of modifications on the same protein, including GPI anchor and palmitoylation (sonic Hedgehog), and myristoylation and palmitoylation [epithelial nitric oxide synthase (eNOS) and the T cell protein tyrosine kinases such as Lck]. The $G\alpha$ subunits of trimeric GTPase enzymes can be myristoylated and palmitoylated, with two or more modifications required for stable association of a cytoplasmic protein with membrane. The G protein γ subunits can be prenylated, typically with the C_{20} geranylgeranyl group and less frequently with the C_{15} farnesyl group.

Glycosyl Phosphatidylinositol (GPI) Anchored Proteins

Many cell surface proteins in lower and higher eukaryotes (Englund, 1993; Udenfriend and Kodukula, 1995) contain GPI anchors that tether them to the outer layer of the plasma membrane (Mayor, 2004). These include the prion protein (Rudd et al., 1999), lymphoid cell surface proteins, folate receptors, and hydrolytic enzymes such as 5'-nucleotidase. Trypanosomatids can have their whole surface covered by one GPI-anchored protein, the variant surface glycoprotein, which protects against immune surveillance by the infected host (Englund, 1993). The mammalian ADP-ribosyltransferases, ART1 and ART2, are also GPI-linked (Okazaki and Moss, 1998), as they act to ADP-ribosylate cell surface proteins such as integrin-α_7. GPI-anchored proteins can also serve as receptors for bacterial cytolytic toxins such as aerolysin from *Aeromonas hydrophila* and clostridial α-toxin (Hong et al., 2002), which allows them to bind to plasma membranes and become proteolytically activated there. There are consensus sequences in eubacterial proteins that suggest GPI anchoring (Eisenhaber et al., 2001).

Proteins that contain GPI anchors have these anchors attached in the ER where the preformed GPI anchor has been built up, and then the GPI protein transits to the cell surface. The dwell time at the cell surface can be controlled by internalization and/or by release of the protein into the extracellular milieu by enzymatic hydrolysis of the lipid anchor by phosphatidylinositol-specific phospholipase C (Figure 7.4). This sloughing of the protein moiety by PLC activity is a unique attribute of GPI-anchored surface proteins. Experimentally, release of a cell surface protein on PI-specific PLC treatment is taken to validate GPI anchoring. The prion protein PrP^C is internalized (Muniz and Riezman, 2000) from cell surfaces where it sits in sphingolipid microdomains, known as lipid rafts, which may be important for protein clustering.

The prior assembly of a glucosyl-mannosyl$_3$ tetrasaccharide chain on a phospholipid scaffold is reminiscent of the *en bloc* preassembly of the tetradecasaccharyl lipid in N-glycosylations discussed in Chapter 10. Indeed, parallel logic is involved in the step-by-step elongation of the GPI chain on the ER membrane by a series of mem-

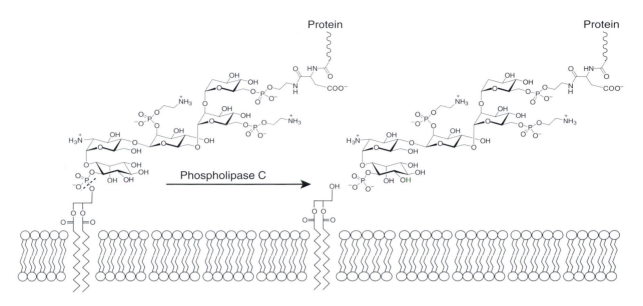

Figure 7.4 GPI-anchored proteins: hydrolytic release of the protein moiety to the extracellular space by action of phosphatidylinositol-specific phospholipase C, leaving the diacylglycerol in the membrane.

brane-associated glucosyl- and mannosyltransferases (Kinoshita and Inoue, 2000). The three mannosyl residues are provided by dolichol-P-mannose, the same donor of the last four mannosyl residues in the $Glc_3Man_9GlcNAc_2$ chain buildup (Maeda et al., 2000).

The initial phospholipid scaffold is provided by phosphatidylinositol, a minor phospholipid, but one crucial in signaling cascades (Berridge, 1993). The PI lipid molecule in the ER membrane is then elaborated by a series of GPI anchor biosynthetic enzymes (Kinoshita and Inoue, 2000), starting with attack by the C6-OH of the inositol ring on the GlcNAc moiety of UDP-GlcNAc to build the disaccharyl-PI in the ER membrane (Figure 7.5, first step). This is followed by deacetylation of the GlcNAc and palmitoylation of the 2-OH of the inositol moiety, exchanging a short-chain acetyl on the glucose for a long-chain C_{16} on the inositol (Figure 7.5, second step). This disaccharyl-PI is subject to mannosylation by dolichol-P-mannose in an α-1,4-linkage. This first mannose is then decorated with phosphoethanolamine, arising by attack of the 2-OH of the mannosyl moiety on the phospholipid phosphatidylethanolamine, releasing the diacylglycerol. This is a harbinger of a subsequent phosphoethanolamine transfer at the last step of GPI anchor assembly.

The second and third mannosyl residues are then enzymatically transferred from another two molecules of Dolichol-P-mannose in α-1,6- and α-1,2-regiospecific linkages to build the full-length tetrasaccharyl-inositol-phospholipid chain (third step of Figure 7.5). The remaining modification is to install one (or two, with the second at C_6 of the just added mannose) more phosphoethanolamine groups at position 6 of the terminal mannose to put in place the amine nucleophile that will participate in the protein transamidation. It is estimated that some 20 gene products are involved in the 10–11 steps of GPI anchor *preassembly* in the ER membrane (Kinoshita and Inoue, 2000) in yeast, trypanosomatid parasites, and humans. As in protein N-glycosylation

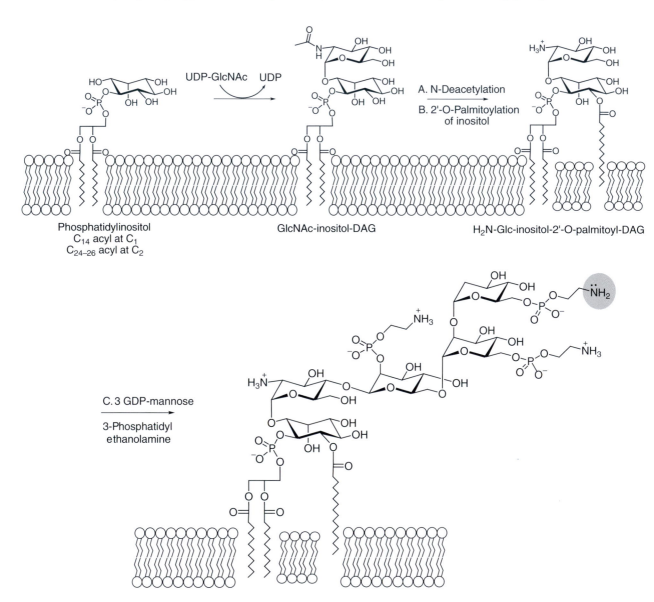

Figure 7.5 Biosynthesis of the GPI anchor: generation of the disaccharyl-PI in the endoplasmic reticulum is followed by **A.** N-deacetylation and **B.** 2′-O-palmitoylation of the inositol ring. **C.** Tris-mannosylation and ethanolamine decoration involves addition of the third and fourth mannosyl residues to build the full length tetrasaccharyl chain of the GPI anchor; installation of the third P-ethanolamine moiety brings in the attacking nucleophile for attachment to the C-terminus of proteins.

tetradecasaccharyl lipid assembly (Chapter 10), flipping between the cytoplasmic and luminal sides of the ER membrane is proposed during GPI anchor assembly.

The attachment of the mature GPI to the C-terminus of proteins to be covalently modified also occurs in the ER. Nascent proteins destined for GPI capture actually have two signal sequences. The first is at the N-terminus and directs the new protein into the ER where the N-terminal signal sequence is cleaved by signal peptidase activity in transit. The second signal sequence is about 10 residues from the C-terminus, where a DAA sequence is recognized in many human proteins and DSS in trypanosome variant surface glycoproteins. The transamidase is multicomponent, is membrane associated, and functions like a cysteinyl protease (Meyer et al., 2000; Ohishi et al., 2001; Ohishi et al., 2000), making an acyl-S-enzyme intermediate (Figure 7.6) with release of the C-terminal peptide. The acyl-S-transamidase is captured specifically by the GPI anchor as cosubstrate with the terminal ethanolamine-P-mannosyl residue acting as amine nucleophile. Amide bond formation installs the GPI anchor at the new C-terminus of the protein and effects an isoenergetic net transamidation. A comparable protein cleavage reaction at a particular peptide bond near the C-terminus occurs in the reactions catalyzed by the bacterial sortase enzymes (Navarre and Schneewind, 1999). Sortases also generate covalent peptidyl-S-enzyme intermediates (Zong et al., 2004) and transfer the protein acyl fragment not to water, but to a cosubstrate amine, in this case provided by the peptidyl cross-bridges of the peptidoglycan layers of Gram-positive bacteria such as *Staphylococcus aureus*.

In analogy with the N-glycoprotein pathway, where there is hydrolytic trimming of the core oligosaccharide in the ER after transfer to the proteins, there are editing steps in GPI-protein maturation (Figure 7.7). One is the removal of the palmitoyl group on the inositol ring (Chen et al., 1998). The fatty acyl chains on the diacylglycerol moiety of the PI lipid can be heterogenous. In GPIs they appear biased toward saturated chains. The second editing step is the addition of extra sugars to the mannose$_3$ core of the GPI anchor.

For example, the GPI anchor of the prion protein could be analyzed after peptidase cleavage and isolation of the GPI anchor attached to the C-terminal peptide. Six resolvable GPI anchors were detected, one of which contained the trisaccharide sialic acid-Gal-GalNAc attached to the first mannose (Stahl et al., 1992). Other forms lacked the sialic acid or the sialic acid-Gal disaccharide or had a fourth mannose residue. Thus, glycosyltransferase-mediated maturation of GPI anchors has analogies to the maturation of the Man$_5$GlcNAc$_2$ core of N-glycan chains of glycoproteins. The heterogeneity of prion protein then encompasses the 52 variants of the two N-glycan chains and the six variants in the GPI anchor. Another GPI-linked protein in neurons, Thy1, has distinct modifications to the mannose core of the GPI anchor, with a GalNAc added to Man$_1$ and another mannose attached to Man$_3$ (Rudd et al., 2002).

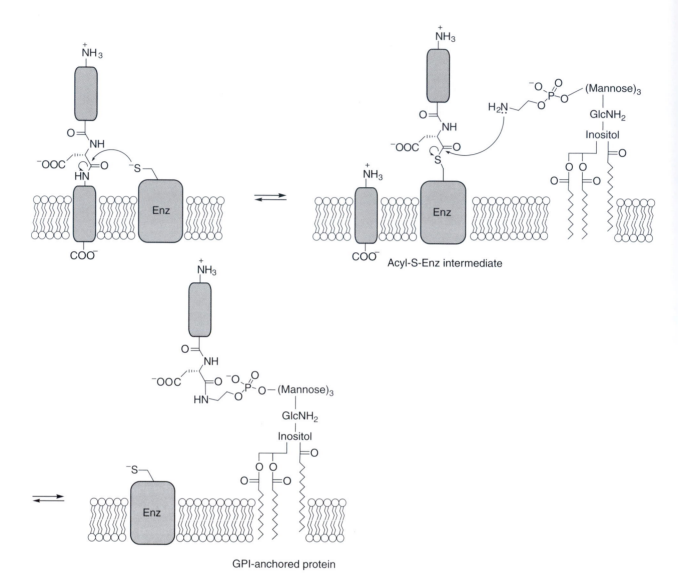

Figure 7.6 Mechanism of the transamidase connecting the GPI anchor to proteins. The acyl-S-enzyme intermediate leads to amide bond formation by acyl transfer to the ethanolamine terminus of GPI.

Transit through the Golgi complex requires cargo receptor proteins in the secretory pathway (Kinoshita and Inoue, 2000). Some free GPI anchors also transit to cell surfaces. Recall from above that the GPI anchors may predispose such modified proteins to associate with cholesterol and sphingolipid rafts as part of the channeling to plasma membranes. Protein sorting may also go on in the lipid rafts as they transit through the secretory pathway (Muniz and Riezman, 2000).

In yeast, GPI-anchored proteins can become covalently attached to β-1,6-glucan strands in the cell wall, like the bacterial sortase-mediated cross-linking of proteins to

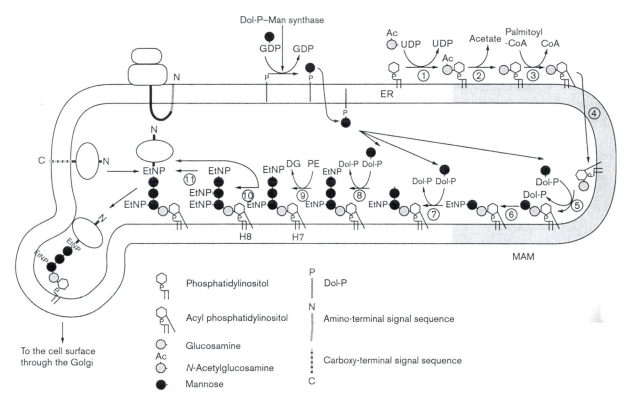

Figure 7.7 Assembly of the preformed GPI anchor (steps 1–10) in the endoplasmic reticulum membranes before transamidative coupling (step 11) to the C-terminus of proteins that then transit through the Golgi complex to the cell surface (reproduced with permission from Kinoshita and Inoue, 2000).

wall components noted earlier (Mazmanian et al., 2001). There are variations in fungi and trypanosomatid elaboration of the GPI anchors, such as the inositol acylation step, which may offer therapeutic targets.

The beginnings of a proteomics approach to the experimental characterization of the subproteome of GPI-anchored proteins has recently been reported (Elortza, 2003). It is based on the prior fractionation of membrane proteins, followed by the release of GPI-anchored proteins back into the aqueous phase by treatment with phosphatidylinositol-specific phospholipase C. The proteins thus freed of their lipid anchoring were concentrated, analyzed by denaturing gel electrophoresis, and subjected to peptide identification by mass spectrometry. From a lipid-raft-enriched human membrane fraction, six of 13 proteins were GPI-anchored. From *Arabidopsis thaliana* plant membranes, 44 (of some 248 predicted in the *Arabidopsis* genome) proteins were identified that fit predicted algorithms for GPI anchoring. Proteomic approaches that target both myristoylation sites and GPI sites have been described (Eisenhaber et al., 2004).

Protein N-Myristoylation

Covalent acylation of proteins with the C_{14} myristoyl group occurs on eukaryotic proteins and some envelope proteins of viruses that replicate in eukaryotic cells on N-terminal glycyl residues (McLaughlin and Aderem, 1995; Resh, 1999). The modification catalyst N-myristoyltransferase (NMT) has been purified from fungal and higher eukaryotic sources and studied for structure and mechanism. NMT catalyzes acyl transfer from myristoyl CoA as the donor to the nucleophilic N-terminal amine of Gly_1 in protein substrates (Figure 7.1) (Johnson et al., 1994). While the more abundant (by 5- to 20-fold) C_{16} palmitoyl CoA will bind to NMT, there is no detectable palmitoyl transfer. In some cases a small amount of C_{12} and unsaturated $C_{14:1}$ and $C_{14:2}$ olefinic acyl chains get transferred by NMT, but usually the acyl CoAs are at low abundance and heterogeneity is low. The specificity for Gly at the N-terminus is absolute and the x-ray structure (Farazi, 2001; Johnson et al., 1994) of NMT complexes rationalizes why no R group other than hydrogen can be accommodated at C_α and accounts for extended site selectivity at GXXXS/T N-terminal sequences.

The N-myristoylation of protein substrates is cotranslational, occurring while the nascent chain is emerging from the peptide exit tunnel in the large ribosomal subunit and before folding has occurred. Because all eukaryotic protein synthesis generates N-terminal methionyl residues, the precursor proteins have MGXXXS/T N-terminal sequences. Before Gly_2 can be N-myristoylated, the Met_1-Gly_2 peptide bond must be cleaved by methionine aminopeptidases (MAP1 or MAP2), also acting cotranslationally (Figure 7.8A). Thus, there are two cotranslational modifications that go in tandem: MAP hydrolysis to liberate the $Gly–NH_2$ as the new amino terminus. Subsequent N-myristoylation by NMT occurs in rapid succession before the nascent protein folds (Figure 7.8B). In proteins with both myristoylation (at Gly_1) and S-palmitoylation (e.g., GAP43 in neurons at Cys_3), palmitoylation occurs after myristoylation.

Among the many classes of N-myristoylated proteins are protein kinases and protein phosphatases (Resh, 1999) (also see Chapter 2). These include Src family tyrosine kinases (Src, Fyn, Lck, and Blk) and others where association of these cytoplasmic enzymes with membranes is relevant to function, and Abl tyrosine kinase. The most well-known protein serine/threonine kinase that gets N-myristoylated is the cAMP-dependent protein kinase A, which gets N-myristoylated on the catalytic subunit (Johnson et al., 2001). The A kinase anchoring protein AKAP18 is also N-myristoylated. The pSer/pThr protein phosphatase calcineurin B is in this category as is a yeast protein phosphatase.

Guanine nucleotide binding proteins with slow GTPase activity are likewise acylated at Gly_1, including $G\alpha_{i1}$ and $G\alpha_o$, as well as several of the ADP-ribosylation factors (Arf-1, -3, -5, -6). Ca^{2+} binding proteins such as recoverin, and neurocalcin, along with membrane- and cytoskeleton-affiliated proteins such as MARCKS and

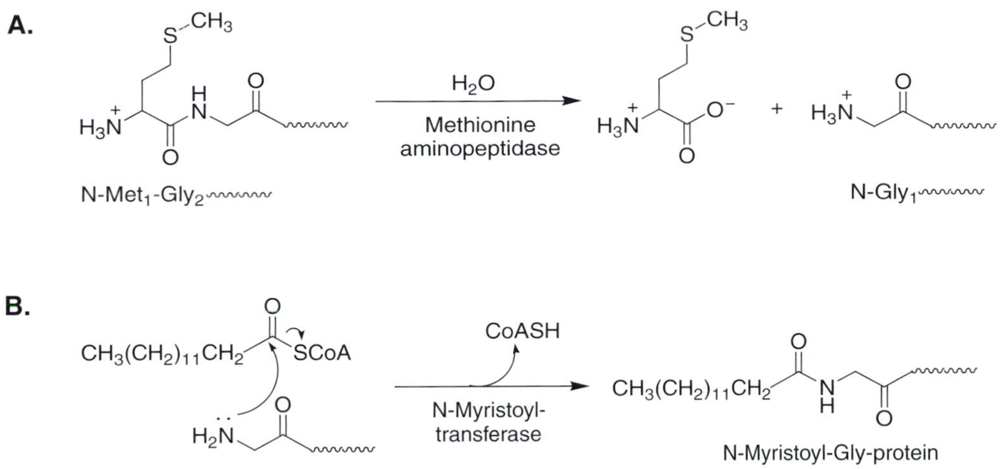

Figure 7.8 Two-stage modification to create myristoylated protein N-termini. **A.** Cleavage of Met_1-Gly_2 bonds by methionine aminopeptidase. **B.** Myristoylation of the newly liberated N-terminal Gly by N-myristoyltransferase.

AnnexinXII, are cotranslationally myristoylated, as are the NADH cytochrome b5 reductase and the signal transduction enzyme nitric oxide synthase (NOS) (Resh, 1999). Viral Gag proteins that are modified include HIVC-1, SIV-1, Moloney murine sarcoma virus, and Friend murine leukemia virus, as well as the HIV-1 Nef protein.

The purpose of N-terminal Gly acylation is presumed to be the addition of a hydrophobic handle that can assist protein association with membrane surfaces. In this context the C_{14} myristoyl chain may have been selected over the more abundant C_{16} palmitoyl chain to titrate the strength of the interaction; the shorter myristoyl chain has less energy of interaction with the membranes and the proteins may have faster "on–off" rates than they would if they were N-palmitoylated. (There are a few rare N-palmitoylated proteins, but these are thought to arise by S-to-N acyl shifts via S-palmitoyl-Cys_2 initial adducts).

It has also been proposed that myristoylated proteins can switch between conformations where the C_{14} hydrophobic acyl group is exposed and can interact with membrane lipids and conformations where the myristoyl chain is sequestered internally within the modified protein and not available for membrane binding (McLaughlin and Aderem, 1995). The x-ray structure of the PKA catalytic subunit shows that the fatty acyl chain is well ordered within a groove in the protein, presumably the membrane-inaccessible state (Johnson et al., 2001) (Figure 7.9A). It can be well-ordered, as in polio VP4, where the myristoyl group is a part of the virion subunit structure; also, Arf-1 (Goldberg, 1998) and recoverin (Ames et al., 1997) have grooves proposed to bind the myristoyl chain. It is proposed that many other N-terminally myristoylated proteins have the chain extended on the protein surface (see recoverin in Figure 7.9B) and disordered in the absence of membranes. In the cAbl kinase, the N-myristoyl group

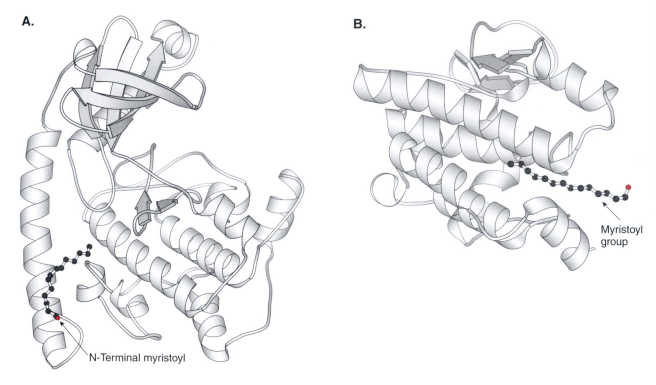

Figure 7.9 Location of covalent myristoyl groups in proteins. **A.** Location of the myristoylated N-terminus of protein kinase A in an internal groove of the protein (figure made using PDB 1CMK). **B.** Myristoyl tail in recoverin (figure made using PDB 1OPJ).

binds in the kinase domain and causes autoinhibition (Nagar et al., 2003). The two posttranslational modifications, phosphorylation of Tyr and N-myristoylation of Gly_1, have been proposed (Hantschel et al., 2003) to control kinase activity.

Evidence has accumulated that the N-myristoyl hydrophobic tail is one signal for membrane association, but that efficient membrane association is driven by a second signal, either an adjacent S-palmitoyl group (see next section below) or by clusters of positively charged side chains. The "myristate plus basic residues" (Resh, 1999) is proposed to have the myristoyl insertion synergized with electrostatic attraction of negatively charged phospholipids head groups with the cationic side chains in the vicinity of the myristoyl-Gly_1 residue. Src has six basic side chains in the first 16 residues, enhancing phospholipid binding by 3000-fold (Resh, 1999). Equivalent basic motifs are found in HIV-1 Gag and Nef proteins, and in MARCKS.

Resh (1999) has summarized three types of myristoyl switches that could drive conformations such that the myristoyl group was available or unavailable to interact with membranes. The first is ligand-mediated myristoyl switching, exemplified by calcium-ion-binding to recoverin, which ejects the myristoyl group from its sequestered site (Ames et al., 1997) in a hydrophobic pocket, exposing it and bringing

Ca^{2+}–recoverin to the ROS membrane (Figure 7.10A), where it inhibits rhodopsin kinase. In Arf-1 with GDP bound, the myristoylated N-terminal helix is in a shallow hydrophobic groove. Replacement of GDP by GTP extrudes a loop, breaking down the pocket, expelling the myristoyl group and leading to subsequent membrane binding (Amor et al., 1994; Goldberg, 1998). The second proposed switch inducer is electrostatic (Figure 7.10B), exemplified by the phosphorylation of MARCKS by protein

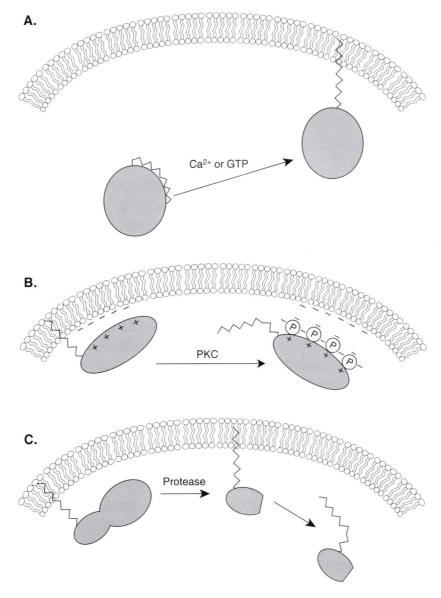

Figure 7.10 Three types of proposed conformational switches driven by myristoyl groups at protein N-termini. **A.** Reorientation of the N-myristoyl group driven by Ca^{2+} ions. **B.** Electrostatic repulsion of a myristoylated protein. **C.** Proteolytic cleavage to reorient N-myristoylated protein.

kinase C (Newton, 2001). This introduces negative charges from the phosphates, which balances the cationic side chain charges and weakens the electrostatic interaction with the anionic phospholipid head groups in the membranes, leading to dissociation of MARCKS from the membranes. The third switch proposed is proteolytic, as exemplified by HIV Gag-1. Gag is translated as a 55-kDa polyprotein (Pr55Gag) and then cleaved by the HIV protease into fragments prior to viral assembly (Figure 7.10C). The N-terminal 17-kDa fragment (p17MA) bearing the N-myristoyl group binds less tightly to membranes than Pr55Gag, presumably a protease-triggered myristoyl switch [but see Tang et al. (2004) for a view of a preequilibrium for exposed and buried myristoyl conformations]. The release of p17MA provides a cytosolic 17-kDa matrix protein for the preintegration complex that translocates to the nucleus (Bukrinskaya et al., 1992).

Protein S-Palmitoylation

The S-acylation of one or two cysteine thiolates in proteins by palmitoyl chains [or occasionally by stearoyl (C_{18}), oleoyl ($C_{18:1}$), or even arachidonyl ($C_{20:4}$) chains] is a pervasive posttranslational modification in eukaryotic cells. The acyl donor is palmitoyl CoA, the next higher homolog of myristoyl CoA, a major product from cellular fatty acid synthases. The ionized thiolate side chain of cysteine residues is the most reactive side chain nucleophile in proteins and the acyl transfer is an energy-neutral transthiolation. To date there is no reliable sequence in protein substrates that predicts S-palmitoylation. Cysteine residues in protein domains that interact with membranes are the targets, reflecting the membrane locale of both palmitoyl CoA and the palmitoyl acyl transferases (PATs). The PATs have been very recalcitrant to solubilization and purification. One catalyst that has been characterized is a palmitoyltransferase that is acting on the Hedgehog proteins and has been named skinny Hedgehog (Chamoun et al., 2001; Lee et al., 2001). In yeast the palmitoyltransferase operating on Ras has been identified by both genetic and biochemical criteria (Lobo et al., 2002; Roth et al., 2002). Traditionally, palmitoylation has been assayed with radioactive palmitoyl CoA, making it difficult to get good stoichiometry. The N-terminal peptide GAP43 has been recently evaluated by mass spectrometry after digestion with trypsin. It still contains Met_1, which is N-acetylated to 100%, providing an internal standard for the bis-palmitoylation at Cys_3 and Cys_4, which was about 30%. This fractional stoichiometry of modification was accompanied by heterogeneity of acylation, with both $C_{18:0}$ stearoyl and $C_{18:1}$ oleoyl groups present at low fractional abundance (el-Husseini Ael and Bredt, 2002).

The generic reaction is shown in Figure 7.1B and, in contrast to protein myristoylation (Figures 7.1A and 7.8B), produces not an acyl amide, but an acyl thioester modification (Bijlmakers and Marsh, 2003) that can go on and come off reversibly. While NMT acts cotranslationally at the N-terminal Gly$_1$ of protein substrates, PATs act posttranslationally and the cysteines can be near the N- or C-termini or at internal regions of the protein. This is illustrated in Figure 7.11. Figure 7.11A shows integral membrane proteins, from single transmembrane (TM) proteins such as the CD8 α-subunit to the seven TM β$_2$-adrenergic receptors that get S-acylated near the transmembrane domain/cytoplasmic domain (TMD/CD) boundary. Figure 7.11B shows the single TM HIV-1 Env protein and the seven TM CCR5 chemokine receptors that are palmitoylated at cysteines distal from the TMD/CD boundary. Figure 7.11C shows cytosolic protein examples, Lck and Gα$_z$, which are both N-myristoylated and S-palmitoylated in the N-terminal region. Figure 7.11D shows a different composition of dual lipidation—namely, palmitoylation and prenylation at the C-terminus of N-Ras and H-Ras. Finally, Figure 7.11E shows two cytosolic proteins where the dual

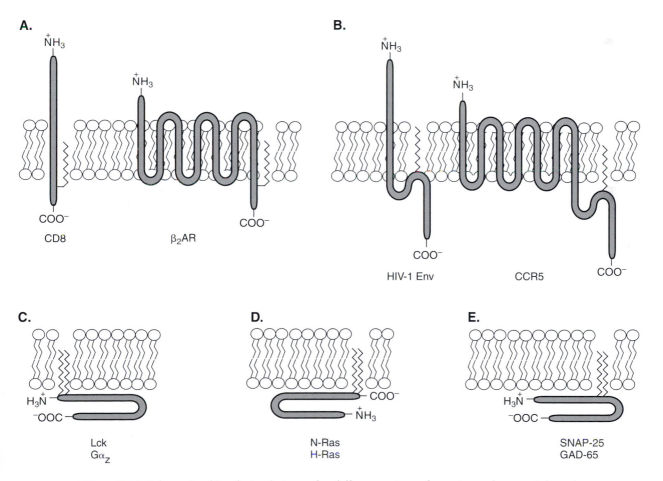

Figure 7.11 Schematic of S-palmitoylation at five different regions of protein cosubstrates (adapted from Bijlmakers and Marsh, 2003). Mono- and dipalmitoylations are shown.

lipidation is bis-palmitoylation, in the central domains of the neuronal proteins SNAP-25 and GAD-65 (glutamate decarboxylase, which generates the neurotransmitter GABA) (Bijlmakers and Marsh, 2003).

For the cytosolic proteins in Figures 7.11C–7.11E, the membrane localization effects of the two lipid chains is an important device for determining a stable membrane cellular address. Of the three lipid modifications (myristoyl, palmitoyl, and prenyl), only the palmitoyl thioester is hydrolytically labile and reversible by thioesterases. Thus, palmitoylation is a dynamic modification and can switch "on and off" the association of cytosolic proteins with specific membranes in cells. Studies on the mannose-6-P protein receptor, which moves 6-P-mannosylated N-glycoproteins to the lysosomal compartments (see Chapter 6), indicted that the two palmitoyl moieties, on Cys_{30} and Cys_{34}, turned over with a two-hour half-life, while the protein turned over with a 40-hour half-life, strongly suggesting selective depalmitoylation, perhaps to control localization or affinity (Schweizer et al., 1996).

When two different types of lipid chains are present in the same proteins, the palmitoylation occurs last. Recall from above that N-myristoylation may be insufficient to obtain stable association of a protein with membranes and can require a subsequent palmitoylation. The N-myristoyl protein is doubtless a better substrate for PATs because of the transient association to the membrane interface where they reside, provided by the N-terminal acyl chain. Likewise, prior prenylation at the C-terminus, as in Ras proteins, increases the probability that the protein will be in the vicinity of membranous PATs for the second modification.

A brief survey of palmitoylated proteins encompasses G-protein-coupled receptors (β_2-adrenergic receptor and α_2-adrenergic receptor, and CCR5), ion channels (Na^+ channel α subunit), protein kinases (Src, Lck, Fyn, and Hck), the $G\alpha$ subunits of heterotrimeric GTPases ($G\alpha_o$, α_z, α_{11}), epithelial nitric oxide synthase, viral envelope proteins, and both CD4 and CD8α. The Src family kinases and $G\alpha$ subunits are typically myristoylated and palmitoylated while some $G\alpha$ subunits are also prenylated to exemplify all three lipid modifications (Figure 7.12) (Bijlmakers and Marsh, 2003).

Several studies have indicated that the C_{16} saturated palmitoyl chain modification selectively concentrates palmitoylated proteins in cholesterol- and sphingolipid-rich lipid rafts (Waheed and Jones, 2002). These rafts can ferry protein cargoes differentially to specific membrane destinations. These rafts can also be preconcentration platforms for assembly of intermediates and complexes such as CDK, CD4, and CD8α on the way to T cell plasma membranes, or for HIV envelope protein sorting for viral assembly, or for transporting neuronal proteins to axons (GAP43) or dendrites (PSD95).

A recent analysis of the occurrence and function of palmitoylated proteins in neurons suggests that palmitoylation is a ubiquitous posttranslational modification that is

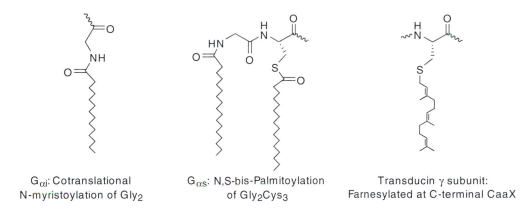

Figure 7.12 Three lipid modifications on some forms of the small GTPases: $G\alpha_i$ N-myristoylation, $G\alpha_s$ N,S-bis-palmitoylation, and transducin γ-S-prenylation.

Table 7.2 S-Palmitoylated Neuronal Proteins

Family	Protein
GPCRs	β_2AR, α_2AR, and $5HT_{4a}R$
Ion channels	Glu6R, nAChR, and Na^+ channel α subunit
Synaptic vesicle proteins	Synaptotagmin-1, synaptobrevin-2, SNAP-25, and GAD-65
Scaffolding proteins	PSD95, Ca^{2+} channel β_{2A} subunit, and AKAP-18
Signaling proteins	$G\alpha_1$, GRK6, RGS4, H-Ras, Fyn, RhoB, and GAP43

central to neuronal function and development (el-Husseini Ael and Bredt, 2002). First is the long list of proteins found both presynaptically and postsynaptically that are palmitoylated. On the presynaptic side these are proteins regulating neurotransmitter release. On the postsynaptic side both ion channels and neurotransmitter GPCRs are palmitoylated, as are key scaffolding proteins that organize complexes both of receptors and their G protein signaling partners, as noted in Table 7.2.

el-Husseini Ael and Bredt (2002) suggest palmitoylation and depalmitoylation of the above proteins and others regulate synaptic transmission. This includes synaptic vesicle fusion, neurotransmitter biosynthesis and release on the presynaptic side, and modification of subunits for both the sodium channels and the voltage-dependent calcium channel. Axonal growth cones contain GAP43 and NCAM140. On the postsynaptic side are the palmitoylated GPCRs (such as the metabotropic mGluR4) and ion channels (such as the kainate receptor GluR6). Palmitoylated PDZ domain proteins include the postsynaptic density proteins PSD95 and PSD93, which interact with NMDA and AMPA receptors to control their localization. The palmitoylation content of all these neuronal proteins may contribute to and regulate sorting, localization, and protein–protein interactions that contribute to neuronal function and development.

Protein S-Isoprenylation

The fourth type of modification that specifically delivers a lipid anchor to proteins is the posttranslational addition of C_{15} or C_{20} isoprenyl groups to the thiolate anion of cysteine side chains within five residues of the C-termini of the protein substrates. The C_{15} chain is a farnesyl unit and the C_{20} is a geranylgeranyl unit, derived from farnesyl-1-diphosphate and geranylgeranyl-1-diphosphate, respectively (Figure 7.13). Isoprenyl chains are built up from the five-carbon building blocks of Δ^2- and Δ^3-isoprenyl-PP by elongations of one isoprenyl group at a time. Thus the C_{10} geranyl-PP arises from one coupling, the C_{15} farnesyl-PP by two elongations, and the C_{20} geranylgeranyl-PP by three elongations. The dolichol-PP groups in yeast and higher eukaryotic N-glycosylation tetradecasaccharyl assemblies, and in GPI-anchor assemblies arise by 12–20 C_5-elongation steps. In cholesterol biogenesis two farnesyl-PP groups are condensed enzymatically head-to-head to yield the C_{30} squalene that is then cyclized and rearranged to cholesterol. The C_{15} and C_{20} isoprenyl-PP groups

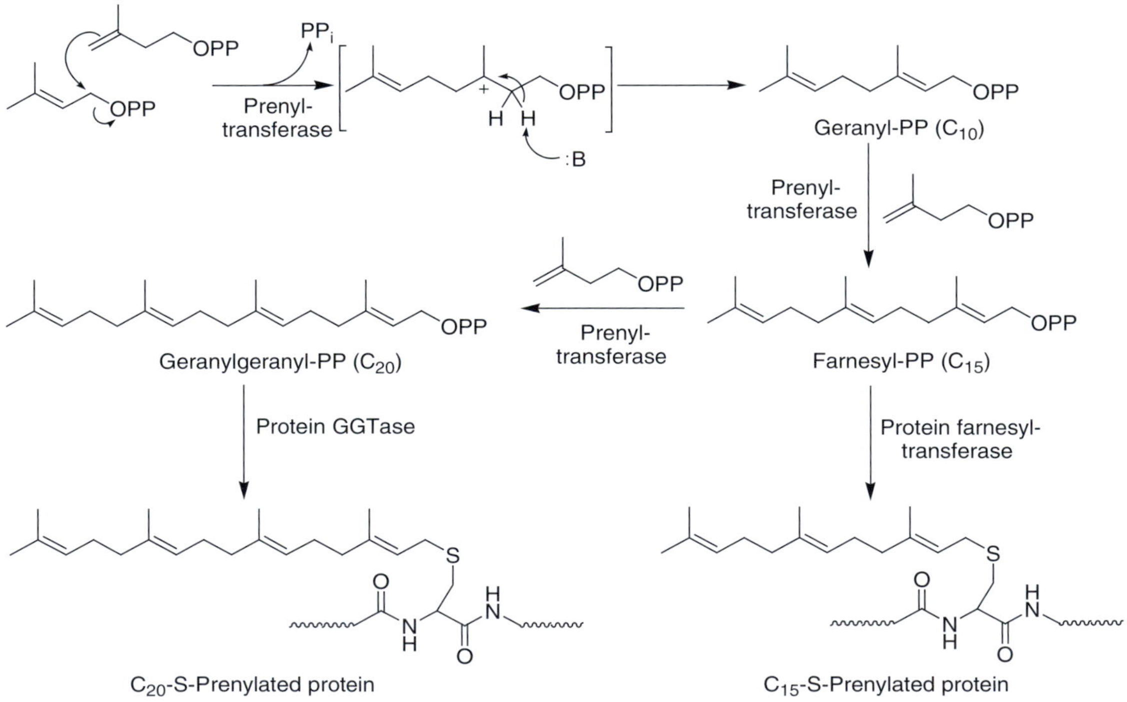

Figure 7.13 Biogenesis of the C_{15} farnesyl-PP and C_{20} geranylgeranyl-PP isoprenyl group donors from C_5 Δ^2- and Δ^3-isoprenyl-PP building blocks.

are thus normal membrane constituents in cells and available both in the ER membranes and partition into the cytoplasm for conscription by the posttranslational modification enzymes.

Early farnesylated proteins to be discovered were sex pheromones from jelly fungi (Glomset et al., 1990), yeast mating type *a* factor, and nuclear lamin (Zhang and Casey, 1996). Interest rose when it became clear that members of the Ras superfamily of small GTPases (Ras, Rho, Rac, and Arf, but not Ran) were prenylated, generally with the C_{20} prenyl chain. Also, rhodopsin kinase and the γ subunit of transducin in the retina are prenylated, but with the C_{15} farnesyl group, presumably reflecting specificity of the protein prenylation enzymes (Figure 7.12). The multisubunit retinal phosphodiesterase is farnesylated on one subunit and geranylgeranylated on another, making the retinal proteome rich in posttranslational prenylations.

The Rab family in the Ras superfamily of GTPases constitutes the largest set (greater than 60 members) (Pereira-Leal and Seabra, 2000) and they are exclusively geranylgeranylated, usually on two adjacent cysteine side chains at the C-terminus of these 20- to 25-kDa proteins. It has been estimated that 2% of eukaryotic proteins are prenylated (Nalivaeva and Turner, 2001), with about 150 protein substrates defined (Roskoski, 2003).

On further examination, the Ras GTPase superfamily members divided into two cysteine sequons that become prenylated. One set has a CaaX motif at the C-terminus. When X is small, such as Ala, Ser, or Met, the C residue is farnesylated, as in Ras itself (Figure 7.14). When X is a terminal Leu residue, as in the Rac and RhoA GTPases, the prenyl donor is the C_{20} isoprenyl-PP and the cysteine thiolate is geranylgeranylated (Pereira-Leal et al., 2001; Zhang and Casey, 1996). The farnesylation reaction is carried out by protein farnesyltransferase (FT), while the C_{20} prenylation is carried out by geranylgeranyltransferase I (GGT-I). G-Protein-coupled receptor kinases, other than the one for rhodopsin noted above and retinal cyclic GMP phosphodiesterases, fall in this CaaL category and are geranylgeranylated.

The second set, found in all of the more than 60 Rab isoforms, has two cysteines, arranged at the carboxy terminus as XXCC, XCCX, CCXX, CCXXX, or XCXC. Both cysteines get geranylgeranylated. The two C_{20} lipid anchors are introduced tandemly by GGT-II (Figure 7.14) to produce the doubly prenylated products.

X-ray structures of FTase (Park et al., 1997), GGT-I (Taylor et al., 2003), and GGT-II (Zhang et al., 2000) have been determined, showing a binding pocket for the isoprenyl-PP in a central cavity of the β subunit of both transferases (Figure 7.15). All three transferases are α,β-heterodimers with a zinc-containing active site in the β subunit. The FT and GGT-I actually share the identical α subunit. The related but distinct β subunits impose specificity for the C_{15} versus C_{20} isoprenyl donor and for the X residue in the CaaX motif of the protein to be farnesylated or geranylgera-

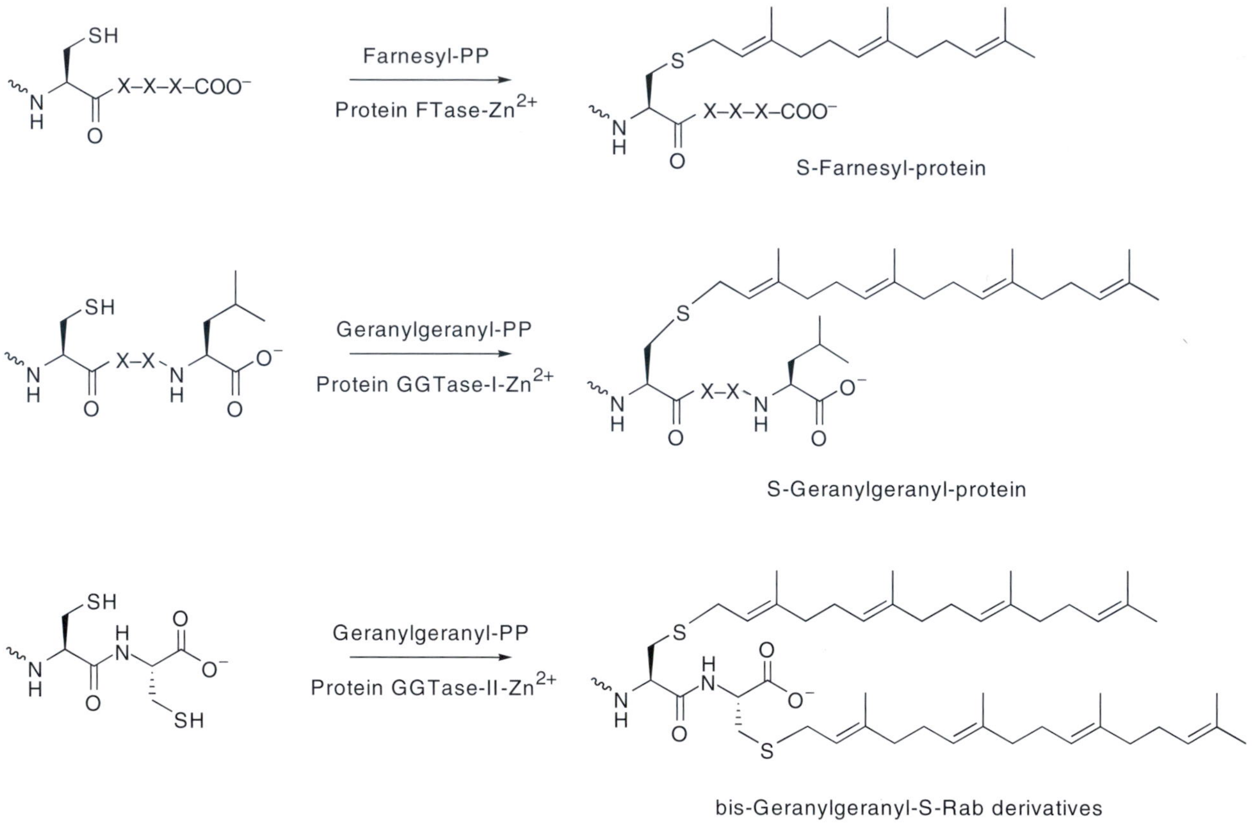

Figure 7.14 Prenylation of Ras superfamily GTPase members at distinct cysteine sequons. CaaX motifs can be farnesylated or geranylgeranylated when X = Leu, whereas bis-geranylgeranylation of CC motifs can occur at the C-termini of Rab proteins.

nylated. The GGT-II enzyme is also related to GGT-I, but has some distinctive substrate recognition properties, as noted below. The prenyltransferases use the active site zinc to coordinate the reactive Cys–S$^-$ of the protein substrate and catalyze thioether bond formation with inversion of configuration at C_1 of the isoprenyl moiety. The transition state may utilize the electrophilicity of a partial allylic cation at C_1 of the farnesyl and geranylgeranyl moieties, but C_1-OPP bond cleavage is incomplete according to the S_N2 stereochemical outcome (Figure 7.16).

Unlike the protein–thioester linkages that arise on S-palmitoylation, the protein–thioether linkages from S-prenylations are stable hydrolytically. Oxidation of the sulfur atom in the thioether could labilize the Cys–S–OH for elimination and this appears to be the mechanism for a prenylcysteine lyase (Digits et al., 2002) thought to act during protein degradation in lysosomes. It is presumed that the most prenylated proteins are catabolized proteolytically with the S-prenyl linkages still intact (Zhang

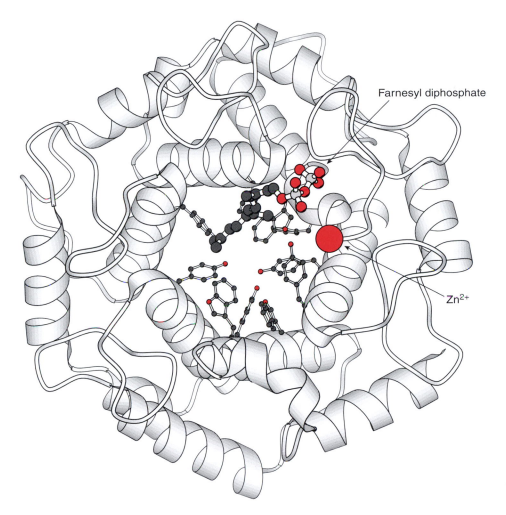

Farnesyl diphosphate

Zn²⁺

Figure 7.15 Architecture of the common β subunit of FTase and GGTase with a bound prenyl-PP ligand (figure made using PDB 1KZO).

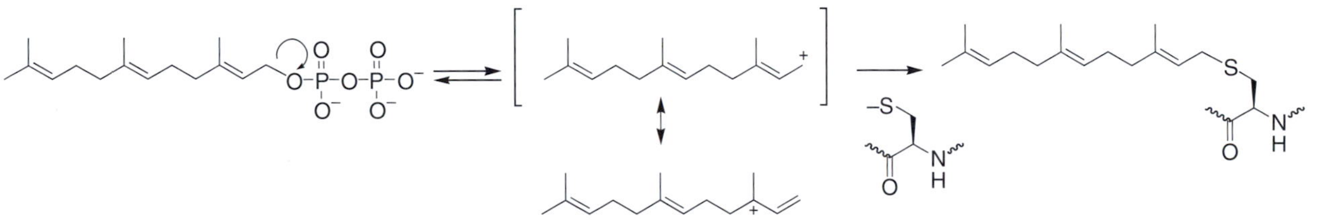

Figure 7.16 Prenyltransferase mechanism: transfer of an incipient allylic cation at C1 of the migrating prenyl unit.

and Casey, 1996). Thus, S-prenylation is thought generally to be an irreversible biological switch in the functions of the intact, folded proteins that get modified.

Recall the dual lipid anchor pattern in the analysis of S-palmitoylated proteins targeting to membranes, where a single long-chain acyl group brought about transient association with membranes. The same appears to be true for the cytoplasmic small GTPases that get prenylated. One farnesyl residue does not provide permanent association with membranes. Subsequent palmitoylation at a nearby cysteine occurs in H-Ras (at Cys_{181} and Cys_{184}) and N-Ras to drive the equilibrium toward membrane association. Alternatively, a cluster of adjacent basic residues can provide electrostatic attraction to the anionic phospholipid head groups as noted above for myristoylated proteins. Some Gα subunits have two of the three lipid anchors: N-myristoyl and S-palmitoyl, but not S-prenyl substituents. The bis-geranylgeranylation of Rab proteins does impart strong membrane targeting to the C-terminal region.

The farnesylated-CaaX and geranylgeranylated-CaaX Ras families of GTPases undergo two additional posttranslational processing steps to finish their maturation. The first is endoproteolytic cleavage by a membrane protease (hRce1) specific for the prenylcysteine side chain, releasing the aaX tripeptide and placing the prenylcysteine as the new carboxy terminus with a COOH group (Figure 7.17). The new free carboxylate of the prenylcysteine is then methylated by the SAM-dependent carboxymethyltransferase pcCMT, also an ER-membrane enzyme, to yield the final processed Ras protein. In total, three consecutive posttranslational modifications reshape the C-terminus of CaaX proteins: prenylation (C_{15} or C_{20}), proteolysis, and O-methylation of the new C-terminus. The methylation is reminiscent of carboxylate methylation in the chemotaxis transmembrane receptors (Chapter 4). In this case, covering the C-terminal carboxylate negative charge by converting it to the more hydrophobic methyl ester increases protein binding by 10- to 100-fold to ER membranes (Seabra, 1998).

The Rab protein CC-bis-geranylgeranylation does not involve associated regional proteolysis. In Rab–CXC, but not Rab–XCC, termini carboxymethylation does occur. The GGT-II-mediated double prenylation of Rab proteins does have its own twists. When first isolated, GGT-II had a third protein component, later recognized as a Rab escort protein (REP1; there is a REP2, too). It turns out that Rab proteins on their own are not prenylated by GGT-II but must be presented by REP as a Rab–REP stoichiometric complex (Pylypenko et al., 2003). Thus, REP escorts newly translated and unprenylated Rabs to the GGT-II (Figure 7.18). In the Rab–REP–GGT-II complex two molecules of geranylgeranyl-PP are utilized, with the N-terminal cysteine being prenylated first and then the adjacent downstream Cys–S$^-$ undergoing alkylation before any dissociation of monoprenylated Rabs (Thoma et al., 2001). The doubly prenylated Rab is escorted away for the GGT-II

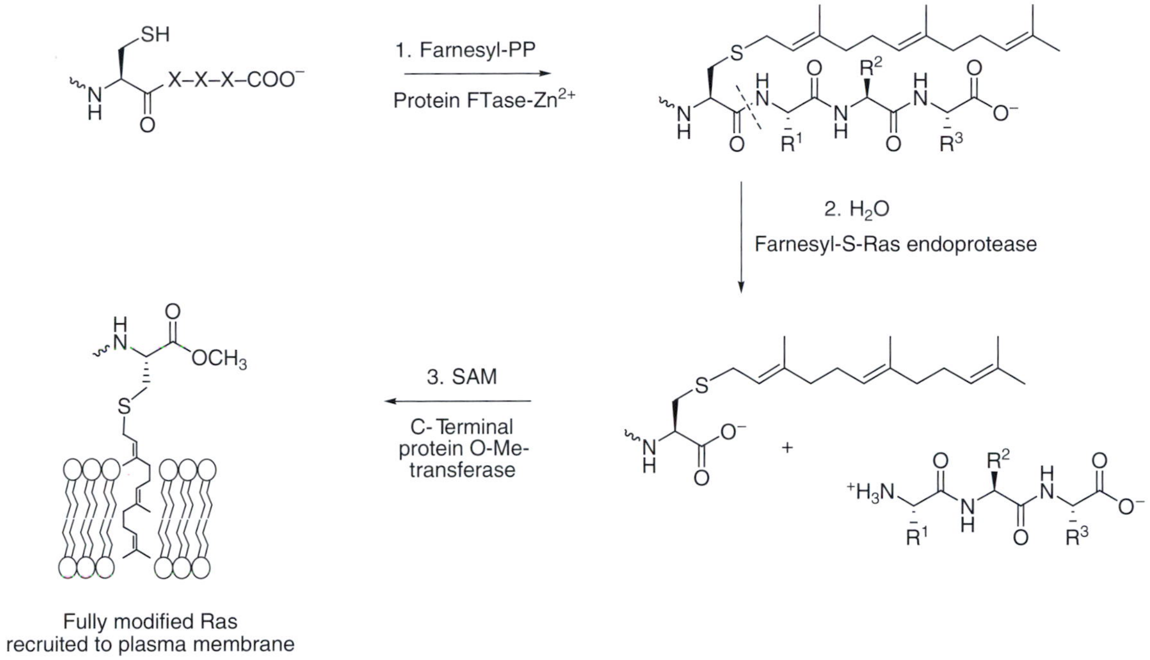

Figure 7.17 Further posttranslational processing of prenylated Ras. Endoproteolytic cleavage at the prenylated cysteine places it at the new C-terminus. SAM-dependent O-methylation caps the negatively charged C-terminal carboxylate as the neutral methyl ester.

active site, still bound to REP, which will deliver a given Rab to many different membrane regions in cells. Rab8 and Rab13 are exceptions in having only a single cysteine at the carboxy terminus, and so they function as monoprenylated proteins (Pereira-Leal and Seabra, 2000). The C_{20} isoprenoid chain, unlike a single C_{15} chain, appears to have enough hydrophobic volume to drive functional association of these Rab isoforms with membranes.

Prenylation can have effects on specific protein–protein interactions as well as protein-lipid membrane interactions (Casey and Seabra, 1996). For example, it is likely that farnesylation of yeast Ras increases its affinity for membranes, thereby presenting a higher effective concentration for interaction with its target enzyme adenyl cyclase by 100-fold (Kuroda et al., 1993), and farnesylation of K-Ras is required for interaction with its partner guanine nucleotide exchange protein SOS. Trimeric G$\alpha\beta\gamma$ GTPases assemble stepwise with $\beta\gamma$ assembling first in the cytosol. The β subunit appears to have equal affinity for γ whether γ is prenylated or not, but the subsequent assembly of α to preformed $\beta\gamma$ and reconstitution of the resting holo enzyme requires the γ subunit to be prenylated (Zhang and Casey, 1996).

The five dozen different Rabs are involved in spatially distinct vesicle formation and cargo transport in cells (Pereira-Leal et al., 2001; Pereira-Leal and Seabra, 2001).

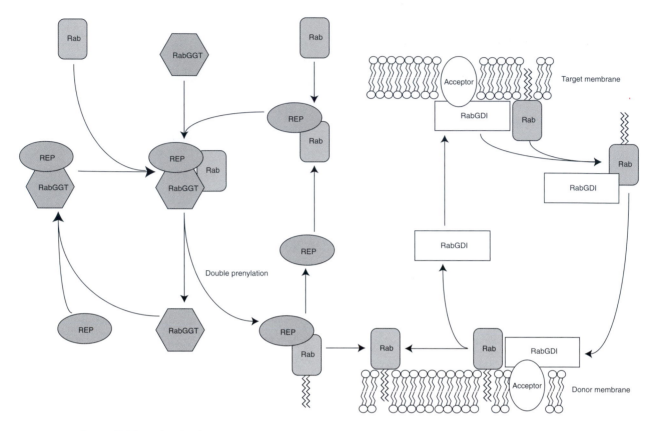

Figure 7.18 Role of Rab escort protein (REP) in presentation of Rab proteins to RabGGT for double geranylgeranyl thioether formation. This modification enables cycling of Rab proteins in GTP hydrolysis and in membrane association by interaction with the GDI partner protein to effect GDP release.

The prenylated Rabs bind to membranes, and recruit partner proteins that direct cargo protein binding and vesicle fusions as proteins move through the secretory pathway, exocytically and endocytically (Schimmoller et al., 1998). The Rabs are then recycled back to their starting membranes to begin another cycle. Particular Rabs localize in distinct subsets of intracellular membranes: Rab1 in the ER and Golgi complex for early steps in secretion, Rab5 in the plasma membrane and early endosomes for early endocytic pathway involvement, and Rab9 to the trans Golgi network where it regulates cycling of mannose-6-P receptors (Seabra, 1998). The GTPase activity of Rabs is essential for these functions.

All the Rab isoforms, and also Rac, Rho, and Ras GTPases, bind both GTP and GDP with subnanomolar K values, so Rabs perambulate in cells as GTP or GDP complexes. The GTP-Rab forms are active while the GDP-Rab forms are inactive for the vesicle and transport functions. The slow GTPase activity of Rabs determines the lifetime of the "on" state, and the switching is mediated by changes in loop regions of the protein between GTP- and GDP-bound states that interact differentially with partner proteins. The favorable thermodynamics of GTP hydrolysis allows the small G proteins to do work. When G-GDP forms accumulate on vesicles, they are extracted from the membrane milieu by another Rab chaperone called GDI (for GDP-dissociation inhibitor). The Rab-GDI complexes partition between cytoplasm and back to fusion-competent vesicles, their starting membrane microenvironment. GDIs have structural homology to REP (Rak et al., 2003; Wu et al., 1996), but while REPs function as initial escorts of Rabs to GGT-II, the GDIs bind only to prenylated Rabs and function in the continuous recycling of components required for semicontinuous vesicle trafficking. GDIs bind only to the GDP-bound forms of Rabs, illustrating how the GDP versus GTP ligands perturb Rab loop conformations for recognition of specific partners in one state or the other.

A prototypical Rab cycle (Figure 7.18, right-hand side) would have Rab-GTP on a donor membrane, fusing into vesicles and having the vesicles and associated cargos move to target membranes and fuse with them, under the aegis of SNARE protein machinery. The GTP-Rab converts to GDP-Rab in this target membrane fusion. Now Rab-GDP is retrieved and extracted out into the cytoplasm by GDI, traveling as the Rab-GDP GDI complex back to the donor membrane. In that membrane milieu, the Rab GEF proteins pry loose the bound GDP and allow it to be replaced by GTP to start the next cycle (Wu et al., 1996).

The bis-geranylgeranylation of Rabs creates sufficient membrane affinity that Rab proteins partition into membrane phases. The structural features that localize particular Rabs to particular subsets of cellular membranes remain to be sorted out. The REP and GDI escorts and recycling proteins represent variant chaperones for directing the location of Rabs before and after the posttranslational prenylations.

Prenylation of proteins in prokaryotes has been rarely reported. The main example to date is in a 43-residue protein KW30, with bacteriocidal activity, produced by *Lactobacillus plantarum* (Kelly et al., 1996). The terminal cysteine is in thioether linkage with a C_{15} farnesyl group, in analogy to the C-terminal placement of prenyl lipid anchors in eukaryotic proteins. The *Lactobacillus* farnesylation enzymatic machinery is as yet uncharacterized.

References

Ames, J. B., R. Ishima, T. Tanaka, J. I. Gordon, L. Stryer, and M. Ikura. "Molecular mechanics of calcium-myristoyl switches," *Nature* **389**:198–202 (1997).

Amor, J. C., D. H. Harrison, R. A. Kahn, and D. Ringe. "Structure of the human ADP-ribosylation factor 1 complexed with GDP," *Nature* **372**:704–708 (1994).

Bairoch, A., and R. Apweiler. "The SWISS-PROT protein sequence database and its supplement TrEMBL in 2000," *Nucleic Acids Res.* **28**:45–48 (2000).

Berridge, M. J. "Inositol trisphosphate and calcium signalling," *Nature* **361**:315–325 (1993).

Bijlmakers, M. J., and M. Marsh. "The on–off story of protein palmitoylation," *Trends Cell Biol.* **13**:32–42 (2003).

Bukrinskaya, A. G., G. K. Vorkunova, and Y. Tentsov. "HIV-1 matrix protein p17 resides in cell nuclei in association with genomic RNA," *AIDS Res. Hum. Retroviruses* **8**:1795–1801 (1992).

Casey, P. J., and M. C. Seabra. "Protein prenyltransferases," *J. Biol. Chem.* **271**:5289–5292 (1996).

Chamoun, Z., R. K. Mann, D. Nellen, D. P. von Kessler, M. Bellotto, P. A. Beachy, and K. Basler. "Skinny Hedgehog, an acyltransferase required for palmitoylation and activity of the Hedgehog signal," *Science* **293**:2080–2084 (2001).

Chen, R., E. I. Walter, G. Parker, J. P. Lapurga, J. L. Millan, Y. Ikehara, S. Udenfriend, and M. E. Medof. "Mammalian glycophosphatidylinositol anchor transfer to proteins and posttransfer deacylation," *Proc. Natl. Acad. Sci. U.S.A.* **95**:9512–9517 (1998).

Digits, J. A., H. J. Pyun, R. M. Coates, and P. J. Casey. "Stereospecificity and kinetic mechanism of human prenylcysteine lyase, an unusual thioether oxidase," *J. Biol. Chem.* **277**:41086–41093 (2002).

Eisenhaber, B., P. Bork, and F. Eisenhaber. "Post-translational GPI lipid anchor modification of proteins in kingdoms of life: Analysis of protein sequence data from complete genomes," *Protein Eng.* **14**:17–25 (2001).

Eisenhaber, B., F. Eisenhaber, S. Maurer-Stroh, and G. Neuberger. "Prediction of sequence signals for lipid post-translational modifications: Insights from case studies," *Proteomics* **4**:1614–1625 (2004).

el-Husseini Ael, D., and D. S. Bredt. "Protein palmitoylation: A regulator of neuronal development and function," *Nat. Rev. Neurosci.* **3**:791–802 (2002).

Elortza, F., T. S. Nuhse, L. J. Foster, A. Stensballe, S. C. Peck, and O. N. Jensen. "Proteomic analysis of GPI-anchored membrane proteins," *Mol. Cell. Proteomics* **2**:1261–1270 (2003).

Englund, P. T. "The structure and biosynthesis of glycosyl phosphatidylinositol protein anchors," *Annu. Rev. Biochem.* **62**:121–138 (1993).

Farazi, T. A., G. Waksman, and J. I. Gordon. "Structures of *Sacchromyces cerevisiae* N-myristoyltransferase with bound myristoyl CoA and peptide provide insights about substrate recognition and catalysis," *Biochemistry* **40**:6335–6343 (2001).

Glomset, J. A., M. H. Gelb, and C. C. Farnsworth. "Prenyl proteins in eukaryotic cells: A new type of membrane anchor," *Trends Biochem. Sci.* **15**:139–142 (1990).

Goldberg, J. "Structural basis for activation of ARF GTPase: Mechanisms of guanine nucleotide exchange and GTP-myristoyl switching," *Cell* **95**:237–248 (1998).

Hantschel, O., B. Nagar, S. Guettler, J. Kretzschmar, K. Dorey, J. Kuriyan, and G. Superti-Furga. "A myristoyl/phosphotyrosine switch regulates c-Abl," *Cell* **112**:845–857 (2003).

Hong, Y., K. Ohishi, N. Inoue, J. Y. Kang, H. Shime, Y. Horiguchi, F. G. van der Goot, N. Sugimoto, and T. Kinoshita. "Requirement of N-glycan on GPI-anchored proteins for efficient binding of aerolysin but not *Clostridium septicum* alpha-toxin," *EMBO J.* **21**:5047–5056 (2002).

Johnson, D. A., P. Akamine, E. Radzio-Andzelm, M. Madhusudan, and S. S. Taylor. "Dynamics of cAMP-dependent protein kinase," *Chem. Rev.* **101**:2243–2270 (2001).

Johnson, D. R., R. S. Bhatnagar, L. J. Knoll, and J. I. Gordon. "Genetic and biochemical studies of protein N-myristoylation," *Annu. Rev. Biochem.* **63**:869–914 (1994).

Juncker, A. S., H. Willenbrock, G. Von Heijne, S. Brunak, H. Nielsen, and A. Krogh. "Prediction of lipoprotein signal peptides in gram-negative bacteria," *Protein Sci.* **12**:1652–1662 (2003).

Kelly, W. J., R. V. Asmundson, and C. M. Huang. "Isolation and characterization of bacteriocin-producing lactic acid bacteria from ready-to-eat food products," *Int. J. Food Microbiol.* **33**:209–218 (1996).

Kinoshita, T., and N. Inoue. "Dissecting and manipulating the pathway for glycosylphosphatidylinositol-anchor biosynthesis," *Curr. Opin. Chem. Biol.* **4**:632–638 (2000).

Lee, J. D., P. Kraus, N. Gaiano, S. Nery, J. Kohtz, G. Fishell, C. A. Loomis, and J. E. Treisman. "An acylatable residue of Hedgehog is differentially required in *Drosophila* and mouse limb development," *Dev. Biol.* **233**:122–136 (2001).

Lobo, S., W. K. Greentree, M. E. Linder, and R. J. Deschenes. "Identification of a Ras palmitoyltransferase in *Saccharomyces cerevisiae*," *J. Biol. Chem.* **277**:41268–41273 (2002).

Maeda, Y., S. Tanaka, J. Hino, K. Kangawa, and T. Kinoshita. "Human dolichol-phosphate-mannose synthase consists of three subunits, DPM1, DPM2 and DPM3," *EMBO J.* **19**:2475–2482 (2000).

Mayor, S., Riezman, H. "Sorting GPI-anchored proteins," *Nat. Rev., Mol. Cell Biol.* **5**:110–120 (2004).

Mazmanian, S. K., H. Ton-That, and O. Schneewind. "Sortase-catalysed anchoring of surface proteins to the cell wall of *Staphylococcus aureus*," *Mol. Microbiol.* **40**:1049–1057 (2001).

McLaughlin, S., and A. Aderem. "The myristoyl-electrostatic switch: A modulator of reversible protein–membrane interactions," *Trends Biochem. Sci.* **20**:272–276 (1995).

Meyer, U., M. Benghezal, I. Imhof, and A. Conzelmann. "Active site determination of Gpi8p, a caspase-related enzyme required for glycosylphosphatidylinositol anchor addition to proteins," *Biochemistry* **39**:3461–3471 (2000).

Muniz, M., and H. Riezman. "Intracellular transport of GPI-anchored proteins," *EMBO J.* **19**:10–15 (2000).

Nagar, B., O. Hantschel, M. A. Young, K. Scheffzek, D. Veach, W. Bornmann, B. Clarkson, G. Superti-Furga, and J. Kuriyan. "Structural basis for the autoinhibition of c-Abl tyrosine kinase," *Cell* **112**:859–871 (2003).

Nalivaeva, N. N., and A. J. Turner. "Post-translational modifications of proteins: Acetylcholinesterase as a model system," *Proteomics* **1**:735–747 (2001).

Navarre, W. W., and O. Schneewind. "Surface proteins of gram-positive bacteria and mechanisms of their targeting to the cell wall envelope," *Microbiol. Mol. Biol. Rev.* **63**:174–229 (1999).

Newton, A. C. "Protein kinase C: Structural and spatial regulation by phosphorylation, cofactors, and macromolecular interactions," *Chem. Rev.* **101**:2353–2364 (2001).

Ohishi, K., N. Inoue, and T. Kinoshita. "PIG-S and PIG-T, essential for GPI anchor attachment to proteins, form a complex with GAA1 and GPI8," *EMBO J.* **20**:4088–4098 (2001).

Ohishi, K., N. Inoue, Y. Maeda, J. Takeda, H. Riezman, and T. Kinoshita. "Gaa1p and gpi8p are components of a glycosylphosphatidylinositol (GPI) transamidase that mediates attachment of GPI to proteins," *Mol. Biol. Cell* **11:**1523–1533 (2000).

Okazaki, I. J., and J. Moss. "Glycosylphosphatidylinositol-anchored and secretory isoforms of mono-ADP-ribosyltransferases," *J. Biol. Chem.* **273:**23617–23620 (1998).

Paetzel, M., A. Karla, N. C. J. Strynadka, and R. E. Dalbey. "Signal peptidases," *Chem. Rev.* **102:**4549–4579 (2002).

Pallen, M. J., A. C. Lam, M. Antonio, and K. Dunbar. "An embarrassment of sortases—a richness of substrates?" *Trends Microbiol.* **9:**97–102 (2001).

Park, H. W., S. R. Boduluri, J. F. Moomaw, P. J. Casey, and L. S. Beese. "Crystal structure of protein farnesyltransferase at 2.25 angstrom resolution," *Science* **275:**1800–1804 (1997).

Pereira-Leal, J. B., A. N. Hume, and M. C. Seabra. "Prenylation of Rab GTPases: Molecular mechanisms and involvement in genetic disease," *FEBS Lett.* **498:**197–200 (2001).

Pereira-Leal, J. B., and M. C. Seabra. "The mammalian Rab family of small GTPases: Definition of family and subfamily sequence motifs suggests a mechanism for functional specificity in the Ras superfamily," *J. Mol. Biol.* **301:**1077–1087 (2000).

Pereira-Leal, J. B., and M. C. Seabra. "Evolution of the Rab family of small GTP-binding proteins," *J. Mol. Biol.* **313:**889–901 (2001).

Porter, J. A., S. C. Ekker, W. J. Park, D. P. von Kessler, K. E. Young, C. H. Chen, Y. Ma, A. S. Woods, R. J. Cotter, E. V. Koonin, and P. A. Beachy. "Hedgehog patterning activity: Role of a lipophilic modification mediated by the carboxy-terminal autoprocessing domain," *Cell* **86:**21–34 (1996).

Pylypenko, O., A. Rak, R. Reents, A. Niculae, V. Sidorovitch, M. D. Cioaca, E. Bessolitsyna, N. H. Thoma, H. Waldmann, I. Schlichting, R. S. Goody, and K. Alexandrov. "Structure of Rab escort protein-1 in complex with Rab geranylgeranyltransferase," *Mol. Cell* **11:**483–494 (2003).

Rak, A., O. Pylypenko, T. Durek, A. Watzke, S. Kushnir, L. Brunsveld, H. Waldmann, R. S. Goody, and K. Alexandrov. "Structure of Rab GDP-dissociation inhibitor in complex with prenylated YPT1 GTPase," *Science* **302:**646–650 (2003).

Resh, M. D. "Fatty acylation of proteins: New insights into membrane targeting of myristoylated and palmitoylated proteins," *Biochim. Biophys. Acta* **1451:**1–16 (1999).

Roskoski, R., Jr. "Protein prenylation: A pivotal posttranslational process," *Biochem. Biophys. Res. Commun.* **303:**1–7 (2003).

Roth, A. F., Y. Feng, L. Chen, and N. G. Davis. "The yeast DHHC cysteine-rich domain protein Akr1p is a palmitoyl transferase," *J. Cell Biol.* **159:**23–28 (2002).

Rudd, P. M., T. Endo, C. Colominas, D. Groth, S. F. Wheeler, D. J. Harvey, M. R. Wormald, H. Serban, S. B. Prusiner, A. Kobata, and R. A. Dwek. "Glycosylation differences between the normal and pathogenic prion protein isoforms," *Proc. Natl. Acad. Sci. U.S.A.* **96:**13044–13049 (1999).

Rudd, P. M., A. H. Merry, M. R. Wormald, and R. A. Dwek. "Glycosylation and prion protein," *Curr. Opin. Struct. Biol.* **12:**578–586 (2002).

Sankaran, K., and H. C. Wu. "Lipid modification of bacterial prolipoprotein. Transfer of diacylglyceryl moiety from phosphatidylglycerol," *J. Biol. Chem.* **269:**19701–19706 (1994).

Schimmoller, F., I. Simon, and S. R. Pfeffer. "Rab GTPases, directors of vesicle docking," *J. Biol. Chem.* **273:**22161–22164 (1998).

Schweizer, A., S. Kornfeld, and J. Rohrer. "Cysteine34 of the cytoplasmic tail of the cation-dependent mannose 6-phosphate receptor is reversibly palmitoylated and required for normal trafficking and lysosomal enzyme sorting," *J. Cell Biol.* **132:**577–584 (1996).

Seabra, M. C. "Membrane association and targeting of prenylated Ras-like GTPases," *Cell Signalling* **10:**167–172 (1998).

Stahl, N., M. A. Baldwin, R. Hecker, K. M. Pan, A. L. Burlingame, and S. B. Prusiner. "Glycosylinositol phospholipid anchors of the scrapie and cellular prion proteins contain sialic acid," *Biochemistry* **31:**5043–5053 (1992).

Tang, C., E. Loeliger, P. Luncsford, I. Kinde, D. Beckett, and M. F. Summers. "Entropic switch regulates myristate exposure in the HIV-1 matrix protein," *Proc. Natl. Acad. Sci. U.S.A.* **101:**517–522 (2004).

Taylor, J. S., T. S. Reid, K. L. Terry, P. J. Casey, and L. S. Beese. "Structure of mammalian protein geranylgeranyltransferase type-I," *EMBO J.* **22:**5963–5974 (2003).

Thoma, N. H., A. Niculae, R. S. Goody, and K. Alexandrov. "Double prenylation by RabGGTase can proceed without dissociation of the mono-prenylated intermediate," *J. Biol. Chem.* **276:**48631–48636 (2001).

Udenfriend, S., and K. Kodukula. "How glycosylphosphatidylinositol-anchored membrane proteins are made," *Annu. Rev. Biochem.* **64:**563–591 (1995).

Waheed, A. A., and T. L. Jones. "Hsp90 interactions and acylation target the G protein Gα 12 but not Gα 13 to lipid rafts," *J. Biol. Chem.* **277:**32409–32412 (2002).

Willert, K., J. D. Brown, E. Danenberg, A. W. Duncan, I. L. Weissman, T. Reya, J. R. Yates III, and R. Nusse. "Wnt proteins are lipid-modified and can act as stem cell growth factors," *Nature* **423:**448–452 (2003).

Wu, S. K., K. Zeng, I. A. Wilson, and W. E. Balch. "Structural insights into the function of the Rab GDI superfamily," *Trends Biochem. Sci.* **21:**472–476 (1996).

Zhang, F. L., and P. J. Casey. "Protein prenylation: Molecular mechanisms and functional consequences," *Annu. Rev. Biochem.* **65:**241–269 (1996).

Zhang, H., M. C. Seabra, and J. Deisenhofer. "Crystal structure of Rab geranylgeranyltransferase at 2.0 Å resolution," *Struct. Fold. Des.* **8:**241–251 (2000).

Zong, Y., S. K. Mazmanian, O. Schneewind, and S. V. Narayana. "The structure of sortase B, a cysteine transpeptidase that tethers surface protein to the *Staphylococcus aureus* cell wall," *Structure (Cambridge)* **12:**105–112 (2004).

Proteolytic Posttranslational Modification of Proteins

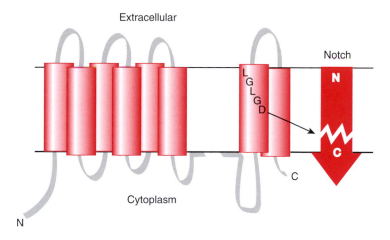

Extracellular

Notch

Cytoplasm

Intramembrane proteolysis of Notch by the transmembrane protease presenilin.

The destruction of proteins by hydrolysis of their primary covalent linkages, the peptide bonds, by proteins acting catalytically as proteases is an essential aspect of homeostasis (Creighton, 1993). The hydrolysis of the peptide backbones of proteins controls the dynamics of protein turnover. Since peptide bonds in proteins are indefinitely stable under physiological conditions, their turnover requires protease action. The proteolytic cleavage of proteins is also posttranslational, by definition, because proteins are cleaved at some time after they are synthesized on and exit from ribosomes. Proteases acting as posttranslational modification catalysts can fulfill a multitude of roles, depending on the time, place, and specificity of proteolytic cleavages.

Among the cellular processes controlled by regulated proteolysis are the removal of unassembled protein subunits and misfolded proteins. At a nominal concentration of 100–150 mg/mL in cytosol (Clausen et al., 2002), proteins that tend to aggregate have the problematic liability of introducing gelation and so the concentration of these aggregation-prone forms need to be carefully controlled. Also removed by proteolysis are short-lived proteins broken down by proteasomes. Antigen processing, programmed cell death, shedding of cell surface proteins, and some signaling pathways are also mediated by proteases.

The hydrolytic cleavage of peptide bonds to the component acid and amine fragments (Figure 8.1A) is thermodynamically favored, with a K_{eq} of about 10^5, and thus represents an irreversible posttranslational modification. Proteolysis therefore can be integral parts of one-way biological switches. These irreversible switches can be *biosynthetic*, as in the cleavage of single-chain proinsulin to the active two-chain protein hormone insulin or in the carving out of multiple-peptide hormones from the single-chain precursor proopiomelanocortin (POMC). A single peptide bond, between Asp_{36} and Tyr_{37}, is cleaved by the protease caspase-1 in the conversion of the

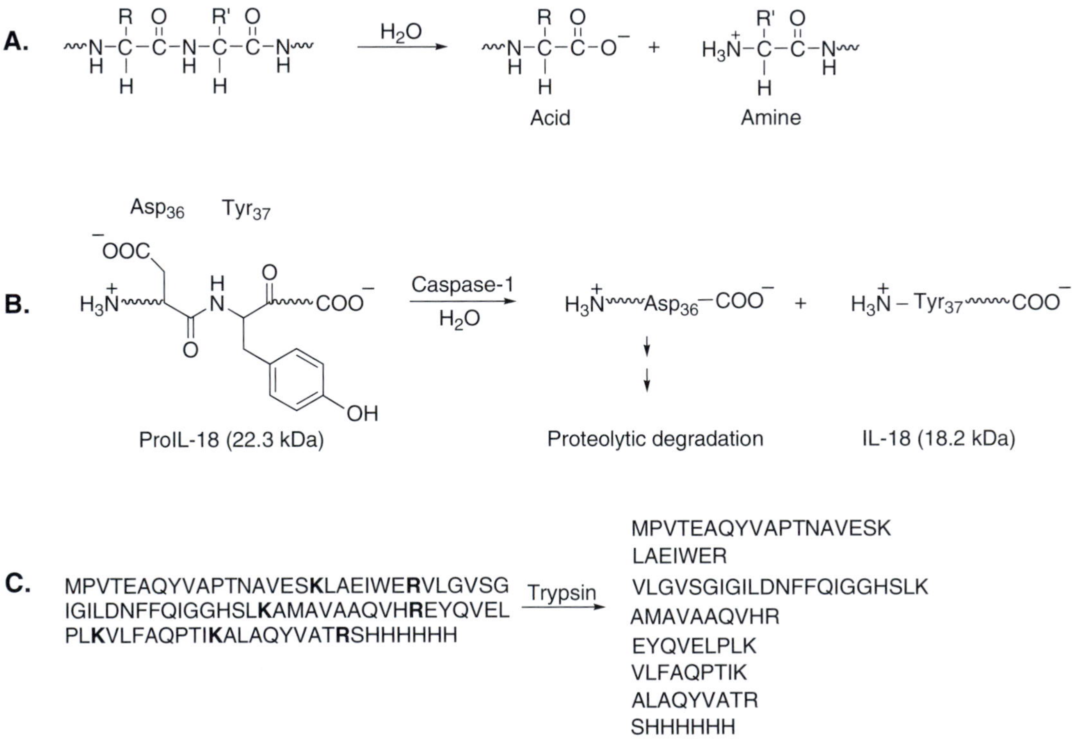

Figure 8.1 **A.** Hydrolysis of the primary covalent linkage in proteins, the peptide bond, is thermodynamically favored. **B.** Limited proteolytic cleavages of proteins can have biosynthetic functions as in ProIL-18 maturation. **C.** Proteolysis can be degradative in cleaving proteins to limit peptides. The cleavage of TycC3 by trypsin is shown. Trypsin cleaves after the Arg (R) or Lys (K) indicated in bold.

precursor form of interleukin-18 to the active cytokine (Figure 8.1B) (Rockwell et al., 2002). Proteolytic cleavages can also be *degradative*, by extracellular digestive proteases such as chymotrypsin, trypsin, and carboxypeptidase in the GI tract, or intracellular when short-lived proteins such as the cyclin subunits of cell-division kinases (Chapter 2) are degraded by proteasomes. Degradation usually involves cleavage of many peptide bonds in protein substrates, all the way to limit peptides (Figure 8.1C). Even small peptides can still contain useful information content, as in the display of foreign peptides as antigens by cells of the immune system.

Controlled proteolysis can also function at essential steps in development, such as the cleavage of Hedgehog and Notch proteins or in the cascade of extracellular proteases acting in tandem to set up dorsal or ventral axes (Ye and Fortini, 2000). Key steps in fertilization involve the proteolytic action of sperm enzymes to penetrate into the egg's cytoplasm. Proteolytic cascades are also at work in defense responses, extracellularly to generate the rapid amplification responses needed for blood coagulation, and also in intracellular defense where virally-infected cells may commit suicide through a cascade of caspases (Thornberry and Lazebnik, 1998).

The function of the more than 11,000 proteases with sequences in the protein databases depend on their specificity and the mechanisms by which peptide bonds are cleaved in protein substrates, on their locations in intracellular or extracellular compartments, and on how long they get to act. Some proteases, such as the proprotein convertases that act on insulin (Figure 8.1B), are highly selective endoproteases, perhaps making a single hydrolytic cut between adjacent basic site residues (e.g., –R–R–) in a small number of protein substrates (Steiner, 1998). Their purpose is to create functionally active protein fragments rather than to degrade their protein substrates. On the other hand, the proteases in both lysosomes and the chambers of proteasomes (see below) have a degradative function and act to digest proteins into small peptides or even constituent amino acids.

Because proteolytic cleavage of even a single peptide bond in a protein is thermodynamically and kinetically irreversible under physiological conditions, proteases are under tight regulation. Excessive proteolytic activity leads to apoptotic cell death intracellularly and tissue destruction extracellularly, as in emphysema.

Thus, the default position is that proteases are "*off*" *in the basal state* or, if "on," they are tightly controlled in time and space. (There is an analogy here to the logic of keeping most protein kinases "off" in the basal unstimulated states; see Chapter 2). Temporal control is generally exercised by the synthesis of proteases in zymogen form. Cells that produce proteases often produce them in inactive single-chain proenzyme forms, activable by proteolysis, either *in cis* or *in trans*. Famous examples of this strategy include the packaging by pancreatic acinar cells of digestive proteases—

trypsinogen, chymotrypsinogen, procarboxypeptidase, and proelastase—as inactive zymogens in secretory granules, releasable by regulated secretion (Figure 8.2A). Once in the pancreatic duct, the zymogens get activated by other active protease molecules, as exemplified for conversion of the single-chain inactive chymotrypsinogen to the three-chain active form of chymotrypsin, held together by disulfide bonds (Figure

A.

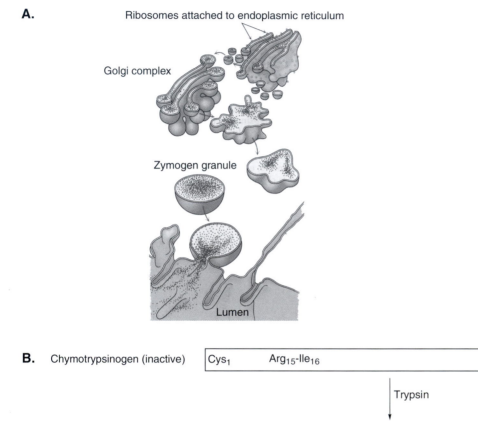

B.

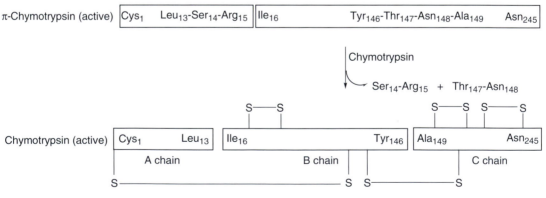

Figure 8.2 Spatiotemporal control of protease activity involves the general strategy of the biosynthesis of proteases as inactive precursors (zymogens = proenzymes). **A.** Secretion of zymogens by an acinar cell of the pancreas (reproduced with permission from Berg et al., 2002, p. 281). **B.** Proteolytic activation of bovine chymotrypsinogen.

8.2B). The caspase cascade within eukaryotic cells is kept in check in part by the synthesis of the proteases as inactive zymogens. Other proteases get activated at some appropriate time on demand, such as prococoonase to cocoonase at the time of insect molting by developmental signals.

Spatial control can be exercised within cells, most often by compartmentalization. The separation of digestive enzymes in lysosomes away from the cytoplasm and nucleus represents one sequestration strategy. The placement of protease active sites within proteasome chambers such that only unfolded proteins can be threaded into the chambers and cleaved represents a distinct protection logic to avoid uncontrolled access of active proteases to protein substrates (Bochtler et al., 1999).

Spatial control of protease action outside cells can be brought about by protease inhibitors (antiproteases) circulating at high concentrations. Collisional interaction of a protease with a tight-binding antiprotease, such as elastase with α1-antitrypsin or thrombin with antithrombin, leads to an inactive complex. The lifetime and spatial distribution of the active protease can be controlled by the concentration of the inhibitor and the stochastic collision frequency. Analogously, there are intracellular protein inhibitors for proprotein convertases and caspases (Hook et al., 1994; Shi, 2002b).

Classification of Proteases

Proteases have been classified by many criteria. One is by the site of action on protein substrates. When proteases attack from one end or the other, the amino terminus, or the carboxy terminus, they are termed exopeptidases, aminopeptidases, or carboxypeptidases, respectively (Figure 8.3A). While removal of one amino acid at a time is typical, removal of a dipeptide, such as a carboxydipeptide, occurs when angiotensin

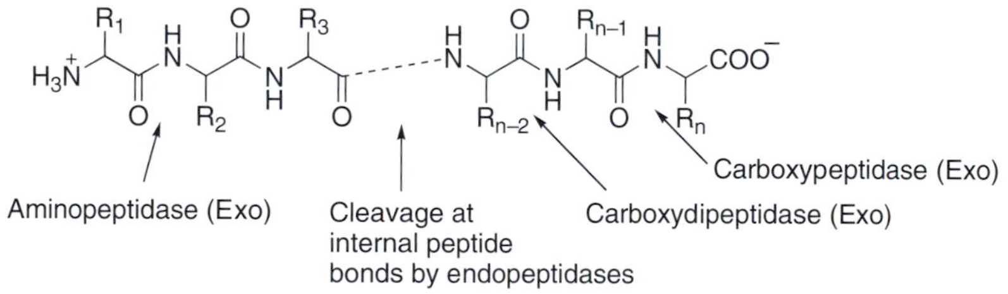

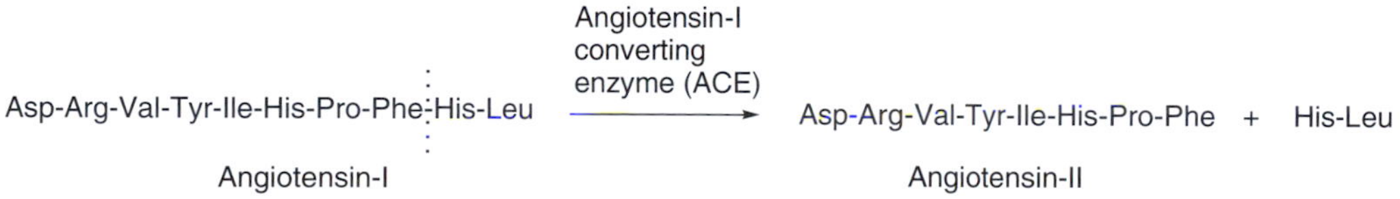

Figure 8.3 **A.** Classification of proteases as exo- and endoproteases. **B.** Angiotensin converting enzyme as a carboxydipeptidase removes two residues at a time from the C-terminus of its substrate.

converting enzyme, a major pharmacologic target in the control of hypertension, converts its decapeptide substrate to an octapeptide product (Figure 8.3B). All other proteases in this classification are endoproteases or endopeptidases, since they cleave at one or more internal peptide bonds of proteins. Since internal peptide bonds comprise all but two linkages in the primary sequence of proteins, these classifications are not very useful for categorizing the vast body of endoproteases.

More useful is the classification of proteases by catalytic mechanism and, in particular, by the active site groups involved in lowering the barriers for the net addition of water to the particular peptide bond to be hydrolyzed. The vast majority (>90%) of proteases fall into four mechanistic categories: serine proteases, cysteine proteases, aspartic acid proteases, and zinc (metallo) proteases (Figure 8.4). From genome sequence analysis there are 553 proteases in the human genome, distributed as follows (Puente et al., 2003): 203 Ser proteases, 143 Cys proteases, 21 Asp proteases, and 186 metalloproteases.

In serine and cysteine proteases the –OH and –SH groups of an active site Ser or Cys, respectively, act as the catalytic nucleophiles (Figure 8.5A) to attack the scissile bond of a bound protein substrate. This cleaves the peptide bond, releases the amine-containing product fragment, and generates a covalent peptidyl-O-protease or peptidyl-S-protease covalent enzyme intermediate (acylation), respectively, that must be hydrolyzed in a second half-reaction (deacylation).

The aspartic acid and zinc proteases do not engage in such covalent acyl-enzyme catalysis but instead lower the energy barriers for direct addition of a water molecule to the protein bound in the active site. The aspartic acid proteases invariably have two Asp residues, one as a general base and one as a general acid, disposed on either side of

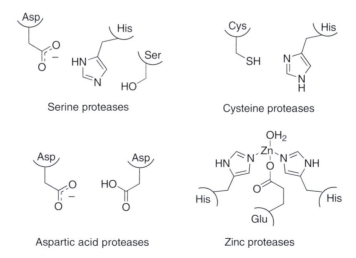

Serine proteases

Cysteine proteases

Aspartic acid proteases

Zinc proteases

Figure 8.4 Classification of proteases by active site machinery: serine proteases, cysteine proteases, aspartic acid proteases, and zinc proteases.

the peptide bond to be cleaved (Figure 8.5B). The zinc proteases tend to coordinate the divalent cation by histidines and a Glu/Asp acidic side chain, with the attacking water as the fourth ligand (Figure 8.5C). The substrate tetrahedral adduct generated during catalysis can be coordinated to zinc to lower activation barriers and speed up rates. The distinction in mechanisms (covalent acyl-enzyme versus direct generation of a tetrahedral adduct by water addition) and the presence or absence of zinc in the

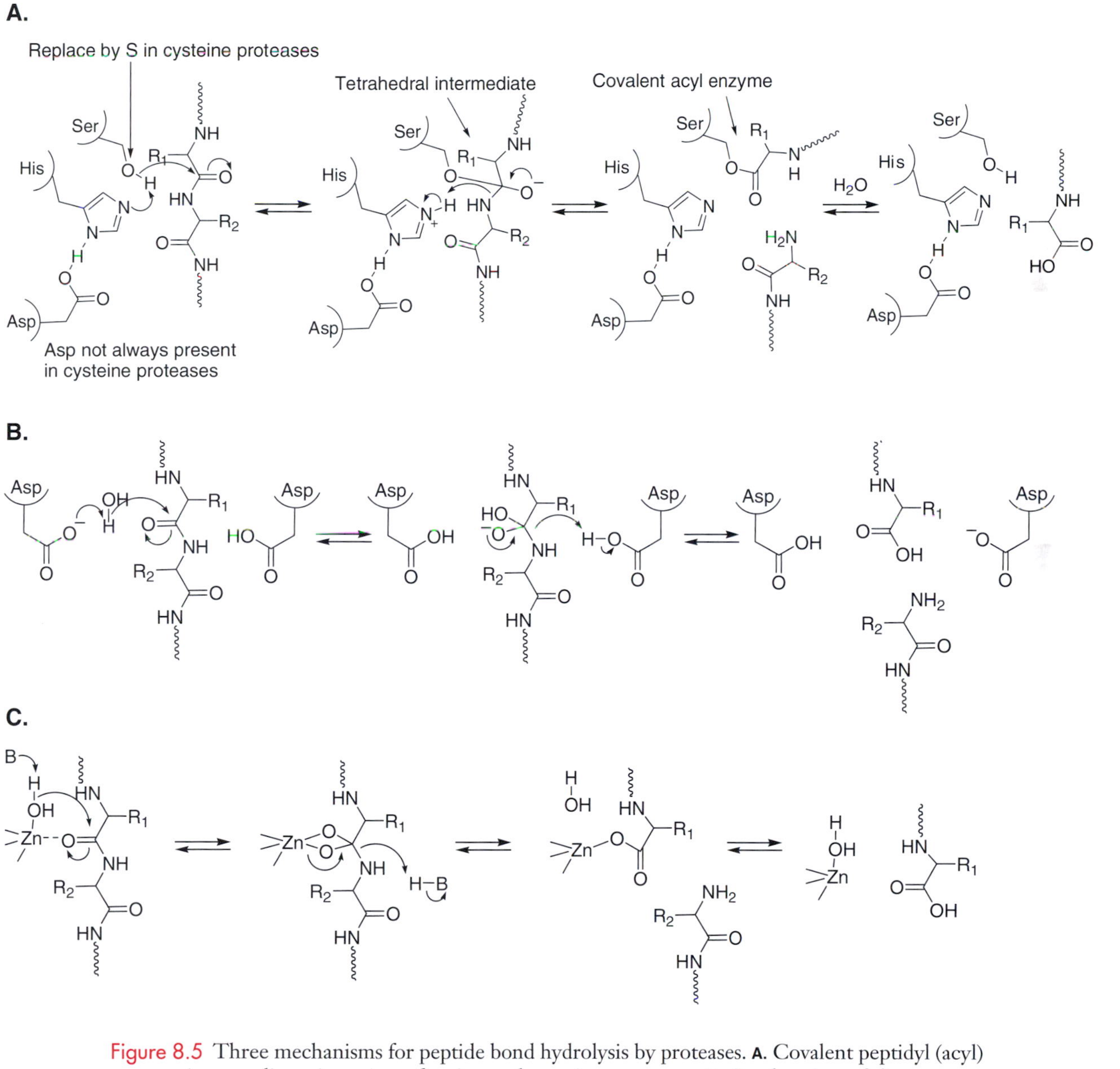

A.

B.

C.

Figure 8.5 Three mechanisms for peptide bond hydrolysis by proteases. **A.** Covalent peptidyl (acyl) enzyme intermediates in action of serine and cysteine proteases. **B.** Acceleration of direct water attack by aspartic acid proteases. **C.** Coordination of protein substrate and water by the active site metal in zinc proteases.

active site has enabled the design of mechanism-specific inhibitors for all four major classes of proteases for therapeutic interventions.

The serine protease superfamily is underrepresented in yeast with only a few examples. There are about 20 in the worm *C. elegans*, and up to several hundred members in *Drosophila*. In one survey (Clausen et al., 2002) it was estimated that 36% of the 11,000 predicted protease sequences are members of the serine superfamily. Another 30% are metalloenzymes, 19% are cysteine proteases (including 11 cathepsins in lysosomes and 11 caspases in the cytoplasm), and 7.5% are aspartic acid proteases. These four categories account for 92.5% of the more than 10,000 sequences, and suggest four main paths for the evolution of catalytic apparatuses in proteins that can hydrolyze peptide bonds efficiently.

Given the clustering of thousands of proteases into the four superfamilies, such as the more than 3000 serine proteases all using the same fundamental mechanisms for rate accelerations in their active sites, how do proteases of the same superfamily achieve specificity for cleavage at particular sites in protein substrates? The answer lies in the creation of specificity pockets within the protease active sites (Branden and Tooze, 1999). Essentially all proteases discriminate based on the side-chain identity of the amino acid side chains at the peptide bond undergoing cleavage. Thus, for the three related serine proteases elastase, chymotrypsin, and trypsin, elastase cleaves at peptide bonds after a small uncharged side chain (e.g., Gly or Ala), chymotrypsin cleaves after hydrophobic side chains (including the aromatic Phe, Trp, and Tyr), while trypsin cleaves after positively charged side chains (Arg and Lys). Examination of the specificity subpocket shows (Figure 8.6) that elastase has a cleft blocked by side chains, so only Gly and Ala from the substrate can fit; chymotrypsin has a deep pocket lined by Gly_{216} and Gly_{226}, allowing bulky hydrophobic side chains to fit; and trypsin has a pocket with Asp_{189}–COO^- at the bottom, providing the electrostatic interaction with the cationic Arg and Lys side chains of bound protein substrates.

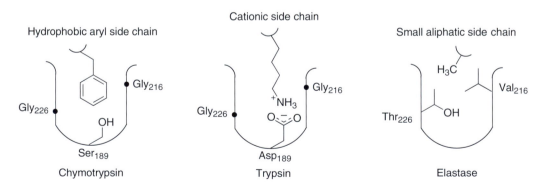

Figure 8.6 Distinct substrate binding specificity pockets enable regioselective cleavage of protein substrates by the serine proteases chymotrypsin, trypsin, and elastase (redrawn from Branden and Tooze, 1999).

Proteases in Prokaryotic Cells:
The Quality Control Inventories and Secretory Functions

We begin with some prototypic bacterial proteases to survey protease functions and protease types and locations in prokaryotic cells as a platform for analyzing additional posttranslational proteolytic events that occur in eukaryotic cells with multiple compartments. Bacterial proteases are found in the cytoplasm, in the plasma membrane, and in the periplasm of Gram-negative bacteria to deal with the turnover of proteins in all three locales (Hengge and Bukau, 2003). The *E. coli* genome is predicted to encode 72 functional proteases (Clausen et al., 2002), 19 of them serine proteases.

The proteases in the cytoplasm deal with protein degradation in normal housekeeping functions for turnover and also to remove unfolded or misfolded proteins before they can accumulate to high enough concentrations to aggregate. Many bacterial cytoplasmic proteases, such as HslV, and the ClpP, are chambered, composed of two rings of six or seven subunits each (Figure 8.7A) (Wang et al., 1997). Archaeal and eukaryotic proteasome scaffolds have similar core architectures. The active sites of the proteases in the subunits face into the chambers and so isolate the proteolytic activity in a separate microenvironment that is unavailable to proteins on the outside. This is the same strategy used by the proteasome. As a result, folded, native proteins in the cytoplasm of prokaryotic or eukaryotic cells are protected from indiscriminate degradation by these double-ring proteases. These proteases often associate with ATPases (Sauer et al., 2004) that are integral parts to these chambered proteases.

For protein substrates to gain access to the internal proteolytic chambers, they must be unfolded and threaded through a small pore to reach the active sites. Energy is required to unfold the proteins and to thread them into the protease chambers. Thus, it is reasonable to find ATPase catalytic domains required for chambered proteases to function. For ClpP, a protease with two heptameric rings (Ogura and Wilkinson, 2001; Schmidt et al., 1999), hexameric ATPases, such as ClpA, ClpC, and ClpX, can associate at one or both ends of the protease (Grimaud et al., 1998), thus forming ClpAP, ClpCP, and ClpXP, respectively. A cutaway diagram of the ATPase (HslU) and protease (HslV) rings is shown in Figure 8.7B for the HslUV protease. These ATPases are part of the Hsp100 family of heat shock proteins, which can function as chaperones (Schirmer et al., 1996). Hexameric Hsp100 ATPases can function in the absence of protease partners to disassemble macromolecular complexes and resolubilize protein aggregates. During protein degradation these ATPase rings bind a target protein and use ATP hydrolysis energy to drive its unfolding and subsequent translocation into the protease chambers. Using the ClpXP ATPase–protease complex, Kenniston et al. (2003) observed the hydrolysis of more than 500 molecules of ATP during the unfolding and translocation of a single 121-residue protein into the ClpP chamber. The combined energy for denaturation and translocation during

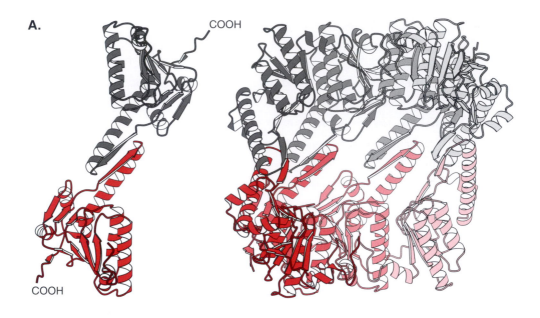

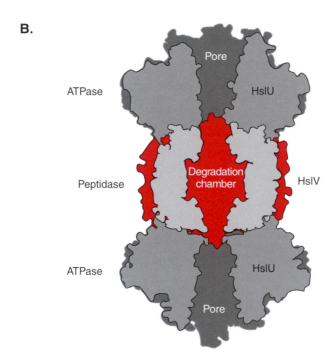

Figure 8.7 **A.** Double hexameric or heptameric rings make up the catalytic chambers in some bacterial proteases. The intra-ring association of ClpP monomers is shown as a ribbon diagram (left). Longitudinal section of a ribbon diagram of ClpP (right) (both figures made using PDB 1TYF). **B.** Cutaway diagram for the pore in HslU and the degradation chamber in HslV (reproduced with permission from Sauer et al., 2004).

degradation clearly represents a serious commitment to housekeeping and stress response functions.

The membrane-bound FtsH protease of *E. coli* also associates with an ATPase–chaperone subunit to unfold membrane or cytoplasmic proteins prior to degradation (Hengge and Bukau, 2003). FtsH substrates include the heat shock σ^{32} subunit of RNA polymerase and the phage λ regulatory protein *cII*, which regulates the lysis/lysogeny state switches for the virus. Degradation of σ^{32} by FtsH occurs continuously under nonstressed conditions and σ^{32} does not accumulate. One hypothesis is that denatured proteins build up under stress conditions, and the chaperone DnaK gets titrated by these accumulating misfolded proteins. Then σ^{32} can escape being bound by DnaK and presented to FtsH. As σ^{32} builds up it can compete for the core RNA polymerase and direct it to promoters for the transcription of stress response genes, including the synthesis of more DnaK. Degradation of another alternate σ subunit, σ^{s}, is controlled by the phosphorylation state of the two-component sensor RssB, which sends σ^{s} to the ClpXP protease (Hengge and Bukau, 2003).

Proteases exported to function in the periplasm of Gram-negative bacteria such as *E. coli* include members of the HtrA superfamily—DegS, DegP, and DegQ. The defining feature of the Deg proteases is the absence of ATPase partners, but the presence of a regulatory/chaperone PDZ domain. DegS has one PDZ domain while DegP and DegQ have two tandem PDZs (Clausen et al., 2002) (Figure 8.8A). The x-ray structure (Krojer et al., 2002) shows that the DegP is hexameric. The 12 PDZ domains form the sides of the chambers, where protein substrates are thought to enter and degraded peptides to exit (Figure 8.8B). The protein can switch between functioning as a chaperone or a protease, based on conformational changes involving the PDZ domains, which may allow the refolding of bound proteins to be explored before irreversible proteolysis is the committed pathway.

Proteases in the mitochondria and chloroplasts of eukaryotic cells contain members of these classes of proteases, reflecting a common bacterial ancestry in evolution (Kaser and Langer, 2000). All the above protease examples are involved in degradation of protein substrates. In chambered proteases, the full-length unfolded protein is threaded in and protease active sites in the chamber act *processively* to make multiple hydrolytic cuts at peptide bonds, releasing only small peptide fragments from the chamber.

A subset of proteins in Gram-negative and Gram-positive bacteria are secreted across the cytoplasmic membrane, a process that is a harbinger of the ER secretion systems in eukaryotes. Almost all proteins secreted by bacteria are processed within 20–30 residues of the N-termini of the preprotein forms by membrane-bound signal peptidases (Paetzel et al., 2002), releasing the N-terminal peptide while the remainder of the protein, intact, passes through the membrane via the Sec YEG secretory pathway (Figure 8.9). Signal peptidase I (SPase I) acts on most proteins, SPase II on

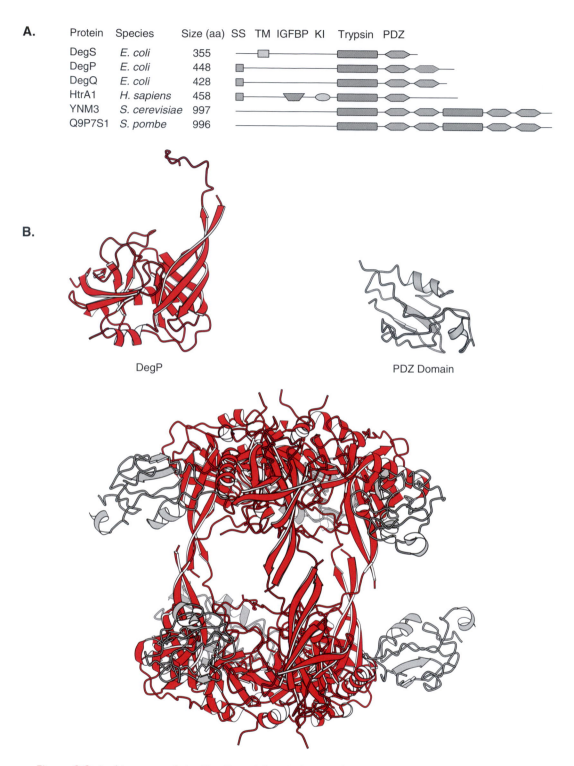

A.

Protein	Species	Size (aa)	SS	TM	IGFBP	KI	Trypsin	PDZ
DegS	*E. coli*	355						
DegP	*E. coli*	448						
DegQ	*E. coli*	428						
HtrA1	*H. sapiens*	458						
YNM3	*S. cerevisiae*	997						
Q9P7S1	*S. pombe*	996						

B.

DegP

PDZ Domain

Figure 8.8 Architecture of the DegP periplasmic bacterial protease. Hexamers with two PDZ domains per subunit create a chambered catalyst for protein degradation. **A.** Domain organization of selected HtrA family members, where SS = signal peptide, TM = transmembrane segments, IGFBP = insulin growth factor binding domain, and KI = Kazal protease inhibitor domain. **B.** Fold of the DegP HtrA domain (left), PDZ domain (right), and the structure of the DegP hexamer (bottom) (all three figures made using PDB 1KY9).

A.

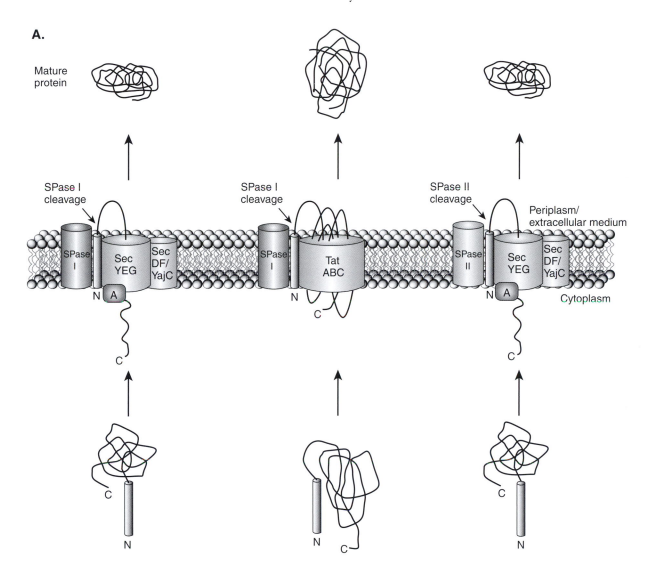

B.

Bacterial signal peptide:

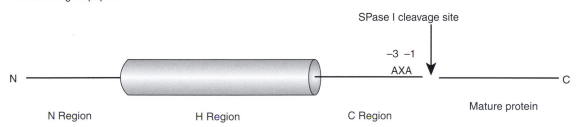

Figure 8.9 Bacterial signal peptidase cleaves the N-terminal region of proteins transiting the cytoplasmic membrane via the Sec YEG secretory pathway. **A.** Protein export across the plasma membrane in bacteria. **B.** Bacterial signal peptide.

lipoprotein signal sequences, and SPIV on prepilins and other protein components of type II secretory pathways (Paetzel et al., 2002). These processing peptidases carry out a controlled clip early in the life cycle of the translocating protein substrates, in contrast to the full degradations performed by ATP-dependent proteases and DegP. Signal sequences at the N-termini of preproteins have multiple parts. There is a central hydrophobic stretch, where substantial residue variation is tolerated, followed by narrower specificity at side chains in the 1- and 3-positions relative to the peptide bond cleaved. The SPases are serine proteases that appear to use a Ser-Lys catalytic diad, with the Ser side chain as the nucleophile and the Lys amino group as the general base (Paetzel and Dalbey, 1997).

About 5% of bacterial cell proteins are turned over each cell divison. In some bacteria that differentiate, the regulated destruction of key control proteins is analogous to the controlled proteolysis of cyclin subunits in cell division. In *Caulobacter crescentus* the two-component regulatory protein CtrA (see Chapter 2 for two-component histidine kinase systems) acts as a master regulator (McAdams and Shapiro, 2003), regulating transcription of 95 genes in 55 separate operons. Consonant with its role in differentiation, there is a sharp temporal onset of synthesis of about 22,000 molecules of CtrA per cell. Then, when its signaling and control role has been executed, CtrA is proteolyzed by a ClpXP type ATPase/chaperone/protease, in just one-half of a predivisioned cell.

Proteasomes and Lysosomes: Sequestration and Degradation

Lysosomes serve as dedicated degradative protease-containing organelles in eukaryotes for controlling the bulk turnover of proteins brought in from the outside via phagocytosis and endocytosis, as well as in the normal autophagocytic reactions of cells (Rouille et al., 2000; Teter and Klionsky, 2000). This is emphasized in Chapter 9, where receptor protein internalizations enabled by the monoubiquitin marking of membrane proteins and TG network sorting are discussed. In a complementary way, proteasomes are famous for using polyubiquitin covalent markers on client proteins for the regulated proteolysis of specific proteins that affect cell cycle progression (cyclins and p21), oncogenesis (p53, Bax, and IκB), gene expression (c-Jun and E2F), memory (protein kinase A regulatory subunit), and protein quality control in the cytoplasm and ER.

The proteasomes of archaea and eukaryotes are chambered proteases that resemble the chambered architecture of the bacterial Clp proteases and also use associated ATPase subunits to recognize and unfold client proteins for threading into the protease chambers. The prototypic proteasomes have a 20S core particle consisting of four stacked heptameric rings (Figure 8.10A). The four rings are comprised of two types of related subunits, α and β, in rings of $\alpha_7\beta_7\beta_7\alpha_7$ (Stock et al., 1996; Voges et al.,

A.

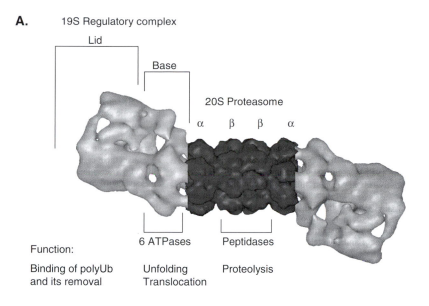

B.

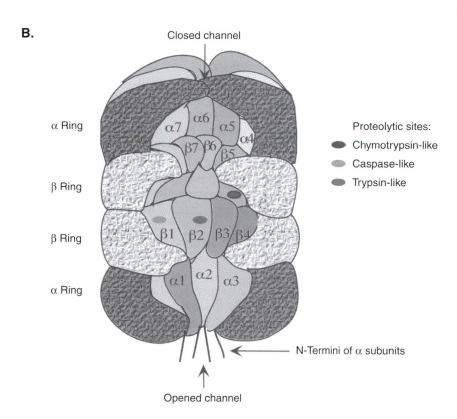

Figure 8.10 Architecture of the 26S and 20S proteasomes of archaea and eukaryotes. **A.** Four rings of αββα heptamers. **B.** All the β subunits in each ring are active proteases in the archaeal proteasome, but in eukaryotes three of the seven β subunits are active, with distinct specificities of peptide bond cleavage. Schematic representation of the cross-section of the eukaryotic 20S proteasome (both drawings reproduced with permission from Kisselev and Goldberg, 2001).

1999). The β_7 rings contain the active sites of the subunits. In the *Thermophila acidophilus* 20S core, all the β subunits are identical, generating 14 active sites in the two $\beta\beta$ rings. In yeast and higher eukaryotes, only three of the seven β subunits in each inner ring are active, generating six protease sites per chamber (Figure 8.10B). Specificity analysis confirms that β_1 subunits cleave after peptide bonds where the side chain is acidic, the small β_2 subunits cleave after basic residues, and β_5 subunits hydrolyze peptide bonds after hydrophobic residues, thus giving broad coverage for the processive cleavage of proteins threaded into the proteasome chamber. The mean peptide product length is 7–9 amino acids, accounting for about 15% of the total peptide products, with a spread of 3–23 residue peptides detectable (Glickman, 2000). The 7–9 residue length is typical for peptides subsequently transported into the ER, bound to MHC class 1 subunits, and presented on immune cells for antigen recognition. Note, in passing, that proteasomes are found in the actinomycete group of bacteria, perhaps by horizontal gene transfer. Little is known about the particular protein substrates of proteasomes in such bacteria, but it has recently been determined that proteasomes are required for the survival of *Mycobacterium tuberculosis* in host phagocytes mounting nitric oxide attacks on the intracellular pathogen (Darwin et al., 2003).

All the β subunits, three per heptameric ring of eukaryotes and the full seven in archaea, are variants of serine proteases, engaging in covalent acyl-enzyme catalysis. The variant feature is that the β subunits that self-assemble into the proteasomal cylinder are proproteins with N-terminal extensions (Kwon et al., 2004). The β subunits that become active engage in autocleavage reactions, as elaborated in the discussion of autocatalysis (Chapter 13), to release an internal threonyl residue, Thr_{76}, as the new N-terminus. This autoprocessing is exemplified for the yeast Doa3 β subunit in Figure 8.11A. This new Thr_1 in the autocleaved β subunits is the active site nucleophile in protease action involving covalent acyl-O-Thr_1 intermediates (Figure 8.11B). The proteasomal autoprocessed β subunits fall into the class of N-terminal nucleophilic catalysts (Chapter 13). The HslV protease of bacteria forms hexameric rings that stack to provide equivalent internal chambers and also uses N-terminal threonine residues as the catalytic nucleophiles in peptide bond hydrolysis.

The heptameric rings of α subunits on each end of the stacked $\beta\beta$ rings serve as gate keepers to the channel. They comprise pores that can be open or shut to allow unfolded protein substrates to pass into the chambers. The pore diameters and passage of proteins through them are controlled by the ATPase subunits in the lids of the proteasomes.

The 19S regulatory particles (Figure 8.12) of the proteasomes contain 17 or more subunits (Pickart and Cohen, 2004), disposed into base and lid units, that can cap the 20S core cylinders at one or both ends. The base of the 19S particle contains nine subunits, six of them with ATPase activity, and at least some of the other three

A.

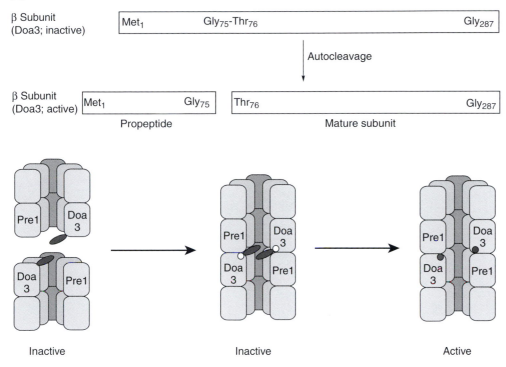

B.

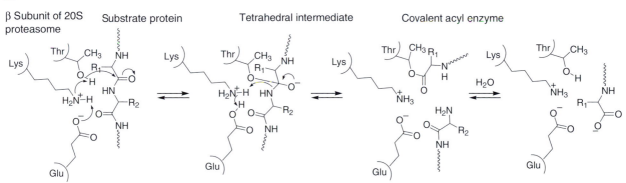

Figure 8.11 Autocleavage of β subunits in the β₇ rings of the proteasome to convert the pro form to the active protease. **A.** The Doa3 propeptide is required for Doa3 incorporation into the 20S proteasome and can function *in trans* (top). Model for autocatalytic proteasome-β-subunit processing and the coupling of active site formation to full proteasome assembly (bottom). Redrawn from Chen and Hochstrasser (1996). **B.** The Thr residue at the newly liberated N-terminus acts as the catalytic nucleophile in acyl enzyme formation during proteolysis of substrates in the chamber.

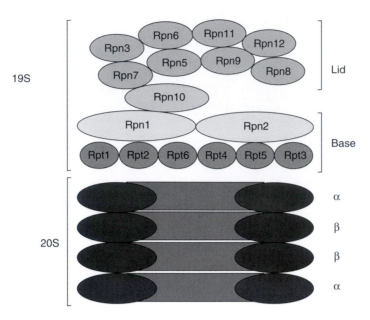

Figure 8.12 Subunit composition of the 19S regulatory particle of the *Saccharomyces cerevisiae* proteasome.

involved in recognition of the polyubiquitin tags that mark client proteins for proteasome-mediated proteolysis, as detailed in Chapter 9. The six ATPase subunits reprise the logic of the hexameric ATPases of the ClpXP, ClpAP, and HluV bacterial proteases noted earlier in this chapter. The 19S ATPases are thought to participate in at least three functions: 1. gating the channel/pore through the α_7 ring, 2. unfolding the client proteins, and 3. threading them through the pore into the protease chambers (Glickman, 2000).

Based on the high energy cost in ATP consumption per protein degraded by the ClpXP protease, comparable energetics are assumed for ATP consumption by the proteasomes. This is in addition to the ATP cost of adding each of the n ubiquitins in poly-U_n chains by the E1 ligase (Chapter 9). You will also learn in the next chapter that the branching polyubiquitin tags need to be removed from the unfolding proteins before they can pass through the α_7 pores, so proteasomes require one or more deubiquitylase (DUB) subunits in the 19S particle.

Proteases in Eukaryotic Cells: The Challenges of the Secretory Pathways

In addition to the posttranslational role of the complete proteolytic digestion of proteins by proteasomes and lysosomes, there are many points in the life cycle of proteins in eukaryotic cells where limited proteolysis is used for protein maturation, the control of protein localization, and the onset and duration of protein signaling. This is particularly well exemplified by proteins passing through the secretory compartments on the way to the Golgi complex and the plasma membrane.

Signal Peptidase

As in bacterial protein secretion, the first step for essentially every protein secreted into the endoplasmic reticulum lumen or membrane involves trimming of the N-terminal signal peptides (median length = 23 residues) by ER membrane signal peptidase action (Paetzel et al., 2002). The subunit composition of ER signal peptidase is more complex than bacterial SPase I. The ER signal peptidase has five subunits (Figure 8.13)—two serine catalytic subunits homologous to bacterial SPI, two glycoproteins, and one that interacts with the Sec61 translocase. The signal peptide is hydrolytically clipped as the membrane-transiting proteins move to the ER lumenal surface. Comparable proteolytic trimming on N-terminal signal sequences occurs as proteins are

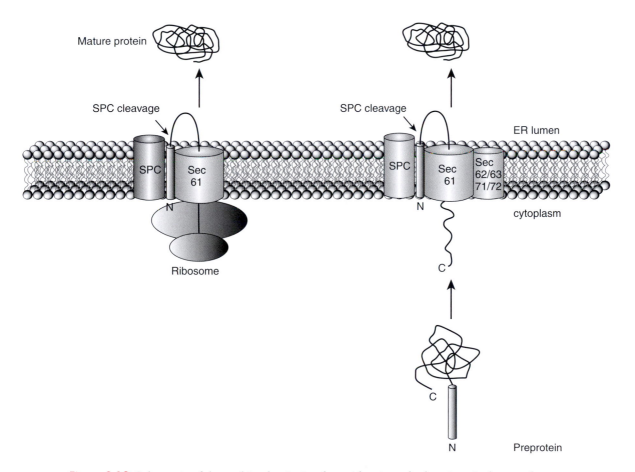

Figure 8.13 Schematic of the multi-subunit signal peptidase in endoplasmic reticular membranes that trims off the N-terminal signal peptides of proteins targeted to the secretory pathway of eukaryotic cells. On the left is cotranslational proteolysis of the N-terminal signal peptide. On the right is posttranslational removal of signal peptide.

imported into mitochondria, chloroplasts, and peroxisomes. In the maturation of insulin, for example, this first proteolytic step converts preproinsulin to proinsulin and helps translocate the pro form of the protein hormone into the secretory pathway.

Proprotein Convertases and Protein Maturation

A second set of proteolytic selective trimming steps occurs in the Golgi network, catalyzed by one or more of a family of seven proprotein convertases (PCs), of which furin is the best studied in higher eukaryotes (Molloy et al. 1999; Nakayama, 1997; Rockwell et al., 2002). Kex2 of yeast was discovered as the founding member of the PC family. Furin is ubiquitously expressed in higher eukaryotic tissues, in the trans Golgi network and endosomal secretory compartments, and also at the plasma membrane. PCs cleave at RR and KR dibasic sites and furin shows extended specificity to cleave at RXR/KR sites and convert proproteins to proteins with mature N-termini. For example, protein hormones such as proinsulin, proparathyroid hormone, pro-$TGF\beta_1$, and pro-$NGF\beta$ are converted to insulin, PTH, $TGF\beta_1$, and $NGF\beta$, respectively (Figure 8.14) (Molloy et al., 1999). Furin has also been implicated in the processing and maturation of pro forms of serum proteins such as proalbumin, proFactorIX, and proprotein C, of cell surface receptors such as the pro forms of the insulin receptor and the Notch receptor, as well as bacterial toxins and viral proteins (Molloy et al., 1999).

The proprotein convertases are Ca^{2+}-dependent serine proteases that themselves undergo a series of controlled proteolysis reactions to control the localization, timing of activation, recycling, and shedding from the plasma membrane. For example, furin is synthesized as a 794-residue inactive preproprotease (Figure 8.15A) (Molloy et al., 1999). The 24-residue N-terminal signal sequence is removed by signal peptidase to deposit profurin in the ER. Autocleavage in the ER with a half-life of about 10 minutes within the 83-residue pro region generates the two-fragment protein with the pro region still associated and inhibitory. Later, in the trans Golgi network, with a half-life of about three hours, a second cleavage occurs in the pro region, which then dissociates to give active furin. The x-ray structure of the catalytic domains of mouse furin and the yeast Kex2 family member have been reported (Henrich et al., 2003; Holyoak et al., 2003). The furin protease molecules can clip substrate proteins in the exocytic arm of the secretory pathway (for maturation purposes, such as the pro forms of receptors or of serum proteins), or can cleave incoming proteins in the endosomes (e.g., bacterial protein toxins and viral proteins), or can cleave proteins while furin is at the cell surface (e.g., the anthrax PA toxin) (Molloy et al., 1999). The efficiency of furin and other proprotein convertases depends on the level of active PC formed, the amounts of substrate proteins moving through the TGN, the catalytic capacity of the PCs [about 12–30 per second for furins (Molloy et al., 1999; Rockwell et al., 2002)],

A.

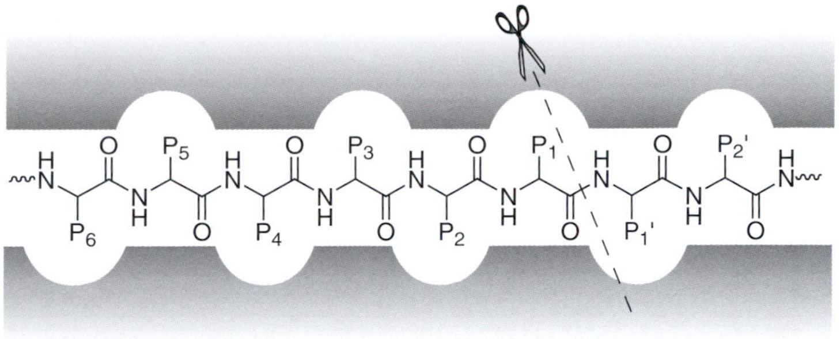

B.

	P6	P5	P4	P3	P2	P1		P1'	P2'
Serum proteins:									
Proalbumin	R	G	V	F	R	**R**		D	A
Profactor IX	L	N	**R**	P	K	**R**		Y	N
Proprotein C	R	S	H	L	K	**R**		D	T
Provon Willebrand factor	S	H	**R**	S	K	**R**		S	L
Hormones and growth factors:									
Pro-β-nerve growth factor	T	H	**R**	S	K	**R**		S	S
BMP-4 precursor	R	R	**R**	A	K	**R**		S	P
ProBNP	T	L	**R**	A	P	**R**		S	P
Proparathyroid hormone	K	S	V	K	K	**R**		S	V
Prosemaphorin D (PCS 1)	K	R	**R**	T	R	**R**		Q	D
ProTGFβ$_1$	S	S	**R**	H	R	**R**		A	L
Cell-surface receptors:									
Insulin pro-receptor	P	S	**R**	K	R	**R**		S	L
Notch-1 receptor	G	G	**R**	Q	R	**R**		E	L
Scatter factor receptor	E	K	**R**	K	K	**R**		S	T
Vitamin B$_{12}$ receptor	L	Q	**R**	Q	K	**R**		S	I
Extracellular matrix proteins:									
BMP-1	R	S	**R**	S	R	**R**		A	A
Human MT-MMP1	N	V	**R**	R	K	**R**		Y	A
Integrin α3-chain	P	Q	**R**	R	R	**R**		Q	L
Profibrillin	R	G	**R**	K	R	**R**		S	T
Stromelysin-3	R	N	**R**	Q	K	**R**		F	V
Bacterial toxins:									
Anthrax toxin PA	N	S	**R**	K	K	**R**		S	T
Clostridium septicum α-toxin	K	R	**R**	G	K	**R**		S	V
Diphtheria toxin	G	N	**R**	V	R	**R**		S	V
Proaerolysin	K	V	**R**	R	A	**R**		S	V
Pseudomonas exotoxin A	R	H	**R**	Q	P	**R**		G	W
Shiga toxin	A	S	**R**	V	A	**R**		M	A
Viral coat proteins:									
Avian Influenza HA (HSN1)	R	R	**R**	K	K	**R**		G	L
Borna disease virus	L	K	**R**	R	R	**R**		D	T
Cytomegalovirus gB	T	H	**R**	T	R	**R**		S	T
Ebola Zaire GP	G	R	**R**	T	R	**R**		E	A
Epstein–Barr virus gB	L	R	**R**	R	R	**R**		D	A
HIV-1 gp160	V	Q	**R**	E	K	**R**		A	V

Figure 8.14 Action of proprotein convertases, cleaving proteins at dibasic sites to liberate mature forms of proteins: proinsulin to insulin, and proPTH to PTH. The P6–P2' cleavage sites are shown in **A.** and the cleavage site sequences for a selected list of proposed furin substrates are shown in **B.** Basic P1 and P4 residues constituting the minimal furin cleavage site are highlighted in bold, while the additional basic P2 and P6 residues generating the consensus furin site are highlighted with a gray background (reproduced with permission from Molloy et al., 1999).

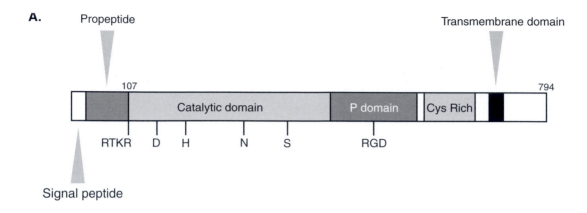

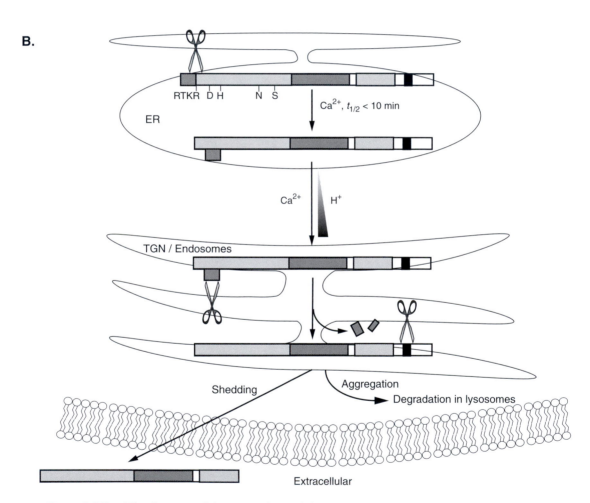

Figure 8.15 A. The domains of the prepro form of the protein convertase furin. B. Controlled proteolytic trimming events in the life cycle of a furin molecule.

and the accessibility of dibasic cleavage sites in the client proproteins. Furin and other PCs have a single transmembrane domain near the C-terminus that anchors them in the secretory compartment membranes. Furin can be cleaved upstream of the TM domain, converting the protease to a lumenal species that can be secreted, a net shedding process that releases furin and other PCs into the extracellular matrix in active forms (Figure 8.15B). The life cycle of furin can thus involve four proteolytic clips, occurring sequentially in time to convert zymogen into active maturation protease, and in space to control the location from cytoplasm to ER to TGN and endosomes to extracellular matrix (Figure 8.15B).

Coupled Ectodomain Shedding and Regulated Intramembrane Proteolysis

In Chapter 9, one mechanism described for the control of the lifetime of transmembrane proteins, including receptor proteins stimulated by ligands, is covalent monoubiquitylation on lysine side chains. This modification sets in motion their internalization and sorting to the TGN, either for recycling or for lysosomal/vacuolar proteolysis. A second route for regulation of the lifetime of proteins anchored at the plasma membrane of multicellular eukaryotes is proteolytic clipping and loss of the ecto domains to the extracellular medium (Blobel, 2000; White, 2003) (Figure 8.16). In some cases this mode of proteolysis generates soluble forms of protein ligands such as cytokines and cytokine receptor ecto domains. Extracellular EGFR fragments are generated this way from membrane-tethered precursors by sheddases in a regulated manner. Also, the pro form of the cytokine tumor necrosis factor is cleaved to the active cytokine (TNFα) by the sheddase TACE (Arribas and Borroto, 2002). Likewise the pro form of TGFα undergoes proteolytic cleavage. Cell adhesion molecules (CAMs), receptors for IL-6, IL-1, and growth hormone, similarly undergo ectodomain shedding, as does amyloid precursor protein (APP).

An early proteolytic shedding activity characterized was TNFα converting enzyme (TACE). This turned out to be one member of a superfamily of transmembrane metalloproteases termed ADAMs (*a d*isintegrase *a*nd *m*etalloprotease), with 33 predicted orfs in the human genome (Arribas and Borroto, 2002). TACE activity is due to the protease ADAM17 (Figure 8.16). The ADAM proteases transit the secretory pathway to the plasma membrane as depicted in the previous section, undergoing signal peptidase trimming and furin-mediated proteolytic processing for activation.

Ectodomain shedding of membrane protein substrates by ADAM processing can be a prelude to one additional set of posttranslational proteolytic steps that can be of great consequence in protein-based signaling and gene transcription. This is exemplified by the life cycle of the protein Notch, involved in juxtamembrane local cell-to-cell

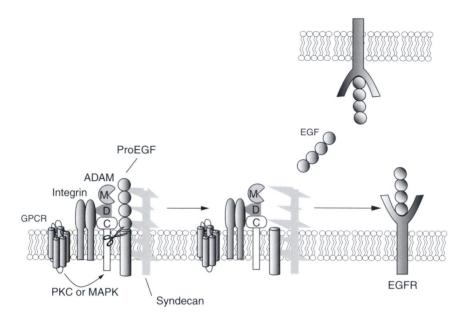

Figure 8.16 Ectodomain shedding on transmembrane proteins by the action of plasma membrane-associated proteases. One such protease is TACE (TNF alpha converting enzyme) also known as ADAM17.

signaling, in cell fate specifications, and in tissue patterning during embryogenesis (Ye and Fortini, 2000).

Notch is translated and emerges from ribosomes as a 300-kDa preproprotein (Figure 8.17). It enters the secretory pathway via signal-peptidase-aided translocation into the ER membrane and then in the TGN undergoes furin-mediated hydrolytic clipping to generate a two-chain heterodimer that stays associated as processed Notch travels to the plasma membrane in secretory vesicles (Figure 8.17A). There are four isoforms of Notch, Notch1–Notch4, in human cells. At the plasma membrane, the extracellular subunit is still associated with the transmembrane subunit. The ligands for the extracellular domain of Notch are themselves transmembrane surface proteins, Jagged and Delta (with multiple isoforms) on neighboring cells (LaVoie and Selkoe, 2003). The lifetimes of Jagged and Notch as signaling ligands are also controlled by ectodomain shedding by one or more ADAM proteases on the presenting cells.

On ligation by Jagged or Delta, the transmembrane subunit of Notch is cleaved (Figure 8.17B) by TACE (ADAM17), acting 12 residues external to the transmembrane spanning region. This hydrolytic cut releases that fragment and the associated extracellular domain. The residual trimmed transmembrane fragment of Notch is now a substrate for proteolysis within the plane of the plasma membrane, a process that has been termed regulated intramembrane proteolysis (RIP) (Brown et al., 2000). The responsible protease is an intrinsic membrane–protein complex, known as pre-

A.

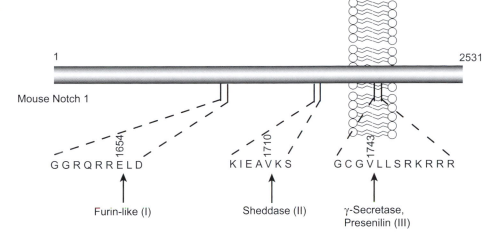

1
2531

Mouse Notch 1

1654
GGRQRRELD

1710
KIEAVKS

1743
GCGVLLSRKRRR

↑
Furin-like (I)

↑
Sheddase (II)

↑
γ-Secretase,
Presenilin (III)

B.

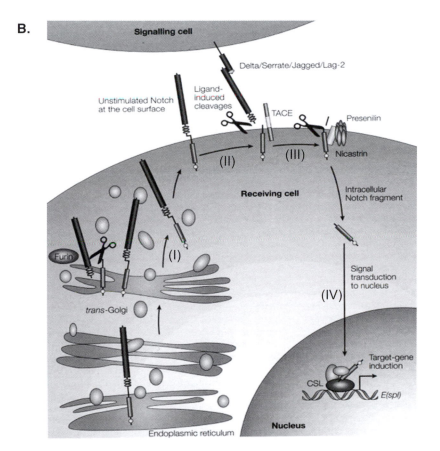

Signalling cell

Delta/Serrate/Jagged/Lag-2

Unstimulated Notch
at the cell surface

Ligand-
induced
cleavages

TACE

Presenilin

(II)

(III)

Nicastrin

Receiving cell

Intracellular
Notch fragment

Furin

(I)

Signal
transduction
to nucleus

(IV)

trans-Golgi

Target-gene
induction

CSL

E(spl)

Endoplasmic reticulum

Nucleus

Figure 8.17 **A.** Proteolytic cleavage sites in Notch. **B.** Proteolytic regulation of Notch-receptor maturation and activation which involves (I) two-stage proteolytic processing of Notch by signal peptidase and furin on its way to the plasma membrane, (II) ectodomain shedding by proteolysis of the transmembrane subunit of Notch on engagement of the protein ligands Jagged or Delta, (III) regulated intramembrane proteolysis of the transmembrane fragment of Notch by the aspartic acid protease presenilin, and (IV) release of the cytoplasmic domain of Notch from the plasma membrane to travel to the nucleus and act as transcription factor (reproduced with permission from Fortini, 2002).

senilin, with an aspartic acid protease catalytic subunit. This cleavage, and perhaps subsequent trimming, liberates the cytosolic fragment of Notch (with three N-terminal residues that had been part of the TM domain). The Notch cytosolic fragment can then migrate to the nucleus and act as a transcription factor for genes that specify developmental responses (Figure 8.17B).

The RIP step constitutes release of a transcription factor from a membrane anchoring site where it had been kept silent. The signaling by Jagged or Delta on one cell via Notch on a neighboring cell involves two regulated proteolysis steps: ligand-induced ectodomain shedding and RIP. The total number of discrete proteolytic events in the life cycle of Notch to generate the active transcription factor fragment that can move to the nucleus is four peptide bond cleavage steps (Figure 8.15).

This two-stage proteolytic signaling for plasma transmembrane proteins appears to be used in multiple situations where a dormant transcription factor or signaling

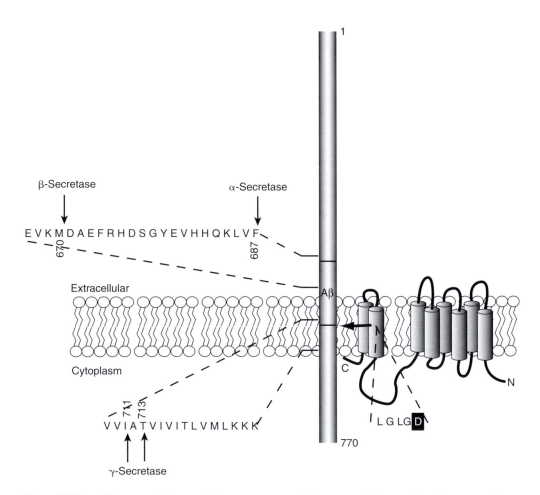

Figure 8.18 A. Cleavage of the amyloid precursor protein by presenilin in regulated intramembrane proteolysis.

protein fragment is activated by intramembrane cleavage (Urban and Freeman, 2002; Weihofen and Martoglio, 2003). The intramembrane cleaving protease responsible for Notch cleavage noted above is a multi-subunit enzyme called presenilin, because it also cleaves the APP precursor of amyloid peptide in Alzheimer's disease (Brown et al., 2000; Urban and Freeman, 2002; Weihofen and Martoglio, 2003) (Figure 8.18A). Presenilin appears to contain four transmembrane subunits—the PS1 subunit with active site aspartates, along with the nicastrin, APH-1, and PEN-2 subunits whose functions are as yet unclear (Fortini, 2002; Haass and Steiner, 2002), but may mediate transmembrane protein–substrate recognition and may enable subcomplexes with distinct selectivities (De Strooper, 2003).

In cholesterol metabolism, the ER membrane-anchored SREBP (sterol response element binding protein) is hydrolyzed by an intramembrane–clipping protease termed site 2 protease (S2P), after a site 1 protease trims the luminal portion of SREBP (Figure 8.18B). The fragment released from the membrane contains a DNA-binding basic helix-loop-helix (bHLH) region that can transit to the nucleus and turn "on" genes with sterol response elements. S2P cleavage also frees the ER membrane

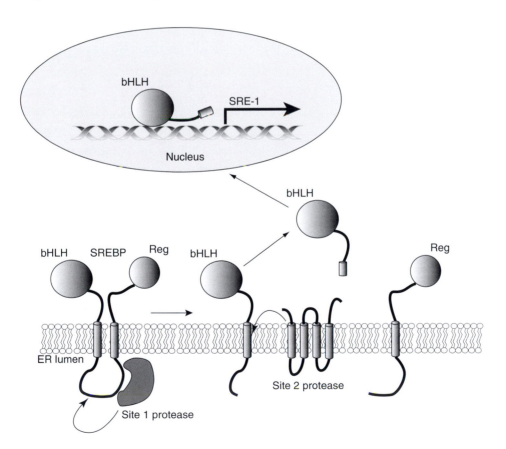

Figure 8.18 B. Schematic of SREBP cleavage by site 1 and site 2 protease.

protein AFT6 to go to the nucleus as a transcription factor in response to unfolded protein accumulation in the ER lumen. These unfolded proteins compete with AFT6 for the chaperone BIP. The AFT6 released from sequestering BIP is then a substrate for S2P.

Analogous RIPs exist in prokaryotes as well. The alternate sigma factor σ^E of *E. coli* is bound at the plasma membrane to the cytosolic domain of RseA. When unassembled outer membrane proteins accumulate in the cell envelope under stress conditions, they bind to the PDZ domain of DegS and activate DegS cleavage of the periplasmic portion of RseA. A second cleavage of RseA then occurs by the protease YaeL in the plane of the membrane (Weihofen and Martoglio, 2003). The σ^E bound to the residual small cytosolic fragment of RseA is then released from the membrane. The RseA portion of this complex is unfolded and degraded by ClpXP, as well as other cytosolic proteases, releasing σ^E to bind the core RNA polymerase and redirect it to promoters for stress-activated genes. The coordinated action of at least four proteases is involved in RseA degradation in this signaling pathway.

In *Drosophila melanogaster* there is a functional protease homolog called Rhomboid (Gallio et al., 2002) that cleaves the protein Spitz in the ER, releasing an EGF-like ligand that can be secreted and can activate the EGF receptor on neighboring cells.

While the function of the distinct protein subunits and associated protein components of presenilin, also known as γ-secretase for its action in APP processing (Figure 8.18), remain to be determined (Fortini, 2002; Haass and Steiner, 2002), it is clear that this is an aspartic acid protease. The S2P members and YaeL, by contrast, are zinc proteases, whereas Rhomboid is a serine protease, establishing that three of the four main mechanisms for protease catalysis have been evolved in these multi-spanning RIP membrane proteases (Weihofen and Martoglio, 2003). The molecular bases for the specificity of the targeting and cleavage of intramembrane proteases for their specific transmembrane target proteins is still under study. Arguments have been advanced that the local sequence may be only one determinant—that hydrophobicity, shape, flexibility for the local unfolding of the protein region (transmembrane helix) to be cleaved, and high lateral mobility (Brown et al., 2000; Weihofen and Martoglio, 2003) may all be important. Such determinants might explain the apparent need in many cases for coupled, prior cleavage of ecto or luminal domains to create flexibility in the TM domain for proteolytic processing in the plane or at the cytoplasmic face of the membranes (Wolfe and Kopan, 2004).

Protease Cascades in Development

The four carefully timed proteolytic cleavages in the life cycle of Notch detailed above represent one manifestation of molecular logic in tandem proteolytic posttranslational signals in cell fate determination. It turns out that the Notch ligands, the isoforms of Jagged and Delta proteins, can also undergo not only ectodomain shedding to regulate lifetime but also the second step of intramembrane proteolysis and nuclear signaling by the liberated cytoplasmic protein fragments after RIP (LaVoie and Selkoe, 2003; Selkoe and Kopan, 2003). Hence, both protein ligands (Delta and Jagged) and the receptors (Notch) are subject to the tandem steps of proteolytic regulation and signaling.

A complementary proteolytic logic is exemplified in the Toll protein pathway for determining the dorsal or ventral axes in *Drosophila* development (Ye and Fortini, 2000), starting with extracellular protease cascades in the perivitelline fluid (Figure 8.19). Four tandem zymogen conversions, proNudel → Gastrulation defective → Snake, and Easter, serve to produce the active form of Easter that then cleaves proSpaetzle to generate Spaetzle, the protein ligand for the transmembrane Toll family receptor. Spaetzle engagement on Toll leads to recruitment of Pelle, a Ser/Thr protein kinase, to the cytoplasmic domain of the Toll receptor. Pelle then phosphorylates Cactus, which is a signal for ubiquitylation and proteolytic degradation, releasing the anchored Dorsal protein. Dorsal is the *Drosophila* homolog of NFκB and can then move to the nucleus and serve as a selective transcription factor.

A third version of proteolytic activation of a protein involved in morphogenesis, cell patterning, and growth is the processing of the Hedgehog protein family (Mullor et al., 2002; Ye and Fortini, 2000) (Figure 8.20). It is noted in Chapter 13 that the 45-kDa single-chain precursor of Hedgehog carries out autoproteolysis at one internal peptide bond via action of the C-terminal half on the N-terminal portion of itself. Furthermore, the N-terminal fragment is transferred not to water as cosubstrate nucleophile but instead to the 3-OH of cholesterol to yield the Hedgehog N-fragment-cholesterol ester (Figure 8.20A). This sterol-tethered protein fragment is restricted to two-dimensional diffusion and local action (Figure 8.20B), both cell autonomously and on neighboring cells, where it can activate Wingless- and TGFβ-signaling pathways (Hall et al., 1997; Porter et al., 1996). Among the targets for the Hedgehog–cholesterol adduct is the protein Cubitus interruptus (Ci), a 155-kDa protein that shuttles between the cytoplasm and nucleus where it activates developmental pathway genes. In resting cells Ci is multiply phosphorylated by protein kinase A, so it is marked for subsequent E3-mediated ubiquitylation and cleavage by the proteasome to a 75-kDa fragment that is a transcriptional repressor. Hedgehog signaling leads to PKA blockade, prevention of Ci155 processing to Ci75, and net transcriptional activation (Figure 8.20B).

These three examples of proteolytic signaling show the versatility and multiple layers of controlled proteolysis in eukaryotic cell protein posttranslational modifications.

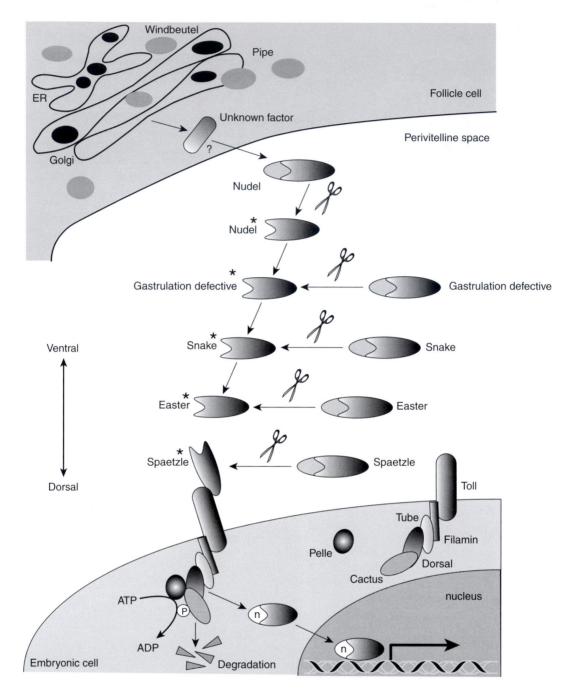

Figure 8.19 Sequential activation of four extracellular proteases in a zymogen to protease cascade during dorsal/ventral axis determination in *Drosophila* (adapted with permission from Ye and Fortini, 2000).

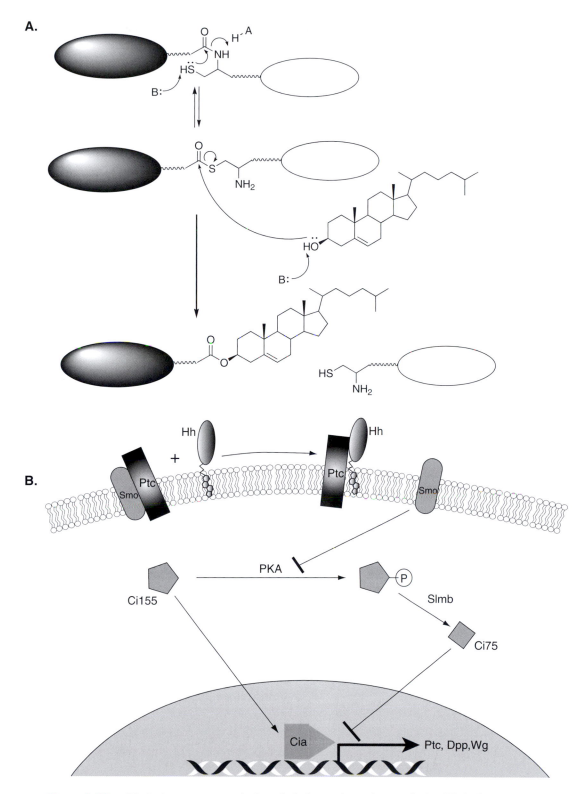

Figure 8.20 **A.** Hedgehog autoproteolysis and cholesterol attachment during Hedgehog maturation. **B.** Model of Hedgehog signaling. Smo has an intrinsic intercellular signaling activity that is repressed by its direct interaction with Ptc within the plasma memberane. This repression is released when Hedgehog (Hh) binds to Ptc, with Smo undergoing a conformational change that allows it to activate its cellular targets.

Apoptotic Cascades and Caspase Regulation

Proteases that employ an active site cysteine thiolate as the catalytic nucleophile to attack the scissile bond of bound protein substrates are represented by two subfamilies in mammalian cells. One is the superfamily of which the thiol protease papain, with its Cys-His-Asp catalytic triad, is the archetype. This family includes Calpain, a Ca^{2+}-activated protease, and the transglutaminase family noted in Chapter 15. The second superfamily of cysteine proteases is a set of 12–15 proteases selective for cleavage at aspartyl residues in protein substrates. These are known as caspases ("C" for the cysteine nucleophile and "Asp" for the selectivity pattern towards substrates). This second family is involved in inflammatory responses (caspase-1) and in programmed cell death (apoptosis) (human caspases-2, 3, 6, 7, 8, 9, and 10) (Thornberry and Lazebnik, 1998).

The caspases active in apoptotic cell death function in hierarchical cascades, with an initiator set (caspases-2, 9, and 10) responsive to different proapoptotic signals, and an effector or executioner set (caspases-3, 6, and 7). As with most proteases, the producing cells make them as zymogens to control their irreversible activity in time and in precise microenvironments. Since the caspase cascade functions intracellularly to generate cell death, there must be a precise balance between activation mechanisms and inactivation or inhibitor mechanisms. Lack of regulation of the balance in favor of activation can lead to inappropriate early cell death (in some neurodegenerative conditions). Lack of regulation in favor of inhibition of the cascade can be a property of cytotoxic-agent-resistant cancer cells.

The zymogen forms, the procaspases, are single-chain polypeptides (Figure 8.21A) that contain an N-terminal propeptide region, a domain that will become the large subunit (20 kDa) and a domain that will become the small subunit (10–12 kDa) (Shi, 2002a). The active forms of the caspases are dimers of heterodimers of the large and small subunits ($\alpha_2\beta_2$), and can have one or two active sites per heterotetramer. The conversion of the zymogen forms to active two-subunit proteases is thought to involve facilitated dimerization of the initiator caspases, those most apical in the cascades. The dimers (e.g., of procaspase-9 or procaspase-2) are thought to have low activity, which can result in propinquity-induced autoactivation, by cleavage at Asp residues between the large and small subunit domains to excise them, and then to trim some of the N-terminal propeptide region. The now active apical caspases in the cascade can activate the zymogen forms of the downstream executioner caspases (Fesik and Shi, 2001; Shi, 2001) by cleavage at Asp residues to generate the two-chain active caspases.

Caspases-2, 8, 9, and 10 have extended pro regions that act as protein–protein recognition domains for the initiation of the caspase cascades when partner proteins provide the signals. Caspases-3, 6, and 7 have much shorter pro regions (Figure 8.21B). They do not engage in regulated interactions with protein partners, and

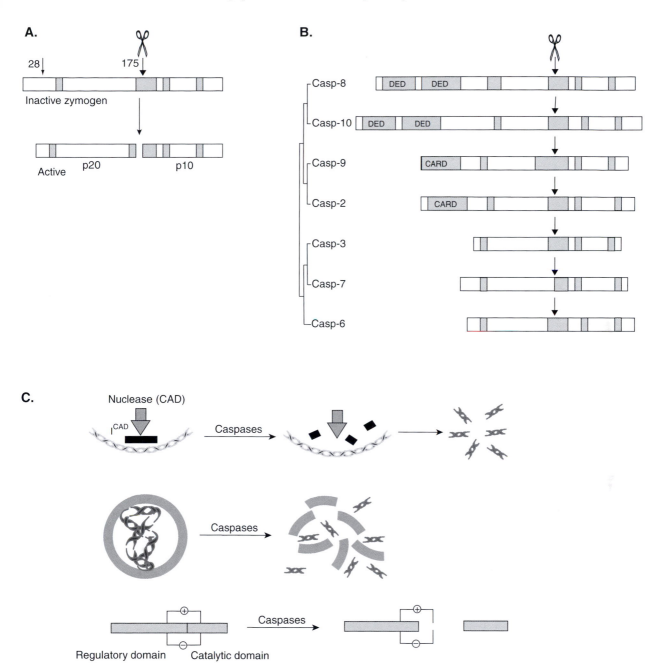

Figure 8.21 **A.** Caspases undergo autocleavage at an internal Asp residue to generate a large and a small subunit from a single-chain proenzyme form. **B.** Schematic diagram of mammalian caspases. **C.** Many proteins essential for cell function are targets of the activated caspases.

instead act as executioner proteases. The activated executioner caspases can, in turn, cleave subsets of proteins that send cells into irreversible programs of cell death when DNA is cleaved, nuclear laminins proteolyzed, cytoskeleton disassembled, and dozens of other proteins fragmented (Figure 8.21C). These include Bcl2, an antiapoptotic protein, DNA-dependent protein kinases, and I^{CAD}, an inhibitor of the caspase-activated DNAse (CAD) that produces the characteristic laddered DNA fragments that are hallmarks of apoptotic cell death. Cleavage of the I^{CAD}–CAD complex on the I^{CAD} releases the CAD to degrade DNA on nucleosomes.

The 10 human caspases show extended site recognition for their protein substrates, starting with the Asp at the cleavage site and extending back four residues towards their N-terminus. For example, studies with combinatorial libraries of peptide substrates indicate that caspases-1, 4, and 5 recognize the sequence WEHD, caspases-2, 3, and 7 prefer to cleave at DEXD, and caspases-6, 8, 9, and 10 cleave at L/VEXD sites in proteins (Thornberry and Lazebnik, 1998). These selectivity differences allow different caspases to act on subsets of cellular proteins for irreversible proteolysis and yet in aggregate provide explosive proteolytic power in apoptotic cascades.

The x-ray structures of several caspases, in active and pro forms, are known and provide some insight into architecture (Figure 8.22) and activation routes (Schweizer et al., 2003; Shi, 2002). In the conversion of procaspase-7 to active caspase-7, for example, three loops rearrange, altering the conformation of the active site cysteine

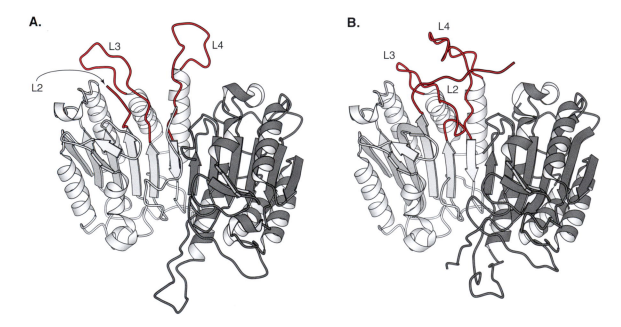

Figure 8.22 Crystal structures of caspase-7 in **A.** the single-chain proenzyme form and **B.** the activated two-chain state (figure made using PDBs 1K88 for **A.** and 1K86 for **B.**).

thiolate from a substrate-inaccessible to substrate-accessible mode. As schematized in Figure 8.21A, the initiator caspases have longer N-terminal propeptide regions (>90 residues) than the executioner caspases (20–30 residues), consistent with the N-terminal regions of the initiator caspases being regions for protein–protein interaction with proteins that carry proapoptotic information.

Two types of domains are indicated: a death effector domain (DED) for caspases-8 and 10, and a caspase recruitment domain (CARD) in caspases-2 and 9. The DEDs are found in proteins [e.g., the Fas-associated protein with death domain (FADD)] carrying information from cell membrane proteins involved in receiving cell death signals from growth factors, while the CARD domains are found in proteins gathering information about cytotoxic agent damage to cells. Procaspase-9 uses its CARD domain to associate with Apaf-1 and cytochrome c in the presence of ATP or dATP to produce the apoptosome (Figure 8.23). In turn, the cytochrome c is released from the

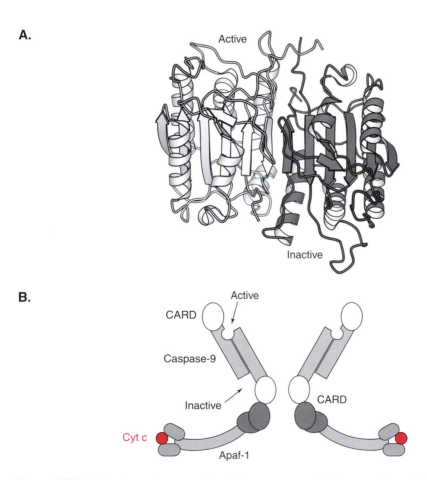

Figure 8.23 **A.** Structure of caspase-9 and the apoptosome. **B.** A proposed model for the activation of caspase-9 (figure made using PDB 1JXQ).

mitochondrion following action of caspase-2, the most apical in this branch of the caspase cascade, on the protein Bid, whose fragment increases the permeability of the outer mitochondrial membrane to effect cytochrome release into the cytoplasm.

These interactions illustrate some of the protein-mediated pathways to zymogen activation in the procaspase hierarchy. The zymogens with long N-propeptides can receive proapoptotic signals from other proteins while the executioner caspases cannot. The zymogen activation can be hierarchical, based on extended site regioselectivity. Procaspase-2 cleaves itself and Bid. The other initiator caspases activate themselves and the zymogen forms of the downstream caspases. The executioner caspases start to proteolyze essential protein targets in cells.

To keep the caspase cascades in check requires constant titration and fine tuning of the caspase activity levels. One way is to control the duration and intensity of the protein-based inputs for turning on the cascades. This can involve compartmentalization of the zymogens from activator proteins, clearly illustrated by using the mitochondrial protein cytochrome c as an activator of cytoplasmic apoptosome formation. Cytochrome c becomes available in the cytoplasm only after the Bcl family member Bid is cleaved by activated caspase-2. Caspase levels can be controlled transcriptionally and by posttranslational modification, including the covalent ubiquitylation (discussed in detail in Chapter 9) that marks them for proteasomal proteolysis.

The other general mechanism for caspase activity control is the use of protein inhibitors selective for this superfamily of cysteine proteases. The concentration and location of such inhibitor proteins is important for controlling inappropriate minibursts of caspase activities. Much information about protein inhibitors—IAPs (inhibitors of apoptosis)—came from initial kinetic and structural studies on viral proteins that have evolved caspase inhibitors to short circuit the cell-death programs that eukaryotic cells initiate when they sense viral infection. Eight mammalian IAPs are known, targeting selectively the initiator caspase-9 and the executioner caspases-3 and 7 (Shi, 2002). The structural element in IAPs that creates the submicromolar potency inhibition of caspases is 80 residues surrounding a zinc atom. These elements are termed baculoviral inhibitor repeats (BIR), and they typically occur in groups of three. When complexed to caspases, they block substrate access to the protease active sites (Figure 8.24) (Shiozaki et al., 2003). Other types of protein inhibitors are also known (FLIPS and ARCS) (Thornberry and Lazebnik, 1998), all presumably integrated to ensure that the basal level of caspase activity is kept very low until sustained or overwhelming activation signals impinge internally or externally on a cell to commit to run the cell-death program.

A.

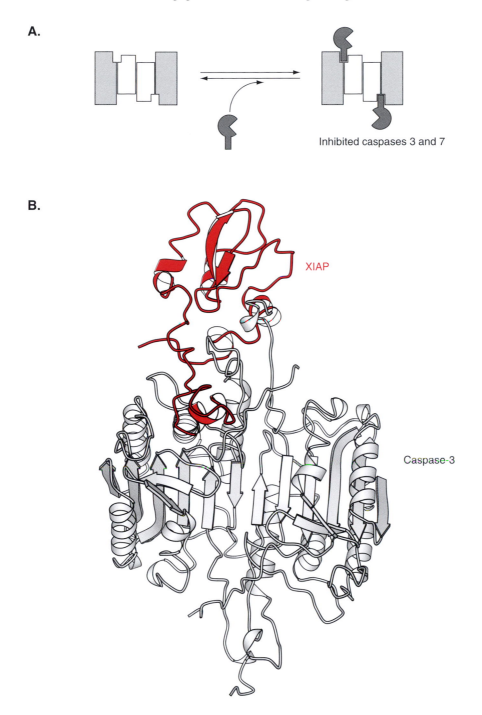

Inhibited caspases 3 and 7

B.

XIAP

Caspase-3

Figure 8.24 **A.** Mechanism of caspase inhibition. **B.** Crystal structure of caspase-3 bound with the BIR2 domain of XIAP (figure made using PDB 1I3O).

References

Arribas, J., and A. Borroto. "Protein ectodomain shedding," *Chem. Rev.* **102:**4627–4638 (2002).

Blobel, C. P. "Remarkable roles of proteolysis on and beyond the cell surface," *Curr. Opin. Cell Biol.* **12:** 606–612 (2000).

Bochtler, M., L. Ditzel, M. Groll, C. Hartmann, and R. Huber. "The proteasome," *Annu. Rev. Biophys. Biomol. Struct.* **28:**295–317 (1999).

Branden, C., and J. Tooze. *Introduction to Protein Structure.* 2nd ed. Garland Publishing: New York (1999) .

Brown, M. S., J. Ye, R. B. Rawson, and J. L. Goldstein. "Regulated intramembrane proteolysis: A control mechanism conserved from bacteria to humans," *Cell* **100:**391–398 (2000).

Chen, P., and M. Hochstrasser. "Autocatalytic substrate processing couples active site formation in the 20S proteasome to completion of assembly," *Cell* **86:**961–992 (1996).

Clausen, T., C. Southan, and M. Ehrmann. "The HtrA family of proteases: Implications for protein composition and cell fate," *Mol. Cell* **10:**443–455 (2002).

Creighton, T. *Proteins: Structure and Molecular Properties.* Freeman: New York (1993).

Darwin, K. H., S. Ehrt, J. C. Gutierrez-Ramos, N. Weich, and C. F. Nathan. "The proteasome of *Mycobacterium tuberculosis* is required for resistance to nitric oxide," *Science* **302:**1963–1966 (2003).

De Strooper, B. "Aph-1, Pen-2, and nicastrin with presenilin generate an active gamma-secretase complex," *Neuron* **38:**9–12 (2003).

Fesik, S. W., and Y. Shi. "Structural biology. Controlling the caspases," *Science* **294:**1477–1478 (2001).

Fortini, M. E. "Gamma-secretase-mediated proteolysis in cell-surface-receptor signalling," *Nat. Rev. Mol. Cell Biol.* **3:**673–684 (2002).

Gallio, M., G. Sturgill, P. Rather, and P. Kylsten. "A conserved mechanism for extracellular signaling in eukaryotes and prokaryotes," *Proc. Natl. Acad. Sci. U.S.A.* **99:**12208–12213 (2002).

Glickman, M. H. "Getting in and out of the proteasome," *Semin. Cell Dev. Biol.* **11:**149–158 (2000).

Grimaud, R., M. Kessel, F. Beuron, A. C. Steven, and M. R. Maurizi. "Enzymatic and structural similarities between the *Escherichia coli* ATP-dependent proteases, ClpXP and ClpAP," *J. Biol. Chem.* **273:**12476–12481 (1998).

Haass, C., and H. Steiner. "Alzheimer disease gamma-secretase: A complex story of GxGD-type presenilin proteases," *Trends Cell Biol.* **12:**556–562 (2002).

Hall, T. M., J. A. Porter, K. E. Young, E. V. Koonin, P. A. Beachy, and D. J. Leahy. "Crystal structure of a Hedgehog autoprocessing domain: Homology between Hedgehog and self-splicing proteins," *Cell* **91:**85–97 (1997).

Hengge, R., and B. Bukau. "Proteolysis in prokaryotes: Protein quality control and regulatory principles," *Mol. Microbiol.* **49:**1451–1462 (2003).

Henrich, S., A. Cameron, G. P. Bourenkov, R. Kiefersauer, R. Huber, I. Lindberg, W. Bode, and M. E. Than. "The crystal structure of the proprotein processing proteinase furin explains its stringent specificity," *Nat. Struct. Biol.* **10:**520–526 (2003).

Holyoak, T., M. A. Wilson, T. D. Fenn, C. A. Kettner, G. A. Petsko, R. S. Fuller, and D. Ringe. "2.4 Å resolution crystal structure of the prototypical hormone-processing protease Kex2 in complex with an Ala-Lys-Arg boronic acid inhibitor," *Biochemistry* **42:**6709–6718 (2003).

Hook, V. Y., A. V. Azaryan, S. R. Hwang, and N. Tezapsidis. "Proteases and the emerging role of protease inhibitors in prohormone processing," *FASEB J.* **8:**1269–1278 (1994.).

Kaser, M., and T. Langer. "Protein degradation in mitochondria," *Semin. Cell Dev. Biol.* **11:**181–190 (2000).

Kenniston, J. A., T. A. Baker, J. M. Fernandez, and R. T. Sauer. "Linkage between ATP consumption and mechanical unfolding during the protein processing reactions of an AAA+ degradation machine," *Cell* **114:**511–520 (2003).

Krojer, T., M. Garrido-Franco, R. Huber, M. Ehrmann, and T. Clausen. "Crystal structure of DegP (HtrA) reveals a new protease-chaperone machine," *Nature* **416:**455–459 (2002).

Kwon, Y. D., I. Nagy, P. D. Adams, W. Baumeister, and B. K. Jap. "Crystal structures of the *Rhodococcus* proteasome with and without its pro-peptides: Implications for the role of the pro-peptide in proteasome assembly," *J. Mol. Biol.* **335:**233–245 (2004).

LaVoie, M. J., and D. J. Selkoe. "The Notch ligands, Jagged and Delta, are sequentially processed by alpha-secretase and presenilin/gamma-secretase and release signaling fragments," *J. Biol. Chem.* **278:**34427–34437 (2003).

McAdams, H. H., and L. Shapiro. "A bacterial cell-cycle regulatory network operating in time and space," *Science* **301:**1874–1877 (2003).

Molloy, S. S., E. D. Anderson, F. Jean, and G. Thomas. "Bi-cycling the furin pathway: From TGN localization to pathogen activation and embryogenesis," *Trends Cell Biol.* **9:**28–35 (1999).

Mullor, J. L., P. Sanchez, and A. R. Altaba. "Pathways and consequences: Hedgehog signaling in human disease," *Trends Cell Biol.* **12:**562–569 (2002).

Nakayama, K. "Furin: A mammalian subtilisin/Kex2p-like endoprotease involved in processing of a wide variety of precursor proteins," *Biochem. J.* **327:**625–635 (1997).

Ogura, T., and A. J. Wilkinson. "AAA+ superfamily ATPases: Common structure—diverse function," *Genes Cells* **6:**575–597 (2001).

Paetzel, M., and R. E. Dalbey. "Catalytic hydroxyl/amine dyads within serine proteases," *Trends Biochem. Sci.* **22:**28–31 (1997).

Paetzel, M., A. Karla, N. C. J. Strynadka, and R. E. Dalbey. "Signal peptidases," *Chem. Rev.* **102:**4549–4579 (2002).

Pickart, C. M., and R. E. Cohen. "Proteasomes and their kin: Proteases in the machine age," *Nat. Rev. Mol. Cell Biol.* **5:**177–187 (2004).

Porter, J. A., S. C. Ekker, W. J. Park, D. P. von Kessler, K. E. Young, C. H. Chen, Y. Ma, A. S. Woods, R. J. Cotter, E. V. Koonin, and P. A. Beachy. "Hedgehog patterning activity: Role of a lipophilic modification mediated by the carboxy-terminal autoprocessing domain," *Cell* **86:**21–34 (1996).

Puente, X. S., L. M. Sanchez, C. M. Overall, and C. Lopez-Otin. "Human and mouse proteases: A comparative genomic approach," *Nat. Rev. Genet.* **4:**544–558 (2003).

Rockwell, N. C., D. J. Krysan, T. Komiyama, and R. S. Fuller. "Precursor processing by kex2/furin proteases," *Chem. Rev.* **102:**4525–4548 (2002).

Rouille, Y., W. Rohn, and B. Hoflack. "Targeting of lysosomal proteins," *Semin. Cell Dev. Biol.* **11:**165–171 (2000).

Sauer, R. T., D. N. Bolon, B. M. Burton, R. E. Burton, J. M. Flynn, R. A. Grant, G. L. Hersch, S. A. Joshi, J. A. Kenniston, I. Levchenko, S. B. Neher, E. S. Oakes, S. M. Siddiqui, D. A. Wah, and T. A. Baker. "Sculpting the proteome with AAA(+) proteases and disassembly machines," *Cell* **119:**9–18 (2004).

Schirmer, E. C., J. R. Glover, M. A. Singer, and S. Lindquist. "HSP100/Clp proteins: A common mechanism explains diverse functions," *Trends Biochem. Sci.* **21:**289–296 (1996).

Schmidt, M., A. N. Lupas, and D. Finley. "Structure and mechanism of ATP-dependent proteases," *Curr. Opin. Chem. Biol.* **3:**584–591 (1999).

Schweizer, A., C. Briand, and M. G. Grutter. "Crystal structure of caspase-2, apical initiator of the intrinsic apoptotic pathway," *J. Biol. Chem.* **278:**42441–42447 (2003).

Selkoe, D., and R. Kopan. "Notch and presenilin: Regulated intramembrane proteolysis links development and degeneration," *Annu. Rev. Neurosci.* **26:**565–597 (2003).

Shi, Y. "A structural view of mitochondria-mediated apoptosis," *Nat. Struct. Biol.* **8:**394–401 (2001).

Shi, Y. "Mechanisms of caspase activation and inhibition during apoptosis," *Mol. Cell.* **9:**459–470 (2002).

Shiozaki, E. N., J. Chai, D. J. Rigotti, S. J. Riedl, P. Li, S. M. Srinivasula, E. S. Alnemri, R. Fairman, and Y. Shi. "Mechanism of XIAP-mediated inhibition of caspase-9," *Mol. Cell* **11:**519–527 (2003).

Steiner, D. F. "The proprotein convertases," *Curr. Opin. Chem. Biol.* **2:**31–39 (1998).

Stock, D., P. M. Nederlof, E. Seemuller, W. Baumeister, R. Huber, and J. Lowe. "Proteasome: From structure to function," *Curr. Opin. Biotechnol.* **7:**376–385 (1996).

Teter, S. A., and D. J. Klionsky. "Transport of proteins to the yeast vacuole: Autophagy, cytoplasm-to-vacuole targeting, and role of the vacuole in degradation," *Semin. Cell Dev. Biol.* **11:**173–179 (2000).

Thornberry, N. A., and Y. Lazebnik. "Caspases: Enemies within," *Science* **281:**1312–1316 (1998).

Urban, S., and M. Freeman. "Intramembrane proteolysis controls diverse signalling pathways throughout evolution," *Curr. Opin. Genet. Dev.* **12:**512–518 (2002).

Voges, D., P. Zwickl, and W. Baumeister. "The 26S proteasome: A molecular machine designed for controlled proteolysis," *Annu. Rev. Biochem.* **68:**1015–1068 (1999).

Wang, J., J. A. Hartling, and J. M. Flanagan. "The structure of ClpP at 2.3 Å resolution suggests a model for ATP-dependent proteolysis," *Cell* **91:**447–456 (1997).

Weihofen, A., and B. Martoglio. "Intramembrane-cleaving proteases: Controlled liberation of proteins and bioactive peptides," *Trends Cell Biol.* **13:**71–78 (2003).

White, J. "ADAMS: Modulators of cell–cell and cell–matrix interactions," *Curr. Opin. Cell Biol.* **15:**598–606 (2003).

Wolfe, M. S., and R. Kopan. "Intramembrane proteolysis: Theme and variations," *Science* **305:**1119–1123 (2004).

Ye, Y., and M. E. Fortini. "Proteolysis and developmental signal transduction," *Semin. Cell Dev. Biol.* **11:**211–221 (2000).

Ubiquitin and Ubiquitin-like Protein Tags

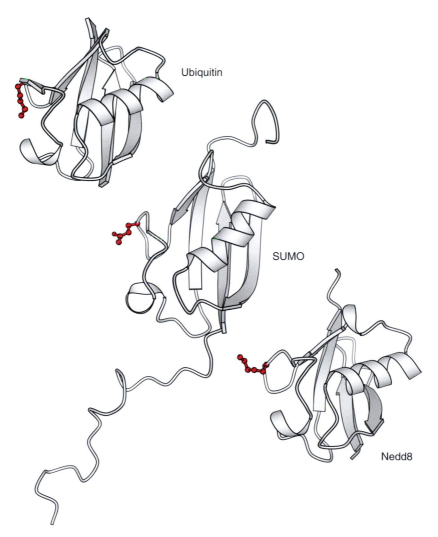

Structures of the information-rich protein tags ubiquitin, SUMO, and Nedd8.

A widely utilized form of posttranslational modification in eukaryotes, but not prokaryotes, is the covalent addition of a small protein, ubiquitin, to the lysine side chains of proteins undergoing modification. Ubiqutin, a 76-residue protein (8 kDa) that folds into an uncommonly stable β structure (Figure 9.1A), was named for its ubiquitous distribution in eukaryotes [see Hershko (1998) and references therein; a ubiquitin fold exists in prokaryotes where it plays a role in sulfur metabolism but not proteolysis]. The C-terminal glycine, Gly_{76}, becomes attached in amide linkage to the ϵ-NH_2 of lysine side chains in client proteins, creating an isopeptide bond (Figure 9.1B) that is resistant to degradation by the normal cast of proteases in cells. Keep in mind, though, that protein ubiquitylation is *reversible* by action of a large number of isopeptidases specific for ubiquitin (and homologs) (Figure 9.1C), known as deubiquitylating enzymes (DUBs) (Wilkinson, 2000).

The posttranslational acylation and deacylation of protein lysine side chains by ubiquitin has a chemical analogy to the acetylation of lysine side chains (e.g., by histone acetyltransferases) discussed in Chapter 6 and the myristoylations of Chapter 7. However, in contrast to the simple C_2 acetyl chain, ubiquitin provides an architecturally complex acyl moiety, allowing recognition by many distinct proteins in eukaryotic cells. The concept of ubiquitin as an *information-rich covalent protein tag* to set up new and expanded protein–protein interactions, *in cis* and *in trans*, has been a useful framework for assessing the diverse functional roles arising from protein ubiquitylations.

In higher eukaryotes ubiquitin is joined by about 10 protein homologs, *ubiquitin-like* proteins (termed Ubls) with the acronyms SUMO, Nedd8, UCRP, FAT10, HAB, ISGN15, Apg8, Apg12, URM1, and An1 (Jentsch and Pyrowolakis, 2000; Schwartz and Hochstrasser, 2003). These Ubls expand the network of protein-based signaling enabled by posttranslational modification of specific sets of proteins recognized by a particular Ubl in isopeptide bond linkage to lysine side chains. Covalent addition of ubiquitin to protein substrates is termed ubiquitination in earlier publications and ubiquitylation in contemporary references. The term "ubiquitylation" will be used in this book (Varshavsky, 1997). The two best studied Ubls are SUMO (*small ubiquitin-like modifier*) and Nedd8 (Figure 9.2A). The terms "sumoylation" and "neddylation" of proteins are used in the literature for the respective posttranslational modification of their protein substrates by these protein tag superfamily members. Different Ubl-isopeptide linkages to proteins are also cleaved by subsets of specific isopeptidases.

A remarkable variation in the transfer of activated Ubls to lysine side chains of proteins is effected by the enzyme(s) processing the Ubl protein Apg8 (also known as Aut7 for its role in autophagy) (Ichimura et al., 2000). This is involved in membrane vesicle formation, which generates autophagosomes in cells (Schlumpberger et al., 1997). The 8-kDa Apg8 protein is attached via its C-terminal glycine carboxylate, not to a protein, but to the ethanolamine head group of the diacylglycerol-based lipid phosphatidy-

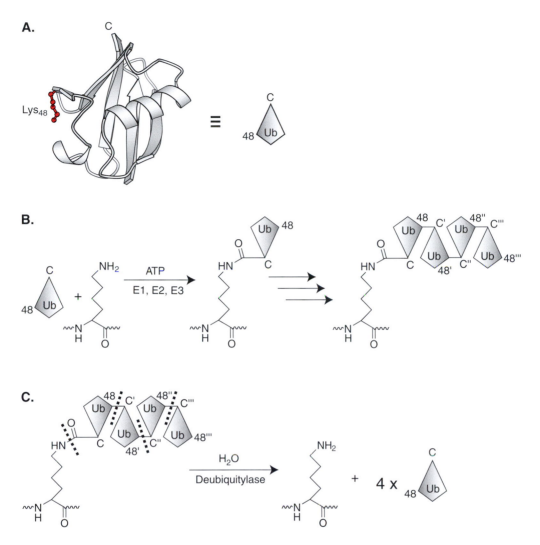

Figure 9.1 **A.** Architectural fold of the 76-residue protein ubiquitin used for posttranslational cova-lent tagging of proteins. The shorthand representation shows the C-terminus at one vertex and the Lys$_{48}$ at another vertex (figure made using PDB 1TBE). **B.** Isopeptide bond formed between C-terminal Gly$_{76}$ of ubiquitin and the ε-NH$_2$ of lysine side chains in proteins undergoing modifica-tion. Polyubiquitylation occurs between Lys$_{48}$ ε-NH$_2$ and the C-terminus of the next Ub molecule. The Ub$_4$ chain shown is required to target such modified proteins to the proteasomes. **C.** Hydrolytic cleavage of the isopeptide bond between ubiquitins and between the most proximal Ub and a tagged protein by deubiquitylase (DUB) action.

lethanolamine (Figure 9.2B). Apg8 is still linked through an amide bond, but now it is linked to a mobile phospholipid. It has been hypothesized that this may be a route for extracting PE lipids from bulk membrane, to form the new protein-containing auto-phagosome vesicles (Ichimura et al., 2000). It is not known if the amide bond is then hydrolyzed to allow each Apg8 protein molecule to recycle and move many molecules

A.

B.

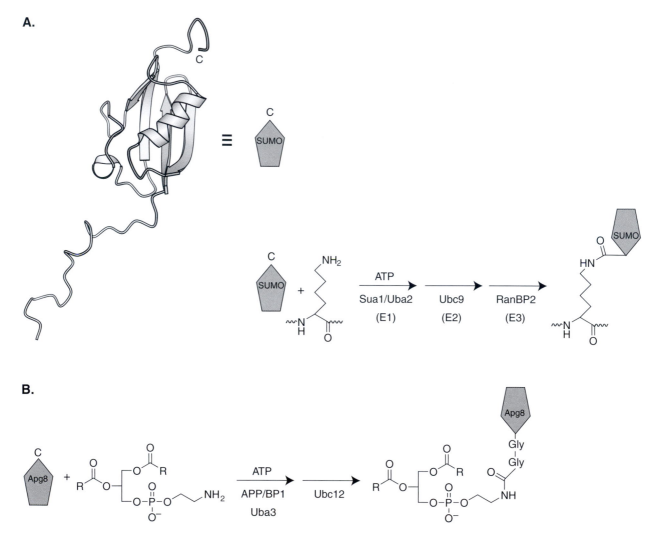

Figure 9.2 Ubiquitin-like protein tags (Ubls): **A.** SUMOylation at lysine side chains to create iso-peptide bonds analogous to ubiquitin linkages (figure made using PDB 1A5R). **B.** Amide linkage between the Ubl Apg8 and the amino group of phosphatidylethanolamine to create a ubiquitylated phospholipid.

of PE into the autophagosomes during vesicle growth. This reaction further demonstrates the versatility of protein tagging strategies and machinery in eukaryotic cells. It is reminiscent of the C-terminal modification of the Hedgehog protein fragment by cholesterol, required for its signaling function (discussed in Chapter 13). One additional variation for ubiquitin tagging is the report that proteasome-mediated proteolytic degradation of the p21 protein inhibitor of cyclin-dependent kinases (Bloom et al., 2003) occurs, not at lysine side chains, but at the N-terminus of p21. This would generate a peptide rather than an isopeptide bond between ubiquitin and p21.

Early on, attention focused on the role of ubiquitin as a protein tag that marked membrane and intracellular proteins for regulated proteolytic destruction by the proteasome (discussed in Chapter 8). Characteristic of proteins sent to the proteasome are polyubiquitin chains, where four tandemly tethered ubiquitins (Ub_4)– are a minimum signal for targeting such marked proteins to the proteasomes (Figure 9.3). The Ub_n chains are built up processively as will be noted below, connected through Lys_{48} residues of successive Ubs. Proteins slated for proteolytic destruction by *polyubiquitylation* include proteins involved in the cell cycle, most famously the cyclin subunits of cell division kinases (Cdks) (Chapter 2); transcription factors, including p53, c-Jun, E2F-1, and NFκB (Herschko and Ciechanover, 1998); enzymes such as the protein kinase A regulatory subunit; HMG CoA reductase, the

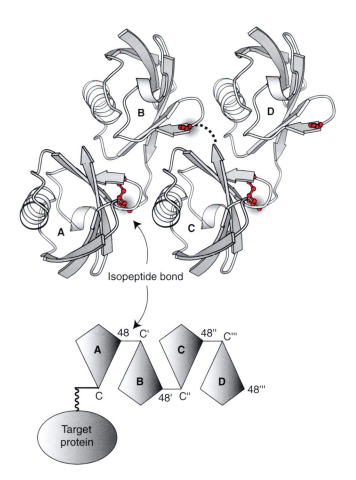

Figure 9.3 Polyubiquitylation: the three-dimensional representation of a tetraubiquitin (Ub_4) chain, with each ubiquitin in isopeptide linkage to the next ubiquitin, is shown. This is the minimal length that sends marked proteins to the proteasome for proteolytic cleavage (figure made using PDB 1TBE).

rate-limiting enzyme in the sterol biosynthetic pathway; and antigens for processing and display on T cells.

More recently, enzymatic *monoubiquitylation* of a diverse range of proteins has been detected, from histones to plasma membrane receptors (Hicke, 2001; Schnell and Hicke, 2003). Monoubiquitylated proteins are generally not routed to proteasomes but are involved in several signaling pathways. Most notable is the internalization of monoubiquitylated membrane receptor proteins and their sorting in the endocytic pathway that leads to transport to lysosomes and proteolytic degradation therein (Cope, 2003). Thus, polyubiquitylation sends proteins to the proteasome for proteolysis while monoubiquitylation and endocytosis can lead to trafficking to the lysosome and destruction by a distinct cadre of proteases (Chapter 8). Each route controls the lifetime and subcellular distribution of distinct subsets of proteins.

The dynamics of monoubiquitylation versus polyubiquitylation are interesting questions for understanding branch points in the use of ubiquitin as a posttranslational protein tag. Recent studies by Li et al. (2003) have shown that the Mdm2 ubiquitin ligase for the p53 transcription factor can effect either mono- or polyubiquitylation of p53. At low ratios of Mdm2/p53 monoubiquitylation predominates and causes the singly tagged p53 to relocalize to the cytoplasm, presumably rendering that p53 molecule inactive as a transcription factor. This is reversible if the single ubiquitin–isopeptide linkage can then be cleaved by deubiquitylases (DUBs), described later in the chapter. At high ratios of Mdm2 ligase to p53, polyubiquitin chains get assembled and p53 is irreversibly removed by subsequent proteolysis by proteasomes in both the nucleus and cytoplasm. Sumoylation and neddylation of proteins rarely, if ever, proceed beyond the addition of one Ubl tag. These modifications are not involved in regulated proteolysis of protein substrates, but rather they are for control of protein location and control of protein function (Seeler and Dejean, 2003).

Consistent with the widespread utilization of ubiquitin and ubiquitin homologs for the tagging of proteins in eukaryotic cells, there are many enzymes, receptors, and adaptor proteins that have ubiquitin recognition domains, required for conferring specificity to trafficking of the tagged proteins to different destinations within cells. The discussion begins with an analysis of the ubiquitin activating machinery and then the subsets of protein domains known to recognize the ubiquitin and Ubl folds and particular surfaces of these protein tags.

Ubiquitin and Ubl-Activating Enzymatic Machinery: E1 and E2 Ligases

Implementation of the strategy of covalent conjugation of ubiquitin and Ubl homologs to the lysine side chains of client/acceptor proteins requires protein machinery for the regiospecific activation of these 8-kDa proteins at the C-terminus and their transfer to subsets of partner proteins. The general solution is to activate the C-terminal carboxylate of ubiquitin and Ubls as thioesters (Figure 9.4A). This is analogous to the logic of using acetyl CoA as acetyl donor in cellular metabolism, including histone acetylations by HATs (Figure 6.4A). Subsequent transfer to lysine amino groups in protein cosubstrates is favored thermodynamically, is catalytically mediated, and proceeds to high stoichiometry (Figure 9.4B).

In contrast to low-molecular-weight acyl CoAs, the thioester form of activated ubiquitin and Ubls are cysteine thioesters in the active site of specific transfer enzymes known as enzymes 2 (E2s) or subsets of enzymes 3 (E3s) of the machinery discussed in the next paragraph. The protein domains of E2 and E3 provide the protein–protein recognition elements for selecting acceptor proteins to be covalently modified on lysine side chains. To make acyl-S-enzyme intermediates (e.g., ubiquityl-S-E2) requires chemical activation of the unactivated, resonance-stabilized form of the C-terminal glycine–COO$^-$, and this is accomplished by spending an ATP molecule to make the ubiquityl-Gly$_{76}$–COOAMP (Figure 9.5A). This is the job of the first enzyme, E1, in the ubiquitylation activation cascade, with two half-reactions—namely, formation of ubiquityl-AMP and then its capture by the thiolate of the active site Cys to generate the acyl-S-enzyme, ubiquityl-S-E1. This acyl

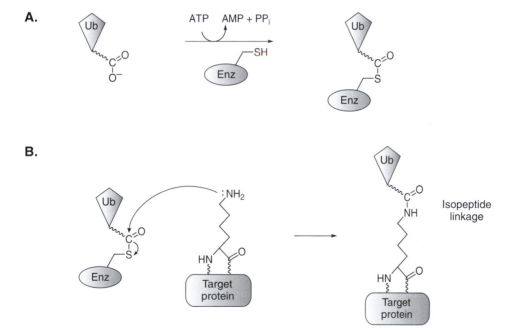

A.

B.

Figure 9.4 **A.** Enzymatic activation of ubiquitin as enzyme-bound acyl thioester to an active site Cys of Enzyme 1. **B.** Thermodynamically favored transfer to make the isopeptide amide bond with lysine residues of proteins that have become targeted for modification.

enzyme now has the requisite thermodynamic activation for the ubiquityl moiety to serve as the acylation reagent.

Enzyme 1 is specific for recognizing the ubiquitin protein substrate and does not recognize any acceptor proteins as cosubstrates. Therefore, the activated ubiquityl moiety is passed from the active site Cys of E1 to the active site Cys–S$^-$ of various E2s, an energy-neutral thioester interchange (Figure 9.5B). There is a separate E1 for making the SUMO-AMP intermediate. NMR and/or x-ray structures of SUMO and Nedd8 (Bayer et al., 1998; Rao-Naik et al., 1998; Whitby et al., 1998) reveal high spa-

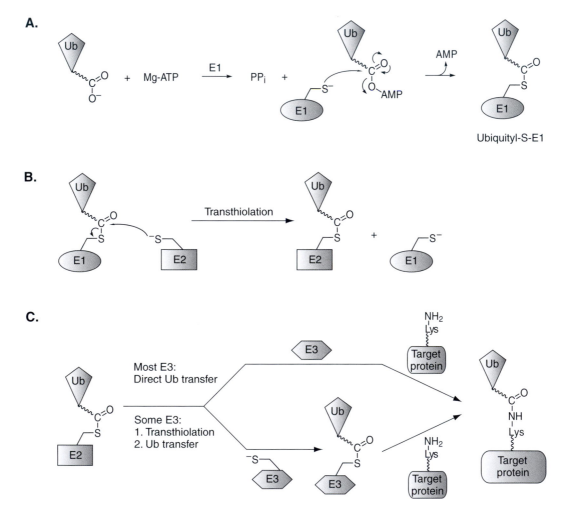

Figure 9.5 Enzymatic activation of ubiquitin Gly$_{76}$ carboxylate (the C-terminal residue) by tandem action of E1, E2, and E3: **A.** ATP-dependent formation of ubiquityl-AMP by E1 and transfer to active site cysteine thiolate to generate ubiquityl-S-E1. **B.** Transthiolation between active site cysteines of E1 and E2 enzymes transfers the ubiquityl group to form ubiquityl-S-E2 isoforms. **C.** E3 action either involves ubiquityl-S-E3 formation (HECT subgroup in the upper branch) or by action as regulatory and/or cosubstrate recruiting subunits for ubiquityl transfers to target proteins by Ub-S-E2 (lower branch).

tial similarity with ubiquitin but with replacement of Arg_{48} by Glu in SUMO and Arg_{72} by Ala in Nedd8; the positive charges are thought to disrupt recognition by the ubiquitin-specific E1. The x-ray structure of the heterodimeric E1 specific for Nedd8 activation shows an adenylation domain, a domain with the catalytic cysteine that becomes neddylated, and a ubiquitin-like domain for interaction with E2s (Walden et al., 2003).

Several E2 variants have had their structures solved by x-ray analysis (Pickart, 2001), revealing some common features for the recognition of the E1 protein by the E2 active site, allowing ubiquityl transfer without its hydrolytic loss. There are 13 E2 proteins predicted in the *Saccharomyces cerevisiae* yeast genome (Weissman, 2001) and substantially more in the human genome, suggesting diversification and specialization for client protein recognition. The E2 family members in yeast are termed Ubc1-13 (*u*biquitin *c*onjugating enzymes) and the human homologs UbcH1, 2, etc. On subsequent purification and assay, it turned out that Ubc9 had no activity with ubiquitin, ATP, and E1, but was involved in sumoylation pathways and so is a SUMO-specific E2 (Seeler and Dejean, 2003). Analogously, Ubc12 is a Nedd8-specific transacylase (Figure 9.6). Some E2s interact directly with particular client protein substrates without the need for an E3 as adaptor. The *Drosophila* protein $TAF_{II}250$, a multimodular protein that is part of the basic transcription factor protein machinery [and recognizes acetylated lysine side chains in histones (see Chapter 6)] contains embedded E1- and E2-like domains and has activity for the monoubiquitylation of histone H1 in nucleosomes (Pham and Sauer, 2000). But, classically, E1 and E2 were insufficient for robust ubiquitylation of acceptor proteins and led to the discovery of additional proteins, defined as enzymes 3 (E3s) (Figure 9.5C) (Hershko and Ciechanover, 1998).

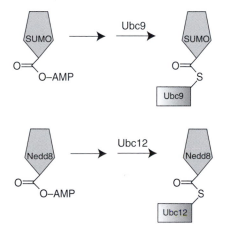

Figure 9.6 Ubc9 is a SUMOyltransferase and Ubc12 is a Neddyltransferase rather than ubiquitin-recognizing E3 enzymes.

Families of E3 Ligases Recognizing Different Partner Proteins

E3s have been harder to characterize than E2s by bioinformatics because they are more diverse, and because many of them are found in multimodular and multisubunit enzyme complexes. Two major subfamilies of E3 ubiquitin ligases (Figure 9.5C) are HECT proteins (*h*omologous to *E6* *c*arboxyl *t*erminus) and RING domain-containing proteins (*r*eally *i*nteresting *n*ew *g*ene) (Pickart, 2001; Weissman, 2001). The HECT E3s tend to follow the catalytic logic of E1s and E2s with active site cysteines and the transfer of the activated ubiquityl chain to the active site of these E3s. The ubiquityl-S-E3 is then the proximal donor to the lysine side chain NH_2 groups of client proteins recognized by E3. The RING domains are typically cysteine-containing zinc binding motifs that are structural elements that sometimes mediate E2–E3 interactions. RING-type E3 proteins are combinations of chaperones and E2 positioning regulatory subunits. They recruit both E2 and protein cosubstrates, to create the proximity for the thermodynamically favored ubiquitylation of lysine side chains, rather than being direct catalytic subunits. There may be a continuum of roles from chaperone to catalyst among the many E3s. While some E3s are monomeric, many of them act in multienzyme complexes as noted below.

One major family that emphasizes the combinatorial possibilities that allow the E1–E2–E3 activation cascade to modify hundreds to thousands of intracellular eukaryotic proteins in a spatially and temporally regulated manner is the SCF E3 ligase complexes. The SCF family of E3 ligases are prototypic of multicomponent E3 protein complexes. An SCF Ub ligase consists of four kinds of protein subunits: a scaffold protein termed a *C*ullin (e.g., Cdc53), a ring domain E2 recruiting subunit termed Roc1/Rbx1/Hrt1, *S*kp1, and one of several *F* box proteins (Cope, 2003). The x-ray structure of the four-component SCF E3 ligase, containing Cul1 (776 residues), Rbx1 (108 residues), Skp1 (163 residues) and the F box[Skp2] protein, has been determined (Zheng et al., 2002), revealing the orientation of the scaffold and the adaptor Skp1, and allowing a model for the interaction with the Ub thioester-linked E2 (Figure 9.7) to be proposed. The E2 ubiquitintransferase is about 60 Å away from the F box subunit that recruits substrate, suggesting a very large cavity in which target proteins can be accommodated for binding and then ubiquityl transfer (Cardozo and Pagano, 2004).

There are 78 human F box proteins predicted from genome sequence analysis (Cardozo and Pagano, 2004), two additional Rbx subunits, and six cullins, suggesting huge combinatorial potential ($78 \times 2 \times 6$) for such E3 complexes in specific recognition of acceptor proteins to be ubiquitylated by SCF E3 ubiquitin ligases (Zheng et al., 2002). For example, Cul2 and Cul5 scaffold proteins use a Skp1-like protein called elongin C and bind not F box proteins but a family of adaptors called BC proteins. Analogously, Cul3 is found in complexes containing BTB box family members instead of F box subunits (Pintard et al., 2003; Xu et al., 2003). The total of F box,

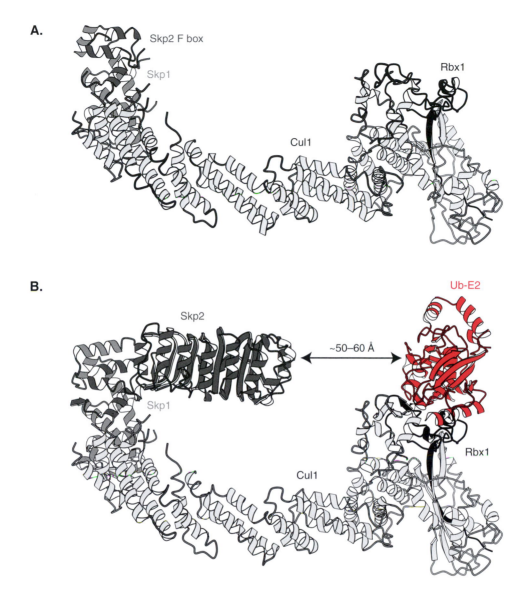

Figure 9.7 **A.** Architecture of the four component SCF E3 ubiquitin ligases (figure made using PDB 1LDK). **B.** Proposed interaction of SCF E3s with ubiquitylated E2 enzymes [coordinates kindly provided by Cardozo and Pagano (2004)].

BC box and BTB box proteins in the human genome is greater than 300 (Xu et al., 2003), suggesting this multipartite superfamily of E3 ligases may be responsible for much of the turnover of short-lived regulatory proteins and protein subunits in eukaryotic cells.

It is noted in Chapter 12 that a four-component SCF-type E3 ligase recognizes the *hydroxylated* version of HIF-1α over the nonhydroxylated form by hydrogen bonding to the HO-Pro$_{564}$ by about 1000-fold. This in turn leads to specific poly-ubiquitylation and proteasomal destruction of the posttranslationally hydroxylated form of the oxygen sensor protein. A number of other F protein components of the SCF E3 ligases recognize *phosphorylated* proteins as a prior set of posttranslational modifications that set proteins up for ubiquitylation. This may be of particular consequence in the destruction of proteins during cell cycle progression. For example, when cyclin-Cdk levels rise in yeast, Sic1 gets phosphorylated at multiple serines or threonines, creating a threshold for recognition by the F box component of a particular SCF E3 ligase (Reed, 2003), triggering Sic1-(Thr–OPO$_3$)$_n$ for polyubiquitylation and proteasome-mediated removal. The structure of the phosphodegron region of the cyclin E subunit recognized by the F box subunit Cdc4 has been determined by crystallography and comprises the pentapeptide sequence LLpTPP (Orlicky et al., 2003) (Figure 9.8).

A third type of posttranslational modification that brings proteins to the attention of a subset of F-subunit-containing E3 ligases is the Man$_9$GlcNAc$_2$ carbohydrate core on *N-glycoproteins* when they show up in the cytoplasm during the ERAD (ER accelerated degradation) protein quality control response. Proteins that have been N-glycosylated (Chapter 10) but cannot be folded in the ER lumen are shipped back to the cytoplasm, tagged, and sent to the proteasome. The Fbs1 and Fbs2 forms of E3 ligases recognize the N-glycan chain on proteins in the cytoplasm and recruit them for polyubiquitylation (Yoshida et al., 2003). Table 9.1, taken from Pickart

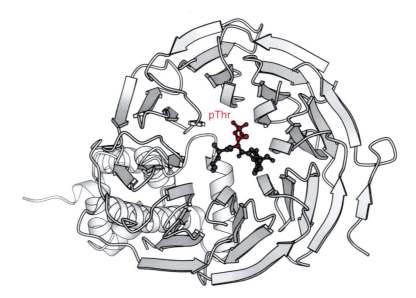

Figure 9.8 SCF-type E3 ligases can recognize partner proteins that have been tagged with a specific posttranslational modification. The phosphorylated pentapeptide LLpTPP is the phosphodegron element in the cell cycle protein Sic1 that allows recognition by the F box subunit of the multicomponent SCF E3 ligase as shown in the cocrystal structure. The phosphoThr residue is in red and the rest of the pentapeptide in black in ball-and-stick representation. The F box subunit is in gray (figure made using PDB 1NEX).

Table 9.1: Prior Modification of Proteins as Signals for Polyubiquitylation

Deacetylation:

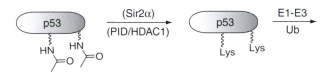

Hydroxylation:

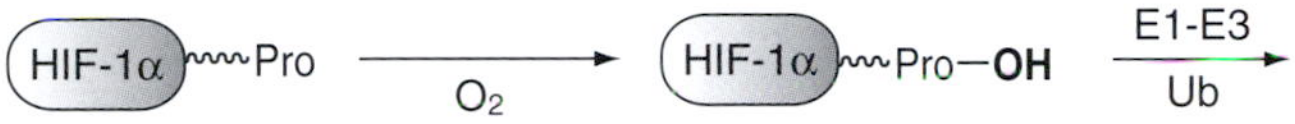

Glycosylation:

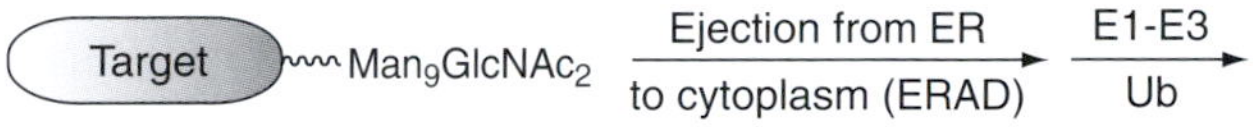

Aminoacylation (N-end rule):

Oxidation (N-end rule):

Phosphorylation:

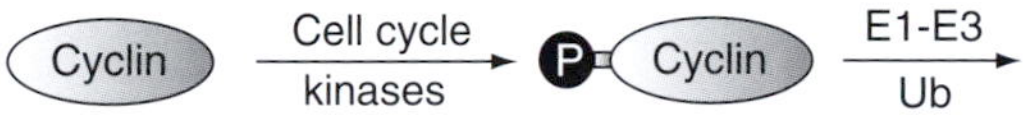

Protein association (or dissociation):

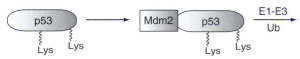

(2004), lists seven different prior posttranslational modifications that set proteins up for recognition and subsequent ubiquitylation by subsets of E3 ligases. In sum, post-translational modification states of proteins are read by particular F box components of multisubunit E3 ligases in the subtly orchestrated control of protein lifetimes by ubiquitylation.

Using mass-tagged versions of Skp1 in yeast (Seol et al., 2001) have identified at least seven distinct Skp1-continuing complexes. The minimum number of proteins in an E2-SCF-client protein substrate complex from the considerations above is six; up to 14 proteins were found in some of the yeast Skp1 complexes, emphasizing the complexity that could permit specificity and regulation. The anaphase-promoting complex that ubiquitylates mitotic cell cycle proteins has 11 protein components (Murray, 2004). The Skp1 subunit in *Dictyostelium* is subject to two further types of posttranslational modifications, hydroxylation of Pro_{143} to hydroxyproline and then glycosylation of that –OH group by three glycosyltransferase activities to create a fucosylgalactosyl-GlcNAc-O-prolyl linkage (Van Der Wel et al., 2002a and 2002b).

For SCF E3 ligases to have optimal activity, the cullin subunit must be modified at Lys_{720} by the ubiquitin homolog Nedd8 (Zheng et al., 2002). A Nedd8-specific E1 and E2 carry out the posttranslational tagging of the cullin subunit to activate SCF E3 ligases to carry out ubiquitylation of specific protein cosubstrates. This is a tandem, double covalent tagging of the E3 *catalyst* (neddylation) and the *protein substrate* (ubiquitylation) in the largest family of E3 ligases (Figure 9.9). The 8-kDa Nedd8 moiety must effect a conformational change in the multicomponent E3 ligase that has a net activating effect.

Exactly how these multicomponent E3 ligases orient incoming proteins for multiple rounds of ubiquitylation is not known (Pickart, 2001). In particular, the mechanism of *processive* elongation of monoubiquitin chains to the minimum of Ub_4 units, required for recognition by the proteasome to target the Ub_4–NH–Lys-protein for

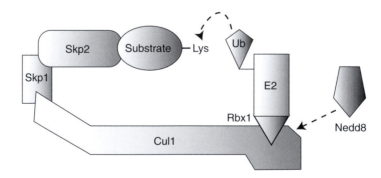

Figure 9.9 Double modification by SCF E3 ligases: neddylation of the cullin subunit by a Nedd8-specific E1–E2 pair; ubiquitylation of client protein by the active neddylated E3.

proteolysis, is not fully clear. Combinations of E3 ligases increase the yield of poly-ubiquitylation of the UNC4 chaperone for myosin in muscle cells (Hoppe et al., 2004). Chain elongation up to Ub_{20} has been detected in *in vitro* E1–E3 reconstitutions.

In addition to the kinetic issue of multiple rounds of transfer of ubiquitin from successive ubiquityl-S-E2 molecules, there is the issue of change in regioselectivity between the first ubiquityl transfer, which modifies a substrate protein lysine side chain, and *all* subsequent transfers which sequentially derivatize the most recently added ubiquitin (Figure 9.10). Ubiquitin has seven lysine residues—K_6, K_{11}, K_{27}, K_{29}, K_{33}, K_{48}, and K_{63}. Of these, K_{48} is the lysine side chain selected for polyubiqui-tin chain growth that sends Ub_n-marked proteins to proteasomes. As shown in Figure 9.11, Lys_{48}, Lys_{63}, and Lys_{29} are on three different faces of the ubiquitin fold and can be regioselected for ubiquitin chain modification by distinct E2–E3 ligase pairs. While both K_{48} and (less commonly) K_{29} are sites for polyubiquitylation and subse-quent escorting to proteaseomes for breakdown, K_{63} can be *mono-* and polyubiquity-lated, and signals differently in membrane protein internalization and/or DNA repair

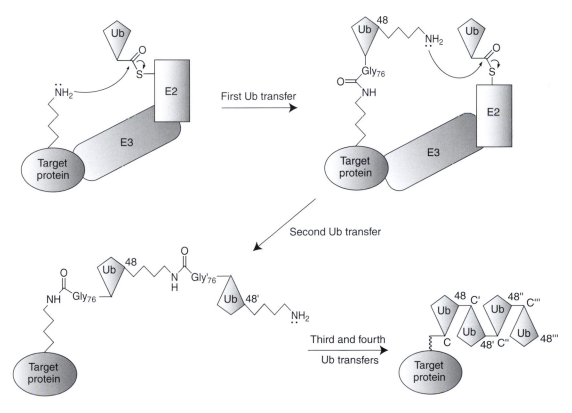

Figure 9.10 Processive elongation of ubiquitin chains on a protein targeted by multicomponent E3 ligases: isopeptide bonds between the C-terminal Gly_{76} carboxyl and the ϵ-NH_2 of Lys_{48} of the pre-ceding ubiquitin in the elongating polyubiquitin chain.

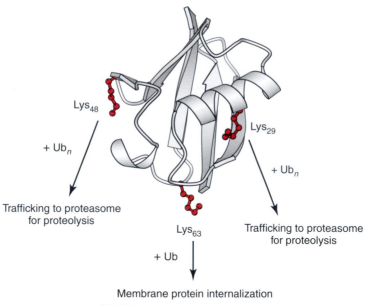

Lys$_{48}$

+ Ub$_n$

Trafficking to proteasome
for proteolysis

Lys$_{29}$

+ Ub$_n$

Trafficking to proteasome
for proteolysis

Lys$_{63}$

+ Ub

Membrane protein internalization
Trafficking to endosomal compartments
Activation of DNA repair processes

Figure 9.11 Ribbon diagram of ubiquitin with side chains of K$_{29}$, K$_{48}$, and K$_{63}$ shown in red ball and stick. In yeast, polyubiquitin chains linkages are found in the order K$_{48}$ > K$_{63}$ > K$_{29}$ (figure made using PDB 1TBE).

(Schnell and Hicke, 2003). The regioselective utilization of different Lys side chains in ubiquitin elongation adds yet another layer of recognition to Ub-based protein signaling. In the activation of the transcription factor NFκB in resting lymphocytes, the protein is anchored, inactive, in the cytoplasm by binding to the inhibitory protein IκB. Interleukin1-mediated signaling (see subsequent section) of TRAF6 (Schwartz and Hochstrasser, 2003) ultimately leads to polyubiquitylation of a phosphorylated form of IκB, and its subsequent destruction by the proteasome and release of NFκB, which moves to the nucleus and binds to promoter regions of target genes. The TRAF6 protein in this signaling pathway is thought to contain K$_{63}$-linked Ub$_n$ chains rather than K$_{48}$-linked ubiquitins.

Recognition of Ubiquitin and Ubl Domains for Protein Trafficking and Sorting in the Cell

The three-dimensional fold of ubiquitin (Figure 9.1) and ubiquitin-like (Ubl) protein domains (Jentsch and Pyrowolakis, 2000) are both recognized and distinguished by many proteins in eukaryotic cells, notably in the proteasome capture of Ub$_n$-proteins by protein receptors in the 19S proteasome lid, in the endocytic sorting of Ub_1-membrane proteins undergoing internalization, and in the change of location of monosumoylated proteins (Schnell and Hicke, 2003). Multiple domains and recognition sequences in chaperones, adaptors, and docking proteins seem to have evolved to interact with Ub and Ubl architectures and enable trafficking to particular subcellular destinations.

Bioinformatic analysis (Upadhya, 2003) has recently suggested that Ubl domains are embedded in up to 62 proteins in addition to the 10 free-standing Ubls thought to be used as bona fide protein tags (Aguilar and Wendland, 2003). For ubiquitin, SUMO, Nedd8, and most of the rest of the Ubls used as covalent protein tags, they are expressed in inactive precursor forms, with C-terminal peptide extensions. Until the C-termini are removed hydrolytically by endopeptidase action, the –GG–COO⁻ end that gets adenylated by ATP via E1 action is unavailable. Ubiquitin itself is typically encoded as five head-to-tail repeats, each Ub domain of which can be removed by DUBs and ubiquitin C-terminal hydrolase (Uch-L1) (Figure 9.12). Analogously, SUMO is expressed in a 101-residue precursor form and needs to have the 25 C-

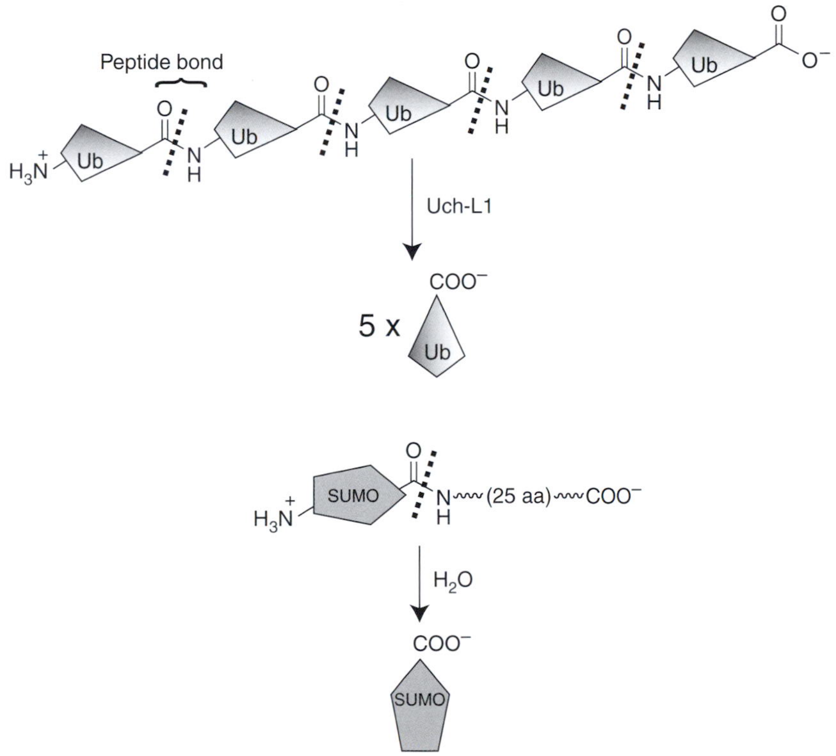

Figure 9.12 Precursor forms of ubiquitin and SUMO are inactive for use in covalent posttranslational modification of proteins until they are processed by proteolytic trimming to yield GG at the newly liberated C-termini for recognition and activation by E1 isoforms.

terminal residues removed by an endoproteolytic clip to reveal the mature 76 residue SUMO, with –GG–COO$^-$ available for activation and protein conjugation (Finley, 2002; Pickart, 2001). The other 62 embedded Ubl proteins do not appear to be excised but rather can be recognized in situ by the chaperone, adaptor, and docking proteins that recognize ubiquitylated proteins.

One such Ubl-containing chaperone protein is Rad23, which is proposed to bind the polyubiquitin motif and ferry Ub_n-proteins to the S5a subunit of the proteasome (Hicke, 2001; Weissman, 2001). Rad23 also contains a 40-residue motif, designated as the ubiquitin binding associated (UBA) domain. Structural analysis of related UBA domains revealed a three-helix bundle, likely to interact with the hydrophobic surface of ubiquitin, containing $Leu_8Ile_{44}Val_{70}$ (Schnell and Hicke, 2003). The S5a subunit of the 19S subunit lid of the proteasome has two repeats of a 20-residue sequence termed a ubiquitin interaction motif (UIM). The UIM domain from several proteins has been characterized for ubiquitin-binding affinity and the UIM from the yeast vacuolar sortin protein Vps27p has been studied by NMR in complex with ubiquitin (Fisher et al., 2003). A third motif, also 40 residues in length, and also forming a three-helix bundle (Kang et al., 2003; Prag et al., 2003; Shih et al., 2003), is the CUE motif (*c*oupling of *u*biquitin conjugation to *E*R degradation) found in proteins involved in the ubiquitylation and retrotranslocation of malfolded proteins out of the ER compartment back into the cytoplasm for transport to the proteasome (Kostova and Wolf, 2003). The UIM and CUE motifs are thought to interact with the same conserved hydrophobic surface on ubiquitin and Ubl domains as the UBA domain. UIM, UBA, and CUE domains probably recognize both Ub_1 and Ub_n chains, perhaps with different affinities and binding lifetimes; these differences are likely to mediate control of the trafficking and intracellular destination of tagged proteins.

The structure of the Ub_4 motif that is the minimal recognition element to direct a marked protein to proteasomes has been solved by NMR (Cook, 1994). It shows a two-fold screw axis (Figure 9.3), allowing recognition of multiple Ub domains by tandemly arrayed UIM or UBA domains. Additional domains found to recognize the three-dimensional fold of ubiquitin tags are UEV domains (E2-based Ub binding site minus the catalytic cysteine), PAZ domains (*p*olyubiquitin *a*ssociated *z*inc finger), and NZF (*n*ovel *z*inc *f*inger) domains (Schnell and Hicke, 2003). The structure of a NZF-Ub complex has been solved by NMR and reveals interaction surfaces (including the I_{44} surface) of ubiquitin that confer recognition (Alam et al., 2004).

The histone deacetylase HDAC6 (Verdin et al., 2003) has a PAZ domain while the vacuolar protein involved in sorting Vps36 has a NZF domain (Schnell and Hicke, 2003). The UEV protein Mms2 interacts with an E2 and directs regiospecificity of subsequent ubiquitylation chain elongation to K_{63}. Polyubiquitylation via Ubc13 action at successive K_{63} residues occurs on PCNA, the processivity-conferring clamp subunit of DNA polymerase in DNA repair (Hicke, 2001; Weissman, 2001), and also in the IκB proteolysis pathway (Schwartz and Hochstrasser, 2003).

A variety of ubiquitin recognition motifs have thus evolved to direct tagged proteins down different pathways. The Rad23 chaperone may use its UBA domain to bind Ub_n-tagged proteins and then its embedded Ubl domain may enable recruitment of the complex to the tandem UIM motifs of S5a for docking at the proteasome (Hartmann-Petersen et al., 2003). Some of these adaptor and binding proteins get monoubiquitylated in the process, including the epsins involved in the endocytosis of receptor proteins. One of the most surprising aspects of protein degradation assisted by CUE domain proteins is that malfolded proteins that have failed multiple cycles of chaperone-assisted folding in the ER are ubiquitylated (Uba1/Ubc7), possibly in the ER, then retrotranslocated through the Sec61 translocon channel. At the ER membrane the p97 ATPase is involved in providing the energy for the retrotranslocation, assisted by Ufd1p and Np14p adaptor proteins, each of which has ubiquitin recognition motifs (Flierman et al., 2003). Once in the cytosol these marked proteins are chaperoned by Rad23 to the proteasome for degradation. If the Ub_n-marked protein is N-glycosylated, Rad23 apparently presents it in the cytoplasm to the protein N-glycosylase to remove the N-glycan carbohydrate chain before the proteasome can thread the substrate polypeptide chain into the proteolytic chamber.

Proteins Marked for Ubiquitylation and Proteasomal Destruction During the Cell Cycle

There are many proteins targeted for ubiquitin-directed destruction at different points in the cell cycle progression [see Reed (2003) and Murray (2004) for reviews] that, in aggregate, commit the cell to traverse unidirectionally through the cycle. The cyclin subunits, for example cyclins A–E, of cyclin-dependent cell division kinases (CDKs) were identified and named for the property of undergoing cycles of synthesis and regulated proteolytic destruction. Also marked for proteolytic degradation by polyubiquitylation are proteins acting as negative regulators of cyclin-CDK holo enzymes (Reed, 2003). These include the Sic1 protein of *S. cerevisiae*, which inhibits B-type cyclin CDKs. Its ubiquitylation and rapid destruction occurs at the G1-S transition in the yeast cell cycle. In mammalian cells, cognate kinase inhibitors p27^{Kip1} and p21^{Cip1} are destroyed by this route. For example, the ternary complex of cyclin E-CDK•p27 accumulates and is inactive in G1. At the G1-S transition, the ubiquitin-mediated destruction of p27 liberates the now active cyclin E-Cdk2 (Chapter 2) and helps drive the transition to S phase by the phosphorylation of protein substrates.

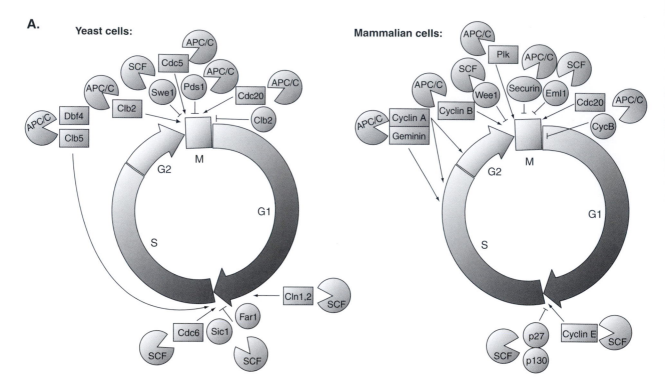

Figure 9.13 **A.** Controlled destruction of regulatory subunits of cyclin-dependent cell division kinases occur at the G1-S transition by SCF-type E3 ligase and at the G2-M boundary by APC/C E3 ligases [redrawn from Reed (2003)].

Figure 9.13A [redrawn from Reed (2003)] shows that those proteins ubiquitylated at the G1-S boundary in both yeast and mammalian cells are substrates for the SCF-type E3 ligases noted above. At the G2-M transition, on the other hand, a set of conditionally short-lived proteins is targeted by a different enzyme called the anaphase promoting complex (APC/cyclosome), a complex multisubunit ubiquitin ligase. The APC/cyclosome E3 ligase can contain an activating subunit, Cdc20, that may also be ubiquitylated and proteolytically broken down. An alternate subunit, Cdh1, is in turn negatively regulated through phosphorylation by protein kinases. SCF forms of E3 ligases are active throughout the cell cycle, and their substrate choices at particular time points appears to depend on the regulated phosphorylation of potential substrates by cell cycle kinases (Murray, 2004). In contrast, the APC/cyclosome multisubunit E3 ligase is regulated by the phosphorylation of its specific subunits rather than through the phosphorylation of target proteins (Figure 9.13B).

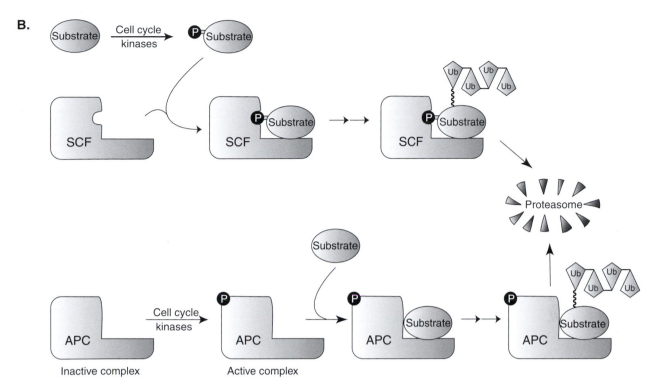

Figure 9.13 B. Cell-cycle dependent action of SCF ligases is regulated by kinase-dependent phosphorylation of target protein *substrates*. By contrast it is the phosphorylation of subunits of the *catalyst* that regulates APC ubiquitin ligase activity.

Reversibility of Ubiquitylation and Ubl-Tagging

Covalent ubiquitin tags can be removed in at least two kinds of circumstances. One is when the polyubiquitylated proteins have been captured by the S5a subunit in the lid complex of the proteasome and are being unfolded by its ATPase and threaded into the chamber containing the multiple protease active sites (Chapter 8). The Rpn11 subunit of the 19S lid has isopeptidase activity capable of cleaving the Ub_n chain. This preserves ubiquitin tags for recycling by Uch-LI back to ubiquitin monomers and also makes sure that the unfolded protein to be degraded can be threaded into the chamber. The bulky Ub_n branch is otherwise likely to impede the threading process or later stages of proteolysis (depending on the point of attachment of the Ub_n chain).

A second set of contexts where cleavage of the ubiquitin CO–NH protein isopeptide bond (Figure 9.14) might occur is for retrieval of proteins before they get chewed up by the proteasome degradation machinery. This could reverse premature degradation of specific proteins. In support of the concept that deubiquitylases (DUBs) might have regulatory functions is the large multiplicity of such predicted isopeptidases, 16

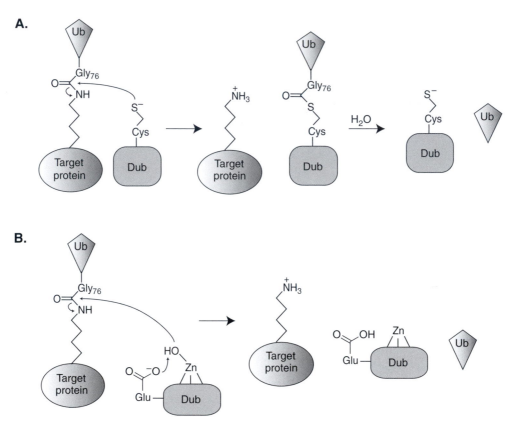

Figure 9.14 Contexts for hydrolytic removal of polyubiquitin chains by deubiquitylases and the regeneration of ubiquitin monomers from Ub_n chains. Two mechanistic subclasses of deubiquitylases cleave isopeptide linkages in polyubiquitin chains and between the proximal ubiquitin and the targeted protein: **A.** Active site cysteine as nucleophile and acyl-S-enzyme intermediate. **B.** Zinc-metalloenzyme strategy for isopeptide bond hydrolysis.

in the yeast genome and about 50 in the human genome. If DUBs had only a Ub_n-chain-removal function at the proteasome it is not obvious why so many would be needed. Four subfamilies of ubiquityl-protein (iso)peptidases have been noted (Gan-Erdene et al., 2003). The Uch group is probably involved in the initial C-terminal activation of proubiquitin and proUbls and is likely not to work on isopeptide links. The 50-member UBP/USP (ubiquitin-specific protease) class is noted above and these are hydrolases with an active site cysteine, most likely generating ubiquityl-S-enzyme intermediates during hydrolysis (Figure 9.14A), analogous to the acyl thio-ester intermediates during ubiquitin activation by E1 and E2. Phenotypic analysis of various USP mutants is underway but already DUBs have been implicated in a plethora of cellular functions, from modulation of gene silencing, neuronal migration, synapse size and strength, chemotaxis, differentiation, and oncogenesis (Gan-Erdenne et al., 2003; Wilkinson, 2000). There are seven human members of a ubiquitin-like pro-

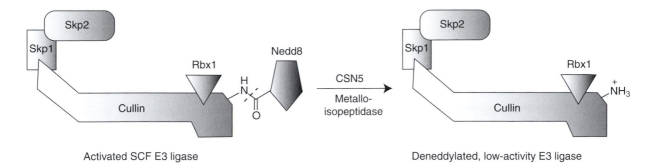

Figure 9.15 The CSN5 subunit of the COP9 signalosome is a zinc-dependent isopeptidase that cleaves the Nedd8-protein from SCF-type E3 ligases, down-regulating the ubiquitylating activity from the activated to basal state.

tease subfamily. There are also multiple examples of metallopeptidases that cleave the Ub_n-protein linkages (Figure 9.14B). The Rpn11 subunit of the proteasome is in this subclass, as is the CSN5 subunit of the COP9 signalosome (Cope, 2003) discussed below.

The COP9 signalosome was first detected in photosignaling pathways of gene regulation in plants (Cope and Deshaies, 2003), but has also been found in other organisms, from worms to humans. It is an eight-subunit, 500-kDa protein complex that has, as one of its functions, the ability to down-regulate the activity of SCF-type multicomponent E3 ubiquitin ligases. The signalosome can achieve this by reversal of the tandem neddylation of the E3 cullin subunit of the catalyst and by deubiquitylation of bound Ub_n-substrate proteins. The deneddylation reaction (Figure 9.15) is direct; the CSN5 subunit is a zinc isopeptidase specific for the Nedd8 acyl group. Removal of the Nedd8 tag converts the E3 ligase back to low basal activity and thereby decreases SCF activity. It is not yet known whether the dozens of possible different SCF E3 ligase complexes are deneddylated equally well by the COP9 signalosome. The CSN subunit is also reported to recruit a DUB, Ubp12, to the COP9-E3 complex. This DUB could deubiquitylate the associated ubiquityl-S-E2 and block the tagging process at that step or Ubp12 might cleave one or more ubiquitin moieties from the Ub_n protein product before it dissociates. Additional deneddylases have also been purified (Wu et al., 2003), suggesting that the full complement of isopeptidases is not yet known.

Doubtless there are specific desumoylases as well, given the signaling functions of sumoylation noted in a later section. For the other Ubl tags not involved in targeting proteins for proteolytic destruction, it is likely that a balance of isopeptide ligase/isopeptide hydrolase activity will be maintained for reversible titration of the functional modifications, akin to kinase/phosphatase and HAT/HDAC posttranslational modification strategies.

Monoubiquitylation of Membrane Proteins During Endocytosis

A system that exemplifies the nonproteosome arm of ubiquitin protein conjugation in protein signaling is the trafficking of cell surface proteins from the plasma membrane to the lysosomes (or the equivalent vacuolar compartment in yeast). Engagement of extracellular ligands by transmembane proteins functioning as receptors in eukaryotes, from yeast to humans, can lead to regulated internalization and subsequent partitioning between recycling back to the cell surface or transfer to multivesicle compartments (Bonifacino and Traub, 2003; Katzmann et al., 2002) on the way to lysosomes for degradation by lysosomal proteases. Transmembrane proteins, so mobilized, include both G-protein-coupled receptors [the STE2 mating factor α receptor was the first example (Hicke, 2001)], such as the β-adrenergic receptor and the chemokine CXCR4 receptor, and the receptor tyrosine kinases, such as growth hormone receptor and epidermal growth factor receptor (EGFR) (Figure 9.16A).

These transmembrane proteins are ubiquitylated on their cytoplasmic domains while still at the plasma membrane. For the epidermal growth factor receptor (EGFR) the signaling/internalization pathway has been well studied (Figure 9.16B). EGF engagement externally drives the dimerization of two EGFR subunits and the resultant *in trans* autophosphorylation of their cytoplasmic domains at several tyrosine residues in the cytoplasmic tail, as detailed in Chapter 2. The pY residues recruit several proteins, often by SH_2 domain engagement on the protein undergoing docking on the pY residue (Chapter 2). Recruitment of the multidomainal Cbl protein via its SH2 domain brings the Ubl and RING domains of Cbl proximal to the cytoplasmic tail of phosphorylated EGFR. The RING domain recruits a ubiquityl-S-E2 that monoubiquitylates EGFR on a specific Lys in its cytoplasmic domain. This single ubiquitin tag is necessary and sufficient for EGFR endocytosis, presumably by recognition of elements of the clathrin-mediated formation of early endosomes (Katzmann et al., 2002). Failure to ubiquitylate EGFR leads to its efficient recycling back to the plasma membrane and continued signaling and prolonged activation of the MAP kinase pathway. Oncogenic versions of EGFR, such as ErbB2, lack the ubiquitylatable lysine and indeed recycle at much higher efficiency (Duan et al., 2003). Endocytosis of the adrenergic receptor uses an *in trans* variant of the monoubiquitylation signaling (Figure 9.16C). Its partner cytoplasmic protein β-arrestin is the protein that gets monoubiquitylated and thereby mediates adrenergic receptor internalization. Analogously, the growth hormone receptor acts *in trans* with monoubiquitylated Epsin 15 (*EGFR pathway substrate 15*) for an early step required for endocytosis.

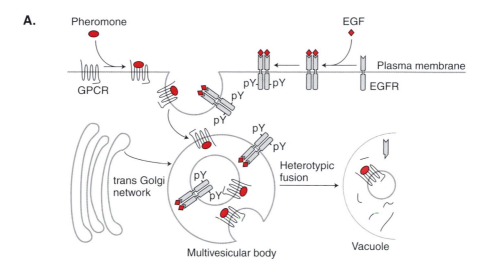

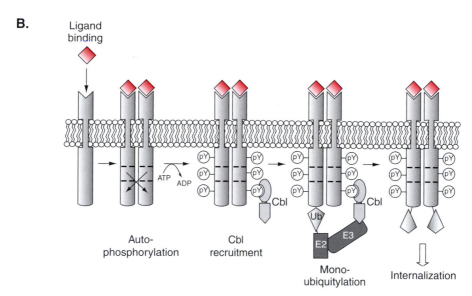

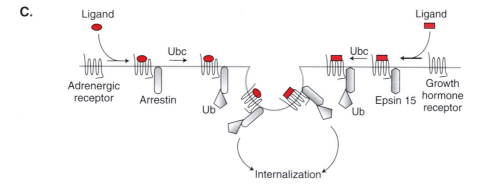

Figure 9.16 A. Internalization of transmembrane receptor GPCRs and receptor tyrosine kinases.
B. Internalization of the EGF receptor: autophosphorylation on tyrosine residues in the cytoplasmic
domain recruits Cbl, which monoubiquitylates the phosphorylated EGFR on one lysine residue.
C. Monoubiquitylation *in trans* on a partner protein is the internalization system for the adrenergic
receptor (arrestin) and the growth hormone receptor (Epsin 15).

The ubiquitin tag on a transmembrane protein or a closely associated partner protein (e.g., arrestin or Epsin) is then "read" by several proteins involved in a second set of sorting steps, as early endosomes get sorted into late endosomes on the way to lysosomes, or are sent back to the plasma membrane. Many of these proteins contain one of the several ubiquitin recognition domains noted above. From yeast genetic analyses there are some 50 genes encoding vacuolar proteins (Vps) that participate in vacuole formation (Katzmann et al., 2002). Of these, Vps23 has a UEV domain, Vps27 a tandem UIM pair of domains, and Hse1 a UIM domain. Vps23, Vps28, and Vps37 form a complex called ESCRT-I (Figure 9.17), presumably recognizing one or more

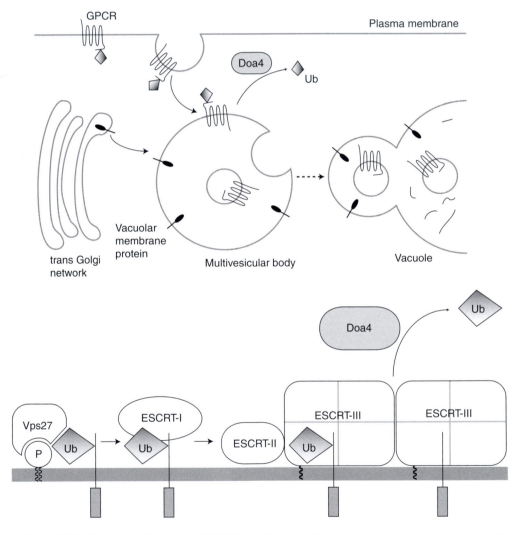

Figure 9.17 Monoubiquitylation of GPCRs marks them for sorting by proteins that contain ubiquitin recognition motifs in the early endosome and trans Golgi network. Assembly of ESCRT complexes to chaperone the monoubiquitylated membrane protein undergoing internalization. Reversal of the monoubiquitylation occurs in the ESCRT-III complex that contains a deubiquitylase (Doa4) that removes the ubiquitin tag hydrolytically before passage of proteins into yeast vacuoles.

of the internalizing, ubiquitylated transmembrane protein components. Several other vacuolar proteins self-associate to produce ESCRT-II and ESCRT-III complexes. When ESCRT-III has been recruited, a DUB, known as Doa4 (degradation of alpha-4), binds to the assembled protein machinery and hydrolytically removes the monoubiquitin tags from the internalizing proteins. Thus, the ubiquitin protein tag is used *reversibly* to initiate endocytosis of many proteins, and is removed before the final passage of the proteins into lysosomes or vacuoles at a point where the tag has provided two stages of sorting instructions.

Small Ubiquitin-like Modifier (SUMO)

Probably the best characterized of the 10 or so ubiquitin homologs that serve as alternate posttranslational tags is the SUMO protein (*s*mall *u*biquitin-like *mo*difier), with one isoform in yeast, three in mammals, and eight or more in plants. As noted in earlier sections of this chapter, SUMO is released from a precursor form by a C-terminal hydrolase (Figure 9.12) and then activated at the newly uncovered $-GG-COO^-$ tail by an E1–E2 pair that is specific for SUMO rather than ubiquitin [a two-subunit E1 (AosI/Uba2) and the E2 (Ubc9)]. Some of the 60 known sumoylation protein targets (Seeler and Dejean, 2003) appear to interact directly with the acylated E2, SUMO-S-Ubc9, for lysine-sumoylation, but there are at least three SUMO E3 ligases recently identified—the polycomb protein, the RanB2 binding protein for the Ran GTPase that is involved in the nucleocytoplasmic shuttling of proteins, and the four isoforms of protein inhibitors of STAT (the transcription factors in interferon signaling pathways).

Given the absence of the lysine corresponding to ubiquitin K_{48} in SUMO, it is thought that polysumoylation is unlikely, and sumoylation does not appear to be used to mark proteins for proteolytic destruction either in proteasomes or lysosomes. Rather, there is a set of SUMO-specific isopeptidases and it is likely to be the balance of sumoylation/desumoylation that determines signaling functions of these protein tags (Bylebyl et al., 2003).

Many of the known sumoylation substrates are nuclear proteins (Rogers et al., 2003), including transcription factors such as p53, CREB, STAT1 and STAT4, GATA-2, coactivators and repressors, p300, and HDAC1 and HDAC4 (see Chapter 6). There are also signaling pathway proteins such as IκBα, Mdm2 (a specific p53-ubiquitin E3 ligase), and Mek1 (Table 9.2) (Seeler and Dejean, 2003). One of the first proteins known to be sumoylated was the Ran GTPase activating protein, RanGAP1. Both RanGAP1 and its sumoylating E3 RanBP2 can localize at nuclear pore complexes. The hypothesis has been proposed that reversible sumoylation by RanBP2 on one side of the nuclear pore and desumoylation on the other side by a SUMO isopep-

Table 9.2: SUMO Substrates Grouped by Function

Transcription Factors	Genome Integrity/Structure
AR	Top1
PR	Top2
GR	PCNA (Sc)
p53	Dnmt3b
p73α	WRN
c-Jun	BLM
CREB	Rad22 (Sp), Rad52
HSF1	Rhp (Sp), Rad51
HSF2	TDG
AP-2	**Signal Transduction**
Lef-1	IκBα
Tcf4	Mdm2
IRF-1	CamKll (Dm)
C/EBP	Mek1
Ttk69 (Dm)	**Nuclear Bodies**
Dorsal (Dm)	PML
Bicoid (Dm)	Sp100
Sp3	HIPK2
c-Myb	Daxx
PDX1	TEL, TEL-AML1
ARNT	**Nuclear Pore Complex**
APA-1	RanGAP1
STAT1, STAT4	RanBP2
GATA-2	**Cytoplasmic**
Transcription Cofactors	Yeast septins
GRIP1	GLUT1, GLUT4
SRC-1	
p300	
DJ-1	
TIF1α	
HDAC1	
HDAC4	
PlASxβ	
PlAS1	
CtBP	
Pc2	

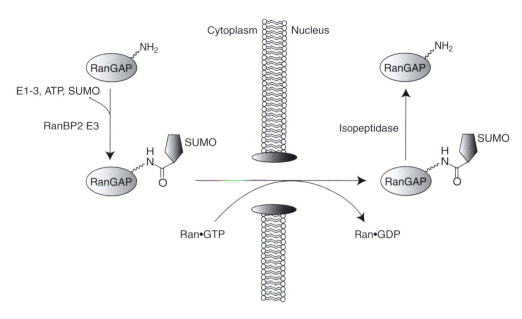

Figure 9.18 RanGAP as SUMOylation substrate and the SUMO-specific E3 ligase RanBP2 have been proposed to participate in shuttling substrate proteins in and out of the nucleus by reversible SUMOylation and deSUMOylation, by a SUMO-specific isopeptidase, at opposite sides of the nuclear pore.

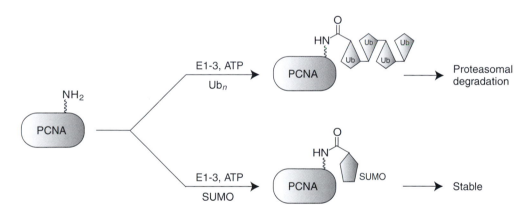

Figure 9.19 PCNA, a protein processivity clamp for DNA polymerase, can be modified on the same lysine residue either by ubiquitin or SUMO, with antagonistic functional outcomes.

tidase may contribute to Ran-mediated shuttling of some proteins in and out of the nucleus (Seeler and Dejean, 2003) (Figure 9.18). There are also cytoplasmic protein substrates for SUMO—namely, the glucose transporters GLUT1 and GLUT4, and (in yeast) the GTPase septins, required for budding.

This cast of sumoylated proteins has led to the suggestion that SUMO tags are used to direct protein location and relocation within the nucleus and between the nucleus and cytoplasm (Melchior et al., 2003). The tags would modify the activity and/or physically sequester the many transcription factors that can be monosumoylated. The premise is that nuclear protein sumoylation is involved in several aspects of monitoring and maintaining genome integrity, including topoisomerases I and II during DNA damage and the processivity protein clamp PCNA, where sumoylation at a specific lysine residue antagonizes polyubiquitylation at the same lysine (Figure 9.19). Sumoylation of histone H4 (residue not yet determined) has been reported to occur with effects on transcriptional repression (Shiio and Eisenman, 2003). Many roles for the SUMO protein tag and the other Ubls that are as yet largely uncharacterized will reveal many additional facets of these versatile 8-kDa protein posttranslational tags (Schwartz and Hochstrasser, 2003).

Ubiquitylation and Other Posttranslational Modifications of the Same Proteins

The timing of the degradation of short-lived proteins in eukaryotic cells (e.g., cyclin subunits of cell division kinases at particular temporal check points) can be controlled by several factors, including the recognition of particular lysine side chains to be modified by E3 ligase machinery. The accessibility of the lysine side chains in some target proteins can be controlled by additional posttranslational modifications, including phosphorylations at nearby side chains, to activate or block ubiquitylation, and also acetylations that would directly compete with those ubiquitylations for covalent tagging of the protein.

The IκBα subunit that anchors the NFκB transcription factor in the cytoplasm of resting lymphocytes is destroyed by ubiquitin tagging and proteasomal proteolysis only after both Ser_{32} and Ser_{36} are phosphorylated by IκBα protein kinases. Then Lys_{21} and Lys_{22} can become polyubiquitylated (Hershko and Ciechanover, 1998), resulting in proteasomal destruction of this inhibitory subunit. On the other hand, the c-Jun transcription factor, when phosphorylated, is blocked from ubiquitylation. Thus, specific E3 ligase machineries can recognize the Ser–OPO_3^{2-} anionic side chains, positively or negatively.

Analogously, the p53 transcription factor can be multiply phosphorylated, on Ser_{20} by the kinase Chk1 and on Thr_{81} by the Jun kinase Jnk (Brooks and Gu, 2003). Both of these phosphorylations disrupt the interaction of p53 with the E3 ligase Mdm2. The result is a longer half-life for p53 since ubiquitylation is blocked (Li et al., 2002b) (Figure 9.20A). The Mdm2 E3 ligase can also be phosphorylated, with the same result of blocking p53 recognition and increasing its half-life *in vivo* (Figure 9.20B). The lifetime of p53 can also be lengthened by the action of one or more deubiquitylating enzymes, one of which is the herpes-virus-associated cellular protein

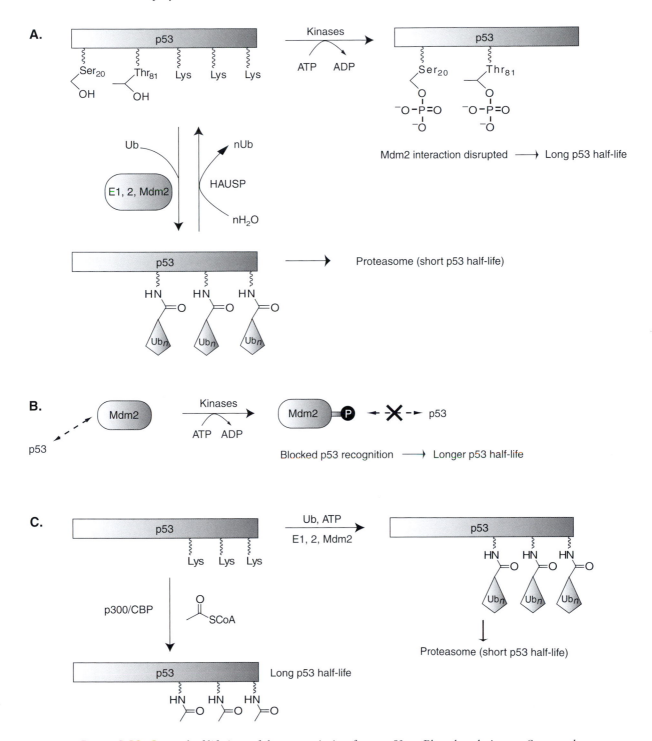

Figure 9.20 Control of lifetime of the transcription factor p53. **A.** Phosphorylation on Ser_{20} and Thr_{81} blocks interaction with the E3 ligase Mdm2. **B.** Phosphorylation on Mdm2 E3 ligase blocks p53 recognition. **C.** Acetylation and ubiquitylation of lysines on p53 protein.

(HAUSP) (Li et al., 2002a), which opposes the action of Mdm2. The balance of Mdm2/HAUSP activity helps to titrate the ubiquitylation level and thereby the half-life of p53 (Brooks and Gu, 2003). Mdm2 can itself become autoubiquitylated or autoneddylated to add an additional layer of modifications (Xirodimas et al., 2004), while its target p53 can also be neddylated or ubiquitylated.

A second example where posttranslational modification of a target protein is a crucial determinant for recognition by an E3 ubiquitin ligase is the HIF-1α subunit of the oxygen sensing system, noted in Chapter 13 (Ivan et al., 2001). The prior hydroxylation of two proline side chains to hydroxyprolines in HIF-1α by a nonheme iron protein hydroxylase enables specific hydrogen bonding of the HO–Pro residues in HIF-1α to the substrate recognition domains of the four-component, cullin-2-containing SCF-type E3 ligase (see Chapter 12). The difference in affinity of SCF ligase for Pro_{542} versus $HO–Pro_{542}$ forms of HIF is at least 1000-fold.

There are cases where lysine residues in proteins with transcription factor and/or signaling functions can be modified with two mututally exclusive posttranslational modifications. The p53 protein is one such example. Near the C-terminus of p53 there are five lysine residues (K_{370}, K_{372}, K_{373}, K_{378}, and K_{382}), any of which can become acetylated (Li et al., 2002b) by the HAT domains of the transcription factor coactivator proteins p300/CBP (also see Chapter 6). These lysines are also the site(s) of ubiquitylation and chain extension to polyubiquitins. If HAT activity is high, ubiquitylation is physically blocked and the lifetime of p53 is increased (Figure 9.20C).

A second case is the Smad7 protein in the signaling pathway for TGFβ cytokines (Gronroos et al., 2002). TGFβ binds to transmembrane receptors that have cytoplasmic serine/threonine kinase domains (Chapter 2). The autophosphorylation of those domains recruits Smad proteins, which form complexes and move to the nucleus as transcription factors. There are also inhibitory Smad proteins, such as Smad7, that act as dominant negative regulators of the pathway. Smad7 is in the nucleus of cells at rest but moves to the plasma membrane upon TGFβ activation and can inhibit signaling by two modes (Izzi and Attisano, 2004). One is by binding to the cytoplasmic domain of the transmembrane receptor Ser/Thr kinase to compete with and block signaling by the other Smads. The second is that Smad7 can be ubiquitylated on two N-terminal lysines and so forms a complex with ubiquitin E3 ligases such as Smurf1 and Smurf2. On translocation from the nucleus to the plasma membrane Smad7 carries Smurf1 with it. The Smurf1 E3 ligase can now monoubiquitylate the receptor and start it down an internalization and degradation pathway of the type noted above for such membrane proteins. The same two N-terminal domain lysines of Smad7 can be N-acetylated by the HAT activity of p300 in the nucleus. This blocks subsequent association with and ubiquitylation by Smurf ligases. In turn, this prevents the negative regulatory actions of Smad7 in the TGF signaling pathway.

The huntington protein, whose mutant forms with longer than normal glutamine repeats cause Huntington's disease, can be analogously sumoylated or competitively ubiquitylated at the same lysine side chains. Sumoylation exacerbates neurodegenerative tendencies while ubiquitylation abrogates them (Steffan et al., 2004).

The Beginnings of a Proteomics Approach to Define the Full Range of Ubiquitylated and Ubl-Modified Proteins

Some 25 years after the initial discoveries of ubiquitin conjugation as a widely used protein posttranslational modification (Hershko et al., 2000), it is still unclear how large is the ubiquitylated proteome, what sets of lysines are modified in target proteins, and which ones are modified by other Ubls. This knowledge base will be required, *inter alia*, to generalize the posttranslational competitions between the acetylation and ubiquitylation of proteins mentioned in the preceding section. A proteomics-based survey has been made in *S. cerevisiae* (Peng et al., 2003) using the expression of an N-His_6-tagged ubiquitin, followed by nickel-NTA affinity column. Proteins eluted from the nickel column were digested with trypsin, and 1075 candidate yeast proteins thereby identified. These were then further screened for a mass increase of 114.1 mass units, arising from a Gly-Gly-Lys peptide fragment from the junction of the ubiquitin C-terminus and the lysine of a target protein, diagnostic for a ubiquitin attachment site. Seventy-two confirmed ubiquitylated yeast proteins were identified. This is probably just the tip of the iceberg thus far, since it is the most abundant proteins that will be most readily detected by the mass spectrometer. Furthermore, some highly unstable proteins known to be ubiquitylated, such as Cln1 and Cln2, were not detected, reflecting low abundance. Conditions of proteasome inhibition will presumably reveal the particularly short-lived ubiquitylated proteins.

Of the 72 yeast proteins, 66 were novel in terms of posttranslational ubiquitylation, adding to the knowledge base of the yeast ubiquitylated proteome. This first pass also identified particular lysines in the sequences of the ubiquitylated proteins, giving a sense of the distribution of the tags across sequence and presumed conformational space of the particular protein targets. Two additional features are worth noting. First, one protein, Emc21, is ubiquitylated at six different lysines, an example of multisite ubiquitylation, known previously only qualitatively (Hershko and Ciechanover, 1998), but without specific sequence information for the lysines modified. Second, ubiquitin was found to participate *in vivo* (in yeast) in polyubiquitin chains at all seven of its lysine residues, with $K_{48} > K_{63}$ as expected, but K_{11} was next most abundant, followed by much lower levels of K_{33}, K_{29}, and K_6 chains. The biological significance of this can be evaluated, for example, by knockouts of particular E2 or E3 ligases in yeast to see what linkages and what target proteins go away. In this regard, a follow-up study (Hitchcock et al., 2003) identified 83 membrane-associated proteins that get ubiquitylated in yeast in the ERAD response for targeting poorly folded and unfolded proteins in the early stages of the secretory system for proteasome degradation, as noted in an earlier section of this chapter. This is the beginning of the definition of the subset of ubiquitylated proteins that are localized to membranes. This approach can also be utilized with His_6-SUMO and any other Ubls expressed in yeast.

References

Aguilar, R. C., and B. Wendland. "Ubiquitin: Not just for proteasomes anymore," *Curr. Opin. Cell Biol.* **15:**184–190 (2003).

Alam, S. L., J. Sun, M. Payne, B. D. Welch, B. K. Blake, D. R. Davis, H. H. Meyer, S. D. Emr, and W. I. Sundquist. "Ubiquitin interactions of NZF zinc fingers," *EMBO J.* **23:**1411–1421 (2004).

Bayer, P., A. Arndt, S. Metzger, R. Mahajan, F. Melchior, R. Jaenicke, and J. Becker. "Structure determination of the small ubiquitin-related modifier SUMO-1," *J. Mol. Biol.* **280:**275–286 (1998).

Bloom, J., V. Amador, F. Bartolini, G. DeMartino, and M. Pagano. "Proteasome-mediated degradation of p21 via N-terminal ubiquitinylation," *Cell* **115:**71–82 (2003).

Bonifacino, J. S., and L. M. Traub. "Signals for sorting of transmembrane proteins to endosomes and lysosomes," *Annu. Rev. Biochem.* **72:**395–447 (2003).

Brooks, C. L., and W. Gu. "Ubiquitination, phosphorylation and acetylation: The molecular basis for p53 regulation," *Curr. Opin. Cell Biol.* **15:**164–171 (2003).

Bylebyl, G. R., I. Belichenko, and E. S. Johnson. "The SUMO isopeptidase Ulp2 prevents accumulation of SUMO chains in yeast," *J. Biol. Chem.* **278:**44113–44120 (2003).

Cardozo, T., and M. Pagano. "The SCF ubiquitin ligase: Insights into a molecular machine," *Nat. Rev. Mol. Cell Biol.* **5:**739–751 (2004).

Cook, W. J., L. C. Jeffrey, M. Carson, Z. Chen, and C. M. Pickart. "Structure of tetraubiquitin shows how multiubiquitin chains can be formed," *J. Mol. Biol.* **236:**601–609 (1994).

Cope, G. A., and R. J. Deshaies. "COP9 signalosome: A multifunctional regulator of SCF and other cullin-based ubiquitin ligases," *Cell* **114:**663–671 (2003).

Duan, L., Y. Miura, M. Dimri, B. Majumder, I. L. Dodge, A. L. Reddi, A. Ghosh, N. Fernandes, P. Zhou, K. Mullane-Robinson, N. Rao, S. Donoghue, R. A. Rogers, D. Bowtell, M. Naramura, H. Gu, V. Band, and H. Band. "Cbl-mediated ubiquitinylation is required for lysosomal sorting of epidermal growth factor receptor but is dispensable for endocytosis," *J. Biol. Chem.* **278:**28950–28960 (2003).

Finley, D. "Ubiquitin chained and crosslinked," *Nat. Cell Biol.* **4:**E121–E123 (2002).

Fisher, R. D., B. Wang, S. L. Alam, D. S. Higginson, H. Robinson, W. I. Sundquist, and C. P. Hill. "Structure and ubiquitin binding of the ubiquitin-interacting motif," *J. Biol. Chem.* **278:**28976–28984 (2003).

Flierman, D., Y. Ye, M. Dai, V. Chau, and T. A. Rapoport. "Polyubiquitin serves as a recognition signal, rather than a ratcheting molecule, during retrotranslocation of proteins across the endoplasmic reticulum membrane," *J. Biol. Chem.* **278:**34774–34782 (2003).

Gan-Erdene, T., K. Nagamalleswari, L. Yin, K. Wu, Z. Q. Pan, and K. D. Wilkinson. "Identification and characterization of DEN1, a deneddylase of the ULP family," *J. Biol. Chem.* **278:**28892–28900 (2003).

Gronroos, E., U. Hellman, C. H. Heldin, and J. Ericsson. "Control of Smad7 stability by competition between acetylation and ubiquitination," *Mol. Cell* **10:**483–493 (2002).

Hartmann-Petersen, R., M. Seeger, and C. Gordon. "Transferring substrates to the 26S proteasome," *Trends Biochem. Sci.* **28:**26–31 (2003).

Hershko, A., and A. Ciechanover. "The ubiquitin system," *Annu. Rev. Biochem.* **67:**425–479 (1998).

Hershko, A., A. Ciechanover, and A. Varshavsky. "Basic medical research award. The ubiquitin system," *Nat. Med.* **6:**1073–1081 (2000).

Hicke, L. "Protein regulation by monoubiquitin," *Nat. Rev. Mol. Cell Biol.* **2:**195–201 (2001).

Hitchcock, A. L., K. Auld, S. P. Gygi, and P. A. Silver. "A subset of membrane-associated proteins is ubiquitinated in response to mutations in the endoplasmic reticulum degradation machinery," *Proc. Natl. Acad. Sci. U.S.A.* **100:**12735–12740 (2003).

Hoppe, T., G. Cassata, J. M. Barral, W. Springer, A. H. Hutagalung, H. F. Epstein, and R. Baumeister. "Regulation of the myosin-directed chaperone UNC-45 by a novel E3/E4-multiubiquitylation complex in *C. elegans*," *Cell* **118:**337–349 (2004).

Ichimura, Y., T. Kirisako, T. Takao, Y. Satomi, Y. Shimonishi, N. Ishihara, N. Mizushima, I. Tanida, E. Kominami, M. Ohsumi, T. Noda, and Y. Ohsumi. "A ubiquitin-like system mediates protein lipidation," *Nature* **408:**488–492 (2000).

Ivan, M., K. Kondo, H. Yang, W. Kim, J. Valiando, M. Ohh, A. Salic, J. M. Asara, W. S. Lane, and W. G. Kaelin, Jr. "HIFalpha targeted for VHL-mediated destruction by proline hydroxylation: Implications for O_2 sensing," *Science* **292:**464–468 (2001).

Izzi, L., and L. Attisano. "Regulation of the TGFbeta signalling pathway by ubiquitin-mediated degradation," *Oncogene* **23:**2071–2078 (2004).

Jentsch, S., and G. Pyrowolakis. "Ubiquitin and its kin: How close are the family ties?" *Trends Cell Biol.* **10:**335–342 (2000).

Kang, R. S., C. M. Daniels, S. A. Francis, S. C. Shih, W. J. Salerno, L. Hicke, and I. Radhakrishnan. "Solution structure of a CUE-ubiquitin complex reveals a conserved mode of ubiquitin binding," *Cell* **113:**621–630 (2003).

Katzmann, D. J., G. Odorizzi, and S. D. Emr. "Receptor downregulation and multivesicular-body sorting," *Nat. Rev. Mol. Cell Biol.* **3:**893–905 (2002).

Kostova, Z., and D. H. Wolf. "For whom the bell tolls: Protein quality control of the endoplasmic reticulum and the ubiquitin-proteasome connection," *EMBO J.* **22:**2309–2317 (2003).

Li, M., C. L. Brooks, F. Wu-Baer, D. Chen, R. Baer, and W. Gu. "Mono- versus polyubiquitination: Differential control of p53 fate by Mdm2," *Science* **302:**1972–1975 (2003).

Li, M., D. Chen, A. Shiloh, J. Luo, A. Y. Nikolaev, J. Qin, and W. Gu. "Deubiquitination of p53 by HAUSP is an important pathway for p53 stabilization," *Nature* **416:**648–653 (2002a).

Li, M., J. Luo, C. L. Brooks, and W. Gu. "Acetylation of p53 inhibits its ubiquitination by Mdm2," *J. Biol. Chem.* **277:**50607–50611 (2002b).

Melchior, F., M. Schergaut, and A. Pichler. "SUMO: Ligases, isopeptidases and nuclear pores," *Trends Biochem. Sci.* **28**:612–618 (2003).

Murray, A. W. "Recycling the cell cycle: Cyclins revisited," *Cell* **116**:221–234 (2004).

Orlicky, S., X. Tang, A. Willems, M. Tyers, and F. Sicheri. "Structural basis for phosphodependent substrate selection and orientation by the SCFCdc4 ubiquitin ligase," *Cell* **112**:243–256 (2003).

Peng, J., D. Schwartz, J. E. Elias, C. C. Thoreen, D. Cheng, G. Marsischky, J. Roelofs, D. Finley, and S. P. Gygi. "A proteomics approach to understanding protein ubiquitination," *Nat. Biotechnol.* **21**:921–926 (2003).

Pickart, C. M. "Mechanisms underlying ubiquitination," *Annu. Rev. Biochem.* **70**:503–533 (2001).

Pickart, C. M. "Back to the future with ubiquitin," *Cell* **116**:181–190 (2004).

Pintard, L., J. H. Willis, A. Willems, J. L. Johnson, M. Srayko, T. Kurz, S. Glaser, P. E. Mains, M. Tyers, B. Bowerman, and M. Peter. "The BTB protein MEL-26 is a substrate-specific adaptor of the CUL-3 ubiquitin-ligase," *Nature* **425**:311–316 (2003).

Prag, G., S. Misra, E. A. Jones, R. Ghirlando, B. A. Davies, B. F. Horazdovsky, and J. H. Hurley. "Mechanism of ubiquitin recognition by the CUE domain of Vps9p," *Cell* **113**:609–620 (2003).

Rao-Naik, C., W. de la Cruz, J. M. Laplaza, S. Tan, J. Callis, and A. J. Fisher. "The rub family of ubiquitin-like proteins. Crystal structure of Arabidopsis rub1 and expression of multiple rubs in Arabidopsis," *J. Biol. Chem.* **273**:34976–34982 (1998).

Reed, S. I. "Ratchets and clocks: The cell cycle, ubiquitylation and protein turnover," *Nat. Rev. Mol. Cell. Biol.* **4**:855–864 (2003).

Rogers, R. S., C. M. Horvath, and M. J. Matunis. "SUMO modification of STAT1 and its role in PIAS-mediated inhibition of gene activation," *J. Biol. Chem.* **278**:30091–30097 (2003).

Schlumpberger, M., E. Schaeffeler, M. Straub, M. Bredschneider, D. H. Wolf, and M. Thumm. "AUT1, a gene essential for autophagocytosis in the yeast *Saccharomyces cerevisiae*," *J. Bacteriol.* **179**:1068–1076 (1997).

Schnell, J. D., and L. Hicke. "Non-traditional functions of ubiquitin and ubiquitin-binding proteins," *J. Biol. Chem.* **278**:35857–35860 (2003).

Schwartz, D. C., and M. Hochstrasser. "A superfamily of protein tags: Ubiquitin, SUMO and related modifiers," *Trends Biochem. Sci.* **28**:321–328 (2003).

Seeler, J. S., and A. Dejean. "Nuclear and unclear functions of SUMO," *Nat. Rev. Mol. Cell Biol.* **4**:690–699 (2003).

Seol, J. H., A. Shevchenko, and R. J. Deshaies. "Skp1 forms multiple protein complexes, including RAVE, a regulator of V-ATPase assembly," *Nat. Cell Biol.* **3**:384–391 (2001).

Shih, S. C., G. Prag, S. A. Francis, M. A. Sutanto, J. H. Hurley, and L. Hicke. "A ubiquitin-binding motif required for intramolecular monoubiquitylation, the CUE domain," *EMBO J.* **22**:1273–1281 (2003).

Shiio, Y., and R. N. Eisenman. "Histone sumoylation is associated with transcriptional repression," *Proc. Natl. Acad. Sci. U.S.A.* **100**:13225–13230 (2003).

Steffan, J. S., N. Agrawal, J. Pallos, E. Rockabrand, L. C. Trotman, N. Slepko, K. Illes, T. Lukacsovich, Y. Z. Zhu, E. Cattaneo, P. P. Pandolfi, L. M. Thompson, and J. L.

Marsh. "SUMO modification of Huntingtin and Huntington's disease pathology," *Science* **304:**100–104 (2004).

Upadhya, S., and A. H. Hedge. "A potential proteasome-interacting motif within the ubiquitin-like domain of parkin and other proteins," *Trends Biochem. Sci.* **28:**280–283 (2003).

Van Der Wel, H., S. Z. Fisher, and C. M. West. "A bifunctional diglycosyltransferase forms the Fucalpha1,2Galbeta1,3-disaccharide on Skp1 in the cytoplasm of dictyostelium," *J. Biol. Chem.* **277:**46527–46534 (2002a).

Van Der Wel, H., H. R. Morris, M. Panico, T. Paxton, A. Dell, L. Kaplan, and C. M. West. "Molecular cloning and expression of a UDP-N-acetylglucosamine (GlcNAc): Hydroxyproline polypeptide GlcNAc-transferase that modifies Skp1 in the cytoplasm of dictyostelium," *J. Biol. Chem.* **277:**46328–46337 (2002b).

Varshavsky, A. "The ubiquitin system," *Trends Biochem. Sci.* **22:**383–387 (1997).

Verdin, E., F. Dequiedt, and H. G. Kasler. "Class II histone deacetylases: Versatile regulators," *Trends Genet.* **19:**286–293 (2003).

Weissman, A. M. "Themes and variations on ubiquitylation," *Nat. Rev. Mol. Cell Biol.* **2:**169–178 (2001).

Whitby, F. G., G. Xia, C. M. Pickart, and C. P. Hill. "Crystal structure of the human ubiquitin-like protein NEDD8 and interactions with ubiquitin pathway enzymes," *J. Biol. Chem.* **273:**34983–34991 (1998).

Wilkinson, K. D. "Ubiquitination and deubiquitination: Targeting of proteins for degradation by the proteasome," *Sem. Cell Dev. Biol.* **11:**141–148 (2000).

Wu, K., K. Yamoah, G. Dolios, T. Gan-Erdene, P. Tan, A. Chen, C. G. Lee, N. Wei, K. D. Wilkinson, R. Wang, and Z. Q. Pan. "DEN1 is a dual function protease capable of processing the C terminus of Nedd8 and deconjugating hyper-neddylated CUL1," *J. Biol. Chem.* **278:**28882–28891 (2003).

Xirodimas, D. P., M. K. Saville, J. C. Bourdon, R. T. Hay, and D. P. Lane. "Mdm2-mediated NEDD8 conjugation of p53 inhibits its transcriptional activity," *Cell* **118:**83–97 (2004).

Xu, L., Y. Wei, J. Reboul, P. Vaglio, T. H. Shin, M. Vidal, S. J. Elledge, and J. W. Harper. "BTB proteins are substrate-specific adaptors in an SCF-like modular ubiquitin ligase containing CUL-3," *Nature* **425:**316–321 (2003).

Yoshida, Y., F. Tokunaga, T. Chiba, K. Iwai, K. Tanaka, and T. Tai. "Fbs2 is a new member of the E3 ubiquitin ligase family that recognizes sugar chains," *J. Biol. Chem.* **278:**43877–43884 (2003).

Zheng, N., B. A. Schulman, L. Song, J. J. Miller, P. D. Jeffrey, P. Wang, C. Chu, D. M. Koepp, S. J. Elledge, M. Pagano, R. C. Conaway, J. W. Conaway, J. W. Harper, and N. P. Pavletich. "Structure of the Cul1-Rbx1-Skp1-F boxSkp2 SCF ubiquitin ligase complex," *Nature* **416:**703–709 (2002).

Protein Glycosylation

The tetradecasaccharyl donor for transfer to Asn residues in protein N-glycosylation.

Covalent glycosylation occurs in a significant set of the proteins that transit through the endoplasmic reticulum in eukaryotic cells (Helenius and Aebi, 2004; Trombetta, 2003; Trombetta and Parodi, 2003). This includes proteins that stay resident in the ER (e.g., HMG CoA reductase in cholesterol biosynthesis) and proteins that go into other organelles of the secretory pathway, including all the way to the plasma membrane, as well as proteins secreted into the extracellular milieu. Both N-glycosylation and O-glycosylation of proteins occur and contain remarkable complexity in the oligosaccharide chains that are added. The N-glycan chains can occupy a volume that is a substantial fraction of the total volume of the protein backbone. N-Glycosylation is localized to the carboxamido nitrogen of Asn residues, in the consensus tripeptide sequence Asn-X-Ser/Thr. The β-OH of the Thr or Ser residue is thought to act via hydrogen bonding to the carboxamido nitrogen to enable a conformation with enhanced nucleophilicity of the otherwise unreactive nitrogen in the $CONH_2$ group of the Asn residues (Figure 10.1A). The glycan chain transferred is a branched 14-mer (tetradecasaccharide). About 30% of the proteins with such predicted N-glycosylation sequons that pass into the ER are actually glycosylated, perhaps reflecting some mixture of accessibility during refolding and the kinetic capacity of the glycosylation posttranslational catalytic enzyme machinery. The O-glycosylation of proteins occurs commonly on both Ser and Thr residues (Figure 10.1B), but there are several examples where posttranslational hydroxylation (Chapter 12) creates 5-hydroxylysine and 4-hydroxyproline side chains that are sites of subsequent O-glycosylation (see Chapter 12). There are even a few examples of C-glycosylation [e.g., of Trp_2 in RNAse (Loffler et al., 1996) (Figure 10.1C)], although these are rare since the requisite carbon nucleophiles on amino acid side chains are difficult to generate.

Oligosaccharide structures, especially the ones with multiple outer branches, are bulky and highly hydrophilic. They clearly alter the physical properties of the protein microenvironments at their local attachment sites, as well as more globally (Rudd and Dwek, 1997). They can be handles for particular subcellular addresses (phosphomannosylation sends proteins to lysosomes) or to assist in protein folding during the secretory pathway, and can function as selective ligands for recognition (Wormald et al., 2002). The N-glycan moiety can be important for the participation in cycles of protein folding in the ER and for the subsequent protease protection for proteins at cell surfaces (Wormald and Dwek, 1999). In contrast, most O-linked glycosylations occur later, in the Golgi complex of the cell.

Proteins that are N-glycosylated on Asn residues have complex carbohydrate chains, with 23 or more separate enzymatic steps involved in the assembly, trimming, and maturation of branched carbohydrate structures. The large number of glycosyltransferases (Gtfs) and glycosidases involved in this posttranslational modification

A.

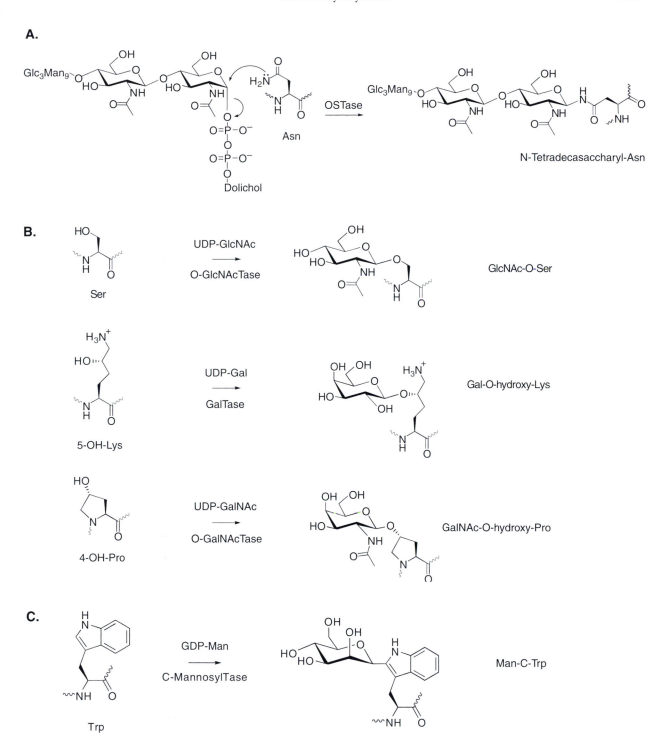

Figure 10.1 Protein glycosylation categorized by protein side chain nucleophiles attacking the activated sugar donor. **A.** Asn N-glycosylation by tetradecasaccharyl-PP-dolichol. **B.** O-Glycosylation of Ser, 5-OH-Lys, and 4-OH-Pro by NDP-hexoses. **C.** C-Glycosylation of Trp by GDP-mannose.

pathway suggests great potential for diversity. The almost bewilderingly complex library of carbohydrate arrays that can be found in N-linked glycoproteins reflects, among other things, the different levels of Gtf expression, the distinct compartmentalization in organelles [the ER and different regions (cis, medial, or trans) of the Golgi complex], the order of glycosyl transfer enzyme encounter with the growing N-glycan chain, and the residence time for multiple processing by the same glycosyltransferase (Gtf) or glycosidase (Kornfeld and Kornfeld, 1976; Trombetta, 2003). Because the more than 23 enzymes that are involved in assembling and tailoring the carbohydrate chains of glycoproteins appear to act as independent, stochastic agents and function without templating, the populations of glycans even at one site of one glycoprotein can be heterogeneous.

Multiple glycoforms of the same protein add to the diversity and molecular complexity of the glycoproteome. On the other hand, all nascent N-glycoproteins in eukaryotes are formed with an almost identical branched tetradecasaccharide. It is in the subsequent trimming and re-addition of monosaccharides that diversity is generated (Trombetta, 2003).

The prion protein, found in neuronal and neuroendocrine cell membranes, shows how the heterogeneity can occur. Much attention has been focused on the normal cellular form of the prion protein (PrP^C) and the disease-causing conformer PrP^{SC} that causes *sc*rapie in animals and related spongiform encephalopathies in humans. Prion protein contains two N-glycosylation sites, at Asn_{181} and Asn_{197} (Rudd et al., 2002), and compositional analysis of both PrP^C and PrP^{SC} indicates the same range of heterogeneity in the glycans. Recent studies suggest that the interplay of disulfide bond formation and N-glycosylation can influence fibrillization in a prion fragment (Bosques and Imperiali, 2003). Glycan analysis by cleavage of the glycan chain from the protein, exoenzyme digestion, HPLC, and mass spectrometry indicated at least 52 glycan isoforms populating the two Asn sites. These all had the common $Man_5GlcNAc_2$ core (see below) but then were elaborated to a mixture of bi-, tri-, and tetraantennary oligosaccharide chains with differing amounts of galactose and terminal sialic acids, among other sugars. Recall from Chapter 7 that the prion protein also has a glycosylphosphatidyl anchor, incorporating five additional oligosaccharides, and the opportunity for diversity in that tether. Figure 10.2 (Rudd et al., 1999a) shows a model of PrP^C with two types of posttranslational modifications: the two N-linked glycan chains and the C-terminal GPI membrane anchor. The N-glycan chains have substantial conformational mobility.

O-Glycoproteins tend to have much simpler carbohydrate structures (Gemmill and Trimble, 1999), such as monosaccharide O-GlcNAc or O-GalNAc substituents as

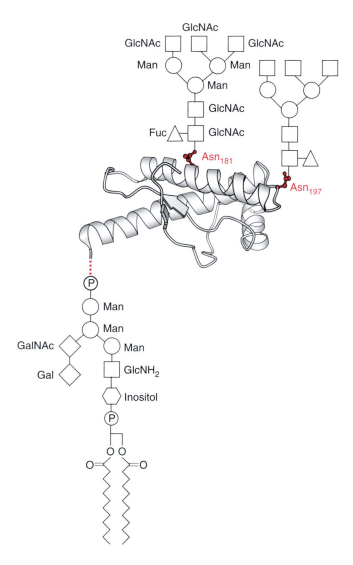

Figure 10.2 Schematic for posttranslational modification of the prion protein: Modification with distinct N-glycan chains on two Asn side chains and at the C-terminus by a GPI anchor give up to 52 isoforms (figure made using PDB 1B10).

well as disaccharide and trisaccharide fucose-containing cores (Okajima et al., 2003) in higher eukaryotes. Figure 10.3A shows the O-linked tetrasaccharide on some of the EGF domains of the extracellular region of the protein Notch. In yeast cells penta-mannosyl-O-proteins are produced (Strahl-Bolsinger et al., 1999) (Figure 10.3B). In this chapter, N-glycosylation is discussed first, followed by O-glycosylation.

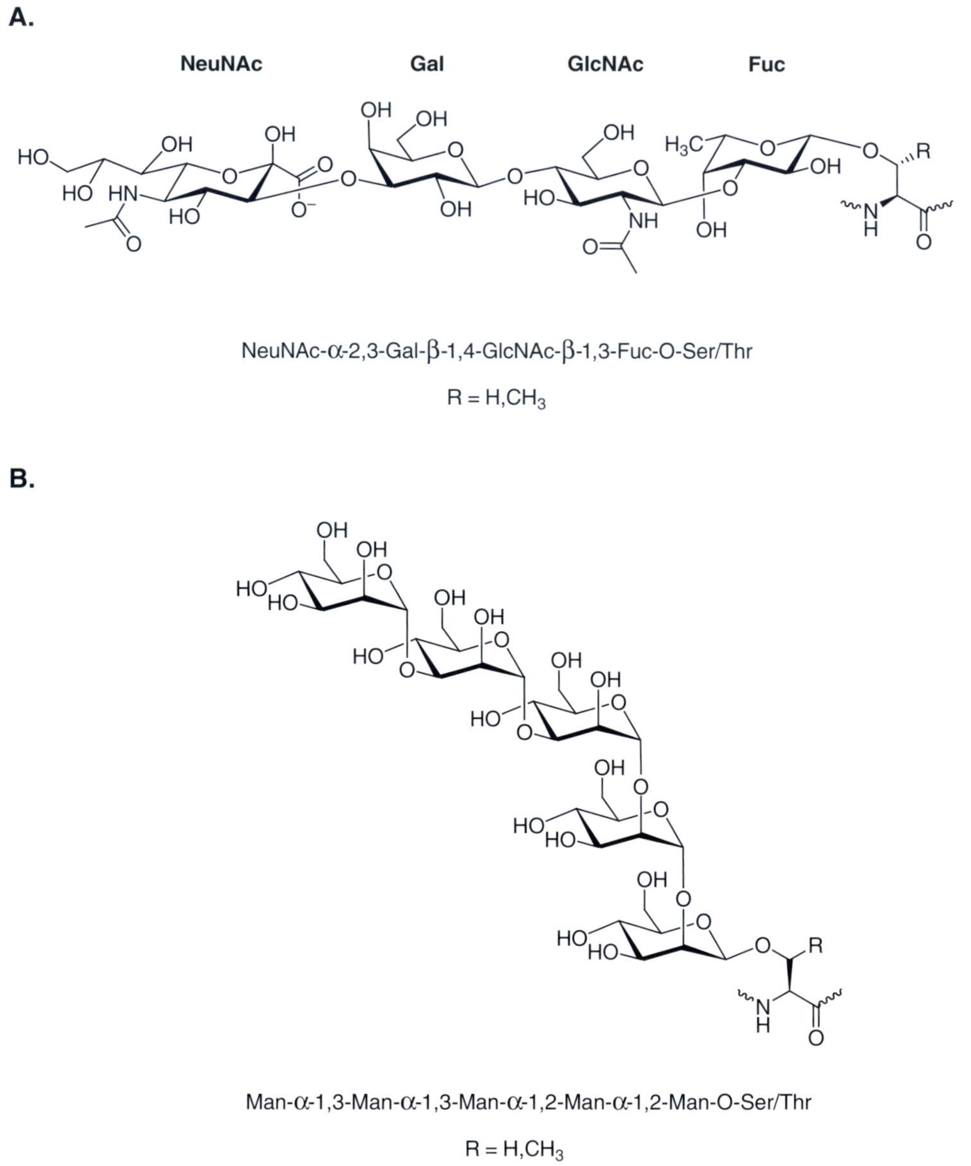
A.
NeuNAc
Gal
GlcNAc
Fuc
NeuNAc-α-2,3-Gal-β-1,4-GlcNAc-β-1,3-Fuc-O-Ser/Thr
R = H,CH3
B.
Man-α-1,3-Man-α-1,3-Man-α-1,2-Man-α-1,2-Man-O-Ser/Thr
R = H,CH3

Figure 10.3 O-Linked glycoproteins have less complex chains. **A.** O-Linked tetrasaccharide on the EGF domain of Notch. **B.** Pentamannosyl-O-linked chains in yeast mannoproteins.

Structural Challenges for the Determination of the Carbohydrate Modifications in Proteins

Structure determination of the glycosyl groups attached to proteins has been challenging as a consequence of the branching of the glycan chains, the configurations of the sugar constituents, and the regiochemistry of the linkage of each monomer unit to

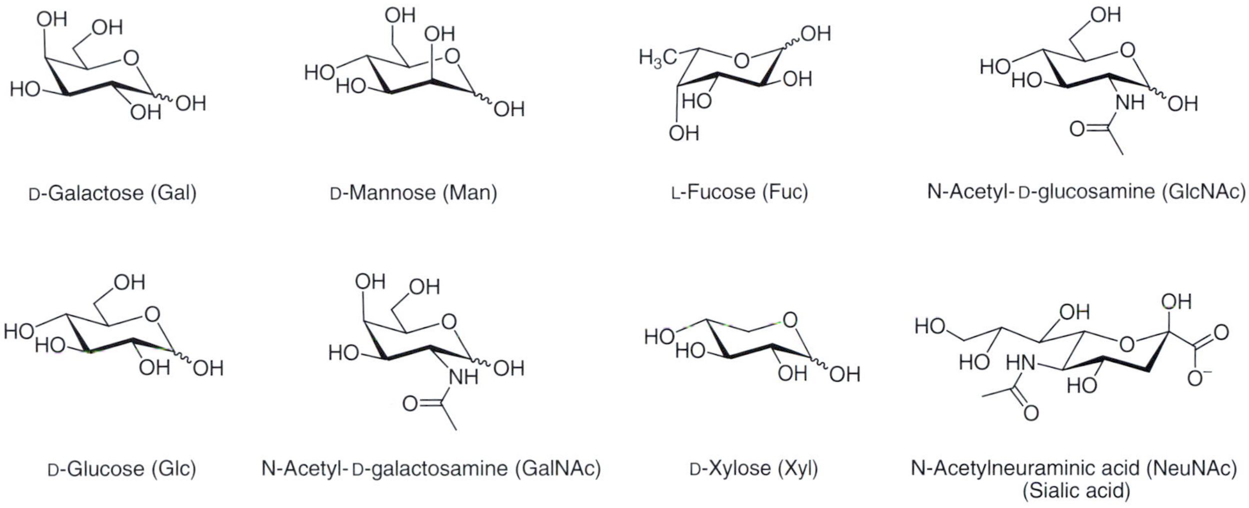

Figure 10.4 Hexoses commonly found as constituents of N-linked glycan chains in glycoproteins.

other sugars (Mechref and Novotny, 2002; Wormald et al., 2002). The monosaccharides that are found in all N-linked glycoproteins are N-acetylglucosamine (GlcNAc) and D-mannose. Other common sugars are D-glucose, which is found in all early intermediates but later trimmed away by glucosidases in the ER and Golgi complex, D-galactose, L-fucose, and N-acetylneuraminic acid, also known as sialic acid (Figure 10.4). In O-linked glycoproteins, glucose, GlcNAc, and the corresponding GalNAc residue are found as well as L-fucose and D-mannose residues (Dell and Morris, 2001).

Historically two enzymes, endoglycosidase H and protein glycanase N, were of particular use in removal of the N-glycan chains, allowing their isolation and subsequent structure determination by degradation (Figure 10.5). The Endo H treatment leaves the proximal GlcNAc attached to the Asn side chain of the protein while glycanase N removes the full N-glycan chain and generates and Asp at the site of the original Asn.

In contemporary contexts mass spectrometry is the dominant structural methodology, along with capillary electrophoresis and HPLC separations of the released glycan fragments (Mechref and Novotny, 2002). Sequential digestion by specific glycosidases (Rudd et al., 1999b and 2001), followed by analysis of the remaining oligosaccharide chain by MS has proven useful, as exemplified by the oligosaccharide analysis of human serum immunoglobulin IgG (Figure 10.6), in which sequential changes in the HPLC migration of the glycan chain can be monitored as hexose residues are removed one at a time by purified glycosidases in panels A–F. FAB-MS can provide composition by the molecular ion and sequence information on the carbohydrate chain by the fragmentation pattern (Dell and Morris, 2001). MALDI-MS is

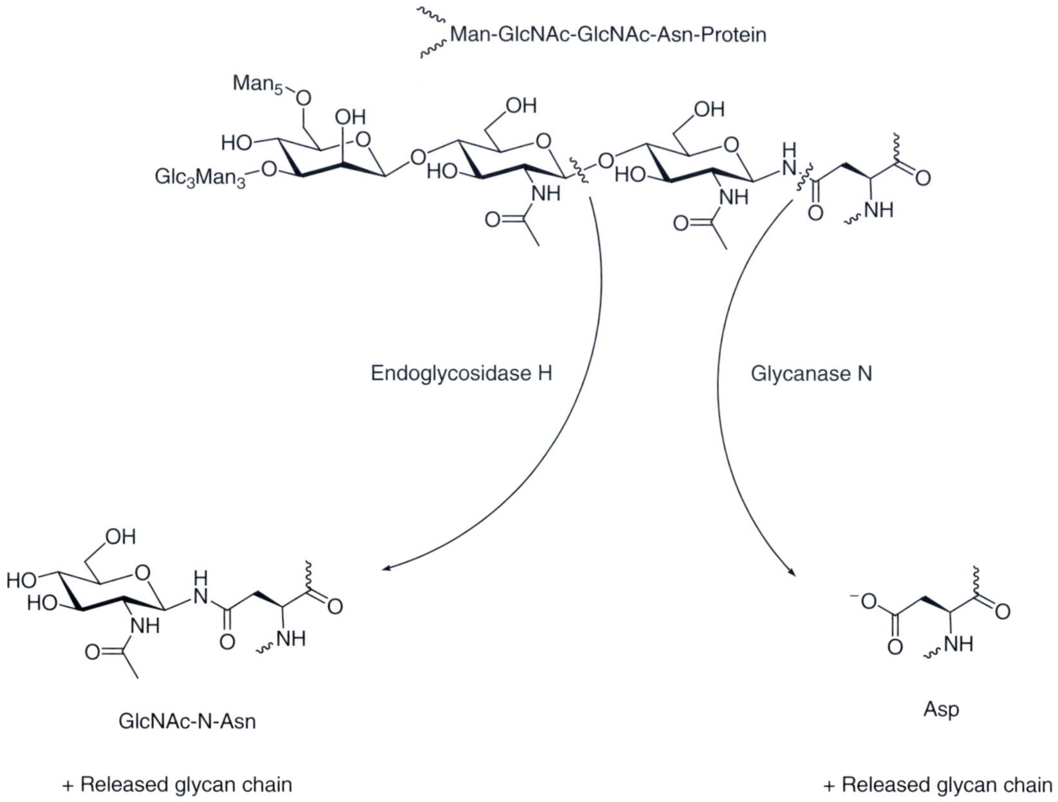

Figure 10.5 Action of endoglycosidase H and glycanase N for cleavage of N-linked glycan chains at the sites of attachment to proteins.

typically effective for the detection of molecular ions; MALDI ionization with quadrupole time-of-flight (Q-TOF) instruments and electrospray-Q-TOF allow fragmentation. Since glycosidic bonds are weaker than peptide bonds, oligosaccharide sequence information can be obtained from the fragmentations (Dell and Morris, 2001; Liebler, 2002).

Electrospray ionization mass spectrometry (ESI-MS) has been used with a quadrupole ion trap detector for MS-MS analysis during multiple stages of fragmentation of branched carbohydrate chains. The MS-MS fragmentations of a $GlcNAc_8Man_3$ chain from ovalbumin that allowed the determination of the identity and connectivity, as 1–6 sugar residues are lost in fragmentations, are shown in Figure 10.7. MS can also be applied to the structure determination of O-linked glycoproteins, although those chains can be susceptible to β-elimination and may need to be studied by soft ionization MS methods (Mechref and Novotny, 2002).

In one case it was possible to show that the Skp1 protein in ubiquitin posttranslational modifications had a hydroxyproline at residue 143 and the hydroxyl group was

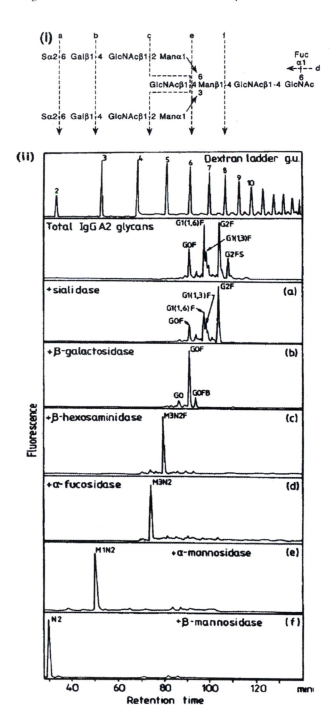

Figure 10.6 Use of sequential glycosidase degradation in the analysis of the complex N-glycan chain in human IgG [reproduced with permission from Guile et al. (1996)].

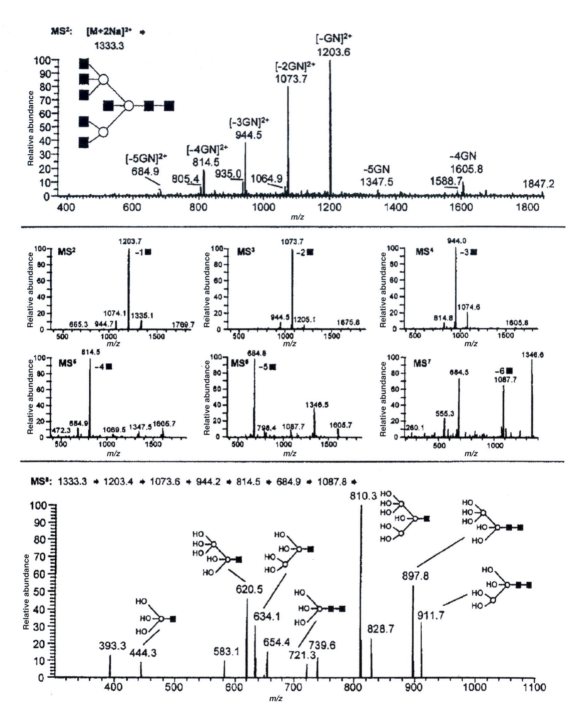

Figure 10.7 ESI/QIT/MS to obtain multiple stages of fragmentation on the GlcNAc$_8$Man$_3$ to determine hexose identity and branch points in the triantennary N-glycan chain from ovalbumin [reproduced with permission from Weiskopf et al. (1998)].

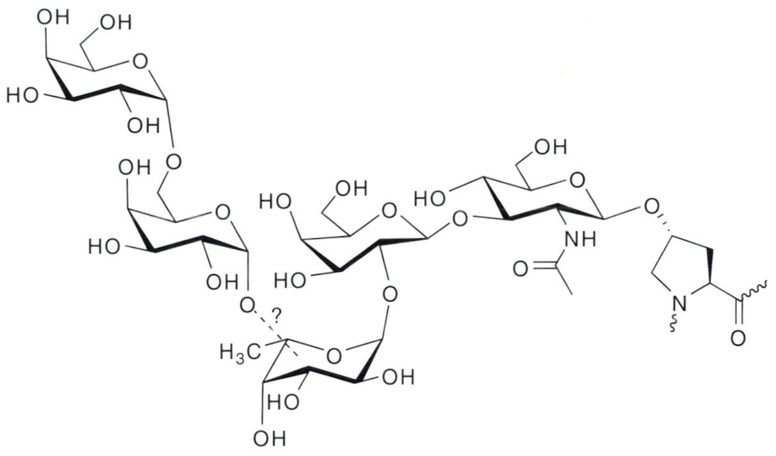

D-Gal- α-1,6-D-Gal-α-1,?-L-Fuc- α -1,2-D-Gal- β-1,3-GlcNAc-4-OH-Pro$_{143}$

Figure 10.8 Hexose chain O-linked to 4-OH-Pro$_{143}$ in Skp1.

linked to a hexasaccharide (Dell and Harris, 2001) (Figure 10.8), a tandem set of two distinct, coordinated types of posttranslational modifications. The loss of 180 mass units during glycan chain fragmentation indicates the release of a hexose, whereas the loss of 209 mass units indicates the loss of an N-acetylhexose in a given cycle. To distinguish whether glucose or mannose (for example) was the hexose released or GlcNAc or GalNAc, however, requires more traditional chromatographic analyses of the released sugars. The regiochemistry of linkages between sugars can be determined by classical permethylation studies with the separation of the permethylated isomers to identify what site had been blocked by connection to adjacent residues. MS methods have also been used on the serum glycoprotein transferrin to evaluate the N-linked oligosaccharide structures and show that defects in the N-glycosylation pathway can be detected by characteristic and identifiable decreases in masses of the glycan chain, reflecting defects in distinct Gtf genes in the pathway [see Dell and Morris (2001) for references]. While these few examples show some facets of the resolving power of mass spectrometric methods for glycoprotein analysis, cataloging the glycoproteome will be challenging because of all the different causes of microheterogeneity.

N-Linked Glycoproteins

N-Linked glycoproteins are found in eukaryotes, from fungi to humans, with much of the enzymatic machinery conserved, and they are also in the Archaea kingdom (Trombetta, 2003). The late stages of glycan chain maturation are distinct because of the different populations of late-stage Gtfs in the different organisms, but the core glycan assembly and trimming reactions in the ER from yeast to humans are strongly related. N-Glycosylation has also recently been detected in the eubacterium *Camphylobacter jejuni* and these findings shed some light on the minimal components of the pathway needed for Asn side chain modifications (Szymanski et al., 2003). It is possible, though, given the limited bacterial distribution that the *Camphylobacter* genes may have arisen by lateral transfer from archaea or fungi, and so not provide independent insight into the evolution of posttranslational glycosylation strategies.

Some prototypic examples of N-linked glycoprotein chains are depicted in Figure 10.9, with two, three, or four outer branches (biantennary, triantennary, or tetraantennary, respectively). The first five sugars, $Man_3GlcNAc_2$, are a conserved core, vestiges of the tetradecasaccharide common intermediate (discussed below) that is the first protein-N-linked glycan chain. The branched tetradecasaccharide gets trimmed back by hydrolytic enzymes to a core that is then elaborated by Gtfs in the Golgi complex to generate the diverse outer branches.

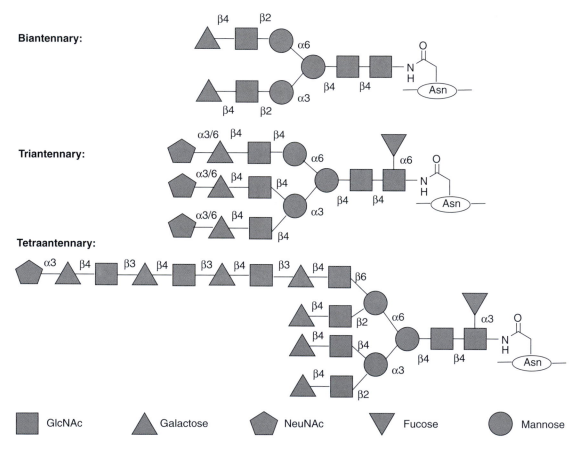

Figure 10.9 Branching in N-linked glycoproteins biantennary, triantennary, and tetraantennary glycan chain examples.

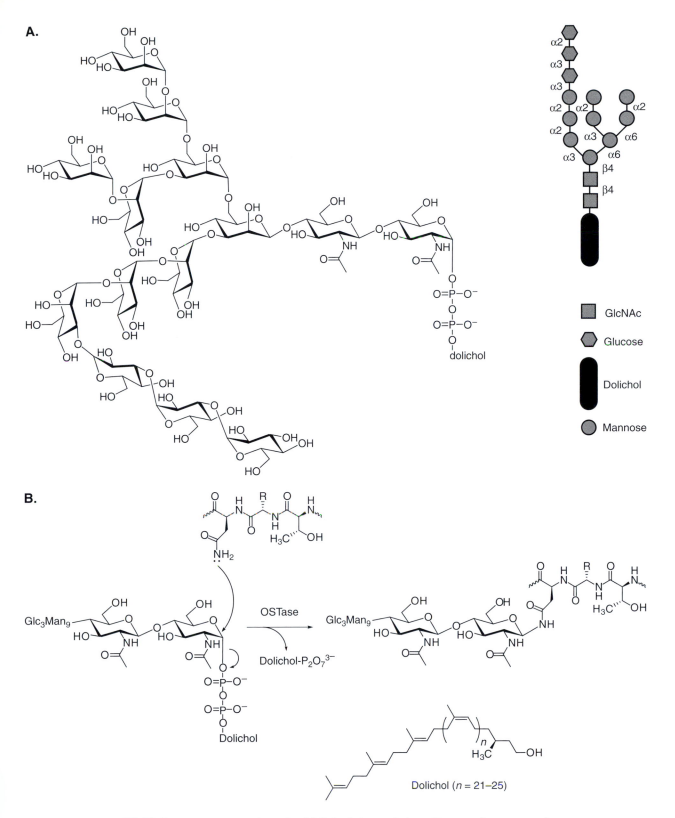

A.

GlcNAc

Glucose

Dolichol

Mannose

B.

Glc₃Man₉

OSTase

Dolichol-P₂O₇³⁻

Glc₃Man₉

Dolichol

Dolichol (*n* = 21–25)

Figure 10.10 Enzymatic maturation of an N-linked glycan chain. **A.** Donor substrate tetradecasaccharyl-PP-dolichol. **B.** Transfer of the oligosaccharyl chain to Asn in a Ser/Thr-X-Asn sequon by action of multicomponent oligosaccharyltransferase (OSTase).

First, consider the enzymatic assembly of the preformed tetradecasaccharyl chain, attached to a long-chain isoprenoid lipid (called dolichol) in the ER membrane (Figure 10.10A) (Burda and Aebi, 1999; Dean, 1999). Then, the N-glycan-forming reaction occurs, catalyzed by a multicomponent oligosaccharyltransferase (OSTase) that is also embedded in the ER membrane. The OSTase transfers the tetradecasaccharyl unit from the lipid carrier to the Asn side chain in the Asn-X-Ser/Thr sequons (Dempski and Imperiali, 2002) (Figure 10.10B).

The chains are trimmed back to a $Man_8GlcNAc_2$ decasaccharyl biantennary chain in the ER compartment and passed to the layers of the Golgi complex for further mannosidase-mediated hydrolytic trimming to the $Man_5GlcNAc_2$ core (van den Elsen et al., 2001). Then, still in the Golgi complex, late-stage Gtfs build the outer branches, including addition of the negatively charged sialic acids on glycoproteins that proceed to the plasma membrane and/or are secreted.

Assembly of the $Glc_3Man_9GlcNAc_2$-PP-dolichol Donor

The initial posttranslational modification reaction in N-glycosylation utilizes a preformed branched oligosaccharyl-lipid as the glycan donor, $Glc_3Man_9GlcNAc_2$-PP-dolichol (Figure 10.10A). The acceptor sites are Asn carboxamido nitrogens in the tripeptide sequon Asn-X-Ser/Thr of proteins during the cotranslational passage across the ER membrane on their way to the ER lumen. The differential and substoichiometric glycosylation of some Asn sequons but not others is thought to reflect the time in which the peptide sequence is available in unfolded form before refolding to the native state. Consistent with this is the observation that Asn residues in N-terminal regions of proteins transiting the ER get glycosylated at higher frequency than those in the C-terminal regions, when global folding of the translocating protein chain may limit access of the later Asn residues to the OSTase complex.

The branched $Glc_3Man_9GlcNAc_2$ tetradecasaccharyl donor is a baroquely complex donor substrate. The identity, sequence, and linkage regioselectivity is conserved from fungi to humans and genetic studies in yeast have elucidated a series of alg (*a*sparaginyl-*l*inked-*g*lycan) genes required for $Glc_3Man_9GlcNAc_2$ assembly on a polyisoprenoid lipid carrier, known as dolichol, with chain lengths of C_{105} to C_{125} (Burda and Aebi, 1999). The isoprene unit proximal to the diphosphate is saturated. Related polyisoprenoid alcohols in bacteria are all unsaturated (known as bactoprenols) with the predominant isomer as the C_{55} undecaprenol. Bacteria use mono- and oligosaccharide-PP-dolichols to assemble peptidoglycan layers and O-antigen layers in cell wall construction and have mechanisms for flipping the sugars from one side of the membrane to the other while attached to the polyisoprenoid anchor (Walsh,

2003). It is likely that the tetradecasaccharyl chain assembled on PP-dolichols for subsequent protein N-glycosylation works off equivalent logic and physical properties.

There are two topographical stages: the first seven monosaccharide residues are added on the *cytoplasmic* face while the next seven sugars are added on the *lumenal* face of the ER. The first seven sugar residues (the two GlcNAcs and five mannoses) are added via UDP-GlcNAc and GDP-mannose donors, respectively, by membrane enzymes whose active sites face the cytoplasm. The first Alg enzyme in the dolichol pathway (Alg70) uses dolichol phosphate as the nucleophile to attack UDP-GlcNAc to release UMP, transfer the GlcNAc-1-P moiety, and create the GlcNAc-PP-dolichol (Figure 10.11) that serves as scaffold for the elongation of the growing glycan. The next six enzymes create glycosidic linkages, releasing UDP and GDP, building the heptasaccharide chain one sugar at a time. This heptasaccharyl intermediate has three mannose residues (d, f, and g) in α3 linkage from the mannose at residue c, and one mannose (residue e) in α6 linkage from mannose c [see Trombetta (2003) for the terminology]. The five mannose units are added in the order c > d > e > f > g.

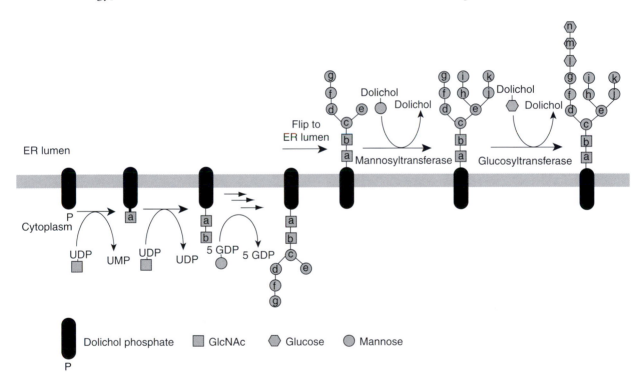

Figure 10.11 Building the oligosaccharyl chain on the polyisoprenoid-phosphate scaffold: GlcNAc-PP-dolichol formation at the cytoplasmic face of the ER membrane is followed by single addition of hexoses from UDP- and GDP-hexose donors to build the heptasaccharyl-PP-dolichol intermediate; flipping to the ER membrane lumenal side precedes stepwise addition of the remaining seven hexose residues from hexosyl-PP-dolichol as donors. Sugars are added in the order a–n.

At the heptasaccharide stage the chain is flipped by the Rft1 protein (Helenius et al., 2002), and the final seven sugar residues (mannoses h, i, j, and k, and glucoses l, m, and n, in that order) are added on the *lumenal* face of the ER membrane. In this phase the sugar donors are not the NDP-sugars, GDP-mannose and UDP-glucose, but rather mannose-PP-dolichol (four times) (Girrbach et al., 2000) and glucose-PP-dolichol (three times), thus building the finished $Glc_3Man_9GlcNAc_2$-PP-dolichol. The choice of isoprenyl lipid sugars rather than nucleotide sugars as hexosyl donors is not because NDP-sugars are absent from the ER lumen, as will be noted below, so there must be some other reason. It does suggest that the growing oligosaccharide chain must be near the membrane surface since the mannosyl- and glucosyl-PP-dolichols are likely to be at the lumen-membrane interface due to the long lipid anchor.

The synthase for dolichol-PP-mannose formation has been characterized as a single protein in yeast and as a three-subunit enzyme in humans (Maeda et al., 2000). It transfers mannose–PO_3^{2-} from GDP-mannose to dolichol, not dolichol–OPO_3^{2-}, and so generates the prenyl-P-hexose, rather than the prenyl-PP-hexose, seen in the first step of the assembly of the tetradecasaccharyl chain. Dolichol-P-mannose is the mannosyl donor, not only in N-glycan elongation, but also in GPI anchor formation, noted in Chapter 7, and also in the tryptophanyl side chain C-mannosylation noted at the end of this chapter, suggesting a converging logic for building saccharide units for posttranslational protein modification.

It is presumed that 14 separate Alg enzymes are required for the incremental build-up of the $Glc_3Man_9GlcNAc_2$-PP-dolichol (Burda and Aebi, 1999; Dempski and Imperiali, 2002). The tetradecasaccharyl chain is the preferred donor for the N-glycation step that occurs next, although Glc_2 and Glc_1 versions of 13-mer and 12-mer glycan chains can be transferred. The $Man_9GlcNAc_2$ 11-mer chain is 10- to 25-fold less good as a donor for OSTase (Karaoglu et al., 2001; Parodi, 2000).

The recent discovery that the intestinal bacterium *Campylobacter jejuni* makes N-linked glycoproteins, where the chain is a heptasaccharide (Benz and Schmidt, 2002; Szymanski et al., 2003) that is presumably flipped as the heptasaccharyl-PP-bacto-prenol from the cytoplasm to the periplasm before transfer of the heptasaccharide to protein, accords with a conserved mechanism for assembling a heptasaccharyl-PP-isoprenoid that has the physical properties that enable membrane flipping.

Oligosaccharyltransferase (OSTase): The Gatekeeper to the N-Glycoprotein Modification Pathway

The OSTase activity (Dempski and Imperiali, 2002), the first committed step in the N-glycosylation posttranslational modification pathway, has been known for decades but has been notoriously difficult to characterize. Up to nine proteins are involved. All are membrane associated (many with multiple predicted transmembrane segments) (Figure 10.12A) and have been difficult to solubilize and purify in catalytically active form. Again, yeast genetics in *S. cerevisiae* have been critical in the genetic definition of the nine subunits, enabling the subsequent biochemical evaluation and demonstration of the equivalent function for many of the human OSTase subunits. The localization of the gatekeeping OSTase (Dempski and Imperiali, 2002) in the ER membrane restricts N-glycosylation to proteins that get into and through the ER compartment of cells. All indications are that N-glycosylation is cotranslational, with Asn-X-Ser/Thr peptide sequences being available for transfer of the intact $Glc_3Man_9GlcNAc_2$ oligosaccharyl chain while the protein is still unfolded during extrusion through the ER membrane. The chemistry of the glycosylation step must be attack by the carboxamido nitrogen of the Asn side chains (Figure 10.12B), despite the low intrinsic nucleophilicity of amide nitrogens. The downstream Ser/Thr is activating, probably by hydrogen bonding, but so presumably is the active site of the

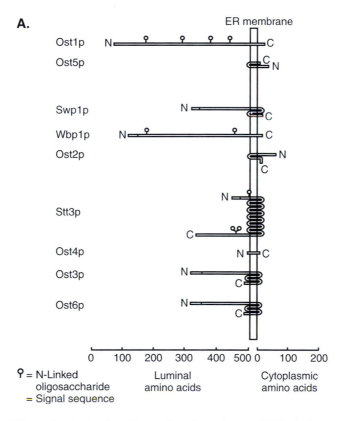

Figure 10.12 **A.** Nine subunits in the oligosaccharyltransferase (OSTase) that transfers the preformed tetradecasaccharyl glycan chain to Asn carboxamide side chains in protein substrates [redrawn from Knauer and Lehle (1999)].

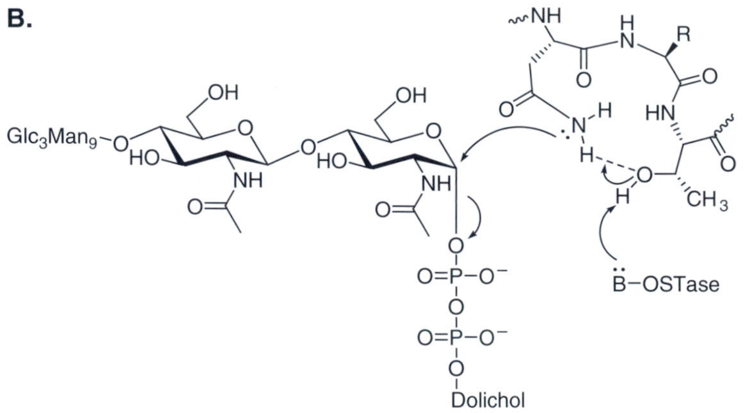

B.

Figure 10.12 **B**. The Asn carboxamide as attacking nucleophile on the $Glc_3Man_9GlcNAc_2$ oligosaccharyl-PP-dolichol donor.

OSTase. The coproduct is dolichol-PP. The polyisoprenoid-OPP is subjected to net pyrophosphatase action. Either the dolichol–OPO_3^{2-} or dolichol–OH then reequilibrates/flips to place the alcohol end on the cytoplasmic face of the ER membrane to salvage the polyisoprenoid carrier to begin the assembly of the tetradecasaccharyl chain again. The molecular basis of the specificity of OSTase for the branched 2500-Da oligosaccharyl donor is not understood, but there are many examples of proteins that act as lectins, binding specifically to one terminal monosaccharide moiety in an oligosaccharyl chain.

In the *C. jejuni* gene cluster there appears to be only a single OSTase subunit, PglB, corresponding to the presumed catalytic subunit, SST3, of yeast and human OSTase. This may suggest that a single subunit can suffice for, in that case, the heptasaccharyl-N-Asn glycosylation enzyme chemistry (Szymanski et al., 2003). This may be a good system to deconvolute the N-glycosylation gatekeeper enzyme mechanism

Hydrolytic Trimming of the Glc_3 Tail in the ER Lumen: Cycles of Deglucosylation/Reglucosylation and Protein Quality Control

Almost immediately after OSTase sets the N-tetradecasaccharyl core in place on Asn side chains, hydrolytic trimming by ER lumen glucosidases commences (High et al., 2000; Parodi, 2000; Roth, 2002). The three terminal glucose residues are removed by two glucosidases (I and II) acting in tandem (Figure 10.13). In competition with removal of the last Glc residue, to yield the $Man_9GlcNAc_2$-Asn-proteins, the Glc-$Man_9GlcNAc_2$-chains bind to both a soluble ER protein chaperone (calreticulin) and

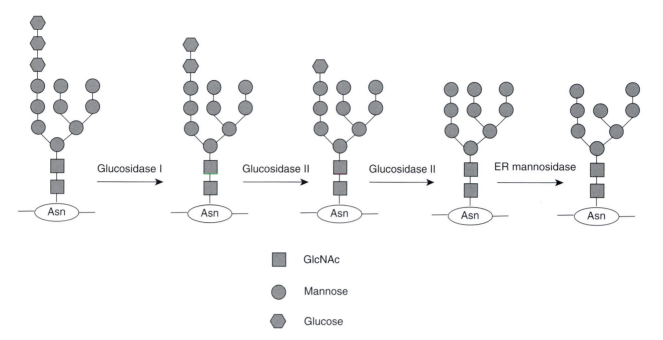

GlcNAc

Mannose

Glucose

Figure 10.13 Trimming of the glucosyl groups of the initial N-glycan chain on nascent glycoproteins by glucosidases in the ER.

the ER membrane-associated protein chaperone (calnexin) (Figure 10.14) (Parodi, 2000). These two ER chaperones behave as lectins, recognizing the terminal glucose of the oligosaccharyl chain. As the glucosidase II removes that last glucose, calreticulin and calnexin have negligible affinity for the mannosyl-terminated oligosaccharyl chain on the nascent glycoproteins, so those deglucosylated glycoproteins are released from the chaperone/protein anchors.

At this point they are substrates for reglucosylation by an ER luminal enzyme UDP-glucose glycoprotein glucosyltransferase (Roth, 2002). This recreates a binding site for calreticulin and calnexin, until glucosidase II acts again to remove the glucose. This cycle of deglucosylation (glucosidase II)/reglucosylation (UDP-glucose protein Gtf) is thought to be involved in protein quality control because UDP-glucose protein Gtf only reglucosylates *unfolded*, non-native nascent glycoproteins (Ellgaard and Helenius, 2003). Thus, this is a cycle to allow partially folded ER proteins time to refold to native structures. Presumably a fraction of the unfolded protein refolds in each deglucosylation/reglucosylation cycle and so quality control is improved at the expense of the net cleavage of a high-energy UDP-glucose bond.

The baroque oligosaccharyl chain on N-glycoproteins may have a positive effect in specifying a subset of proteins that, having been extruded through the ER membrane in an unfolded state, is given more time to achieve a refolded, native state

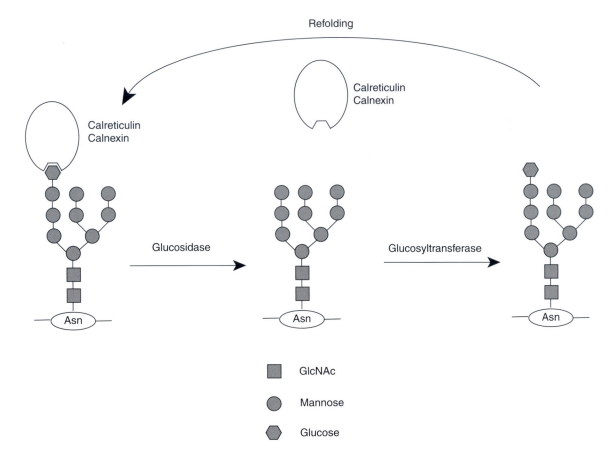

Figure 10.14 Binding of the $Glc_1Man_9GlcNAc_2$ chain to the chaperone proteins calnexin and calreticulin. There is a lack of affinity of the $Man_9GlcNAc_2$ forms of the glycoproteins for the chaperones. Reglucosylation of the $Man_9GlcNAc_2$ chains on glycoproteins in the ER by UDPG glycoprotein glucosyltransferase enables multiple cycles of chaperone-assisted folding to increase the yield of refolded glycoproteins in this compartment.

(Roth, 2002). This is achieved by using the partially trimmed oligosaccharyl chain as the handle and lectin-specific monosaccharide recognition logic for chaperoning, driven by the thermodynamics of UDP-glucose net hydrolysis. The protein secretory pathway may have selected for N-glycosylation as a handle to increase the refolding yield of many classes of proteins. If N-glycoproteins do not refold after multiple cycles, they are shipped back into the cytoplasm, the oligosaccharide chain is removed by protein glycanase N (Figure 10.5), the protein is tagged with ubiquitin, and it is sent to the proteasome (Chapter 9) for proteolysis. Only well-folded $Man_9GlcNAc_2$-N-Asn proteins are sent on for further processing in the Golgi complex. Some mannosidase activities reside in the ER, resulting in further trimming of $Man_9GlcNAc_2$

chains to $Man_8GlcNAc_2$ chains (Figure 10.13). The x-ray structure of the yeast ER α-1,2-mannosidase has been determined (Vallee et al., 2000). This enzyme is itself an N-glycoprotein and in the crystals the N-glycan from one protein molecule extends into the active site of its neighbor, giving a high-resolution view of the bound $Man_8GlcNAc_2$ ligand (Figure 10.15) (Vallee et al., 2000). This is one of the best examples of a biologically relevant conformation for an N-linked glycan chain.

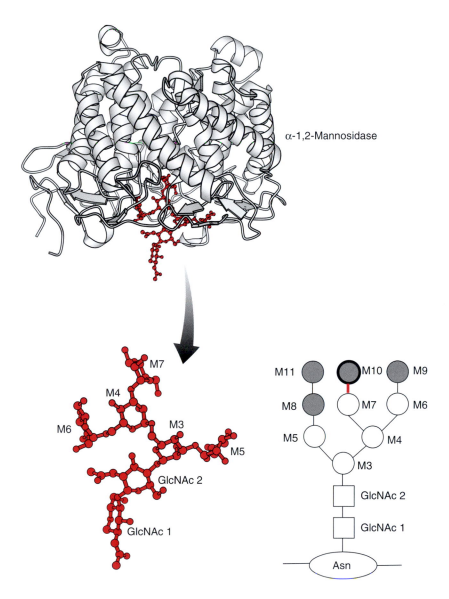

Figure 10.15 High resolution view of a $Man_8GlcNAc_2$ N-glycan chain in the dimer of α-1,2-mannosidase (figure made using PDB 1DL2).

Trimming and Elongation of the Oligosaccharyl Chains in the Golgi Complex

Hydrolytic trimming by mannosidases [see van den Elsen et al. (2001) for the x-ray structure of Golgi α-1,2-mannosidase II] continues in the cis elements of the Golgi complex, where one or more of these enzymes trims the oligosaccharide to the core pentasaccharide $Man_3GlcNAc_2$ that is the common determinant on all the finished N-linked oligosaccharide chains in higher eukaryotes (Figure 10.16) (Roth, 2002; Trombetta, 2003).

In the medial and trans elements of the Golgi complex, the trimming has been completed and there are Gtfs expressed and localized in those subcompartments that build the outer branches of the glycan back up to the diverse final products. For example, in Figure 10.16 the formation of a typical biantennary chain is schematized, with the sequential addition of GlcNAc and Gal from the UDP-hexose substrates to each of the two branches symmetrically by two Gtfs. Many more complex examples could be detailed where tri- and tetraantennary chains are built up by elaborating the two terminal mannosyl residues in the $Man_3GlcNAc_2$ core in the Golgi complex, typically at C_2 and C_6 of one mannose branch and C_2 and C_4 of the other mannose [see Kornfeld and Kornfeld (1976)].

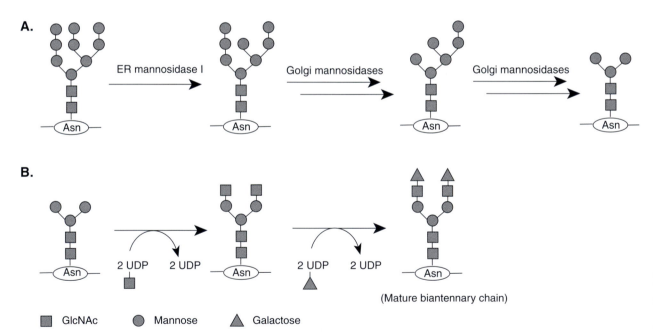

Figure 10.16 N-Glycan chain processing in the Golgi complex. **A.** Mannosidase-mediated trimming of the N-glycan chain to the core pentasaccharide $Man_3GlcNAc_2$. **B.** Building the core pentasaccharide chain back up: enzymatic generation of a mature biantennary chain with 9 sugar residues by the action of four distinct glycosyltransferases.

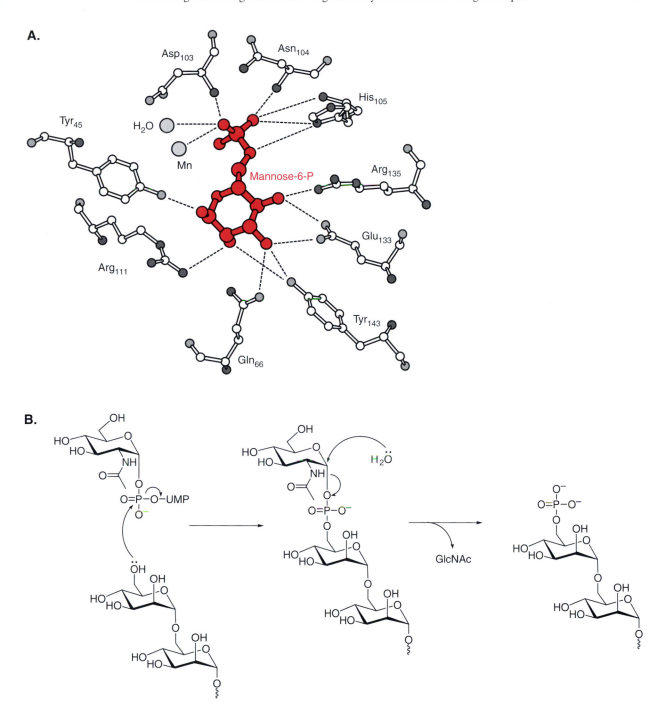

Figure 10.17 **A.** Phosphorylation of mannosyl residues on N-glycan chains of glycoproteins to be targeted to lysosomes (figure made using PDB 1M6P). **B.** Transfer of GlcNAc-P- to mannose followed by hydrolytic removal of the GlcNAc.

For the simple biantennary N-oligosaccharide noted in Figure 10.16B there must be three trimming enzymes in the ER and at least two in the Golgi complex to get to the $Man_5GlcNAc_2$ that is found in hybrid N-glycan structures and then all the way to the $Man_3GlcNAc_2$ core. The additional four remodeling Gtfs in the Golgi complex for this example make nine participating enzymes to go from the original $Glc_3Man_9GlcNAc_2$ scaffold, set in place by OSTase action, to the finished biantennary chain on the glycoprotein. This is after the prior action of the 14 Alg enzymes and OSTase, for at least 24 enzymatic activities to make the rather simple N-linked biantennary glycoprotein posttranslational modification exemplified.

In addition to other Gtfs to remodel the $Man_5GlcNAc_2$ core to more complex finished oligosaccharyl chains, there are even more enzymatic modifications known on the carbohydrate portions of the N-linked glycoproteins in higher eukaryotes. A partial list, just to exemplify the combinatorial prospects for such posttranslational modifications, involves sulfation of mannosyl and GlcNAc residues, acetylation of sialyl groups, and most famously phosphorylation of mannosyl residues [see Kornfeld and Kornfeld (1976)]. The phosphomannosyl residues are epitopes for the trafficking proteins, which send these phosphoglycoproteins to lysosomes and include many lysosomal hydrolases. Enzymatic phosphorylation can happen on any of the five mannoses of the oligosaccharyl core (Figure 10.17) and a given mannose can be bisphosphorylated. The x-ray structure of the mannose-6-P receptor proteins that recognize the phosphorylated mannoses has been determined (Olson et al., 1999; Roberts et al., 1998) and reveals the basis for recognition of these 6-P-mannosyl-N-glycan chains on proteins being chaperoned to lysosomes. Figure 10.17A shows the extensive network of hydrogen bonds between the mannose-6-P moiety and the chaperoning mannose-6-P receptor proteins. The phosphorylation enzymology is also a bit rococo. Each phosphoryl group is put on the mannose as part of a GlcNAc-P addition from UDP-GlcNAc to a mannose–OH, and then the GlcNAc moiety is removed by a subsequent hydrolytic enzyme (Figure 10.17B), leaving the PO_3^{2-} group behind (von Figura and Hasilik, 1986).

The fact that 9–12 Gtfs participate in the elaboration of the $Man_3GlcNac_2$ core to mature N-glycan branched chains without any obvious templating creates much opportunity for heterogeneity. Different cell types and different cells within a tissue (as well as different organisms) express distinct complements of remodeling or extending Gtfs, leading to variety in the outer branches of N-linked glycoproteins. Yeast, where so much genetics has been applied, make high-mannose-containing-proteins where the mature carbohydrate chains (Dean, 1999; Girrbach and Strahl, 2003; Lussier et al., 1999) have mannosyl units that can number over 100 sugars.

O-Glycosylation of Proteins

Most O-glycosylations of proteins occur by the participation of the hydroxyl side chains of threonine and serine residues as nucleophiles to build the sugar-O-protein linkage. In higher eukaryotes, disaccharide and branched trisaccharide chains are typical for O-linked glycoproteins, while in yeast unbranched tetra- and pentamannosyl chains are produced, and it is estimated in those fungi that O-glycosylation is actually more common than the N-glycosylation of proteins (Strahl-Bolsinger et al., 1999). There are a few cases where posttranslational modification of other amino acid side chains occurs, most notably 4-hydroxyprolines and 5-hydroxylysines (see Chapter 12), and those can be subsequently glycosylated. For example, 5-OH-lysyl residues (Figure 10.1B) in collagen chains can be O-glycosylated to create the Glc-Gal-O-hydroxyLys-disaccharide linkages required for the efficient secretion of the collagen chains from fibroblasts.

In higher eukaryotes, the sugar nucleotides, such as UDP-Glc, UDP-GlcNAc, UDP-Gal, UDP-GalNAc (Ten Hagen et al., 2003), and GDP-Fuc, are the typical donors for O-linked glycoproteins, but in yeast where the O-mannosylation of proteins occurs, the mannosyl residue proximal to the Thr/Ser side chain is donated from the mannosyl-PP-dolichol donor (Girrbach and Strahl, 2003) (Figure 10.18A). There are seven such protein mannosyltransferases (Pmt1–Pmt7) in the endoplasmic reticulum of *S. cerevisiae* with overlapping sets of protein cosubstrates. Human homologs of Pmts are known and the O-mannosylation of muscle proteins are known (Yoshida et al., 2001). Defective O-glycosylation of dystrophoglycan is responsible for some muscle wasting syndromes, where a Ser-O-Man-GlcNAc-Gal-sialic acid tetrasaccharide chain is normally built by posttranslational modification (Manya et al., 2004).

In yeast, subsequent elongation of the single O-mannosyl residue to the tetra- and pentamannosyl chains occurs in the Golgi complex with GDP-mannose as the donor by α-1,2-mannosyltransferases (Figure 10.18B) (Lussier et al., 1999). O-Glycosylation has also been detected in the same eubacterium, *C. jejuni*, where N-glycosylation occurs (Szymanski et al., 2003). The protein target for O-glycosylation is the central domain of the flagellin, where 19 Ser/Thr hydroxyls are glycosylated with an anionic sugar known as pseudaminic acid.

Enzymatic addition of a single O-linked GlcNAc residue to mammalian proteins by O-GlcNAc transferase (OGTase) has drawn recent attention in part because it is reversible by O-GlcNAcase (Figure 10.19) action, and may have regulatory roles in controlling gene activity and/or signaling pathways (Vosseller et al., 2002). Sites of GlcNAc addition to Thr/Ser side chains occur, for example, in the transcription factors

A.

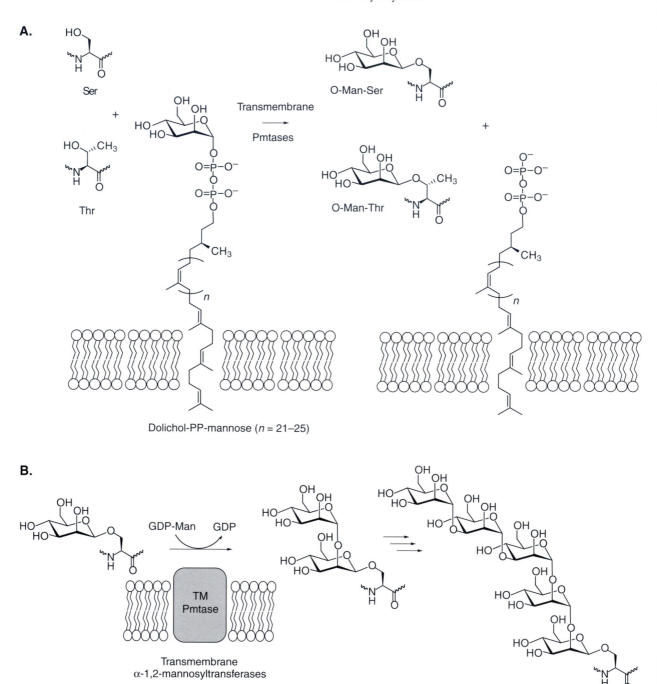

Figure 10.18 **A.** Mannosyl groups on the glycolipid mannosyl-PP-dolichol are the donors for adding the first mannosyl unit in protein O-mannosylation in yeast. **B.** Subsequent mannosyl groups are added enzymatically from GDP-mannose by α-1,2-mannosyltransferases.

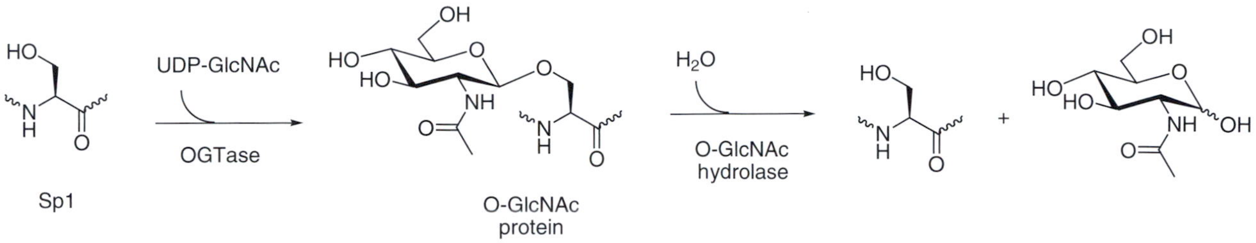

Figure 10.19 Reversible addition of an O-GlcNAc residue to proteins by OGTase and its hydrolytic reversal by O-GlcNAcase.

Sp1, CREB (Lamarre-Vincent, 2003), and Pax6, in a histone deacetylase, in the C-terminal domain of RNA polymerase, and in the epithelial nitric oxide synthase. It has been proposed that the monosaccharide posttranslational modification may block protein–protein interactions or alter phosphorylation patterns and, thus, may impact transcriptional activity. Recently, it has been demonstrated that O-GlcNAc modification of the Rpt2 ATPase subunit in the 19S cap of the proteasome inhibits proteolysis, including Sp1 degradation, and a regulatory role for connecting this glycosylation to intracellular proteolysis has been suggested (Zhang et al., 2003). Detection of GlcNAc-Ser peptide fragments by mass spectrometry [e.g., Chalkley and Burlingame, (2003)] have made it possible to begin the rapid mapping of such O-glycosylation sites. The serum response factor (SRF) transcriptional activator protein has thereby been demonstrated to be modified at three sites—Ser_{283}, Ser_{313}, and Thr_{401}. In the same mass spectrometric characterization, four phosphorylation sites of SRF at Ser_{83}, Ser_{103}, Ser_{224}, and Ser_{435} could be mapped. The extent of heterogeneity and functional consequences of these two types of posttranslational modifications distributed over seven distinct Ser/Thr residues has yet to be established, but it reflects a prototypic challenge for functional proteomics of modified proteins.

There are 24 predicted UDP-GalNAc-O-protein transferases in the human genome, suggesting overlapping sets of specificities for mucin-type proteins with Ser and Thr side chains to be modified by O-GalNAc (Ten Hagen et al., 2003).

A separate set of O-linked glycosylations of cell surface receptors such as Notch occur at two conserved sites in EGF repeat domains (Moloney et al., 2000b). EGF domains in the transmembrane protein Notch are extracellular, spanning about 40 residues with six thiols oxidized to three disulfide bonds to impose architecture. Fucosylation occurs at a Ser/Thr (or occasionally a 5-OH-Lys) residue just before the third Cys, while O-linked glucosylation occurs at serine consensus sites between the first and second Cys (Figure 10.20A). O-Linked glucosylation can be followed by the addition of two α-1,3-linked xylose residues to generate O-linked trisaccharide chains

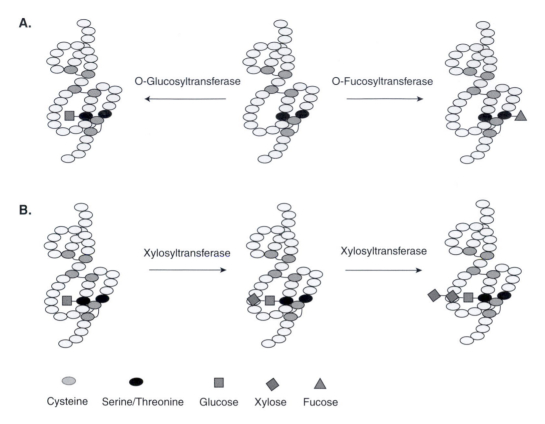

Figure 10.20 **A.** O-Fucosylation and O-glucosylation at S/T sites in the EGF domains of Notch. **B.** Tandem addition of α-1,3-linked xylosyl residues to the glucose moiety to create O-linked trisaccharide chains in the Notch extracellular domain.

(Shao et al., 2002) (Figure 10.20B). Meanwhile, the O-linked fucosyl sugars on the EGF domain repeats of Notch are substrates for elongation by GlcNAc, catalyzed by the Fringe protein in *Drosophila* and the Radical Fringe, Lunatic Fringe, and Manic Fringe homologs in mammalian cells (Fortini, 2000; Moloney et al., 2000a) during transit through the Golgi complex as Notch moves to the cell surface. This O-linked disaccharide can then be further elongated by two Gtfs to give a sialyl-α-2,3-Gal-β-1,4-GlcNac-β-1,3-fucosyl tetrasaccharide chain O-linked to Ser_{224} (Figure 10.3A). Thus, both tri- and tetrasaccharide chains of distinct sequence and with distinct regiospecificity of attachment to the EGF domains are elaborated during the passage of Notch through the secretory system on its way to the cell surface.

This Fringe-initiated processing modulates Notch activation and signaling pathways in some developmental contexts, such as in the establishment of dorsal/ventral polarity (Bruckner et al., 2000; Moloney et al., 2000a). The Notch ligands, the proteins Delta and Jagged, also are O-fucosylated and those can also be elongated by the GlcNAc transferase activity of Fringe proteins (Panin et al., 2002). The glycosylations, in particular the initial O-linked fucosylations, have been found to be required for Notch signaling (Haines and Irvine, 2003) and recognition of the Notch EGF domains by the protein ligands Serrated and Jagged (Okajima et al., 2003), although the molecular bases of the recognition enabled by the sugars has yet to be determined.

O-Glycosylation of Rho and Ras Family GTPases by Clostridial Toxins

Several pathogenic bacteria secrete proteins that act as exoenzymes towards proteins in host cells. Many of theses are ADP-ribosyltransferases, discussed in Chapter 11. Some protein toxins such as the A and B toxins from *Clostridium difficile*, which causes pseudomembranous colitis in the GI tract in humans, and the *Clostridium sordelli* lethal toxin, which causes diarrhea and enterotoxemia in animals (Herrmann et al., 1998; Vetter et al., 2000), carry out the O-glycosylation of a small set of host proteins. The *C. difficile* toxin A and the *C. novyi* α-toxin add a hexose moiety (glucose and Glc-NAc, respectively) to Thr_{37} in the Rho family of GTPases involved in control of the actin cytoskeleton. Rho family members glycosylated include Rho, Rac, and Cdc42. Thr_{37} is in the effector loop, coordinates Mg^{2+} in the $Mg^{2+} \bullet GDP$ complex, and when glycosylated disrupts function of all three of these related GTPases, with consequent depolymerization of actin networks.

Analogously, the *C. sordelli* lethal toxin targets only Rac in the Rho family, but also modifies the corresponding Thr_{35}–OH in Ras, Ral, and Rap members of the Ras GTPase family (Vetter et al., 2000) (Figure 10.21). The O-glucosylation of Ras occurs only in the GDP state, not the GTP state, and prevents the posttranslationally modified effector loop from achieving the conformation required to interact

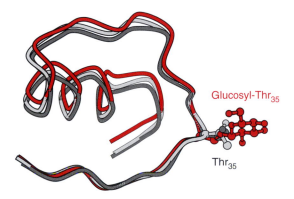

Figure 10.21 O-Glucosylation of Thr_{35} in the effector loop of the Ras GTPases by the *C. sordelli* lethal toxin acting as an O-glucosyltransferase [coordinates kindly provided by I. Vetter (Vetter et al., 2000)].

effectively with partner proteins. These partners include the Ras GTPase activating protein (RAS-GAP), and most notably the downstream Raf protein kinase (see Chapter 2) of one of the parallel MAP kinase pathways in eukaryotic cells. Glucosylation

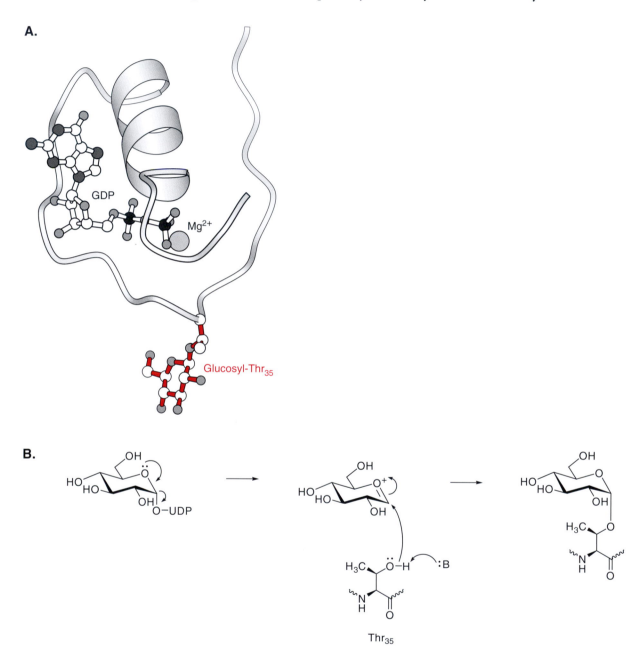

Figure 10.22 **A.** X-ray structure of O-glucosylated Ras shows the C_1-α configuration of the glucosyl residue [coordinates kindly provided by I. Vetter (Vetter et al., 2000)]. **B.** Postulated mechanism for the side chain –OH of Thr_{35} capturing the transferring glucosyl oxocarbenium ion from below.

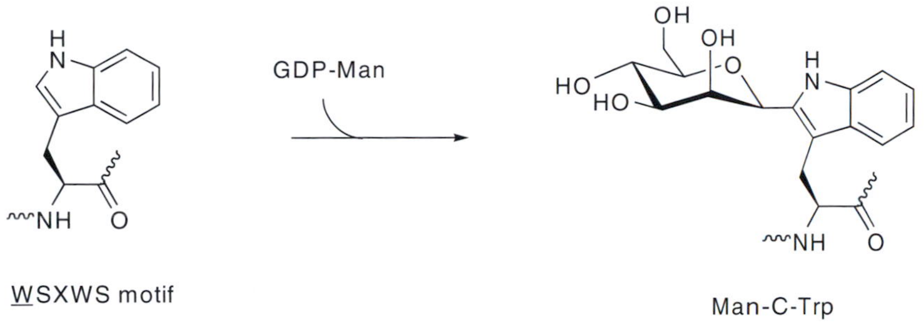

Figure 10.23 Mannosyl transfer from GDP-mannose to the enamine tautomer of the indole side chain of tryptophan residues to effect C-mannosylation.

of Ras at Thr_{35} lowers the catalytic efficiency for recognition of Raf kinase by a factor of 10^5 (Herrmann et al., 1998), shutting down this signaling pathway and leading to host cell death (Vetter et al., 2000). Determination of the x-ray structure of O-glucosylated Ras (Vetter et al., 2000) shows that the glucose group on Thr_{35} has the α-configuration at C_1 (Figure 10.22A) (Geyer et al., 2003), the same as the starting α-configuration in the donor substrate UDP-glucose, consistent with an S_N1 type oxocarbenium ion intermediate. Capture by the OH side chain of Thr_{35} is constrained to occur with net retention from below the plane of the transferring glucosyl oxocarbenium ion (Figure 10.22B).

Protein C-Glycosylation

Recently a small number of proteins have been detected to have mannose linked via its C_1 to C_2 of the indole ring of tryptophan at specific sites in proteins (Furmanek and Hofsteenge, 2000), including Trp_7 of human $RNAse_2$ and Trp_{235} in a highly conserved WSXWS motif in the erythropoietin receptor, in the analogous first tryptophan of this conserved motif in IL-12β, and also in thrombospondins. C-Mannosyltransferase activity has been detected in organisms from *C. elegans* to humans. The sugar donor is GDP-mannose and the carbon nucleophile is presumably the enamine tautomer of the indole ring of tryptophan, an electron-rich site in the indole ring (Figure 10.23). The purpose of the C-mannosylation reaction is not yet known; it could have a role in directing proteins to the cell surface for secretion or could have a structural/conformation-determining role, such as in the conserved WSXWS motif of cytokine receptors. Thrombospondin repeat domains contain both C-mannose and O-fucosylglucose disaccharide links at sites thought to be important for protein–protein recognition (Gonzalez de Peredo et al., 2002).

An Approach to Cotranslational Synthesis of Glycoproteins

The research groups of Wong and Schultz (Zhang et al., 2004) have recently described a novel approach to the direct biosynthetic incorporation of glycosyl-amino acids during protein biosynthesis. A tyrosyl-tRNA-synthetase from a methanogenic bacteria and a cognate tRNA with an anticodon (CUA) that suppresses the stop codon UAG were co-evolved to recognize GlcNAc-O-serine and install it at the 3′-end of the suppressor tRNA. This GlcNAc-O-Ser-tRNA$_{CUA}$ was then able to deliver the glucosylserine at UAG encoded sites in mRNAs translated in *E. coli*, including at residue 4 of a mutant myoglobin gene that had Gly$_4$ converted to the stop sequence TAG (Figure 10.24). Production of the GlcNAc-O-Ser$_4$ mutant myoglobin occurred at 1 mg/liter and the product was characterized by mass spectrometry. To get penetration of the GlcNAc-O-Ser into cells, the tetra-O-acetyl-GlcNAc-O-serine was used, relying on intracellular esterases to remove the sugar protecting O-acetyl groups. Furthermore, the GlcNAc-O-Ser$_4$ site could be elongated to a Gal-GlcNAc-O-Ser$_4$ by a galactosyltransferase with UDP-Gal as donor. This result opens the way to the generation of O-glycoproteins and prospectively some N-glycoproteins with sugars at any desired site, with homogenous occupancy in any given protein.

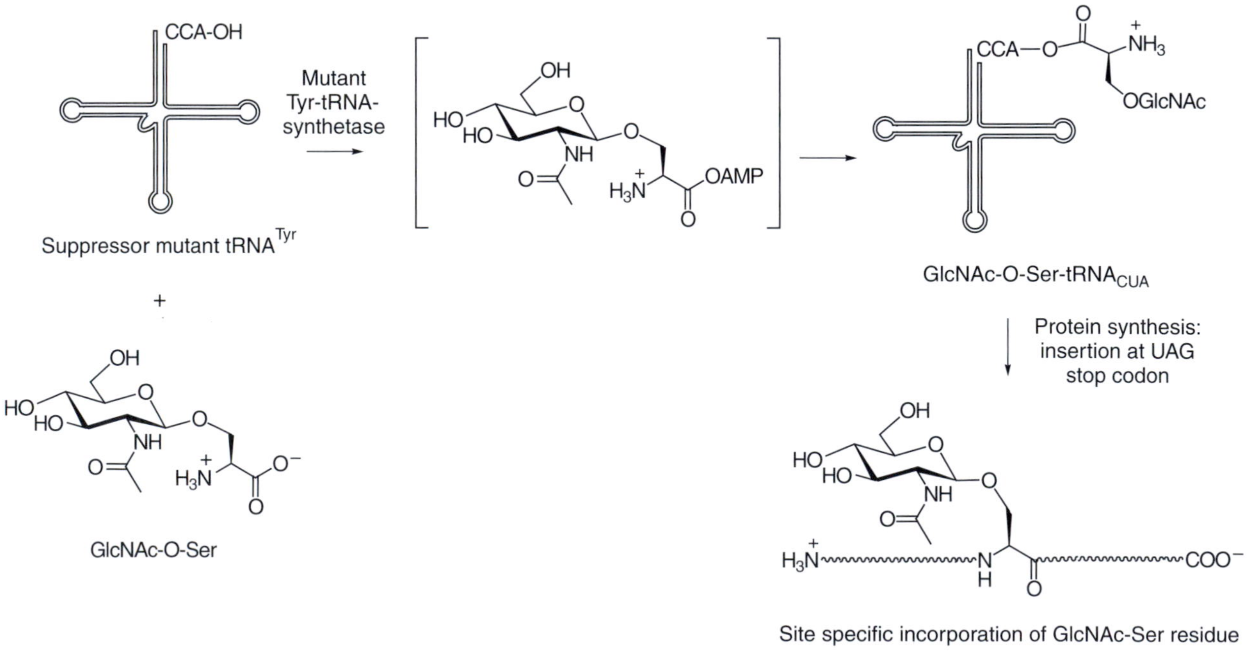

Figure 10.24 Site specific incorporation of the GlcNAc-Ser residue by suppressor tRNA insertion at a UAG stop codon.

References

Benz, I., and M. A. Schmidt. "Never say never again: Protein glycosylation in pathogenic bacteria," *Mol. Microbiol.* **45:**267–276 (2002).

Bosques, C. J., and B. Imperiali. "The interplay of glycosylation and disulfide formation influences fibrillization in a prion protein fragment," *Proc. Natl. Acad. Sci. U.S.A.* **100:**7593–7598 (2003).

Bruckner, K., L. Perez, H. Clausen, and S. Cohen. "Glycosyltransferase activity of Fringe modulates Notch-Delta interactions," *Nature* **406:**411–415 (2000).

Burda, P., and M. Aebi. "The dolichol pathway of N-linked glycosylation," *Biochim. Biophys. Acta* **1426:**239–257 (1999).

Chalkley, R. J., and A. L. Burlingame. "Identification of novel sites of O-N-acetylglucosamine modification of serum response factor using quadrupole time-of-flight mass spectrometry," *Mol. Cell. Proteomics* **2:**182–190 (2003).

Dean, N. "Asparagine-linked glycosylation in the yeast golgi," *Biochim. Biophys. Acta* **1426:**309–322 (1999).

Dell, A., and H. R. Morris. "Glycoprotein structure determination by mass spectrometry," *Science* **291:**2351–2356 (2001).

Dempski, R. E., Jr., and B. Imperiali. "Oligosaccharyl transferase: Gatekeeper to the secretory pathway," *Curr. Opin. Chem. Biol.* **6:**844–850 (2002).

Ellgaard, L., and A. Helenius. "Quality control in the endoplasmic reticulum," *Nat. Rev. Mol. Cell. Biol.* **4:**181–191 (2003).

Fortini, M. E. "Fringe benefits to carbohydrates," *Nature* **406:**357–358 (2000).

Furmanek, A., and J. Hofsteenge. "Protein C-mannosylation: Facts and questions," *Acta Biochim. Pol.* **47:**781–789 (2000).

Gemmill, T. R., and R. B. Trimble. "Overview of N- and O-linked oligosaccharide structures found in various yeast species," *Biochim. Biophys. Acta* **1426:**227–237 (1999).

Geyer, M., C. Wilde, J. Selzer, K. Aktories, and H. R. Kalbitzer. "Glucosylation of Ras by *Clostridium sordellii* lethal toxin: Consequences for effector loop conformations observed by NMR spectroscopy," *Biochemistry* **42:**11951–11959 (2003).

Girrbach, V., and S. Strahl. "Members of the evolutionarily conserved PMT family of protein O-mannosyltransferases form distinct protein complexes among themselves," *J. Biol. Chem.* **278:**12554–12562 (2003).

Girrbach, V., T. Zeller, M. Priesmeier, and S. Strahl-Bolsinger. "Structure-function analysis of the dolichyl phosphate-mannose: Protein O-mannosyltransferase ScPmt1p," *J. Biol. Chem.* **275:**19288–19296 (2000).

Gonzalez de Peredo, A., D. Klein, B. Macek, D. Hess, J. Peter-Katalinic, and J. Hofsteenge. "C-Mannosylation and O-fucosylation of thrombospondin type 1 repeats," *Mol. Cell. Proteomics* **1:**11–18 (2002).

Haines, N., and K. D. Irvine. "Glycosylation regulates Notch signalling," *Nat. Rev. Mol. Cell Biol.* **4:**786–797 (2003).

Helenius, A., and M. Aebi. "Roles of N-linked glycans in the endoplasmic reticulum," *Annu. Rev. Biochem.* **73:**1019–1049 (2004).

Helenius, J., D. T. Ng, C. L. Marolda, P. Walter, M. A. Valvano, and M. Aebi. "Translocation of lipid-linked oligosaccharides across the ER membrane requires Rft1 protein," *Nature* **415:**447–450 (2002).

Herrmann, C., M. R. Ahmadian, F. Hofmann, and I. Just. "Functional consequences of mono-glucosylation of Ha-Ras at effector domain amino acid threonine 35," *J. Biol. Chem.* **273:**16134–16139 (1998).

High, S., F. J. Lecomte, S. J. Russell, B. M. Abell, and J. D. Oliver. "Glycoprotein folding in the endoplasmic reticulum: A tale of three chaperones?" *FEBS Lett.* **476:**38–41 (2000).

Karaoglu, D., D. J. Kelleher, and R. Gilmore. "Allosteric regulation provides a molecular mechanism for preferential utilization of the fully assembled dolichol-linked oligosaccharide by the yeast oligosaccharyltransferase," *Biochemistry* **40:**12193–12206 (2001).

Kornfeld, R., and S. Kornfeld. "Comparative aspects of glycoprotein structure," *Annu. Rev. Biochem.* **45:**217–237 (1976).

Lamarre-Vincent, N., and L. Hsieh-Wilson. "Dynamic glycosylation of the transcription factor CREB: A potential role in gene regulation," *J. Am. Chem. Soc.* **125:**6612–6613 (2003).

Liebler, D. C. *Introduction to Proteomics.* Humana Press: Totowa, NJ (2002).

Loffler, A., M. A. Doucey, A. M. Jansson, D. R. Muller, T. de Beer, D. Hess, M. Meldal, W. J. Richter, J. F. Vliegenthart, and J. Hofsteenge. "Spectroscopic and protein chemical analyses demonstrate the presence of C-mannosylated tryptophan in intact human RNase 2 and its isoforms," *Biochemistry* **35:**12005–12014 (1996).

Lussier, M., A. M. Sdicu, and H. Bussey. "The KTR and MNN1 mannosyltransferase families of *Saccharomyces cerevisiae*," *Biochim. Biophys. Acta* **1426:**323–334 (1999).

Maeda, Y., S. Tanaka, J. Hino, K. Kangawa, and T. Kinoshita. "Human dolichol-phosphate-mannose synthase consists of three subunits, DPM1, DPM2 and DPM3," *EMBO J.* **19:**2475–2482 (2000).

Manya, H., A. Chiba, A. Yoshida, X. Wang, Y. Chiba, Y. Jigami, R. U. Margolis, and T. Endo. "Demonstration of mammalian protein O-mannosyltransferase activity: Coexpression of POMT1 and POMT2 required for enzymatic activity," *Proc. Natl. Acad. Sci. U.S.A.* **101:**500–505 (2004).

Mechref, Y., and M. V. Novotny. "Structural investigations of glycoconjugates at high sensitivity," *Chem. Rev.* **102:**321–369 (2002).

Moloney, D. J., V. M. Panin, S. H. Johnston, J. Chen, L. Shao, R. Wilson, Y. Wang, P. Stanley, K. D. Irvine, R. S. Haltiwanger, and T. F. Vogt. "Fringe is a glycosyltransferase that modifies Notch," *Nature* **406:**369–375 (2000a).

Moloney, D. J., L. H. Shair, F. M. Lu, J. Xia, R. Locke, K. L. Matta, and R. S. Haltiwanger. "Mammalian Notch1 is modified with two unusual forms of O-linked glycosylation found on epidermal growth factor-like modules," *J. Biol. Chem.* **275:**9604–9611 (2000b).

Okajima, T., A. Xu, and K. D. Irvine. "Modulation of Notch-ligand binding by protein O-fucosyltransferase 1 and fringe," *J. Biol. Chem.* **278:**42340–42345 (2003).

Olson, L. J., J. Zhang, Y. C. Lee, N. M. Dahms, and J. J. Kim. "Structural basis for recognition of phosphorylated high mannose oligosaccharides by the cation-dependent mannose 6-phosphate receptor," *J. Biol. Chem.* **274:**29889–29896 (1999).

Panin, V. M., L. Shao, L. Lei, D. J. Moloney, K. D. Irvine, and R. S. Haltiwanger. "Notch ligands are substrates for protein O-fucosyltransferase-1 and Fringe," *J. Biol. Chem.* **277:**29945–29952 (2002).

Parodi, A. J. "Protein glucosylation and its role in protein folding," *Annu. Rev. Biochem.* **69:**69–93 (2000).

Roberts, D. L., D. J. Weix, N. M. Dahms, and J. J. Kim. "Molecular basis of lysosomal enzyme recognition: Three-dimensional structure of the cation-dependent mannose 6-phosphate receptor," *Cell* **93:**639–648 (1998).

Roth, J. "Protein N-glycosylation along the secretory pathway: Relationship to organelle topography and function, protein quality control, and cell interactions," *Chem. Rev.* **102:**285–303 (2002).

Rudd, P. M., C. Colominas, L. Royle, N. Murphy, E. Hart, A. H. Merry, H. F. Hebestreit, and R. A. Dwek. "A high-performance liquid chromatography based strategy for rapid, sensitive sequencing of N-linked oligosaccharide modifications to proteins in sodium dodecyl sulphate polyacrylamide electrophoresis gel bands," *Proteomics* **1:**285–294 (2001).

Rudd, P. M., and R. A. Dwek. "Glycosylation: Heterogeneity and the 3D structure of proteins," *Crit. Rev. Biochem. Mol. Biol.* **32:**1–100 (1997).

Rudd, P. M., T. Endo, C. Colominas, D. Groth, S. F. Wheeler, D. J. Harvey, M. R. Wormald, H. Serban, S. B. Prusiner, A. Kobata, and R. A. Dwek. "Glycosylation differences between the normal and pathogenic prion protein isoforms," *Proc. Natl. Acad. Sci. U.S.A.* **96:**13044–13049 (1999a).

Rudd, P. M., T. S. Mattu, N. Zitzmann, A. Mehta, C. Colominas, E. Hart, G. Opdenakker, and R. A. Dwek. "Glycoproteins: Rapid sequencing technology for N-linked and GPI anchor glycans," *Biotechnol. Genet. Eng. Rev.* **16:**1–21 (1999b).

Rudd, P. M., A. H. Merry, M. R. Wormald, and R. A. Dwek. "Glycosylation and prion protein," *Curr. Opin. Struct. Biol.* **12:**578–586 (2002).

Shao, L., Y. Luo, D. J. Moloney, and R. Haltiwanger. "O-glycosylation of EGF repeats: Identification and initial characterization of a UDP-glucose:protein O-glucosyltransferase," *Glycobiology* **12:**763–770 (2002).

Strahl-Bolsinger, S., M. Gentzsch, and W. Tanner. "Protein O-mannosylation," *Biochim. Biophys. Acta* **1426:**297–307 (1999).

Szymanski, C. M., S. M. Logan, D. Linton, and B. W. Wren. "Campylobacter—a tale of two protein glycosylation systems," *Trends Microbiol.* **11:**233–238 (2003).

Ten Hagen, K. G., T. A. Fritz, and L. A. Tabak. "All in the family: The UDP-GalNAc:polypeptide N-acetylgalactosaminyltransferases," *Glycobiology* **13:**1R–16R (2003).

Trombetta, E. S. "The contribution of N-glycans and their processing in the endoplasmic reticulum to glycoprotein biosynthesis," *Glycobiology* **13:**77R–91R (2003).

Trombetta, E. S., and A. J. Parodi. "Quality control and protein folding in the secretory pathway," *Annu. Rev. Cell. Dev. Biol.* **19:**649–676 (2003).

Vallee, F., F. Lipari, P. Yip, B. Sleno, A. Herscovics, and P. L. Howell. "Crystal structure of a class I alpha1,2-mannosidase involved in N-glycan processing and endoplasmic reticulum quality control," *EMBO J.* **19:**581–588 (2000).

van den Elsen, J. M., D. A. Kuntz, and D. R. Rose. "Structure of Golgi alpha-mannosidase II: A target for inhibition of growth and metastasis of cancer cells," *EMBO J.* **20**:3008–30017 (2001).

Vetter, I. R., F. Hofmann, S. Wohlgemuth, C. Herrmann, and I. Just. "Structural consequences of mono-glucosylation of Ha-Ras by Clostridium sordellii lethal toxin," *J. Mol. Biol.* **301**:1091–1095 (2000).

von Figura, K., and A. Hasilik. "Lysosomal enzymes and their receptors," *Annu. Rev. Biochem.* **55**:167–93 (1986).

Vosseller, K., K. Sakabe, L. Wells, and G. W. Hart. "Diverse regulation of protein function by O-GlcNAc: A nuclear and cytoplasmic carbohydrate post-translational modification," *Curr. Opin. Chem. Biol.* **6**:851–857 (2002).

Walsh, C. T. *Antibiotics: Actions, Origins, Resistance.* ASM Press: Washington (2003).

Wormald, M. R., A. J. Petrescu, Y. L. Pao, A. Glithero, T. Elliott, and R. A. Dwek. "Conformational studies of oligosaccharides and glycopeptides: Complementarity of NMR, x-ray crystallography, and molecular modelling," *Chem. Rev.* **102**:371–386 (2002).

Zhang, F., K. Su, X. Yang, D. B. Bowe, A. J. Paterson, and J. E. Kudlow. "O-GlcNAc modification is an endogenous inhibitor of the proteasome," *Cell* **115**:715–725 (2003).

Zhang, Z., J. Gildersleeve, Y. Y. Yang, R. Xu, J. A. Loo, S. Uryu, C. H. Wong, and P. G. Schultz. "A new strategy for the synthesis of glycoproteins," *Science* **303**:371–373 (2004).

11

ADP-Ribosylation of Proteins from NAD as Donor

The eukaryotic protein synthesis elongation factor EF2 has one His residue converted to diphthamide that is then targeted for ADP-ribosylation by diphtheria toxin.

A variety of eukaryotic proteins are observed to undergo posttranslational ADP-ribosylation. Most of the known protein ADP-ribosyltransferases are secreted as exoenzymes by toxogenic bacteria and taken up into the cytoplasm of host cells where specific sets of proteins get posttranslationally modified (Aktories, 2000). Representative members of the set of bacterial toxins with ADP-ribosyltransferase activity are shown in Table 11.1. Collectively, these enzymes act as virulence factors targeting eukaryotic cell proteins that can cause cell death (EF2 blockade and protein synthesis cessation), altered cell morphology (actin), and disruption of eukaryotic cell signaling pathways (GTPase inhibition).

It is not clear if there are many targets inside the bacterial cell for ADP-ribosylation, although the regulation of nitrogenase activity by reversible ADP-ribosylation has been

Table 11.1 ADP-Ribosylation by Bacterial Toxins at Different Residues in Target Proteins

ADP-Ribosylation Catalyst	Residue Modified	Protein Targeted
P. aeruginosa exotoxin S	Arg	Ras GTPase
V. cholerae cholera toxin	Arg	G_s GTPase
E. coli lethal toxin	Arg	G_s GTPase
C. botulinum C2 toxin	Arg	Actin
B. cereus VIP2 toxin	Arg	Actin
C. botulinum C3 toxin	Asn	Rho family GTPases
S. aureus C3 toxin	Asn	Rho family GTPases
B. pertussis toxin	Cys	G_i GTPase
C. diphtheria toxin	Diphthamide	EF2
P. aeruginosa exotoxin A	Diphthamide	EF2

described (Grunwald et al., 1995 and 2000). The iron–sulfur protein that is the reductant for nitrogenase is ADP-ribosylated on Arg_{100} with loss of electron transfer ability to nitrogenase. The fractional level of ADP-ribosylation can be controlled via a balance of ADP-ribosyltransferase and ADP-ribose hydrolase activity. Other than this case, it may be that the major role of protein ADP-ribosyltransferases in bacteria is to modulate bacterial–eukaryotic cell interactions.

There are a few mammalian ADP-ribosyltransferases known. Their functions, however, are incompletely understood. Some of them cyclize NAD to a derivative cyclic ADP-ribose involved in the mobilization of intracellular calcium ions (Lee, 2001). Other eukaryotic NAD-cleaving enzymes are in the sirtuin family (discussed in Chapter 6) and have histone deacetylase activity, with the acetyl group ending up at the 2′-hydroxyl of the ADP-ribose fragment as NAD is cleaved during the reaction.

The universal donor of the ADP-ribosyl group is the coenzyme molecule NAD. NAD has two different kinds of functions in eukaryotic metabolism: (1) the redox function as the source of hydride ions in redox metabolism, and (2) the ADP-ribosyl group transfer reactions discussed here and in Chapter 6 (Walsh, 1979). Both attributes of NAD reactivity are a consequence of the nicotinamide heterocycle connected to C_1 of the ADP-ribosyl moiety of NAD (Figure 11.1). The dihydro form of NAD (NADH), with the aromatic nicotinamide ring reduced to the dihydroaromatic species, is a thermodynamically activated donor of one of the two prochiral hydrogens at C_4 as a hydride ion. The high driving force ($E^{o'} = -320$ mV) arises from the favorable energy of rearomatization upon hydride transfer. Meanwhile, NADH is kinetically stable in cells and so, with its congeneric NADPH, serves as the main diffusible source of redox currency in cellular metabolism (Walsh, 1979).

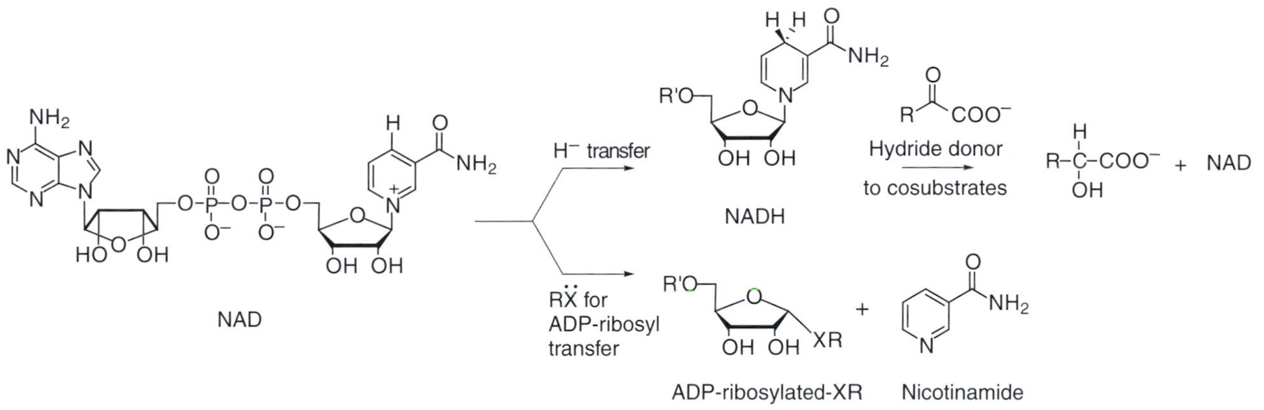

Figure 11.1 Coenzyme NAD with ADP-ribose and nicotinamide moieties joined through a C_1-N linkage. The dihydro form (NADH) is a hydride donor from the dihydropyridine ring. The C_1–N bond of NAD can be cleaved in ADP-ribosyl transfer reactions.

The second mode of reactivity occurs in the *oxidized* form of the coenzyme (NAD), and involves net displacement of the nicotinamide by a nucleophilic atom of a cosubstrate, as free nicotinamide is released and the incoming nucleophilic group is ADP-ribosylated. Because there is a net positive charge on the nicotinamide quaternized nitrogen in NAD, the nicotinamide moiety is activated for departure in a low-energy transition state, generating an ADP-ribose oxacarbenium ion with carbonium ion character at C_1 and the ribose oxygen. This powerful electrophilic species, generated in the ADP-ribosyltransferase active sites, can be captured by a range of nucleophiles. These include typically strong nucleophilic side chains, such as the thiolate of cysteine residues and the imidazole nitrogen of a modified histidine side chain (diphthamide) in elongation factor 2 (Table 11.1), and also weak nucleophiles such as the carboxamide of Asn and the guanidino group of Arg side chains of target protein cosubstrates. In the latter cases hydrogen bonding of the cosubstrate Asn or Arg residues to side chains acting as bases in the ADP-ribosyltransferase activates the carboxamido nitrogen (Asn) or the guanidine nitrogen (Arg). Almost all the bacterial ADP-ribosyltransferases exhibit some net NAD glycohydrolase activity, reflecting a default and presumably adventitious capture mode of the ribooxacarbenium ion by water as the alternate (weak) nucleophile in the active site. This default hydrolysis is a measure of the reactivity of the ADP-ribosylating species (Figure 11.2). Most protein ADP-ribosyltransferases exhibit some basal or competing NAD glycohydrolase activity, consistent with the oxacarbenium ion character of the transferring ribosyl group and a default path for the capture by water.

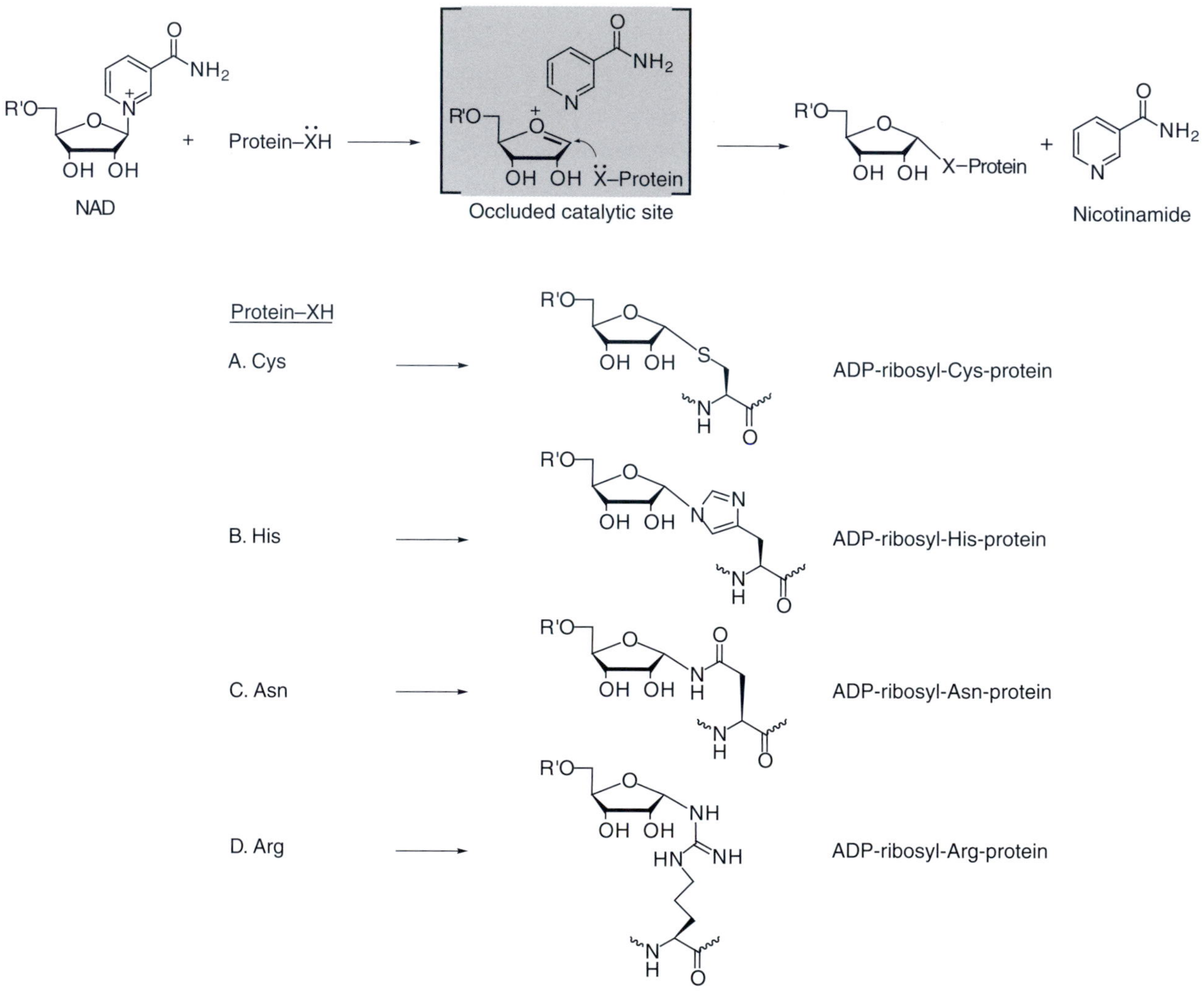

Figure 11.2 Posttranslational modification of protein side chains by ADP-ribosylation at Cys, His, Asn, and Arg side chains.

The terminology "glycohydrolase" and the inferred ribooxacarbenium ion character of the reactions in Figure 11.2 emphasize that this is *glycosyl transfer* chemistry of a pentose (ribose) rather than of the hexoses (e.g., glucose) covered in Chapter 10. It is useful to note the formal similarity to the activation of the C_1 of glucose in UDP-glucose, in that case by the α-UDP moiety, and the activation of C_1 of the ribose in NAD, in this case by the β-nicotinamide moiety (Figure 11.3). The sugar transfers from UDP-glucose thus lead to attachment of the nucleophilic atom to C_1 of glucose and the sugar transfers from NAD in parallel lead to attachment to C_1 of the transferring ribosyl moiety. In the former case the transferring hexosyl moiety is

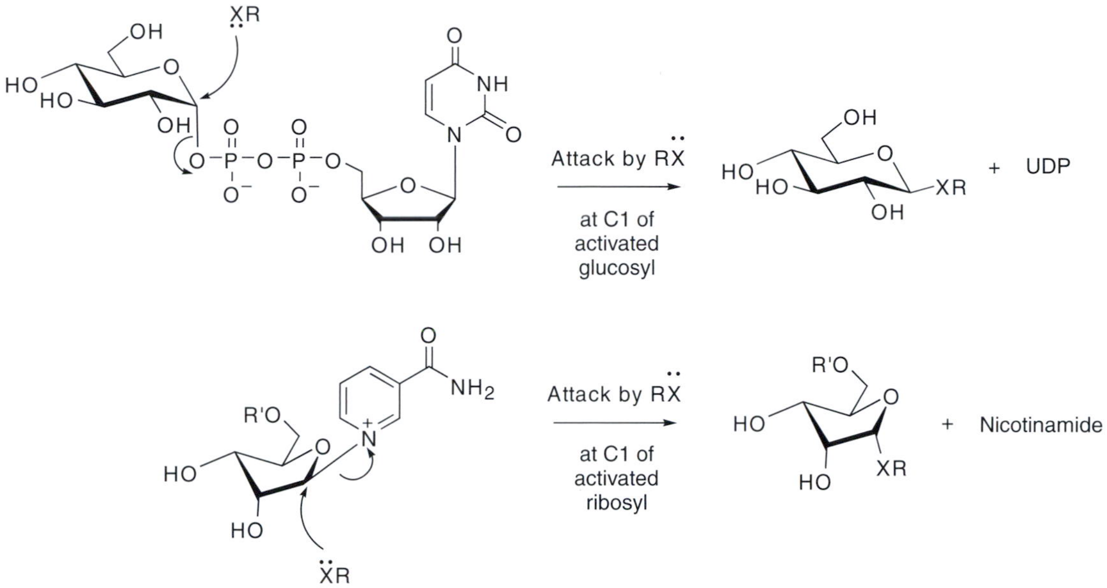

Figure 11.3 Activation of UDP-glucose and NAD for glycosyl transfer reactions: glucosyl transfer via capture at C_1 versus ADP-ribosyl capture at C_1.

the monosaccharide; in the latter case the ribosyl moiety is substituted at C_5 by the ADP substituent.

Posttranslational Formation of Diphthamide in Elongation Factor 2 and its ADP-Ribosylation

One group of bacterial exoenzymes that acts as toxins to mammalian cells includes diphtheria toxin from *Corynebacterium diphtheriae* and exotoxin A from *Pseudomonas aeruginosa* (Collier, 2001; Moss, 1990). They have a two-domain A–B organization with the B domain acting as the ligand for a specific receptor on the host cell membrane, inducing receptor internalization to early endosomes. Then, upon acidification, the B domain oligomerizes and allows the A domain to escape into the cytoplasm where it acts as an NAD-dependent ADP-ribosyltransferase to modify one single protein, eukaryotic elongation factor 2 (eEF2), required for protein synthesis. Diphtheria toxin is first synthesized as a single-chain 60-kDa precursor that is cleaved proteolytically to a 21-kDa A chain still tethered covalently to the 39-kDa B chain by an interchain disulfide bond. The S–S bond is reductively cleaved after internalization and acidification of lysosomes when the B subunit forms an oligomeric pore through which the catalytic A subunit can get to the cytoplasm and its host protein target.

The site of ADP-ribosylation is residue 715 in the eEF2 chain. In nascent eEF2 residue 715 is a histidine but it undergoes a series of three posttranslational steps to convert the imidazole moiety into the residue known as diphthamide, named because it is the target of diphtheria toxin action. Although diphthamide's existence and structure have been known for more than two decades, its enzymology of posttranslational modification is not completely understood (Liu and Leppla, 2003). It is known that the C_2 of the imidazole ring of His_{715} acts as a nucleophile towards cosubstrate SAM in the first enzyme catalyzed step (Han et al., 2001). Rather than attacking the one-carbon CH_3 fragment attached to the trigonalized sulfonium center of SAM, the C_2 of the imidazole ring attacks $C\gamma$ of the methionyl substituent, making a new C–C bond, transferring the aminocarboxypropyl fragment, and releasing 5'-thiomethyl-adenosine (Figure 11.4). Three more molecules of SAM are then consumed, in three

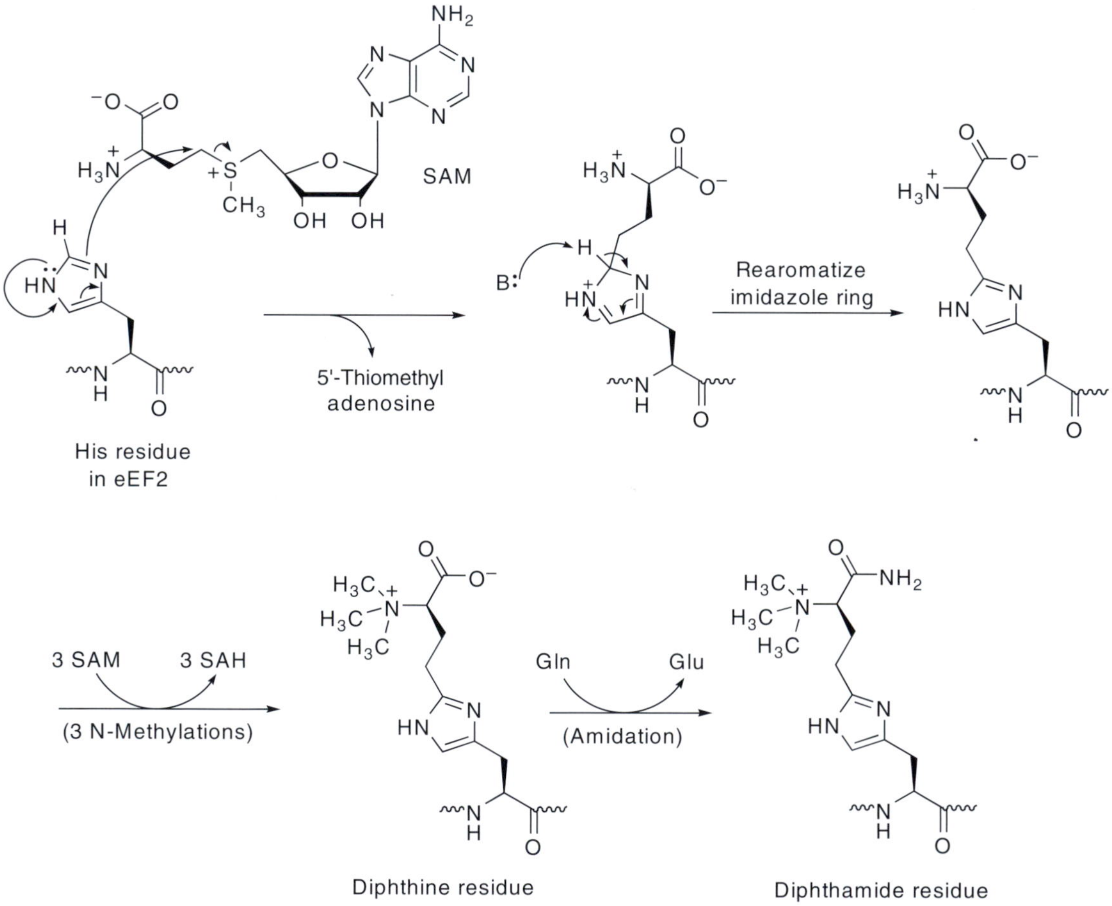

Figure 11.4 Conversion of His_{715} in proeEF2 to diphthamide in active eEF2 by aminocarboxypropyl transfer from SAM, transfer of three CH_3 fragments from three molecules of SAM, and Gln-dependent amidation of diphthine to diphthamide.

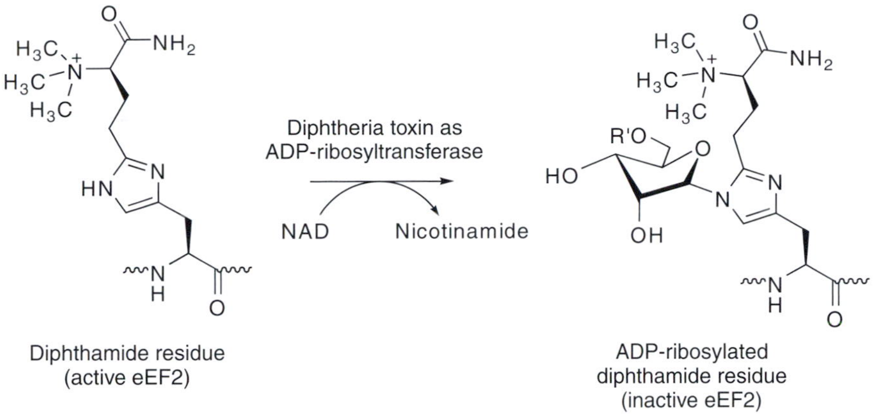

Figure 11.5 The diphtheria toxin catalytic subunit catalyzes ADP-ribosylation of N_1 of the imidazole ring of diphthamide in eEF2.

tandem conventional transfers (Chapter 5) of the activated CH_3 group to the amino moiety of the aminocarboxypropyl substituent on residue 715 to yield the adducted residue known as diphthine. The third type of modification is ATP-dependent amide bond formation, analogous to Glu to Gln or Asp to Asn formation, to cover the charge on the carboxylate of the diphthine side chain and produce the neutral diphthamide. As far as is known, only the single protein eEF2 undergoes the His to diphthamide posttranslational modification, and while its function in protein synthesis is not yet understood, this modification is essential for efficient protein synthesis in eukaryotic cells.

When diphthamide is targeted for ADP-ribosylation on N_1 of its imidazole ring (Figure 11.5) by the catalytic A subunit of diptheria toxin or exotoxin A, the resultant modified eEF2 is nonfunctional. Protein synthesis stops and cell death will ensue in the GI tract epithelial cells that are the proximal targets of *C. diphtheriae*.

Cholera Toxin and Related Toxins that Catalyze ADP-Ribosylations of an Arginine Residue in the Alpha Subunit of the Trimeric GTPase G_s or G_i

The exotoxin of *Vibrio cholerae* has analogies to diphtheria toxin in its A–B didomain structure, but this is a two-subunit AB_5 system (Aktories, 2000). The B pentamer binds to the cell surface glycolipid GM1 preparatory to internalization into endosomes, where the oligomeric B subunits form a pore to allow the A subunit access to the cytoplasm. The ADP-ribosyltransferase activity of the cholera toxin A subunit is

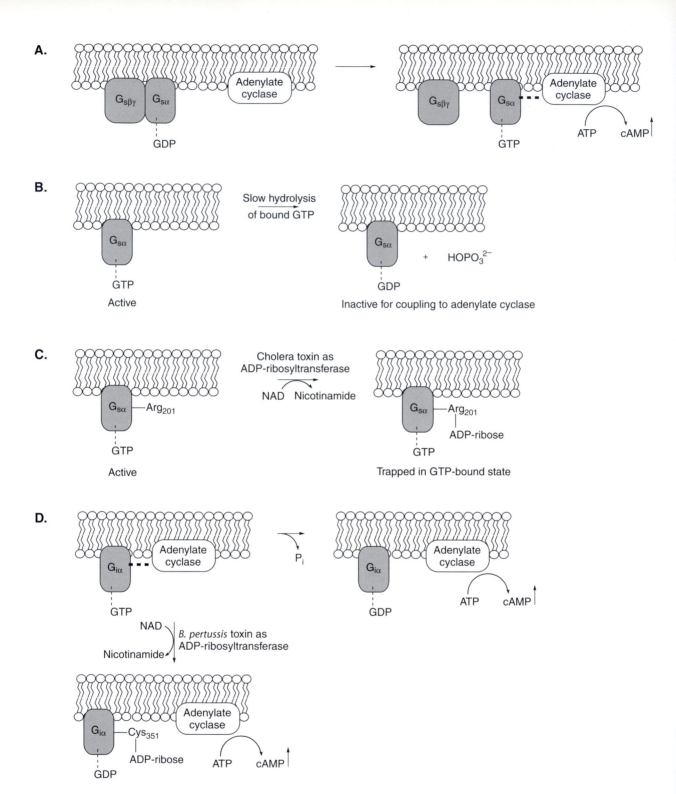

Figure 11.6 Cholera toxin action on $G_{s\alpha}$. **A.** Signaling of the $G_{s\alpha} \bullet$ GTP complex to adenylate cyclase elevates cyclic AMP levels in cells (cAMP ↑). **B.** Hydrolysis of bound GTP to GDP generates a nonsignaling loop conformer in the G•GDP complex. **C.** ADP-ribosylation of Arg_{201} in $G_{s\alpha}$ by cholera toxin traps $G_{s\alpha}$ in the GTP-bound state. **D.** ADP-ribosylation of $G_{i\alpha}$ on Cys_{351} by *B. pertussis* toxin blocks inhibition of adenylate cyclase.

specific for a single arginine side chain, Arg_{201}, in the α subunit of the trimeric GTPase protein G_s. G_s is a conditional GTPase that signals in its GTP-bound form to adenylate cyclase. The duration and intensity of the signal to stimulate the downstream partner adenylate cyclase is controlled by G_s turning off, as bound GTP is slowly hydrolyzed to bound GDP (Figure 11.6A). The switch from bound nucleotide triphosphate to nucleotide diphosphate alters the conformation of an effector loop in the $G_{s\alpha}$ subunit such that it no longer recognizes its downstream partner adenylate cyclase (Figure 11.6B). Lipid modifications of both trimeric G proteins and monomeric small G proteins with conditional GTPase activities are described in Chapter 7.

When cholera toxin A binds selectively to the α subunit, $G_{s\alpha}$, the GTPase presents the guanidino terminus of Arg_{201} to the bound NAD cosubstrate. The toxin activates the transfer of the ADP-ribosyl moiety as the ribooxacarbenium ion while using an active site residue as a base to deprotonate the weakly basic guanidinium group to act as a competent nucleophile (Figure 11.6C). The covalent addition of the bulky ADP-ribosyl group to the Arg side chain blocks the ability of $G_{s\alpha}$ to hydrolyze bound GTP to GDP. Thus, the modified $G_{s\alpha}$ signals, without turn-off mechanism as the enzyme–GTP complex, to adenylate cyclase. The sustained elevation of cAMP turns on protein kinase A (Chapter 2) inappropriately and for extended duration. One result in epithelial cells is excess secretion of monovalent ions into the gut, followed by water to cause the massive watery diarrhea typical of cholera.

In studies to characterize the ADP-ribosylation activity of cholera toxin, it was observed that proteins in host cells would speed up the reaction. These were termed ADP-ribosylation factors (ARFs) that turned out to be conditional GTPases of the Ras superfamily, normally involved in vesicle budding and the movement of protein secretory cargo. It is known that the ARFs increase affinity of the cholera A subunit for the G protein target but the exact mechanism is unclear. There are six ARF isoforms in human cells (Moss, 1995).

Other bacterial toxins that transfer ADP-ribosyl moieties to Arg residues of target proteins by equivalent mechanism are lethal toxin from *E. coli*, the C2 toxin from *Clostridium botulinum*, and the VIP2 toxin from *Bacillus cereus* (Table 11.1). While the *E. coli* enzyme also ADP-ribosylates the G_s GTPase, the *botulinum* and *cereus* toxins target actin molecules and disrupt cytoskeletal architecture (Han et al., 2001).

ADP-Ribosylation of a Cys Residue in a Mammalian Target Protein

The catalytic ADP-ribosyltransferase subunit of *Bordatella pertussis*, the causative agent of whooping cough, is also part of an AB_5 heterohexamer (Aktories, 2000). The A subunit, on extrusion from endosomes into the cytoplasm via the B_5 porin, also targets a trimeric GTPase protein α subunit coupled to adenylate cyclase, but this time the ribosylation target is inhibitory GTPase $G_{i\alpha}$ (Table 11.1). The G_i•GTP complex normally inhibits adenylate cyclase. Its ADP-ribosylation prevents the modified G_i subunit from blocking adenylate cyclase and again, sustained high activity sets off the cAMP–protein-kinase-A cascade with deleterious consequences to the host cell (Figure 11.6D). In contrast to the Arg side chains targeted by the preceding subclass and the Asn side chains of the next subclass (bacterial toxins with ADP-ribosyltransferase activity), the thiolate of ionized Cys residues is a powerful nucleophile. The deprotonation of the thiol to thiolate form of G_i substrate is presumably catalyzed by the pertussis toxin.

Clostridium botulinum Toxin C3: ADP-Ribosylation of Asn$_{41}$ of Rho GTPases

Continuing the theme of bacterial toxins with ADP-ribosylation activity towards host small G protein family members with conditional GTPase activities coupled to signal transduction pathways, there is the case of the C3 protein toxin elaborated by

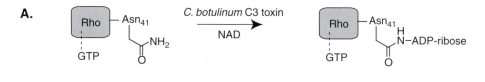

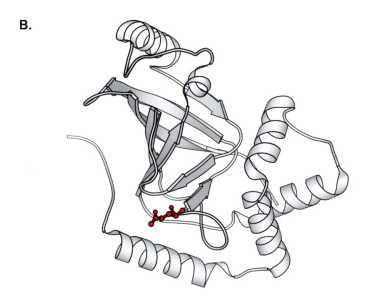

Figure 11.7 Structure of C3 and reaction schematic of ADP-ribosylation of Rho. **A.** Schematic representation of the C3-catalyzed ADP-ribosylation of Rho. **B.** Crystal structure of the C3 exoenzyme with Gln$_{212}$ as red ball and stick (figure made using PDB 1G24).

strains of *Clostridium botulinum*, the organism responsible for botulism outbreaks. A comparable C3 toxin from *Staphylococcus aureus* acts equivalently. The *botulinum* C3 single-chain protein is secreted and undergoes signal-protease-mediated cleavage (Chapter 8) of the N-terminal 40 residues to yield a 211-residue protein with both ADP-ribosyltransferase activity towards the Rho family GTPases as well as weak NAD glycohydrolase activity. The side chain of Rho proteins modified is the carboxamido group of Asn_{41} in a β strand adjacent to and just outside the effector loop (residues 32–40) (Figure 11.7A). The x-ray structure of C3 has been determined (Han et al., 2001) (Figure 11.7B) and suggests that Gln_{212} in the active site is important for the orientation of the Asn_{41} amide side chain of Rho and perhaps stabilizes the isoamide form required for the nitrogen to function as a nucleophile. Thus, the bacterial ADP-ribosyltransferases have learned how to create microenvironments in their active sites that can selectively activate otherwise poorly nucleophilic side chains (Arg of G_α subunits and Asn of Rho) sufficiently to be captured by the reactive ribooxacarbenium ion species generated on NAD fragmentation. The ADP-ribosylation of Rho interferes with partner protein recognition of the effector loop, the typical protein–protein recognition motif in this GTPase superfamily, and leads to the net depolymerization of actin molecules and the breakdown of the cytoskeletal framework.

ADP-Ribosyltransferases in Eukaryotic Cells

Subsequent to the study of the prokaryotic toxins with protein ADP-ribosyltransferase activities towards specific partner proteins, NAD-cleaving ADP-ribosyltransferases were detected and cloned from mammalian cells. Most are arginine-specific transferases in their partner protein selection (Okazaki and Moss, 1998). Five isoforms have been detected (ART1–ART5), some of which have been posttranslationally processed to fix a GPI anchor (Chapter 7) covalently to the C-terminus of the transferase domain. These forms then are ecto enzymes facing the outside environment at the plasma membrane. The integrin α_7 is a substrate for modification, consistent with the predicted orientation of ART1 (Okazaki and Moss, 1998). There is some indication that ADP-ribosylation of Arg side chains could be reversible because ADP-ribosylarginine hydrolase hydrolytic activity has been detected (Figure 11.8). However, there is as yet no understanding of the biological role of protein mono-ADP-ribosylation in mammalian cells. Chapter 13 summarizes the information on poly-ADP-ribosylation.

Recall from Chapter 6 a family of enzymes, sirtuins, where yeast Sir2 is the prototype, participates in the deacetylation of acetylated forms of histones by cleaving NAD stoichiometrically. A mechanism was detailed (Figure 6.10A) in which ADP-

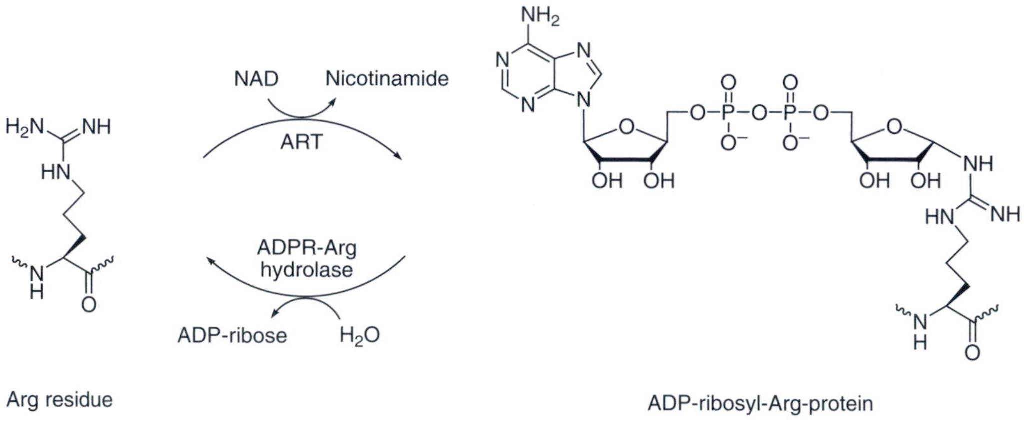

Figure 11.8 Reversible ADP-ribosylation and enzymatic hydrolysis of Arg residues in eukaryotes.

ribosyl transfer from NAD occurs to the acetyl moiety of the acetylated histone, generating peptidyl-O-alkylamidate intermediates, which unravel to 2′O-acetyl-ADP-ribose and the deacetylated histone proteins. The sirtuins should be thought of in the categorization of mammalian protein ADP-ribosyltransferases, specifically where transfer is to a weak nucleophile, in some analogy to the weak protein nucleophiles noted above for Arg and Asn side chains. In the sirtuin family of transformations, the initial ADP-ribosylated protein product is unstable and is then subject to further rearrangement.

Additional Bacterial Toxins with Catalytic Activity Towards the Ras Superfamily of GTPases

Recall near the end of Chapter 10 that some clostridial toxins have enzymatic activity towards Rho family GTPases that involves the O-glucosylation of Thr_{35} in the effector loops of those GTPases. (Recall also the analogy between glucosyl transfer from UDP-glucose and ribosyl transfer from NAD mentioned at the opening of this chapter). That is, glucosylation occurs only six residues away from the ADP-ribosylation of Asn_{41} by the C3 toxin of *C. botulinum* discussed above. Thus, some clostridial species pack a double-barreled posttranslational modification gun for two types of glycosylation in and around the crucial effector loop of this subfamily of the Ras GTPase superfamily.

A third variation in posttranslational modification logic is found yet again in bacterial toxins acting on Rho GTPase host targets. The cytoxic necrotizing factor (CNF-1) of *E. coli* (Buetow et al., 2001; Schmidt et al., 1997) has protein glutaminase

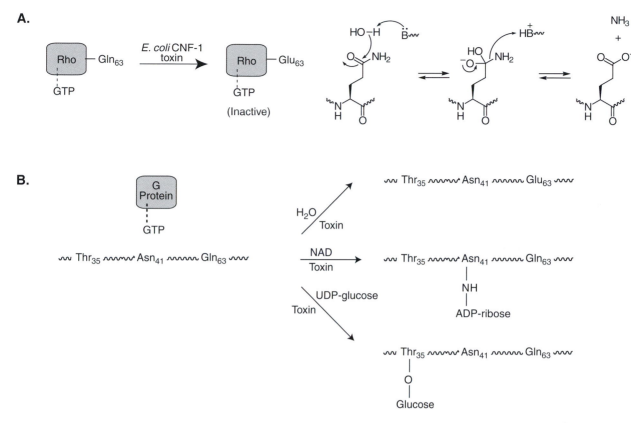

Figure 11.9 **A.** Hydrolytic deamidation of Gln_{63} at the active site of Rho GTPase by the CNF-1 toxin of *E. coli*. **B.** Three types of posttranslational modifications at or near the effector loop or the active site of Rho GTPases by three bacterial toxins with distinct enzymatic activities.

activity, specifically towards Gln_{63} of Rho (Figure 11.9A), thus hydrolyzing the amide to the free carboxylate side chain in the product Glu_{63}. Establishment of this activity took sophisticated detective work, including mass spectrometry of the active site peptide to detect the single dalton difference in mass between active Gln_{63} RhoA and the inactive Glu_{63} product of CNF-1 action. The cellular phenotype was once again depolymerization of actin fibrils, consistent with the known structural biology and mechanistic knowledge of the GTPase catalytic cycle of Rho family GTPases, where the Gln_{63} is thought to be a critical side chain for the activation of a water molecule that attacks bound GTP and converts it to the GDP form. Thus, as shown in Figure 11.9B, there are three posttranslational modifications enacted by three types of catalytically active protein toxins that modify a side chain of the Rho G proteins. Glucosylation, ADP-ribosylation, and deamidation that occur in and around the active site and/or effector loops alter the catalytic cycle or recognition of partner proteins in these critical small GTPases and disrupt important signaling pathways.

References

Aktories, K., and I. Just (eds.). *Bacterial Protein Toxins*. Springer: Berlin (2000).

Buetow, L., G. Flatau, K. Chiu, P. Boquet, and P. Ghosh. "Structure of the Rho-activating domain of *Escherichia coli* cytotoxic necrotizing factor 1," *Nat. Struct. Biol.* **8:**584–588 (2001).

Collier, R. J. "Understanding the mode of action of diphtheria toxin: A perspective on progress during the 20th century," *Toxicon* **39:**1793–1803 (2001).

Grunwald, S. K., D. P. Lies, G. P. Roberts, and P. W. Ludden. "Posttranslational regulation of nitrogenase in *Rhodospirillum rubrum* strains overexpressing the regulatory enzymes dinitrogenase reductase ADP-ribosyltransferase and dinitrogenase reductase activating glycohydrolase," *J. Bacteriol.* **177:**628–635 (1995).

Grunwald, S. K., M. J. Ryle, W. N. Lanzilotta, and P. W. Ludden. "ADP-ribosylation of variants of *Azotobacter vinelandii* dinitrogenase reductase by *Rhodospirillum rubrum* dinitrogenase reductase ADP-ribosyltransferase," *J. Bacteriol.* **182:**2597–2603 (2000).

Han, S., A. S. Arvai, S. B. Clancy, and J. A. Tainer. "Crystal structure and novel recognition motif of rho ADP-ribosylating C3 exoenzyme from *Clostridium botulinum*: Structural insights for recognition specificity and catalysis," *J. Mol. Biol.* **305:**95–107.

Lee, H. C. "Physiological functions of cyclic ADP-ribose and NAADP as calcium messengers," *Annu. Rev. Pharmacol. Toxicol.* **41:**317–345 (2001).

Liu, S., and S. H. Leppla. "Retroviral insertional mutagenesis identifies a small protein required for synthesis of diphthamide, the target of bacterial ADP-ribosylating toxins," *Mol. Cell* **12:**603–613 (2003).

Moss, J., and M. Vaughan. "Structure and function of ARF proteins: Activators of cholera toxin and critical components of intracellular vesicular transport processes," *J. Biol. Chem.* **270:**12327–12330 (1995).

Moss, J., and M. Vaughn. *ADP-Ribosylating Toxins and G Proteins: Insights Into Signal Transduction*. ASM Press: Washington (1990).

Okazaki, I. J., and J. Moss. "Glycosylphosphatidylinositol-anchored and secretory isoforms of mono-ADP-ribosyltransferases," *J. Biol. Chem.* **273:**23617–23620 (1998).

Schmidt, G., P. Sehr, M. Wilm, J. Selzer, M. Mann, and K. Aktories. "Gln 63 of Rho is deamidated by *Escherichia coli* cytotoxic necrotizing factor-1," *Nature* **387:**725–729 (1997).

Walsh, C. T. *Enzymatic Reaction Mechanisms*. Freeman: San Francisco (1979).

Posttranslational Hydroxylation of Proteins

Hydroxylation of Pro$_{564}$ in HIF-1α recruits a ubiquitin E3 ligase (pVHL) that controls HIF proteolysis.

The posttranslational hydroxylation of amino acid side chains in proteins is relatively uncommon. However, because such hydroxylations occur in the most abundant eukaryotic protein, collagen, this modification has been intensively studied for decades and the genesis and function of this rare modification is reasonably well understood. In the triple-helical structure of collagen, proline residues are crucial and abundant and up to half of the prolines have been processed to 4-OH-prolines (Figure 12.1) (Myllyharju and Kivirikko, 1997). Depending on the type of collagen, this can involve 10–150 residues per 1000 (Vranka, 2004). In triple-helix assembly in the endoplasmic reticulum of fibroblasts, Gly-4-OH-Pro amide N–Hs on one strand hydrogen bond to the C=O of Gly-4-OH-Pro dipeptides on neighboring strands. The 4-OH-Pro side chains point away from the helix and hydrogen bond with solvent (Figure 12.2). A second regioisomer of OH-Pro, 3(S)-OH-Pro, is found at about

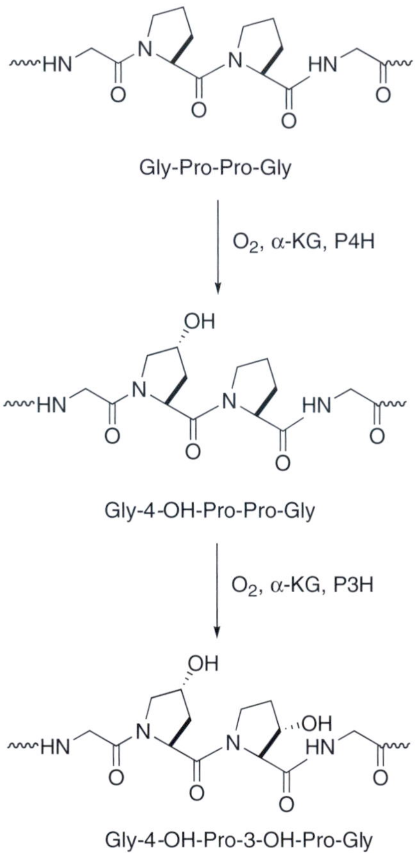

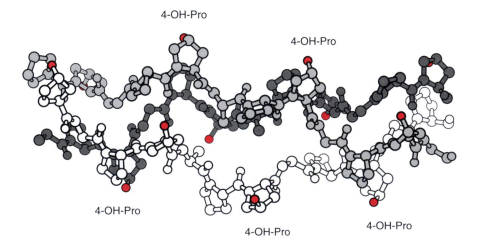

Figure 12.1 Proline residue hydroxylation to 4-OH-prolines and 3-OH-prolines in collagen maturation via posttranslational monooxygenase catalysis by prolyl hydroxylases. (P4H = prolyl-4-hydroxylase; P3H = prolyl-3-hydroxylase).

Figure 12.2 The 4-OH-Pro residues stabilize the register of triple helices that make up collagen protofibrils. Several but not all the 4-OH-Pro residues are labeled (figure made using PDB 1CGD).

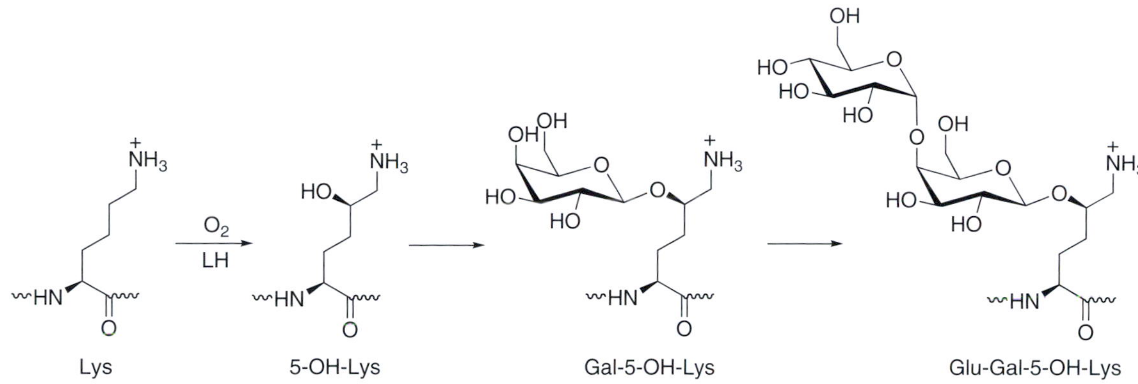

Figure 12.3 Posttranslational hydroxylation of the C_5-CH_2 group in lysine side chains of collagen, setting this locus up for subsequent O-glycosylations.

Figure 12.4 β-Hydroxylation of **A.** Asn and **B.** Asp side chains in proteins.

10% the frequency of 4-OH-Pro (Figure 12.1) in collagen chains. In general, it is likely that the introduction of the additional OH substituents on protein side chains, including creation of the β-OH-Asn residues noted later in the chapter, enable new hydrogen bonding interactions to facilitate protein–protein recognition.

A second type of side chain, lysine, is also hydroxylated at C_5 in collagens and is relevant to both secretion and function in the extracellular matrix. Some of the 5-OH-lysyl side chains are then sites for subsequent O-glycosylation (Figure 12.3) by UDP-galactose and then by UDP-glucose during passage through the secretory pathway. The sugars are necessary for normal function and these carbohydrate moieties can be important immuno-determinants in T cell recognition.

A third type of protein side chain to be hydroxylated is Asn/Asp at the β-CH_2 group to produce the β-OH-Asn/Asp side chains (Figure 12.4). These have been found in secreted proteins in epidermal growth factor-like domains in blood coagulation factors (VII, IX, and X), in thrombomodulin, and in the LDL receptor. Consensus sequences for aspartyl β-hydroxylation at tetrapeptides D-D/N-D/N-Y/F are also found in Notch receptors and in ligands of the Axl family of receptor tyrosine kinases.

Interest in prolyl and asparaginyl residue hydroxylation has intensified in recent years with finding that both types of posttranslational hydroxylations occur as the molecular mechanism for oxygen sensing between hypoxic and normoxic tissues. As noted below, the same subunit (HIF-1α) of the hypoxia inducing factor can be hydoxylated at two proline residues for positive signaling and at one asparaginyl residue for negative signaling.

Protein side chain hydroxylation is irreversible. There are no known enzymes that reverse the action of the hydroxylases. Thus, the hydroxylated residues persist for the lifetime of the specific protein. Indeed, they can be determinants of protein lifetime by signaling to ubiquitylation protein machinery for covalent marking and proteolytic destruction by the proteasome.

Nature of Protein Hydroxylation Reactions

The posttranslational hydroxylation of proline [to 4(R)-OH-Pro and to 3(S)-OH-Pro] residues, lysine (to 5-OH) residues, and asparaginyl [to 3(S)-OH] residues is both regiospecific and stereospecific. In collagen Pro residues in Pro-Gly sequences and Lys residues in Lys-Gly sequences are recognized by their respective hydroxylases. All three modifications occur at CH_2 groups. This is distinct from all the other types of posttranslational modifications noted in this book, which occur at electron-rich, nucleophilic sites of amino acid side chains. This is a strong indication that the chemistry of the hydroxylating modification is distinct from all the other modifications. Indeed, these are formal insertions of one oxygen atom from cosubstrate O_2 into unactivated carbon centers (Figure 12.5). This ability to oxygenate unactivated sites is the hallmark of enzymatic oxygenases, utilizing an oxygen intermediate activated by orbital overlap with a metal ion (typically iron).

Iron-containing oxygenases usually bind stoichiometric amounts of Fe^{II} or Fe^{III} tightly, either in the equatorial plane of heme cofactors, or in a coordination microenvironment where two histidines (through their imidazole side chains) and one carboxylate (Asp/Glu) provide the platform for (Fe^{II}) iron ligation (Figure 12.6). It is members of the second family of oxygenases that are involved in the protein side

Figure 12.5 Posttranslational hydroxylations of protein side chains occur at unactivated methylene loci that do not react as attacking nucleophiles.

chain hydroxylations. These nonheme iron oxygenases constitute a large superfamily, most of which act on small molecule substrates (Ryle and Hausinger, 2002) rather than the proteins that are the subject of this chapter. The mechanistic framework for this superfamily has come mostly from studies on such oxygenases acting on nonprotein substrates.

Typically, these enzymes utilize O_2 as the oxygen donor and also a molecule of α-ketoglutarate as cosubstrate in each catalytic turnover (Figure 12.7A). *In vitro* and probably *in vivo*, some of these enzymes get autoxidized during catalytic turnovers from the active Fe^{II} state to the inactive Fe^{III} oxidation level. The Fe^{III} can be reduced back to the Fe^{II} form by one-electron donors such as vitamin C, which gets oxidized to one-electron-oxidized ascorbate (Hegg and Que, 1997).

During the course of the reaction the α-ketoglutarate gets decarboxylated to succinate and CO_2, with one atom from O_2 ending up in succinate and the other in the hydroxylated product (e.g., β-OH-asparaginyl-protein). This oxygen-labeling pattern initially suggested a peroxysuccinate intermediate (arising from O_2 addition to the C_2 ketone of α-KG, followed by decarboxylation), validated by several other lines of study, arising from reaction of O_2 and α-KG in the iron microenvironment of the hydroxylase active site. The peracid could be a precursor to a high-valent oxoiron species ($Fe^{IV}{=}O$) that is the actual oxygen transfer agent (Figure 12.7B). Such high valent oxoiron reagents have the reactivity to insert into unactivated C–H bonds and to create the C–OH products. The stereo- and regioselectivity of oxygen insertion arises from the specific binding and orientation of the protein side chains to be hydroxylated relative to the $Fe^{IV}{=}O$ iron center.

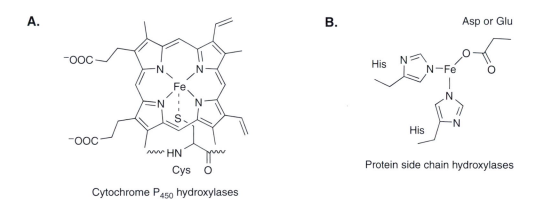

A.

Cytochrome P$_{450}$ hydroxylases

B.

Asp or Glu

Protein side chain hydroxylases

Figure 12.6 **A.** Heme scaffold for Fe^{II} in cytochrome P450 hydroxylases. **B.** Two His, one carboxylate scaffold for Fe^{II} in nonheme iron-containing hydroxylases. The number of water ligands (not shown) varies during the catalytic cycle (see Figure 12.7B).

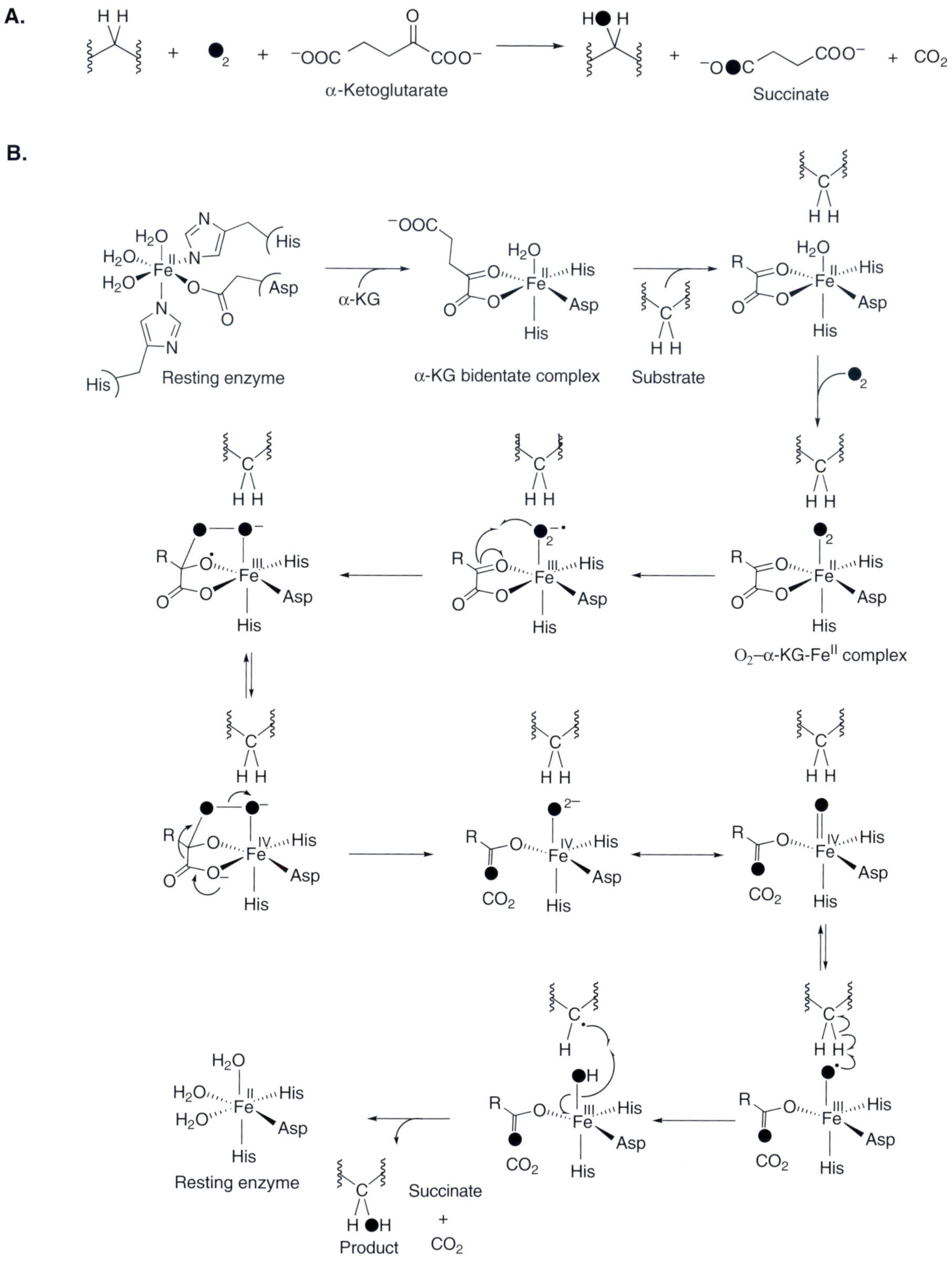

Figure 12.7 **A.** Reaction for nonheme iron-containing protein monooxygenases. **B.** Role of the α-ketoglutarate (α-KG) cosubstrate in oxygen activation; decarboxylation of α-KG and oxygen transfer to substrate; and high-valent oxoiron species as reagent that inserts an oxenoid type oxygen into unactivated C–H bonds (the Asp ligand is replaced by Glu in several enzymes).

Structure and Function of the Protein Hydroxylases

The prolyl hydroxylases and the lysyl hydroxylases (up to three isoforms) (Heikkinen et al., 2000; Pirskanen et al., 1996, Valtavaara et al., 1997; Vranka, 2004) involved in collagen processing are located in the endoplasmic reticular membranes of fibroblasts, with active sites facing the lumenal space where the procollagen molecules are being processed. The lysyl hydroxylases are α_2 dimers, while the 4-OH-prolyl hydroxylase is an $\alpha_2\beta_2$ heterotetramer with the α subunit the catalytic entity and the β subunit a protein with multiple activities as a chaperone and as a protein disulfide isomerase (PDI) (see Chapter 4) (Kivirikko and Pihlajaniemi, 1998). The prolyl 3-hydroxylase has the

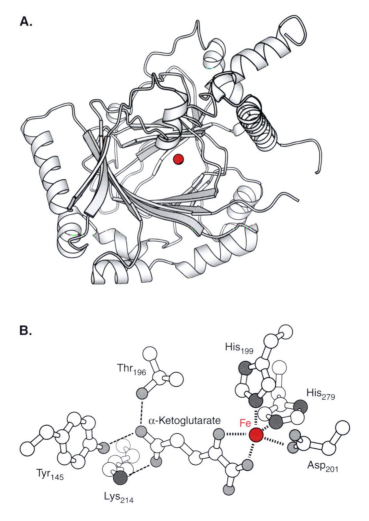

Figure 12.8 **A.** X-ray structure of FIH-1 monoxygenase that hydroxylates Asn$_{803}$ of HIF-1α. The active site Fe is represented as a red sphere (figure made using PDB 1MZE). **B.** Cocrystallization with α-KG confirms anticipated ligation to the active site Fe. Carbons are colored white, iron is colored red, oxygens are colored light gray, and nitrogens are colored dark gray (figure made using PDB 1MZF).

hallmarks of the nonheme iron hydroxylase family but does not use PDI as a regulatory subunit. The endoplasmic reticulum location is consistent with the hydroxylases acting on procollagen during its transit through the secretory compartments. One of the lysine hydroxylase isoforms, form 3, appears to be a bifunctional enzyme (Heikkinen et al., 2000), with associated glucosyltransferase activity that acts two steps later in the processing cascade, on the galactosyl-5-OH-lysyl collagen to create the final Glu-Gal-5-OH-Lys decorations (Figure 12.3). Analogously, the Asp/Asn β-hydroxylase acting on EGF-like domains is in the ER and works on proteins that end up at the cell surface or secreted to the extracellular milieu. These membrane-associated hydroxylases have been difficult to characterize structurally (Dinchuk et al., 2000).

The structure of a bacterial Fe^{II} monooxygenase catalyzing the hydroxylation of free proline to *cis*-3-OH-Pro has been solved as a model for the collagen prolyl-4-hydroxylase catalytic subunit (Clifton et al., 2001). Most relevant is the x-ray structure of the hydroxylase FIH-1 (factor inhibiting HIF-1) that carries out the posttranslational hydroxylation of Asn_{803} on human HIF-1α (Figure 12.8A) (Dann et al., 2002). It has the fold characteristic of other members of the 2-His-1-carboxylate Fe-oxygenase family (Ryle and Hausinger, 2002) and the $H_{199}XD_{201}\ldots H_{279}$ conserved motif is indeed the ligand set for Fe^{II}. Cocrystallization with α-ketoglutarate confirms ligation of the C_1-carboxylate oxygen and the C_2-keto oxygen to the fourth and fifth coordination sites for iron (Figure 12.8B). The axial sixth coordination site is presumably the O_2 binding site initially for the oxygen activation phase of the reaction. The binding locus of the asparaginyl peptide is not yet determined.

Collagen Processing

Collagen, the most abundant protein in vertebrates, occurs in 19 related forms from six collagen genes combined in different chain proportions into triple helical arrays. The subunit length of collagens varies from 600 to 3000 amino acids with type I collagen around 1000 residues. Much of the mature collagen is triple helical, due to repeats of $(Gly-X-Y)_n$ where the Gly residues fit at the interface of the rope-like array of the triple helix (Figure 12.9). Different collagens have different lengths of non-G-X-Y repeat sequences that affect their architecture and function. All collagens are extracellular, structural proteins and transit through the secretory pathways of producing cells, such as fibroblasts.

The primary translation product, preprocollagen, undergoes up to nine types of posttranslational modifications (Figure 12.10A). The presequence directs the nascent protein to the endoplasmic reticulum and is removed by signal peptidase action (Chapter 8) as it passes into the ER lumen. The 4-hydroxylation of about 50% of the proline residues, particularly those in the Y position of the G-X-Y repeat, occurs in the ER by the heterotetrameric prolyl 4-hydroxylase. Prolyl 3-hydroxylase is thought to favor G-P-P-G tetrapeptide sequences and to act at the second prolyl residue *after* the first has been converted to 4(*R*)-OH-Pro (Vranka, 2004) (Figure 12.1). The net result is to yield a Gly-4(*R*)-OH-Pro-3(*S*)-OH-Pro-Gly sequence in that region of the processed collagen. While 4(*R*)-OH-Pro residues stabilize helical regions, the 3(*S*)-OH-Pro are thought to destabilize them. Thus, 3-OH-Pro residues are found in highest amounts in type IV and type V collagen and may contribute to the nonhelical meshwork structure they adopt in basement membranes (Vranka, 2004).

The hydroxylation of some lysine residues at C_5 adjacent to the C_6–NH_2 group is catalyzed by the ER membrane lysine hydroxylases. Both are members of the 2-His-1-carboxylate Fe^{II} α-KG-requiring oxygenases. The 5-OH-Lys residues also occur in the Y position of the G-X-Y repeating triad and point out from the long axis of the triple helix. These hydroxyl groups are the attacking nucleophiles towards UDP-galactose to produce O-galactosyl-Lys residues. Some of these galactosyl moieties in turn are glucosylated while the procollagen is still in the ER/Golgi compartments. Mutations in the lysyl hydroxylase are known and cause a connective tissue disease

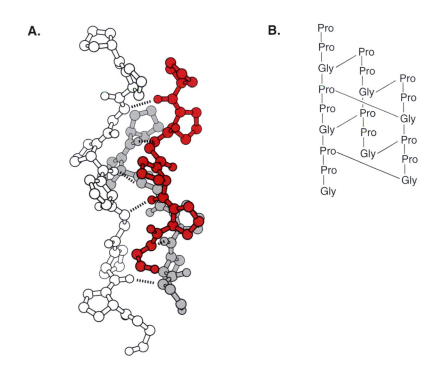

Figure 12.9 Triple helical array in collagen, with critical role for side chain-less Gly residues in the G-X-Y motifs. **A.** Structure of model peptides with G-P-P repeats. Strands are in white, gray, and red (figure made using PDB 1A3J). **B.** Schematic representation of the structure described in **A.**

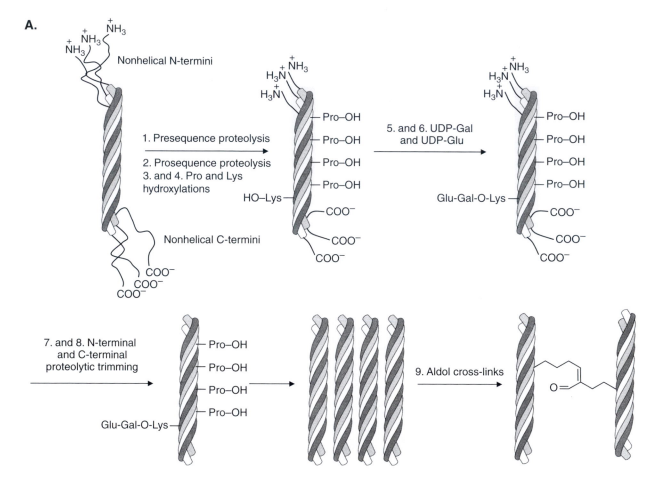

Figure 12.10 **A.** Preprocollagen undergoes up to nine posttranslational modifications, including pre- and prosequence proteolysis, hydroxylation at Pro and Lys residues, glycosylation at HO-Lys residues by galactosyl, then glucosyl moieties, proteolytic trimming at N- and C-terminal nonhelical extensions, and intermolecular cross-linking.

pathological condition termed Ehlers–Danlos syndrome, indicating the importance of this hydroxylation process to folding and maturation.

Triple helix formation is then completed and the processed procollagen transits through the vesicular compartments and is secreted. The nonhelical regions at both the N- and C-termini need to be trimmed proteolytically and there are plasma membrane proteases of the ADAM family (Chapter 8) that effect the N-terminal proteolysis and a separate protease for C-terminal trimming. Additional posttranslational modifications can occur in the extracellular space, including oxidation of the CH_2NH_2 side chains of lysines on the same or separate collagen chains, followed by either imine formation or aldol condensations (Figure 12.10B) to covalently connect the chains by these lysine-based cross-bridges.

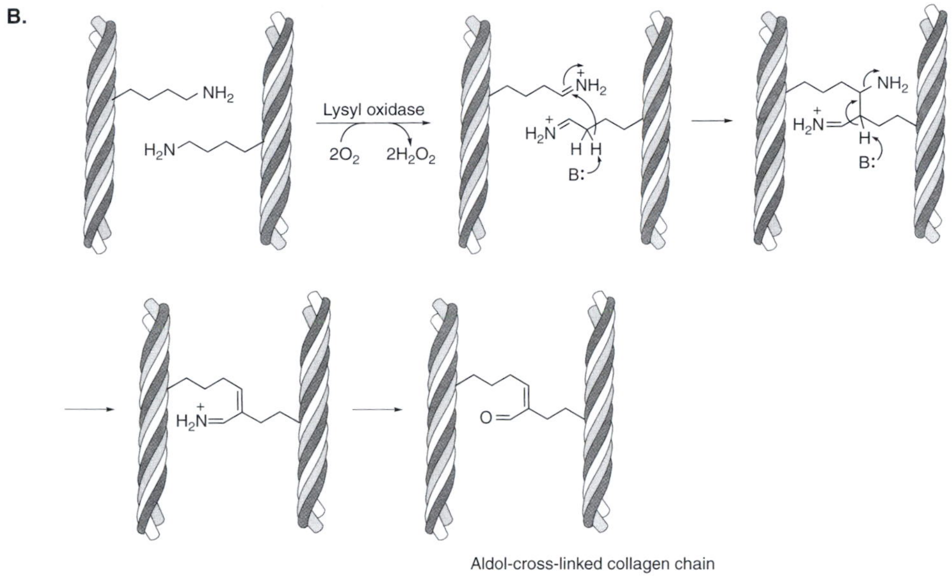

Figure 12.10 **B.** In the extracellular space lysine ϵ-NH$_2$ groups of collagen can be oxidized to the imine, and condensed to aldol type products in intermolecular cross-linkings.

Oxygen Sensing

Many tissues have the ability to respond to hypoxia by transcriptional regulation of subsets of genes. The response has been particularly well scrutinized with regard to angiogenesis in which new blood vessels are induced in response to hypoxia in tissues, both in physiologic and pathophysiologic contexts (Pugh and Ratcliffe, 2003; Semenza, 1999). The ubiquitously expressed heterodimeric transcription factor HIF (hypoxia inducible factor) is acutely sensitive to normoxic versus hypoxic conditions, with net induction of activity at low partial pressures of O$_2$. HIF acts as a transcriptional activator to turn on such genes as erythropoietin, which increases the number of red cells in the blood, thereby increasing the capacity to deliver more O$_2$ to peripheral tissues. Microarray analyses in response to global hypoxia indicate hundreds to thousands of genes in mammalian tissues may be induced by hypoxia through HIF transcriptional regulation (Schofield and Ratcliffe, 2004). HIF-1 is the major isoform and has two subunits, α and β, where levels of β are constant and constitutive; levels of the α subunit vary due to its selective destruction by the proteasome. Thus, the level of functional HIF transcription factor and its consequent activity to upregulate genes for erythropoietin and vascular endothelial growth factor (VEGF), among others, is controlled by the lifetime of the HIF-1α subunit (Figure 12.11).

The proteolysis of HIF-1α is carried out by the proteasome, so the lifetime is controlled by the covalent tagging of the α subunit by ubiquitylation (see Chapters 8

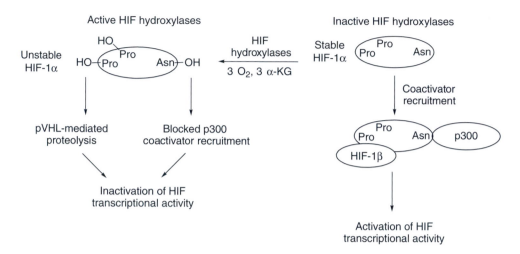

Figure 12.11 Activity of the HIF transcription factor that upregulates erythropoietin and VEGF gene transcription is controlled by the lifetime of the HIF-1α subunit.

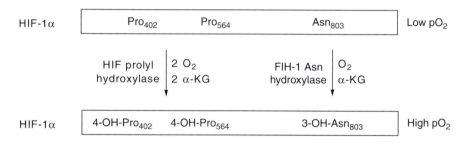

Figure 12.12 Posttranslational hydroxylation of HIF-1α at two Pro and one Asn by two specific hydroxylases. The level of hydroxylation is controlled by pO_2.

and 9). A particular ubiquitin E3 ligase complex, one of whose components is the Von Hippel Lindau (pVHL) tumor suppressor protein, recognizes HIF-1α and enables the protein–protein recognition required for ubiquitylation and subsequent targeting of polyubiquitylated-HIF-1α to the proteasome.

The tuning of the controlled destruction of the HIF-1α to act as an oxygen sensor occurs by control of the hydroxylation state of three residues in HIF-1α—namely, Pro_{402}, Pro_{564}, and Asn_{803}—by two distinct α-ketoglutarate-dependent hydroxylases (Figure 12.12). The HIF prolyl hydroxylase was detected (Bruick and McKnight, 2001) by cloning five human orfs predicted to be homologs of the catalytic subunit of collagen prolyl-4-hydroxylase and demonstrating activity on a peptide fragment of HIF-1α as assayed by binding to the pVHL protein. The Asn hydroxylase FIH-1 was first detected as an HIF-associated protein by yeast two-hybrid assays, observed to have the iron oxygenase 2-His 1-carboxylate motif, and then shown to have Asn hydroxylase activity on peptide fragments from the carboxy-terminal transactivating domain of HIF-1α. The hydroxyproline residues in the N-terminal region and the hydroxyasparagine in the C-terminal region of HIF-1α operate independently, but perhaps with synergy. At low pO_2 there is not enough O_2 to enable the hydroxylases to modify the three residues, so HIF-1α will be produced in the nonhydroxylated form. At the normoxic O_2 level, there is enough pO_2 for both oxygenases to function at high enough rates to produce the trihydroxylated form of HIF-1α.

The OH–Pro$_{402}$ and OH–Pro$_{564}$ residues interact with a specific region of pVHL as determined by x-ray analysis of pVHL with HIF-1α peptide fragments containing either Pro or 4(R)-OH-Pro at residue 564 (Hon et al., 2002; Min et al., 2002). Two hydrogen bonds were made from the 4-OH of the OH–Pro side chain to the pVHL side chains (Figure 12.13). Direct binding assays validated that the 4-OH group of OH–Pro$_{564}$ thus provided about 4 kcal/mol of favorable interaction energy, sufficient for a 1000-fold difference in binding recognition (Hon et al., 2002). If OH–Pro$_{402}$ has comparably favorable energetics of interaction, this would explain how prolyl hydroxylation can be an "on/off" switch for recognition by pVHL and the ubiquitylating complex.

The C-terminal region of HIF-1α interacts with the transcriptional coactivator protein p300 through its CH1 domain. NMR analysis of the Asn$_{803}$ peptide of HIF-1α with the p300 domain has led to a model (McNeill et al., 2002) where the C-terminal tail of HIF-1α is unstructured in solution but becomes helical when complexed with the CH1 domain of p300 (Figure 12.14). NMR analysis of the product β-OH-asparaginyl residue from FIH-1 action on a 19-residue peptide from HIF-1α (788–806) suggests that the product is the 2(S),3(S)-*threo*-hydroxyasparagine.

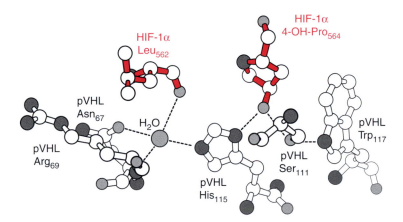

Figure 12.13 Interaction of the 4-OH-Pro$_{564}$ residue of the HIF-1α peptide fragment (red) with pVHL. Carbons are colored white, oxygens are colored light gray, and nitrogens are colored dark gray (figure made using PDB 1LQB).

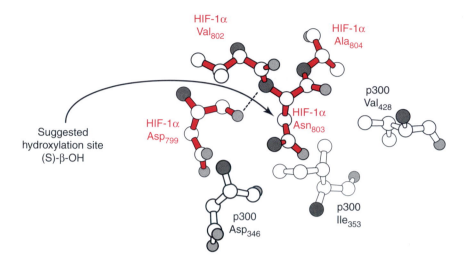

Figure 12.14 Proposed interaction of the β-OH-Asn$_{803}$ side chain (red) with the CH1 domain of the transcriptional coactivator p300. Carbons are colored white, oxygens are colored light gray, and nitrogens are colored dark gray (figure made using PDB 1L3E).

Substitution of the pro-(S) β-H by the OH in the (S)-β-OH-Asn$_{803}$ would introduce a steric clash with the branched side chain of Ile$_{353}$ on p300. In turn, this may disrupt the *i* to *i* + 3 hydrogen bond between D$_{799}$ and N$_{803}$ in the helical conformer of HIF-1α. Both disruptions would lead to reduced affinity for p300 and attenuation of the transcriptional activation by HIF-1α.

It appears that the introduction of a single oxygen to convert a CH$_2$ to a CHOH moiety at two distinct residues in HIF-1α has "on/off" switching effects for the interaction with two partner proteins (Figure 12.15). The 4(R)-OH side chain in OH–Pro$_{564}$ has an attractive effect for the pVHL protein by 1000-fold, setting up recruitment of an E2 ubiquitin ligase for the next stage of posttranslational modification, polyubiquitylation, to lead to shorter lifetime. The introduction of the 3(S)-OH in Asn$_{803}$ in a distinct domain of HIF-1α has the opposite effect—namely, disruption of the interaction with the specific transcriptional coactivator. Taken together, a parallel hydroxylation strategy with two regiospecific posttranslational modification hydroxylases provides oxygen sensing. The Pro$_{564}$ and Pro$_{402}$ hydroxylations lead to rapid destruction of the HIF-1α subunit by proteolysis, leaving HIF-1β nonfunctional for recognition by Epo and VEGF promoters. The Asn$_{803}$ hydroxylation disconnects the p300 circuitry to HIF. The combination of HIF-1α hydroxylations means both a reduction in efficiency as a transcriptional activator and programmed destruction of one subunit.

The prolyl and asparaginyl hydroxylases are the *proximal oxygen sensors*, using the O$_2$ directly as substrate. It will be of interest to see if standard hyperbolic enzyme

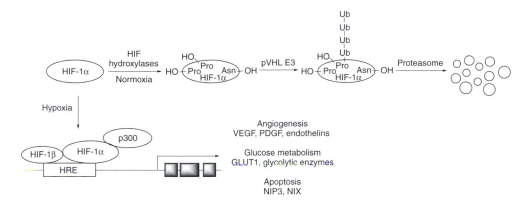

Figure 12.15 Hydroxylation at three residues of HIF-1α separately affects the interaction with two partner proteins, pVHL E3 ubiquitin ligase and p300 transcriptional coactivator [adapted from Bhattacharya and Ratcliffe (2003)].

kinetics are followed or whether there will be allosteric regulation of these sensor catalysts. The ambient pO_2 is then proportional to the fraction of HIF-1α bearing one, two, or three hydroxylated amino acids. The prolyl and asparaginyl hydroxylases are themselves subject to polyubiquitylation and proteasomal degradation by the action of E3 ligases (Chapter 9) of the Siah1a/2 family. Siah is induced under hypoxic conditions and ensures enhanced degradation of the two hydroxylases (Nakayama et al., 2004). As a result, unmodified HIF-1α levels rise. The ubiquitylation/hydroxylation interplay thus occurs at multiple levels of protein control in oxygen sensing.

While it is unclear if other proteins will be analogously regulated by hydroxylation, bioinformatics analysis suggests that the Jumonji family of transcriptional regulators (Aravind, 2001) have the 2-His 1-carboxylate consensus sequence, thus raising the possibility that they will have oxygenase domains.

There are examples in processing of bioactive peptides released from protein precursors where the hydroxylation of C-terminal glycine residues occurs prior to loss of the elements of glyoxalate and retention of the amino group of the Gly residue as part of the $n - 1$ residue as C-terminal amide. This posttranslational machinery is discussed in Chapter 15.

References

Aravind, L., and E. Koonin. "The DNA repair protein AIKB, EGL-9, and leprecan define new families of 2-oxoglucarate- and iron-dependent dioxygenases," *Genome Biol.* **2**:1–7 (2001).

Bruick, R. K., and S. L. McKnight. "A conserved family of prolyl-4-hydroxylases that modify HIF," *Science* **294**:1337–1340 (2001).

Clifton, I. J., L. C. Hsueh, J. E. Baldwin, K. Harlos, and C. J. Schofield. "Structure of proline 3-hydroxylase. Evolution of the family of 2-oxoglutarate dependent oxygenases," *Eur. J. Biochem.* **268**:6625–6636 (2001).

Dann, C. E., III, R. K. Bruick, and J. Deisenhofer. "Structure of factor-inhibiting hypoxia-inducible factor 1: An asparaginyl hydroxylase involved in the hypoxic response pathway," *Proc. Natl. Acad. Sci. U.S.A.* **99**:15351–15356 (2002).

Dinchuk, J. E., N. L. Henderson, T. C. Burn, R. Huber, S. P. Ho, J. Link, K. T. O'Neil, R. J. Focht, M. S. Scully, J. M. Hollis, G. F. Hollis, and P. A. Friedman. "Aspartyl beta-hydroxylase (Asph) and an evolutionarily conserved isoform of Asph missing the catalytic domain share exons with junctin," *J. Biol. Chem.* **275**:39543–39554 (2000).

Hegg, E. L., and L. Que, Jr. "The 2-His-1-carboxylate facial triad—an emerging structural motif in mononuclear non-heme iron(II) enzymes," *Eur. J. Biochem.* **250**:625–629 (1997).

Heikkinen, J., M. Risteli, C. Wang, J. Latvala, M. Rossi, M. Valtavaara, and R. Myllyla. "Lysyl hydroxylase 3 is a multifunctional protein possessing collagen glucosyltransferase activity," *J. Biol. Chem.* **275**:36158–36163 (2000).

Hon, W. C., M. I. Wilson, K. Harlos, T. D. Claridge, C. J. Schofield, C. W. Pugh, P. H. Maxwell, P. J. Ratcliffe, D. I. Stuart, and E. Y. Jones. "Structural basis for the recognition of hydroxyproline in HIF-1 alpha by pVHL," *Nature* **417**:975–978 (2002).

Kivirikko, K. I., and T. Pihlajaniemi. "Collagen hydroxylases and the protein disulfide isomerase subunit of prolyl 4-hydroxylases," *Adv. Enzymol. Relat. Areas Mol. Biol.* **72**:325–398 (1998).

McNeill, L. A., K. S. Hewitson, T. D. Claridge, J. F. Seibel, L. E. Horsfall, and C. J. Schofield. "Hypoxia-inducible factor asparaginyl hydroxylase (FIH-1) catalyses hydroxylation at the beta-carbon of asparagine-803," *Biochem. J.* **367**:571–575 (2002).

Min, J. H., H. Yang, M. Ivan, F. Gertler, W. G. Kaelin, Jr., and N. P. Pavletich. "Structure of an HIF-1alpha -pVHL complex: Hydroxyproline recognition in signaling," *Science* **296**:1886–1889 (2002).

Myllyharju, J., and K. I. Kivirikko. "Characterization of the iron- and 2-oxoglutarate-binding sites of human prolyl 4-hydroxylase," *EMBO J.* **16**:1173–1180 (1997).

Nakayama, K., I. J. Frew, M. Hagensen, M. Skals, H. Habelhah, A. Bhoumik, T. Kadoya, H. Erdjument-Bromage, P. Tempst, P. B. Frappell, D. D. Bowtell, and Z. Ronai. "Siah2 regulates stability of prolyl-hydroxylases, controls HIF1alpha abundance, and modulates physiological responses to hypoxia," *Cell* **117**:941–952 (2004).

Pirskanen, A., A. M. Kaimio, R. Myllyla, and K. I. Kivirikko. "Site-directed mutagenesis of human lysyl hydroxylase expressed in insect cells. Identification of histidine residues and an aspartic acid residue critical for catalytic activity," *J. Biol. Chem.* **271**:9398–9402 (1996).

Pugh, C. W., and P. J. Ratcliffe. "Regulation of angiogenesis by hypoxia: Role of the HIF system," *Nat. Med.* **9**:677–684 (2003).

Ryle, M. J., and R. P. Hausinger. "Non-heme iron oxygenases," *Curr. Opin. Chem. Biol.* **6**:193–201 (2002).

Schofield, C. J., and P. J. Ratcliffe. "Oxygen sensing by HIF hydroxylases," *Nat. Rev. Mol. Cell Biol.* **5:**343–354 (2004).

Semenza, G. L. "Regulation of mammalian O_2 homeostasis by hypoxia-inducible factor 1," *Annu. Rev. Cell Dev. Biol.* **15:**551–578 (1999).

Valtavaara, M., H. Papponen, A. M. Pirttila, K. Hiltunen, H. Helander, and R. Myllyla. "Cloning and characterization of a novel human lysyl hydroxylase isoform highly expressed in pancreas and muscle," *J. Biol. Chem.* **272:**6831–6834 (1997).

Vranka, J. A., L. Y. Sakai, H. P. Bichinger. "Prolyl 3-hydroxylase-1, enzyme characterization and identification of a novel family of enzymes," *J. Biol. Chem.* **279:**23615–23621 (2004).

13

Protein Automodification Reactions

Intramolecular autocatalytic rearrangement of a specific peptide bond to an ester bond sets up
both autocleavage and protein splicing outcomes.

There are several types of protein automodification reactions known, defined in this chapter as autocatalyzed covalent changes. The protein acts enzymatically on itself, *in cis* or *in trans*, and needs to be in the native, folded state to accelerate this chemical transformation. This requirement for the folded state of automodifying proteins most probably reflects the need for precise conformer orientation where a strained or destabilized conformer can be populated to initiate the reaction.

One group of automodifications is carried out by enzymes that normally carry out a posttranslational modification on some other protein as substrate, but in this case catalyze electrophilic fragment transfer to one of their own amino acid side chains. These automodifications mirror the catalytic turnover process, but occur stoichiometrically on the catalyst. Autophosphorylation of protein kinases is in this category.

A second type of automodification occurs in some enzymes carrying out redox transformations of substrates, such as the oxidation of amine or alcohol substrates to aldehyde products. These enzymes act in priming steps to generate radical species that oxidize the electron-rich aromatic side chains of tyrosine and tryptophan residues

349

by one-electron steps to *ortho-* and *para-*quinones [e.g., the tyrosyl side chain to TOPA quinone (TPQ) (Mure, 2004)]. The semiquinone radical intermediates on the way to the quinones can also form C–C bonds by electron abstraction from neighboring side chains, resulting in cross-linking in tryptophan tryptophanylquinone (TTQ) and lysine tyrosylquinone (LTQ) group formation in the active sites of amine dehydrogenases and amine oxidase, respectively. These newly created quinone derivatives of aromatic amino acids in these enzyme active sites serve as the electron sinks for the redox catalytic turnovers that follow (Okeley and van der Donk, 2000).

A third type of automodification involves the intramolecular attack of an amino acid nucleophilic side chain or the amide NH in a peptide bond of the folded protein on a nearby peptide bond. The resulting adduct can rearrange in various ways. One set of outcomes leads to the green fluorescent and red fluorescent chromophores of green fluorescent protein (GFP) and DsRed, respectively (Tsien, 1998). A second outcome can be peptide bond fragmentation. This can involve capture by H_2O, a net hydrolysis as occurs in the formation of the pyruvoyl-enzyme aspartate β-decarboxylase. Deconvolution of the intermediate can instead involve capture by some ROH other than HOH. In the case of the protein called Hedgehog, this means net peptide bond cleavage and transfer of the N-terminal protein fragment to cholesterol to produce the membrane-anchored protein C-terminal cholesterol ester. In other proteins the initial adducts are resolved in net transpeptidations with removal of intervening sequences (inteins) in the remarkable *in cis* excision/religation cycle of protein splicing reactions (Paulus, 2000).

Autophosphorylation of Protein Kinases

Protein kinases are discussed as the most abundant class of posttranslational modification catalysts in Chapter 2. The single aspect of PKs reemphasized here is their automodification reactions. Both major families of protein kinases, the protein serine/threonine kinases and the protein tyrosine kinases, display autophosphorylation properties. This is sufficiently general that a common initial assay strategy for *in vitro* protein kinase activity has been to add γ-[^{32}P]-ATP to a putative protein kinase and look for covalent attachment of the ^{32}P-radioactivity to the protein (e.g., by acid precipitation of the protein). The (sub)stoichiometric autophosphorylation can be detected with picomolar quantities of protein kinases.

As expected, the serine/threonine kinases autophosphorylate on one or more of their own serine or threonine side chains while PTKs maintain their chemoselectivity, autophosphorylating on tyrosines. Thus, the autophosphorylations are viewed as internal examples of the chemospecific protein phosphorylation capacities of the kinase (Johnson, 2001).

These autophosphorylations can be within the PSK and PTK active sites or outside of the catalytic microenvironments. The autophosphorylation of threonine residues in activation loops, such as Thr_{197} in the cAMP-activated protein kinase A or Tyr_{416} in the Src tyrosine kinase, leads to autoactivation of the kinases towards external substrates by reconfiguring the orientation of bound ATP and/or phosphorylatable protein substrates. Autophosphorylations are often accelerated by specific ligand binding and are part of the pathways for turning protein kinases from the "off" state to the "on" state (Kobe and Kemp, 1999). Structural analysis of both serine/threonine and tyrosine kinases (Chapter 2) has revealed that phosphorylation of Thr or Tyr convert mobile loops into fixed architectural elements that contribute to the optimally formed active site geometries. The protein kinases at rest are inefficient phosphoryl transfer catalysts, but even a slow autophosphorylation event can be completed in a few seconds. The phospho form of the PK may then catalyze phosphoryl transfer to external protein substrates at much higher rates, perhaps 100- to 1000-fold faster. Kinase autophosphorylation can also increase the affinity of the protein kinase for either the ATP phosphoryl donor substrate, an effector such as cAMP, or for the protein partner to be phosphorylated. In addition to autophosphorylations for the regulation of protein kinase activity, rounds of protein kinase phosphorylation are fundamental in signal transduction pathways that employ hierarchical cascades of protein kinases, such as MAP kinase pathways.

While autophosphorylations can be *in cis* for many serine and threonine protein kinases and for the Src, Jak, Fak, Abl, and Btk class of intracellular tyrosine kinases, the large number of transmembrane receptor tyrosine kinases (RTKs), the transmembrane histidine kinases of bacterial two-component regulatory systems, and the transmembrane TGFβ receptor Ser kinase all autophosphorylate *in trans*. Ligand (e.g., epidermal growth factor) binding to the extracellular domain of an RTK (such as the EGF receptor) will induce dimerization and bring the two intracellular catalytic domains in apposition. Now the catalytic site of one RTK chain can cross-phosphorylate the second chain, not only in the active site region but at other regions as well. Indeed, as noted in Chapter 2, EGFR transphosphorylates its partner subunit in the homodimeric complex at multiple tyrosines, creating a set of pY residues whose different sequence contexts recruit partner proteins to initiate signal. In RTK function, transphosphorylation at multiple Y residues is essential for partner protein recruitment. Other examples of *supra*stoichiometric autophosphorylation of protein kinases have been reported, including the human cytomegaloviral PK UL97, which autophosphorylates at least nine residues in its N-terminal 200 residues (Back et al., 2002).

As noted in Chapter 2, protein phosphorylation is reversible, including the autophosphorylation of protein kinases. The protein serine/threonine phosphatases and protein tyrosine phosphatases control the lifetime of the phospho forms of proteins by hydrolysis and so in part control the duration and intensity of phosphoprotein signaling circuitry.

Auto-ADP-Ribosylation

A second example of covalent automodification occurs in some of the enzymes that catalyze ADP-ribosyl transfers from NAD, as noted in Chapter 11. Self-modification by mono-ADP-ribosylation has been reported for ADP-ribosyltransferase-2 (ART-2) expressed as an ecto enzyme in differentiated T cells (Maehama et al., 1995). However, the most well-known example is self ADP-ribosylation of the enzyme poly-ADP-ribose polymerase (PARP). PARP occurs in several isoforms, many of which associate with chromosomes and specifically spindles or spindle poles (Chang et al., 2004). The best substrate for PARsylation appears to be the PARP isoforms themselves. The role of distinct PARPs species, both in the inactive non-PARsylated and active, PARsylated state, are under active investigation for roles in cell division and in gene silencing and activation (Ju et al., 2004; Kim et al., 2004).

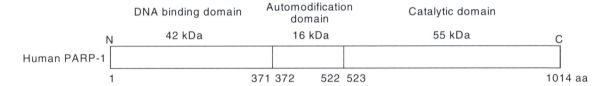

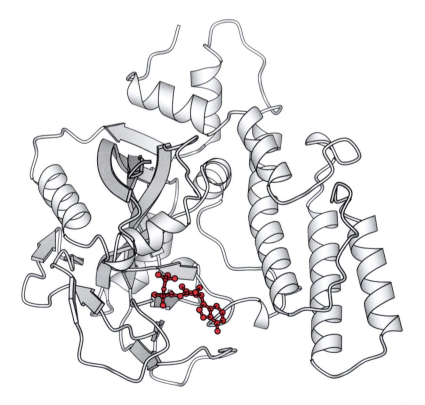

Figure 13.1 Multidomainal organization of the poly-ADP-ribose polymerase (PARP) and x-ray structure of the catalytic domain with bound ADP in ball and stick (figure made using PDB 1A26).

The 113-kDa multidomainal PARP has a 40-kDa C-terminal catalytic domain just downstream of a 20- to 22-kDa domain called the automodification domain that is the site of covalent ADP-ribose chain transfers (Ruf et al., 1998) (Figure 13.1). The poly-ADP-ribose chain that gets built up on DNA binding proteins, including PARP itself, occurs in three phases: initiation, elongation, and branching. The first ADP-ribose moiety, from NAD, is attached as a glutamyl ester to the automodification domain of PARP (Figure 13.2A). The next 20–50 ADP-riboses are in 2′–1′ linkage (Figure 13.2B), where the 2′-OH of the terminal ADP-ribose acts as a *nucleophile* to attack an incoming NAD. Then, a branch point can occur where the terminal ADP-ribose acts as an *electrophile* at C_1'. Now two chains can be elongated from this branch point through elongation to the 2′-OHs of the riboses in each branch (Figures 13.2C and D).

The transfer of the first ADP-ribosyl moiety is formulated as a nucleophilic attack by a Glu γ-carboxylate side chain in the automodification domain on C_1 of the ribosyl moiety of the nicotinamide mononucleotide portion of NAD. The transfer is mediated by the catalytic domain, whose structure has been solved by x-ray analysis (Ruf et al., 1998). This results in the release of free nicotinamide and the capture of the ADP-ribosyl moiety in ester linkage to the attacking PARP side chain carboxylate, with inversion of the stereochemistry from β to α. The reaction is energetically favorable since the nicotinamide moiety in NAD is quarternized and bears a positive charge. This is a net

$$NAD + X \longrightarrow ADP\text{-}5'\text{-ribose-}1'\text{-}X$$

X = Glu residue: initiation
X = 2′-OH of adenine ribose: elongation
X = 2′-OH of nicotinamide ribose: branching

A. Initiation:

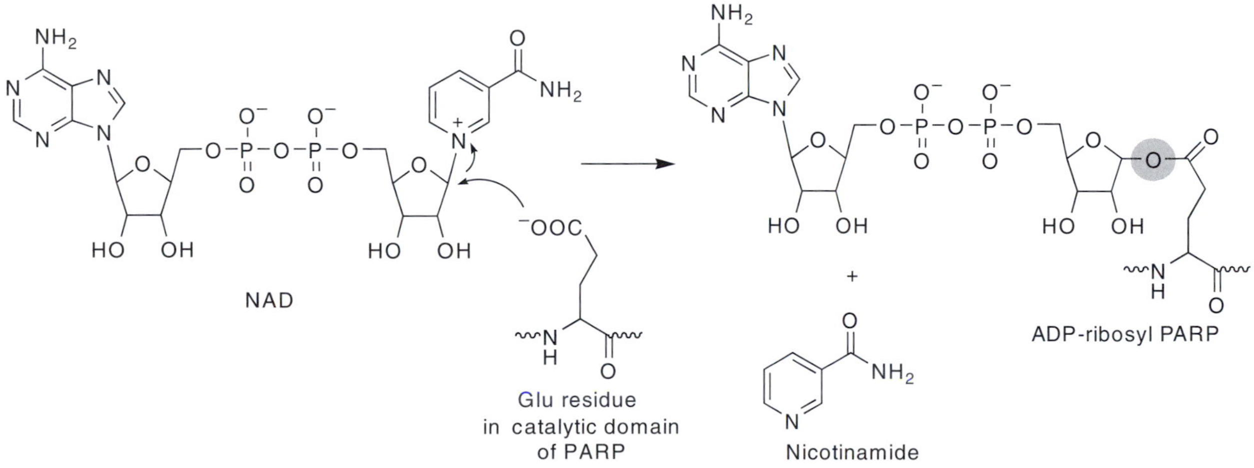

Figure 13.2 Autocatalytic growth of the poly-ADP-ribose chain on a Glu residue in the active site of the catalytic domain of PARP. **A.** Initiation phase: the first NAD molecule utilized.

B. Elongation:

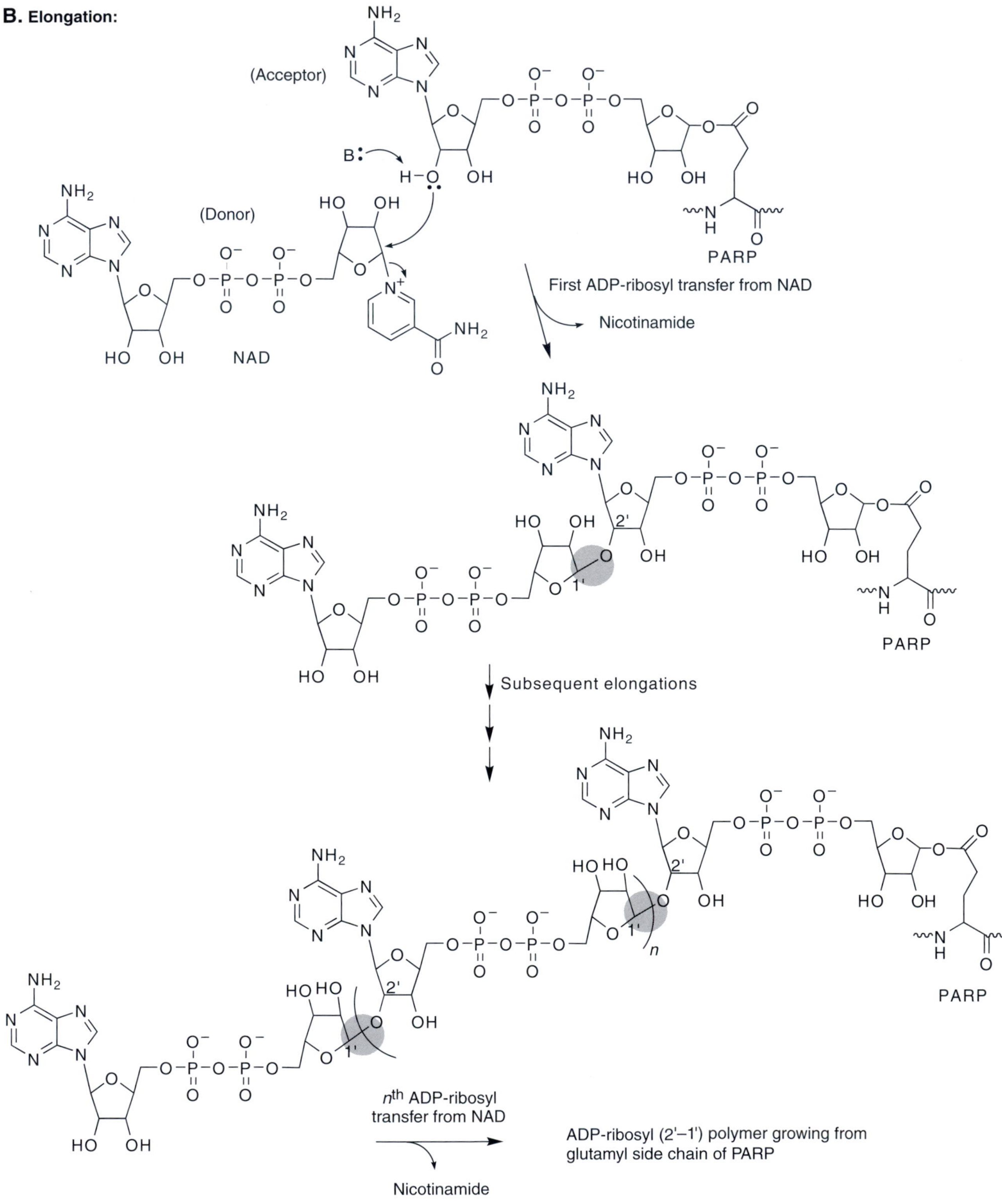

Figure 13.2 B. Elongation by 20–50 ADP-ribosyl units by attack of the 2′-OH of the ADP moiety of the growing chain on the nicotinamide-ribosyl bond of NAD molecules to form 2′-1′-O-ribosyl linkages.

C. Branching:

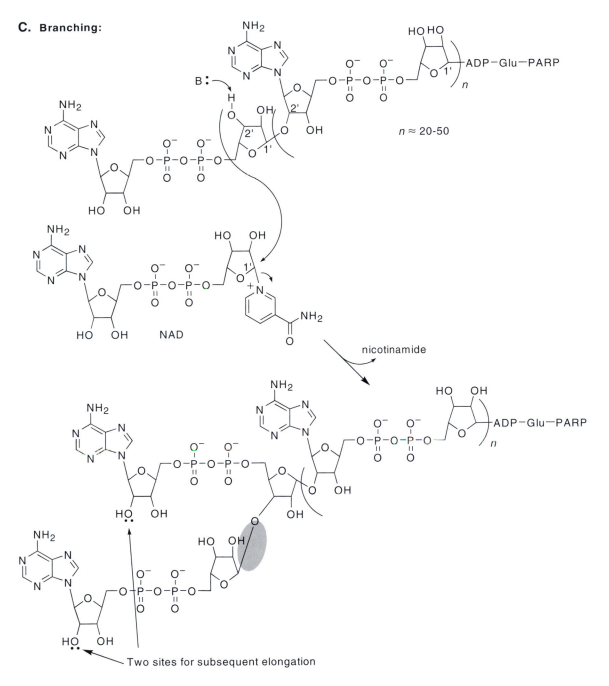

Figure 13.2 **c.** Branch point formation via attack of the alternative ribose 2'-OH; extension can now proceed at two sites of chain growth.

D. Elongation versus branching:

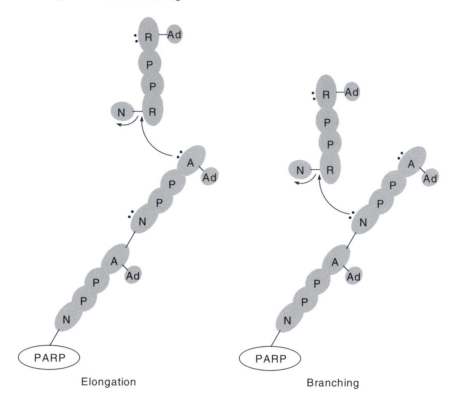

Elongation   Branching

Figure 13.2 **D.** Schematic of different locations of nucleophile within the growing ADP-ribose chain for elongation versus branching (redrawn from Ruf et al., 1998).

ribosylation (glycosylation) of the automodification domain of PARP (Figure 13.2A). This reaction bears strong analogy to the mono-ADP-ribosylations catalogued in Chapter 11.

The iterated series of subsequent ADP-ribosylations that next extend and elongate the chain, with NAD cosubstrate fragmentation in each elongation, build off this mono-ADP-ribosylated PARP, but use a different nucleophile, the 2′-OH of the ribosyl moiety at the far end of the just added ADP-ribosyl unit. NAD can be viewed as a cosubstrate with *two* nucleotide monophosphate portions, AMP and nicotinamide mononucleotide (NMN), joined in diphosphate linkage. There is a ribosyl moiety with an anomeric $C_1′$–N bond and a 2′-OH in each half. In the elongation reactions, it is the same $C_1′$ of the ribosyl moiety of the NMN portion that is transferred as the electrophilic fragment, releasing nicotinamide. The result in the first elongation step is ADP-ribose 2′-1′-ADP-ribosyl 1′-glutamyl-PARP (Figure 13.2B). This elongating, autocatalytic chemistry can be iterated indefinitely to produce (2′–1′ connected-ADP-ribose)$_n$ chains where n = 20–50, each time creating a 2′–1′ linkage with α stereochemistry at that $C_1′$.

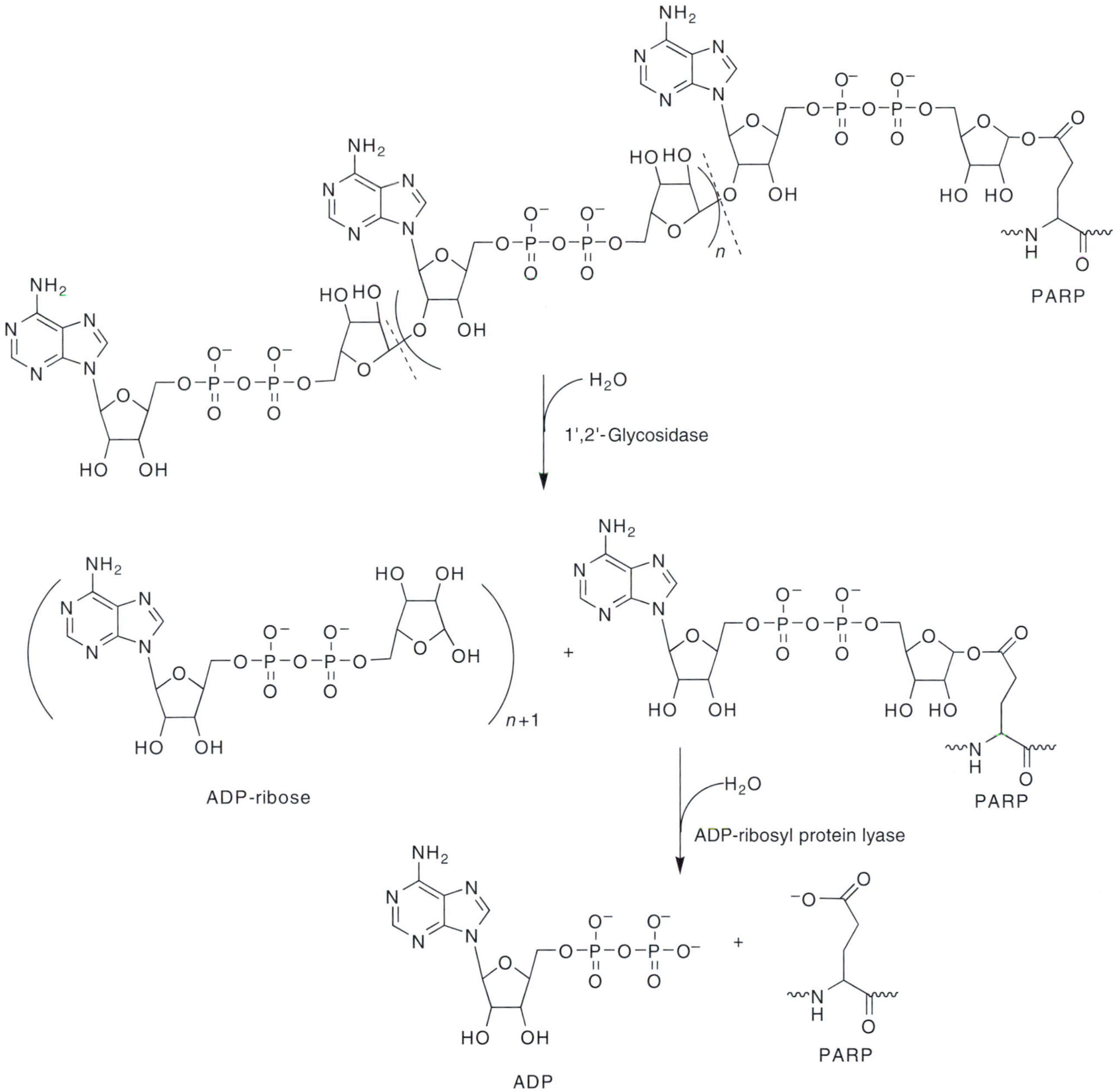

Figure 13.3 Reversal of the poly-ADP-ribose polymerization process by glycosidase action at the 2'-1' linkages followed by removal of the PARP-proximal ADP-ribose by action of ADP-ribosyl protein lyase on the ester bond to the glutamyl side chain of PARP.

For ADP-ribosyl chains of this average length, a third regiochemical variation of the ADP-ribosylation addition occurs, creating a branch point. It has been proposed that the last ADP-ribosyl unit in a chain of that length pivots 180° around the 2′–1′ bond and now brings the 2′-OH of the ribose moiety derived from the NMN end of the last NAD in apposition to the next incoming NAD donor. Capture of the ADP-ribosyl C_1′-carbon by that 2′-OH creates the branchpoint (Figures 13.2 C and D). At such a branch two ADP-ribosyl termini are created, each with a 2′-OH in the AMP portion, allowing for subsequent elongations to occur on both arms of the fork.

Analogous to kinase autophosphorylation, PARP auto-poly-ADP-ribosylation is reversible, with two types of reversing enzymes required, one that cleaves the 2′–1′ glycosidic linkages between ADP-ribosyl units and one, ADP-ribosyl protein lyase, that cleaves the last ADP-ribosyl ester bond to the glutamyl side chain of the PARP automodification domain (Figure 13.3).

Automodification of Aromatic Side Chains of Proteins by Oxidation

A variety of enzymes that carry out redox transformations, such as alcohols to aldehydes and sugars to keto sugars, were observed to require a soluble quinonoid cosubstrate. As the substrate alcohol is oxidized, the quinone cosubstrate is reduced by the corresponding two electrons to the hydroquinone (quinol) oxidation state. The first quinonoid cofactor discovered was the soluble molecule [4,5-dihydro-4,5-dioxo-1H-pyrrolo-2,3-quinolone-2,7,8-tricarboxylate] known as PQQ (pyrrolo-quinolone-quinone) (Stites et al., 2000). The orthoquinone gets readily reduced to the quinol through the one-electron semiquinone (Figure 13.4). PQQ is formed from the two amino acids Glu and Tyr (Gomelsky et al., 1996). Covalently tethered quinonoid prosthetic groups, derived from the oxidation of amino acid side chains of enzymes, were subsequently detected (Mure, 2004).

The best known proteins with quinonoid posttranslational modifications are amine oxidases. Enzymes from both bacteria and eukaryotes oxidize aliphatic and aromatic primary amines to aldehydes (Figure 13.5) through such tethered quinones. The cosubstrate is typically molecular oxygen, undergoing a two-electron reduction to H_2O_2. The aldehyde products arise secondarily from hydrolysis of the primary imine products. Amine oxidases were known to bind stoichiometric amounts of copper ions but also to contain an additional electron sink in the active site. This prosthetic group was identified as a TOPA quinone (TPQ) side chain, arising by a six-electron post-translational modification of a particular tyrosine side chain in an autocatalytic oxidation process (Figure 13.6). The inactive proenzyme forms of these amine oxidases use

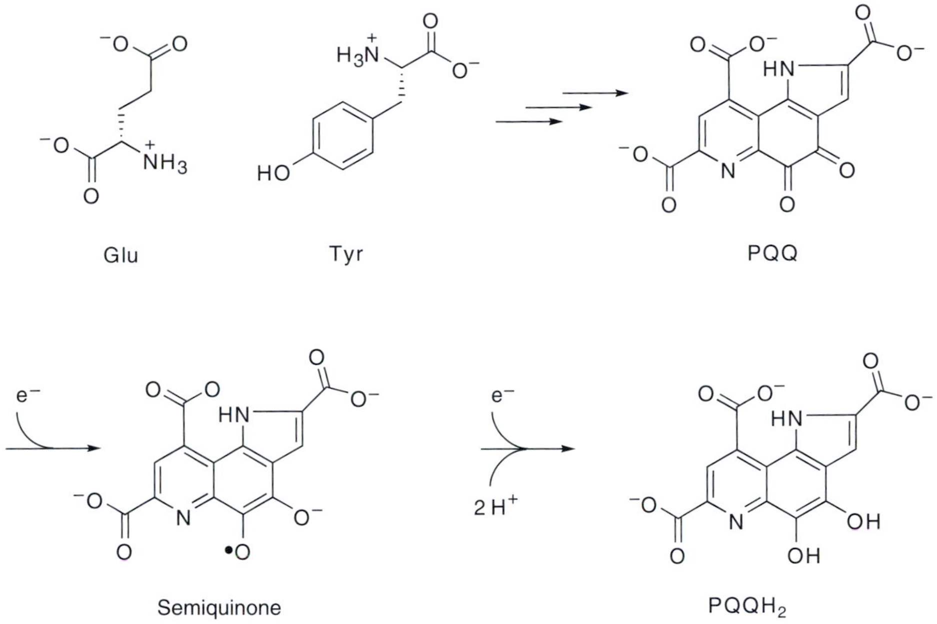

Figure 13.4 PQQ and the orthoquinone–hydroquinol redox cycle.

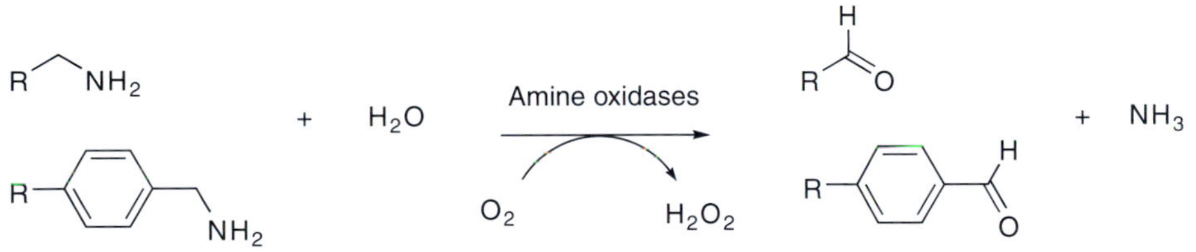

Figure 13.5 Amine oxidase-mediated conversions of amines to aldehyde products with O_2 as reducible cosubstrate, via copper and covalent quinonoid cofactor participation.

the bound Cu^{II} to activate oxygen and carry out the slow autoxidation of Tyr_{382} to TPQ (Wilce et al., 1997). The tyrosyl side chain oxidation is thought to proceed as depicted in Figure 13.6, where Cu^{II} binds to the active site Tyr phenolate ion. Reduction by one-electron transfer to Cu^{I} facilitates electron transfer from copper to O_2. Radical coupling of the Tyr radical and Cu^{I}–O_2^- creates an adduct in which O–O bond cleavage generates the orthoquinonoid DOPA quinone (DPQ) residue. The DPQ side chain can undergo Cu-directed water addition at C_3 and rearomatization to the trihydroxyphenylalanine side chain (TOPA). The TOPA side chain is then oxidized by a second molecule of O_2 to the TOPA quinone, the active form of the amine oxidases. This is a six-electron autoxidation sequence: Tyr→DOPA, quinone→TOPA→TPQ (Mure et al., 2002). It consumes two molecules of O_2 and one molecule of water. Of the

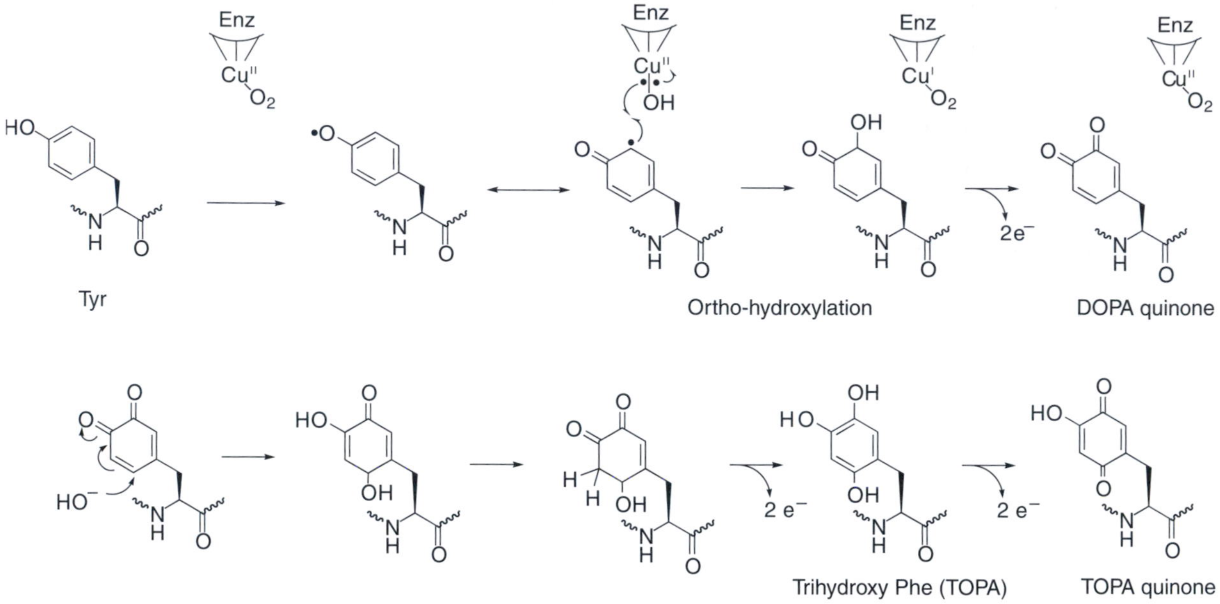

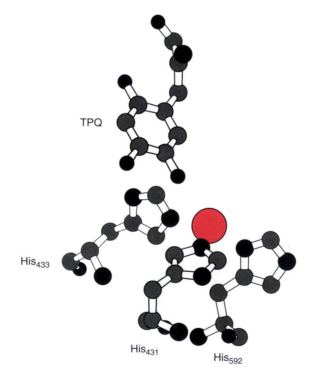

Figure 13.6 The tethered TOPA quinone (TPQ) coenzyme arises from posttranslational autocatalytic modification of an active site tyrosine residue by a six-electron oxidation of the tyrosine side chain, generating the active form of the enzyme.

Figure 13.7 Structure of the ligand sphere of the active site Cu in amine oxidase, depicting the location of TPQ and the three His ligands (figure made using PDB 1AV4).

two oxygens introduced in the Tyr side chain, oxygen-5 comes from the first O_2, oxygen-3 comes from H_2O, and the TOPA to TPQ oxidation consumes the second O_2. X-ray structures of three amine oxidases (Wilce et al., 1997) clearly show the trioxygenated tyrosine (TPQ) residue close to the Cu with three His ligands (Figure 13.7).

The function of the TPQ prosthetic group in catalysis is to serve as a two-electron sink for electrons and the amino group removed from substrate amines during the half-reaction that oxidizes the substrate and reduces the TPQ (Figure 13.8) (Mure, 2004; Mure et al., 2000). Once the aldehyde product has departed, the aminoquinol form of the TPQ side chain is reoxidized by O_2, probably by one-electron pathways.

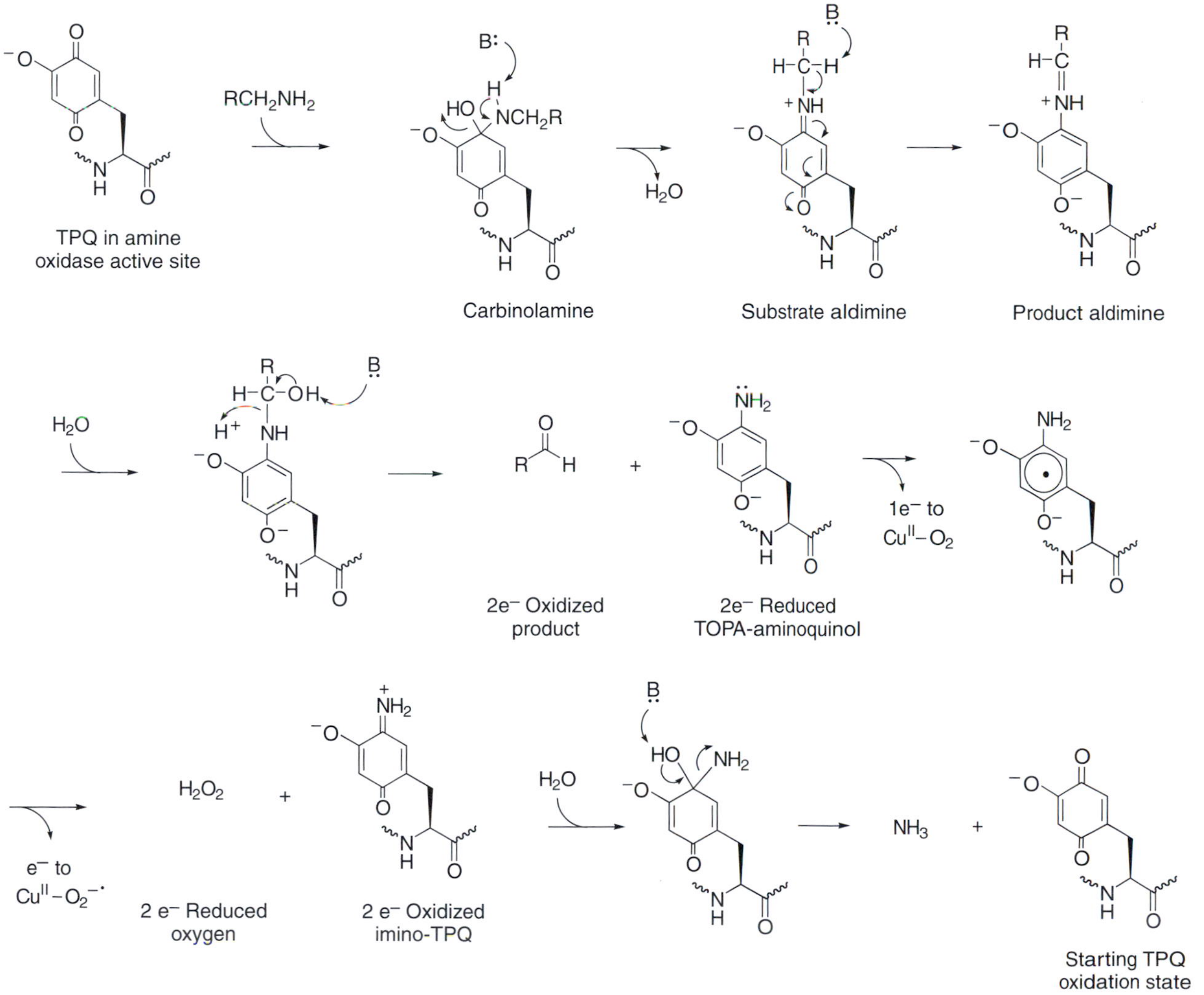

Figure 13.8 Reaction cycle for amine oxidase, involving reversible reduction and reoxidation of the covalently bound TPQ cofactor.

In the enzyme reductive half-reaction, substrate RCH_2NH_2 bound at the active site reacts with C_5 of the quinonoid side chain to form the carbinolamine and then the aldimine. Now C–H cleavage has a low-energy barrier since the carbanion that results can be stabilized by delocalization through the TPQ system. The quinol protonates on C_3 and hydrolysis of the product imine yields the two-electron reduced form of the enzyme as the neutral aminoquinol. Reoxidation is proposed to involve direct one-electron transfer steps from aminoquinol to O_2, without the participation of Cu^I as an electron conduit, to yield the imino-TPQ and H_2O_2 (Mure, 2004). The free NH_3 is released by hydrolysis to reset the initial state of the amine oxidase.

The autocatalytic conversion of a standard side chain, the phenol of tyrosine, to the TPQ to create a *ketone* (as part of a hydroxyquinone) electron sink in the enzyme for imine formation with amine-containing substrates is logic parallel to the autocatalytic generation of the ketone moiety in the pyruvoyl enzymes, discussed in a subsequent section of this chapter. Both represent posttranslational routes to introduce electrophilic moieties into the side chains of amino acid residues to facilitate imine formation with amine-containing substrates.

Analogous oxidation of a tyrosyl residue to a quinone is observed in lysyl oxidase, a secreted enzyme involved in cross-linking of collagen and elastic fibers in extracellular spaces. In lysyl oxidase the prosthetic group is an adduct of lysine and DOPA quinone (Figure 13.9) (Duff et al., 2003). Less is known about the posttranslational mechanism here but it is anticipated that an analogous oxidation of an active site Tyr to DOPA quinone proceeds and this is captured at C_3 by the ϵ-NH_2 of a neighboring lysine side chain to create the LTQ (lysine tyrosylquinone) prosthetic group

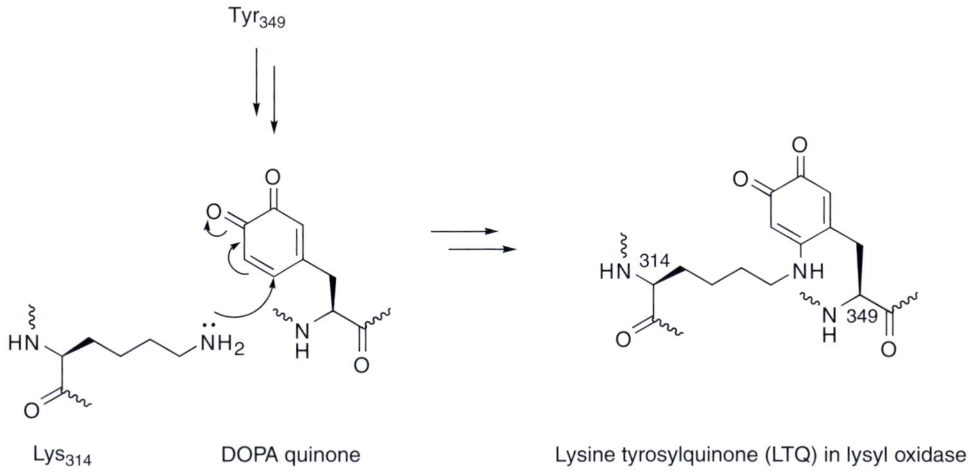

Figure 13.9 Covalently attached quinone prosthetic group at the active site of lysyl oxidase: an adduct between a DOPA quinone residue, formed via controlled autoxidation, and a lysine side chain.

(Okeley and Van der Donk, 2000). In terms of a sequence of posttranslational modifications, lysyl oxidase is translated as a 46-kDa proenzyme form with an ER signal sequence that is removed by signal peptidase (Chapter 8). During transit through the Golgi complex the protein becomes N-glycosylated (Chapter 10) and then packaged into secretory vesicles. In these vesicles, copper ion is bound in the active site and then the autocatalytic modification to form LTQ ($Lys_{314}Tyr_{349}$) proceeds. At the cell surface the enzyme is cleaved in a third proteolytic step, presumably by an ADAM type protease, to a 30-kDa mature enzyme that is active for oxidation of lysyl side chains in collagen and elastin fibers in the extracellular matrix.

The action of lysyl oxidase creates posttranslational modifications in the proteins on which it works. For example, collagen chains can have multiple lysine side chains oxidized to the C_6 aldehydes by this catalyst. If two such aldehyde side chains are in proximity they can undergo aldol-type C–C bond formations, followed by dehydration to create olefinic cross-links (Figure 13.10). If the cross-links are between two different collagens this covalent connection will produce stiffening cross-links to the fibers.

Additional quinonoid prosthetic groups can arise by oxidation of tryptophan side chains. One example is the TTQ (tryptophan tryptophanylquinone) side chain found in bacterial methylamine dehydrogenases (Figure 13.11) (Mure, 2004; Okeley and van der Donk, 2000). One intuits analogous conversion of a tryptophan residue to an

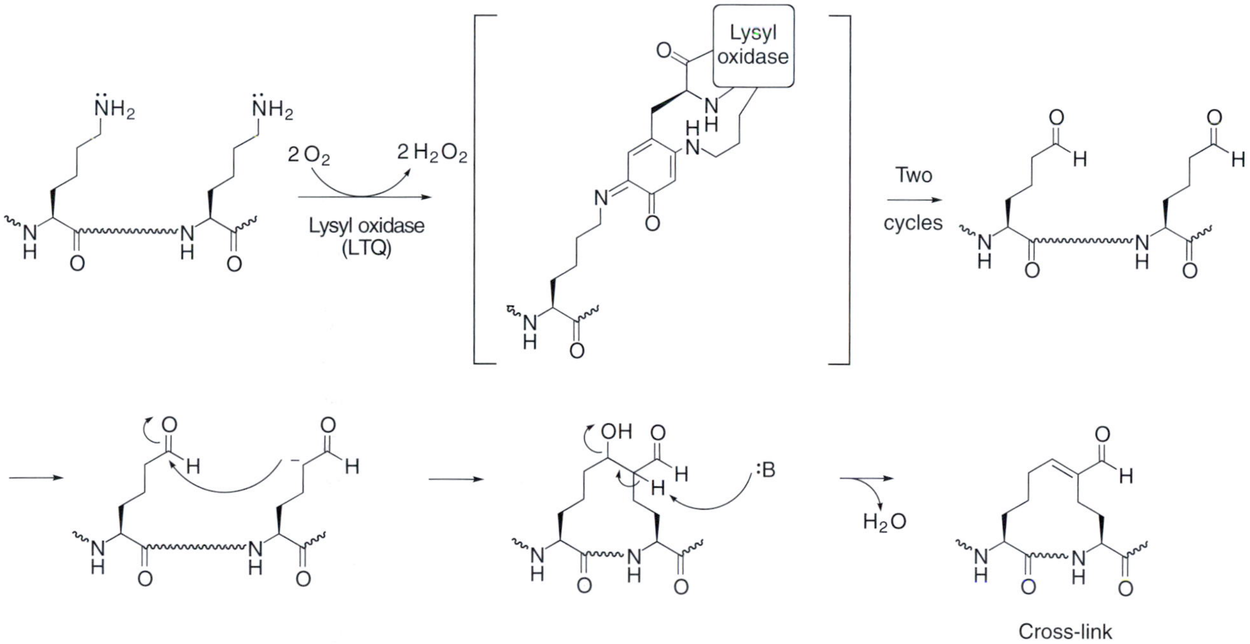

Figure 13.10 Lysyl oxidase action on lysine side chains of extracellular collagen fibrils generates aldehydes that can undergo aldol condensations and subsequent dehydrations to conjugated products.

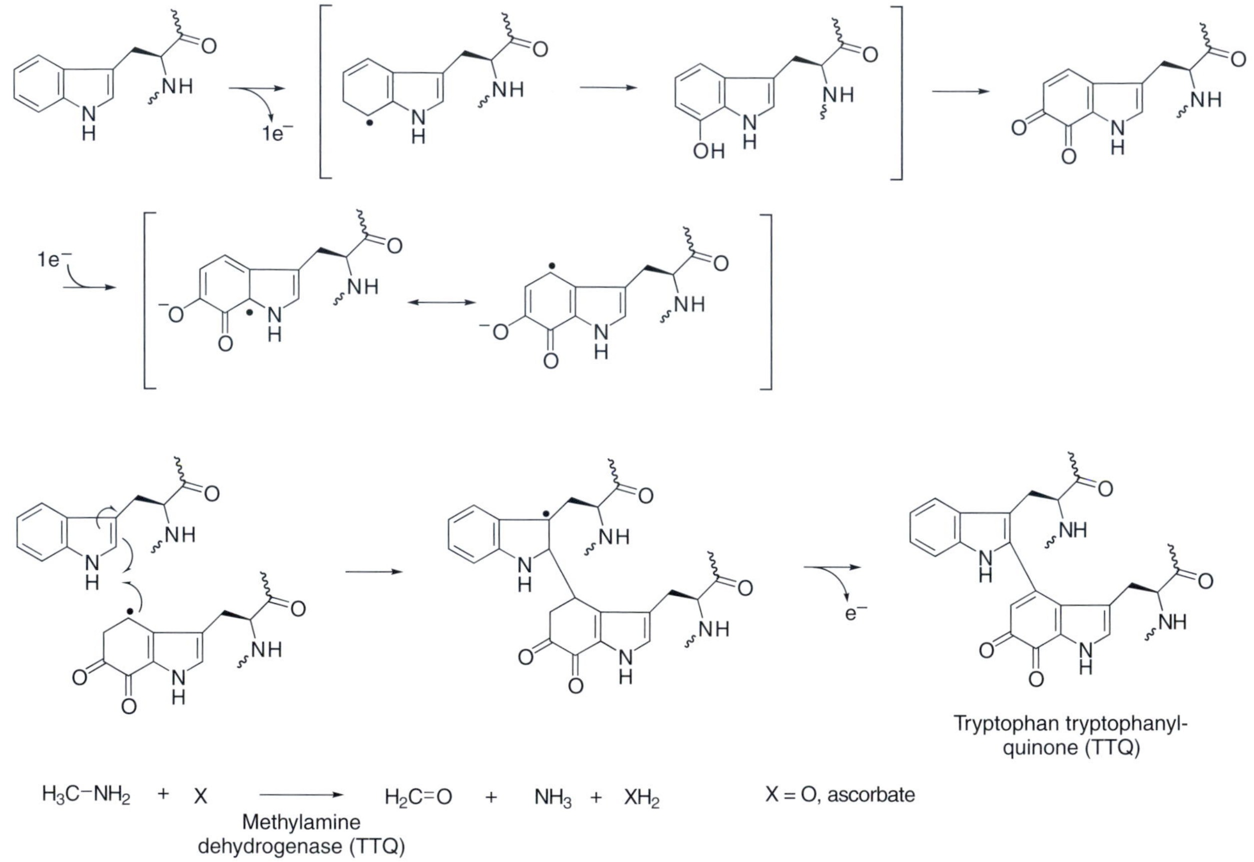

Figure 13.11 Formation of tryptophan tryptophanylquinone (TTQ) in bacterial methylamine dehydrogenase by radical coupling of Trp and a Trp-quinone produces the active form of a methylamine dehydrogenase.

orthoquinonoid form. That could convert to the semiquinone and engage in radical coupling, in this case with an additional nearby tryptophan side chain. The constrained TTQ group in the active site has the orthoquinone functionality to react with methylamine for its oxidation to formaldehyde and ammonia by a mechanism akin to that of the TPQ and LTQ groups. TTQ formation is not autocatalytic, with at least some steps under the control of one or more external enzymes (Pearson et al., 2004).

Comparable covalent capture of a tryptophanylquinone, now by a cysteinyl thiol radical, would yield the CTQ prosthetic group (Figure 13.12A) found in the γ subunit of the amine dehydrogenase from *Pseudomonas* strains (Datta et al., 2001; Vandenberghe et al., 2001). Remarkably this small 8-kDa subunit has two other cysteine cross-links, thioether linkages to Cβ of an aspartyl residue and to Cγ of a glutamyl residue (Figure 13.12B). In both the TTQ- and CTQ-containing enzymes it appears

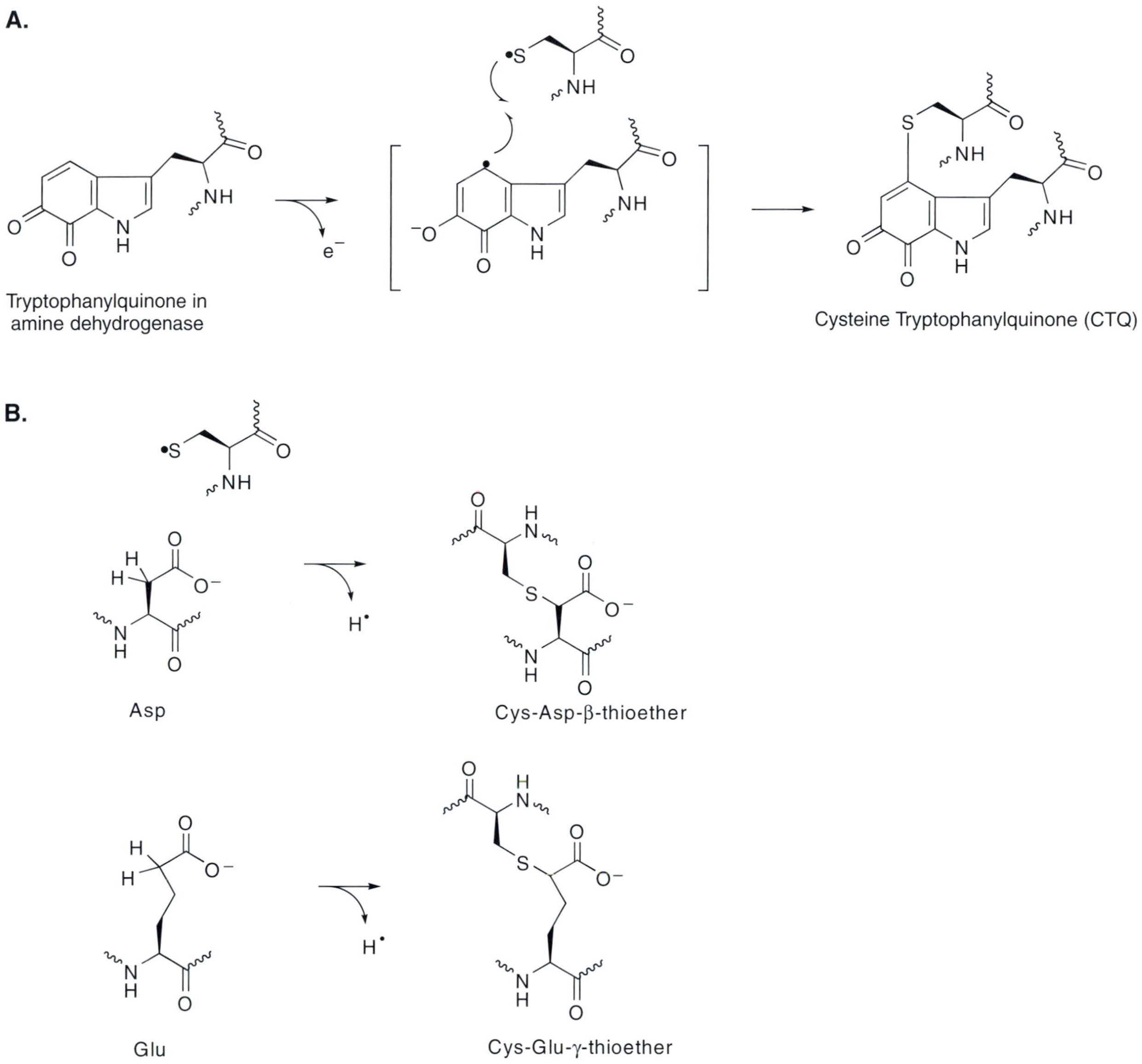

A.

Tryptophanylquinone in
amine dehydrogenase

Cysteine Tryptophanylquinone (CTQ)

B.

Asp

Cys-Asp-β-thioether

Glu

Cys-Glu-γ-thioether

Figure 13.12 Thiyl radicals on cysteine side chains. **A.** Capture of tryptophanylquinone in an amine dehydrogenase to form a CTQ cofactor in the active enzyme. **B.** S–C bond formation to the β-CH$_2$ of an Asp residue or the γ-CH$_2$ of a Glu residue generates cysteine thioether cross-links in the same amine dehydrogenase subunit.

that these posttranslational modifications are not autocatalytic, but instead require a gene encoded in the same operon that acts in a SAM and Fe–S cluster-dependent manner to generate radical intermediates that engage in the cross-linking. The CTQ is presumed to be yet another quinonoid electron sink for amine dehydrogenation catalysis.

Automodification of the Peptide Backbone in Proteins

A small number of protein automodifications are known in which a region of the peptide backbone connectivity is altered. Some of these can result in the cleavage of a specific peptide bond in a particular protein, and these will be taken up in the next part of this chapter. Others involve cyclization/dehydration reactions to create a novel ring structure in the self-modified protein. Related transformations occur in proenzyme-to-enzyme autoconversions for the deamination of the primary amino acids histidine and phenylalanine, and in the generation of protein fluorophores by marine organisms.

In each set of proteins, a three-residue loop in the folded structure of the pro form of the protein engages in an autocyclization to form a five-membered cyclic intermediate (Figure 13.13A) (Baedeker and Schulz, 2002a; Tsien, 1998). That intermediate can be trapped by a subsequent dehydration step, setting up the five-ring chromophore that serves as an electron sink for amino acid deamination catalysis (Figure 13.13B). An alternate trapping route involves one or more subsequent oxidative steps to create the green and red fluorophores of green fluorescent protein and DsRed, respectively (Figure 13.13C). The outcome in the second step is dictated by the side

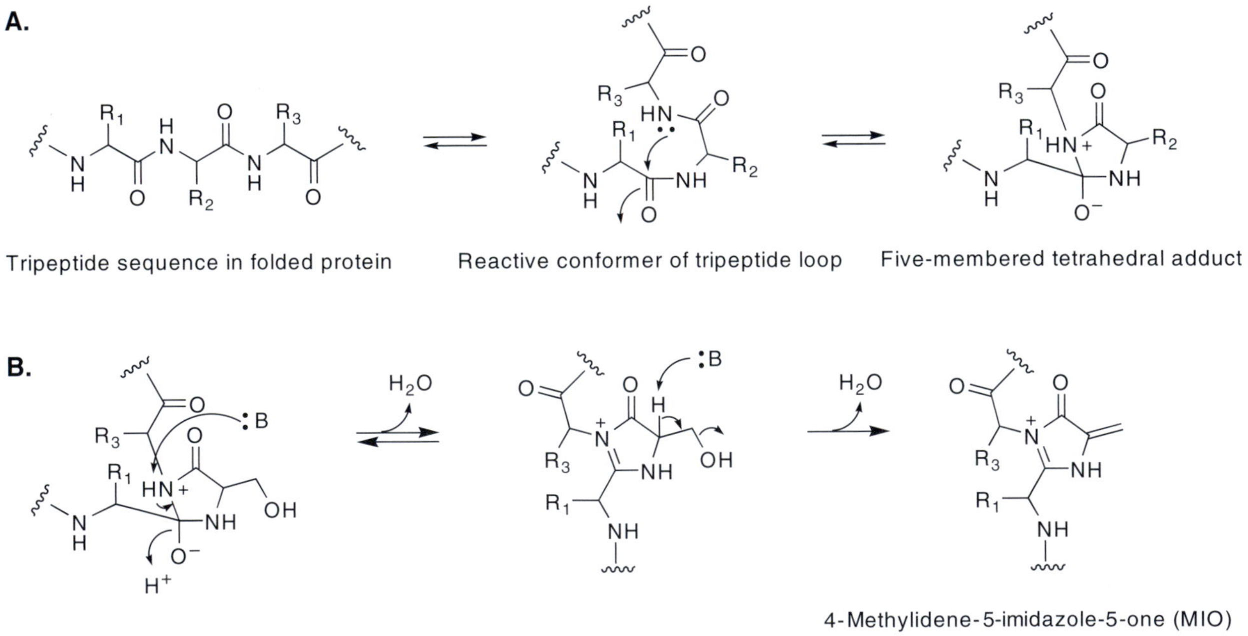

A.

Tripeptide sequence in folded protein   Reactive conformer of tripeptide loop   Five-membered tetrahedral adduct

B.

4-Methylidene-5-imidazole-5-one (MIO)

Figure 13.13 **A.** Automodification of three residues in tripeptide-containing loop to form cyclic five-membered rings as initial tetrahedral adducts in proteins. **B.** Autocatalytic formation of 4-methylidene-5-imidazole-5-one (MIO) from AlaSerGly via the tetrahedral adduct in the precursor form of histidine deaminase.

C.

Fluorophore of green
fluorescent protein (GFP)

Figure 13.13 **c.** Further conversion of the cyclized adduct in GFP to the flurophore by O_2-mediated oxidation to extend the conjugation of the chromophore.

chain of the middle amino acid residue in the tripeptide sequence undergoing automodification. The prosthetic group 4-methylidene-5-imidaziole-5-one (MIO) forms when serine is the middle residue (Figure 13.13B), while the extended chromophore in the fluorescent proteins arises by low-energy autoxidation of the benzylic CH_2 center in a tyrosyl side chain. It is still poorly understood how the particular tripeptidyl sequences only in these proteins, during the late stages of folding into native structure, are subject to autocyclization. The conformational strain of the native structure, anchored at each end of the loop, and acid–base catalysis by a well-localized water molecule have been suggested to be crucial determinants (Baedeker and Schulz, 2002a). The oxidative second step in formation of the GFP and DsRed structures can be very slow, with half-lives of hours.

Autocyclization of a Peptidyl Triad in Histidine Ammonia Lyase and Phenylalanine Ammonia Lyase to Create the Active Form of the Catalyst

From enzymatic mechanistic studies over a 30-year period from the 1960s into the 1990s it was known that the deamination of the two common amino acids, histidine and phenylalanine, to urocanate and cinnamate, respectively (Figure 13.14), involved some unusual cofactor chemistry. The net elimination of the elements of ammonia across Cα and Cβ of Phe and of His require both the cleavage of the Cα–NH_2 and Cβ–H. It seemed unlikely that the NH_3^+ substituent could be cleaved off first, yet the Cβ–H would be an unactivated hydrogen and pose a high-energy barrier for C–H cleavage. A variety of studies had indicated that an electron sink was present in both histidine ammonia lyase (HAL) and phenylalanine ammonia lyase (PAL) and it

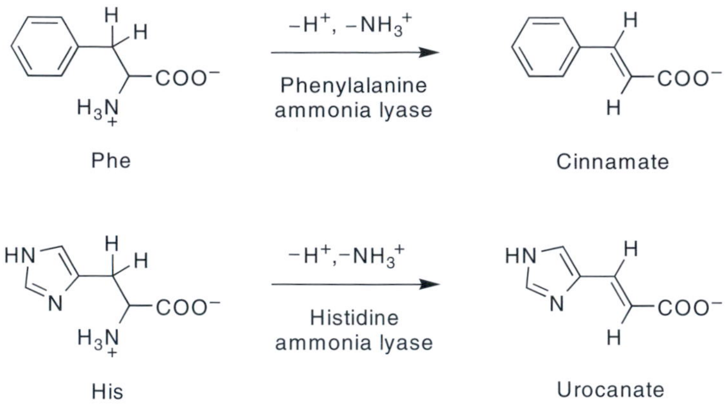

Figure 13.14 Reaction stoichiometries for histidine to urocanate and phenylalanine to cinnamate and free ammonia, catalyzed by His and Phe ammonia lyases (HAL and PAL, respectively).

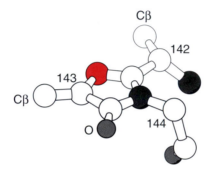

Figure 13.15 X-ray analysis of the MIO prosthetic group covalently attached to the active site of HAL (figure made using PDB 1GKJ).

was widely assumed that this was a dehydroalanyl residue, arising from posttranslational dehydration of a serine residue, Ser$_{143}$ in the *Pseudomonas putida* HAL (Langer et al., 1994). The dehydroalanyl moiety was imputed the role of an electron sink to polarize the substrate and activate it for the α,β-elimination chemistry.

When the x-ray structure of the *P. putida* HAL was determined in 1999 (Schwede et al., 1999), the electron density around residue 143 could not be fit to a dehydroalanyl residue. Instead, a modified cyclized form of an electron sink, the MIO moiety (Figure 13.15), was deduced from the x-ray structure, along with a proposal for MIO formation. Most remarkably, it is thought that prosthetic group formation is initiated by the poorly nucleophilic peptide backbone NH of Gly$_{144}$ as the attacking nitrogen atom on the peptide carbonyl of Ala$_{142}$. A well-ordered water molecule, water 197 in the x-ray structure, is positioned to help deprotonate the amide N–H. The resultant cyclic tetrahedral adduct can then lose water, creating the C=N linkage in the newly formed imidazolone ring. The Ser$_{143}$ Cα–H is now between two carbonyl groups and can yield a stabilized Cα carbanion on proton abstraction, from which loss of the β-OH converts the Ser_{143} side chain into the α,β-methylene form of the prosthetic group, the 4-methylidene-5-imidazole-5-one (MIO) (Figure 13.13B). Thus, the change in backbone peptide connectivity comes from the intramolecular cyclization with a peptide NH as the unlikely nucleophile, followed by two tandem dehydrations to trap out the initial cyclic adduct. Noted below are comparable first cyclization steps in fluorophore formation in GFP and DsRed automodifications.

The autocatalytic formation of the MIO has converted the AlaSerGly tripeptide loop in the inactive proenzyme into a compact electron deficient heterocycle in the active enzyme. The MIO in the active site of Phe ammonia lyase or His ammonia lyase serves as the electrophilic cofactor for reversibly accepting electrons from His or Phe during deamination. For example, His deamination has been proposed to occur by attack of the π electrons of the double bond of the imidazole ring of bound substrate on the terminal carbon of the methylene moiety of the MIO group (Baedeker and Schulz, 2002b) to create a covalent tether between substrate and the now transiently aromatic prosthetic group. More recently, Calabrese et al. (2004) have proposed an E1cb mechanism with the eneone of the MIO moiety attacked by the bound substrate amino group as shown in Figure 13.16. The resulting enolate should be an energetically accessible intermediate. An active site His residue acting as base for the removal of the C_3 pro-(S) hydrogen would generate the urocanate product and the amino group tethered to the MIO enolate, which can readily expel NH_3 to regenerate

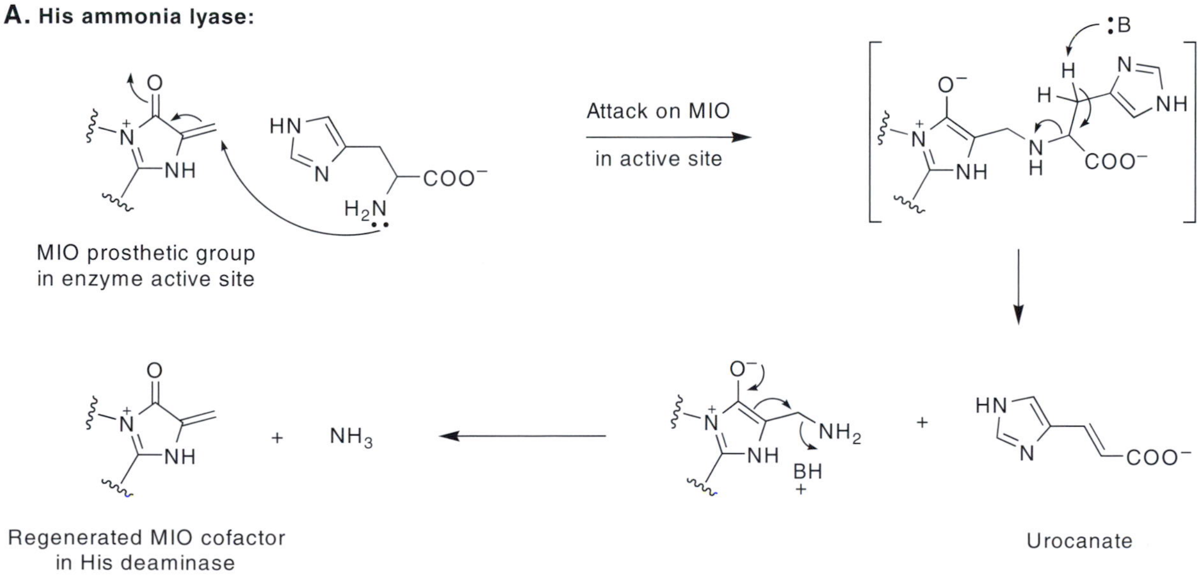

A. His ammonia lyase:

Figure 13.16 Role of MIO as an electron sink in catalysis. **A.** Histidine ammonia lyase.

B. Phe ammonia lyase:

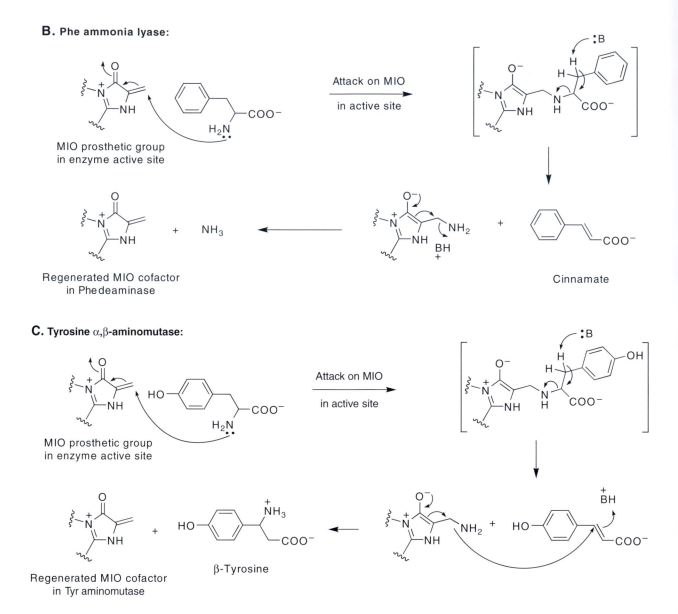

Figure 13.16 **B.** Phenylalanine ammonia lyase. **C.** Tyrosine aminomutase. All three involve an amino-MIO adduct as intermediate.

the starting MIO for the next catalytic cycle (Figure 13.16A). A related mechanism has been proposed for the phenylalanine deamination to cinnamate (Figure 13.16B).

The MIO group does not have an easily identifiable chromophore, so it cannot be readily discerned except by x-ray crystallography or mass spectrometry of a peptide fragment of the active site. It is unlikely that the local loop architecture of AlaSerGly tripeptide sequences in other proteins will cyclize often. However, a recent report on β-tyrosine formation in the biosynthetic cluster for the enediyne antitumor antibiotic

C1027 (Christenson et al., 2003) indicates that the cluster contains a gene for a novel tyrosine aminomutase. This enzyme autocatalytically generates the MIO prosthetic group, which then serves as an electron sink for the chemistry variant from the His and Phe ammonia lyases: the internal transfer of the amino group from Cα of the substrate to Cβ of the amino acid product. A scheme in which the itinerant amino group is stored reversibly in covalent attachment to the MIO cofactor and p-hydroxycinnamate is formed transiently before being recaptured by ammonia attack at the β-carbon, suggests an expanded role for the MIO electrophilic prosthetic group (Figure 13.16C).

Autocyclization of a Peptidyl Triad to Create the Fluorophore in Green Fluorescent and Red Fluorescent Chromoproteins

The remarkable mechanism of intramolecular attack of a peptide nitrogen on the peptide carbonyl group two residues upstream to form a cyclic five-membered tetrahedral adduct is reprised in the autocatalytic maturation of green fluorescent protein (emission maximum at 506 nm) from the jellyfish *Aquorea victoria* (Tsien, 1998). In this instance the tripeptide is $Ser_{65}Tyr_{66}Gly_{67}$. The residue in the middle is now a tyrosine rather than a serine (Figure 13.17), creating 4-(p-hydroxybenzylidene) imidazole-5-one. The attacking nucleophile is again the amide nitrogen of a glycine resi-

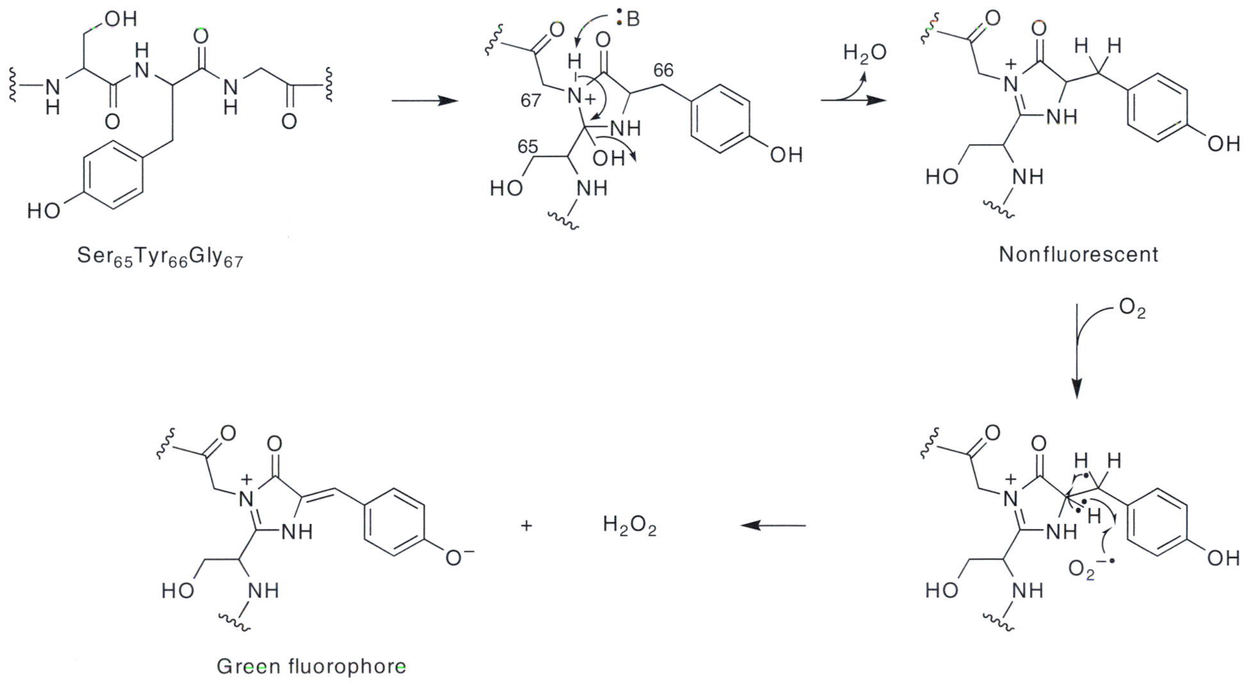

Figure 13.17 Detailed mechanism for the autoconversion of $Ser_{65}Tyr_{66}Gly_{67}$ in the GFP precursor to the green fluorophore in mature GFP.

due. The methylene hydrogens on Cα of glycyl residues provide the smallest possible steric hindrance to any rotation that the peptide NH has to undergo to initiate cyclization. The initial cyclic adduct has been proposed to lose water to accumulate in the forward direction as the imidazolone. As in the MIO case this adduct is neither a significant visible chromophore nor fluorescent. The Tyr cannot readily lose water the way the Ser side chain does in MIO formation to increase the conjugation in the prosthetic group. Instead, the benzylic hydrogens at Cβ of the Tyr$_{66}$ side chain can undergo one-electron oxidation by O$_2$. The resultant radical is delocalized into the phenolic system. A second electron transfer reduces the superoxide to peroxide and creates the sp^2 methine oxidation state of the Tyr side chain, extending the conjugation, creating the green color and the fluorophore. This oxidative conversion constitutes the irreversible trapping of the rearranged backbone. X-ray analysis (Figures 13.18A and B) has indicated that Arg$_{96}$ is placed to enhance the nucleophilicity of the Gly$_{67}$ nitrogen in the cyclization reaction and that Glu$_{222}$ carboxylate can stabilize the tetrahedral adduct involved in the dehydration step (Barondeau et al., 2003).

The autocatalytic cyclization/dehydration/oxidation sequence of GFP maturation requires the 238-residue protein to have folded into its native conformation, a β barrel structure termed a β can (Figure 13.18C) due to its cylindrical architecture formed by β sheet secondary structural elements. The cylinder is formed by eleven strands in a β barrel threaded in the interior by a helix containing the tripeptide from which the chromophore self-assembles. The conjugated cyclic prosthetic group forms within the can and the microenvironment is a constraint both for the fluorophore emission maximum as well as for the quantum yield (~0.7). The slow step in GFP folding and fluorophore formation appears to be the oxidative step (the last steps in Figure 13.17), with a half-life of 20–80 minutes (Tsien, 1998). An argument has been made from x-ray studies of the Y66L mutant of GFP that, in this mutant and by extension in the wild type GFP, the dehydration/oxidation steps in the fluorophore are reversed from the above mechanism. In this alternative for fluorophore autocatalysis, the initial cyclic carbinolamine is first oxidized to create the planar C=N double bond and loss of water is the last step (Rosenow et al., 2004).

The physiological role of this pigmented protein in the jellyfish has been debated, with speculation that it could be involved in light harvesting for energy or in photoprotection. The molecular biology and biotechnology interest in GFP has arisen because of its use as a sensor and as a marker of protein location when expressed in cells, especially as a fusion protein tag to monitor the localization of the fusion partners (Tsien, 1998). Because the fluorophore forms autocatalytically after the β can structure has formed, the GFP moiety is a robust general protein fluorescent signal. To allow for the simultaneous monitoring of two or more GFP-tagged proteins, such as for fluorescence resonance energy transfer (FRET)-based distance measurement, there has been

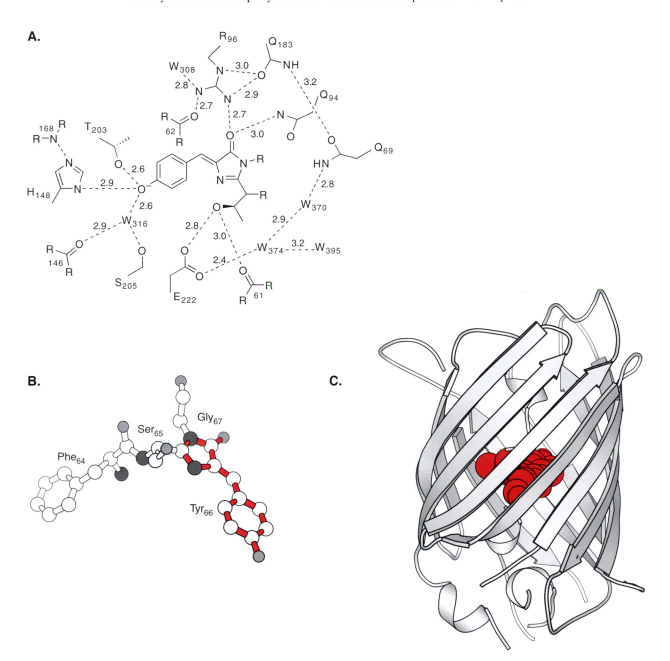

Figure 13.18 Structure of the flurophore in GFP. **A.** Active site microenvironment [redrawn from Tsien (1998)]. **B.** Architecture of the Phe$_{64}$Gly$_{67}$ modified tetrapeptide element in mature GFP (figure made using PDB 1GFL). **C.** Orientation of the fluorophore (shown in red) to the β can structure of GFP (figure made using PDB 1EMA).

much interest in engineering GFP to emit at different wavelengths. This in turn has led to mutagenesis programs to produce enhanced versions with yellow (YFP), cyan (CFP), and other engineered enhancements (Figure 13.19A). These all have the same triad that undergoes automodification to the identical chromophore. It is the protein

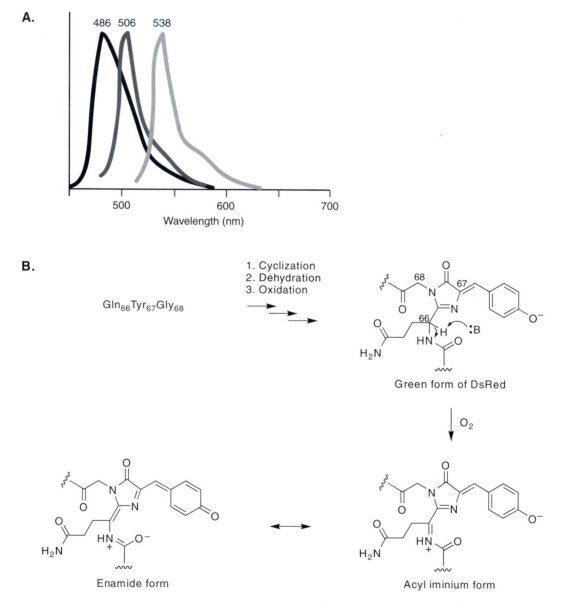

Figure 13.19 Emission spectra of fluorescent protein variants. **A.** Cyan fluorescent protein (CFP) (486 nm), GFP (506 nm), and yellow fluorescent protein (YFP) (538 nm) [adapted from Verkhusha and Lukyanov (2004)]. **B.** Generation of the red fluorescent form of DsRed by way of the green intermediate.

microenvironment surrounding the fluorophore that is being tuned, for instance by the interaction with charged side chains of Arg_{96} and Glu_{222}.

A variety of GFP-like fluorescent proteins have subsequently been characterized from reef anthozoa, bringing the total to over two dozen (Labas et al., 2002; Matz et al., 1999; Shagin et al., 2004). Four natural color variants have been detected: green, yellow, orange-red, and purple-blue, all with related tripeptide-derived fluorophores embedded within comparable β can protein architectures. Some variants replace the aromatic side chain of Tyr_{66} with Trp, His, or Phe. In *A. victoria* GFP the phenol of Tyr_{66} can be dissociated to the phenolate anion to increase the conjugation and electron density available in the system. The phenolate anion forms are yellow, the indole versions of Trp_{66} are cyan, and the imidazole in His_{66} generates blue fluorescent proteins (Tsien, 1998). The reddish-yellow coral fluorescent proteins can have the Ser_{65} replaced by Asn, Gln, or Lys (Matz et al., 1999). Genetic engineering with an expanded genetic code has allowed the replacement of Trp_{66} in ECFP with 4-amino-Trp in the backbone of the enhanced cyan fluorescent protein isoform to create a gold (GD) variant (GDFP) (Hyun Bae et al., 2003).

In the most well-characterized red fluorescent protein, from the coral *Discosoma sp.*, termed the DsRed protein, the tripeptide $Gln_{66}Tyr_{67}Gly_{68}$ undergoes comparable autocyclization and accumulates first as a green fluorescent precursor. It slowly transforms (over days at room temperature) by a second oxidative process to a red version with a more extended chromophore. The postulate that the red fluorophore was an oxidized acyl imine version of the Gln_{66} residue (Figure 13.19B) was validated by x-ray determination, which established a planar geometry at the Gln_{66} $C\alpha$, consistent with the sp^2 hybridization of an imine linkage (Yarbrough et al., 2001). The replacement of Ser in the first position by Gln likely leads to a stable acyl imine; the Ser side chain might instead undergo elimination to a dehydroalanine form. The conversion of the green to red chromophore can be formulated as a second oxygen-mediated series of one-electron oxidations to cleave the $C\alpha$–H bond and oxidize the C–N single bond to the C=N double bond. The x-ray structure of the DsRed tetramer has a mixture of 50% red and 50% green prosthetic groups.

Again, mutagenesis, directed and random, has been performed on DsRed, generating "timer" variants that accelerate the green-to-red maturation and can be monitored *in vivo*, as well as some rapid versions where the half-life for red fluorophore formation is accelerated 10- to 15-fold, generating maturation half-lives of under an hour (Bevis and Glick, 2002; Terskikh et al., 2000). DsRed is tetrameric, in contrast to monomeric GFPs, and this oligomerization state limits DsRed fusions to partner proteins to follow the localization of such fusion proteins in cells. Therefore, Campbell et al. (2002) engineered in mutations that disrupted the subunit contacts to create a monomeric version of the DsRed fluorescent protein.

An intriguing variation of tripeptide autocyclization occurs in the β can protein from *Actinaria sulcata*, leading to a nonfluorescent red-purple chromoprotein (λ_{max} = 595 nm) (Martynov et al., 2001). The chromophore structure (Figure 13.20) indicates cyclization has occurred to form a six-membered ring rather than a five-membered

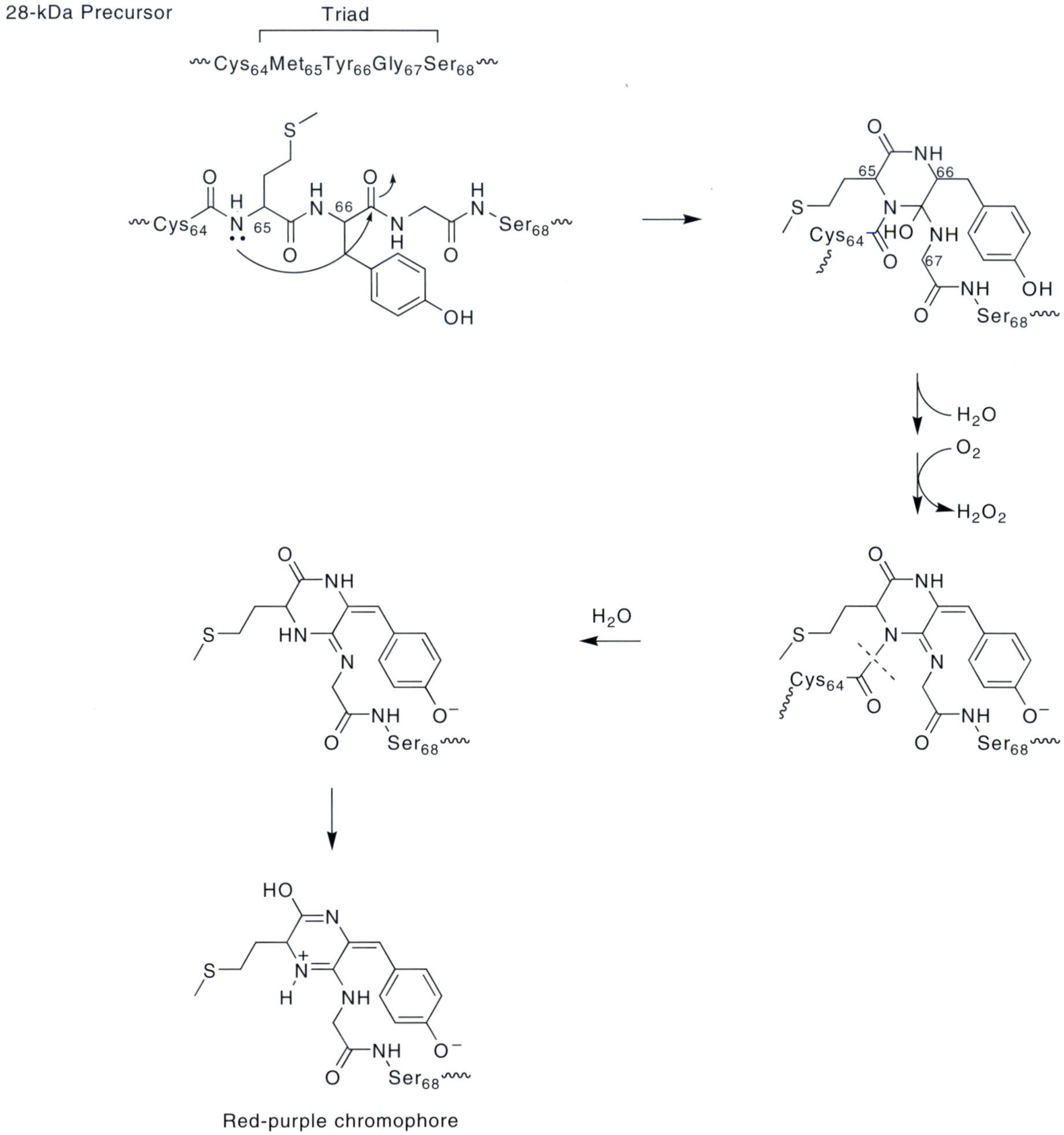

Figure 13.20 The red-purple nonfluorescent chromophore in the *Actinaria sulcata* chromoprotein: cyclization of the tripeptide loop precursor to a six-ring structure rather than a five-ring structure and reversal of polarity in ring construction. The amide of residue 65 is the attacking nucleophile.

ring and has been attended by the cleavage of one of the peptide bonds anchoring the tripeptide loop. One could formulate this autocatalytic cyclization as occurring in the opposite direction, not with the Gly_{67} peptide NH as the attacking nucleophile on the residue 65 carbonyl, but rather with the amide of residue 65 attacking the residue 66 carbonyl (Figure 13.20). At some point in the maturation process the now strained amide 64–65 linkage is hydrolyzed to create the indicated chromophore as the 28-kDa protein autocleaves to an 8-kDa and a 20-kDa chromophoric fragment. This presages the internal peptide bond cleavages seen in the maturation of Hedgehog protein and in protein splicing, where initial tetrahedral adducts are trapped not by some dehydration or desaturation second step but by fragmentation of the C–N bond and the formal disconnection of the peptide backbone.

A Different Route to a Posttranslationally Modified Photoactive Yellow Protein

A different route to photoactive yellow proteins that act as photoreceptors in light-driven proton pumping cycles is found in the purple halophilic bacterium *Ectothiorho-dospira halophila* (Meyer, 1985). The photoactive yellow protein (PYP) from *E. halophila* mediates negative blue light phototropism. The photosensor is a posttranslationally introduced 4-hydroxycinnamoyl chromophore attached in thioester linkage to the side chain of a cysteinyl residue of the 14-kDa PYP (Figure 13.21). The 4-hydroxycinnamoyl (*p*-coumaryl) group is probably derived from tyrosine deamination by an MIO-type deaminase noted above, and then is activated as the CoA thioester and captured in a transthiolation reaction by the Cys of the apo PYP (Borgstahl et al., 1995). It is possible but unlikely that this chromophore tethering is autocatalytic. The photocycle of PYP has been studied and bears analogy to the more celebrated photocycle of bacteriorhodopsin, also a chromophoric protein in halo-

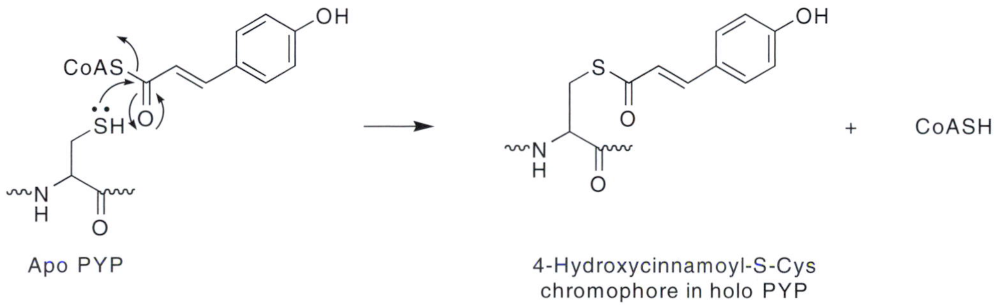

Apo PYP   4-Hydroxycinnamoyl-S-Cys chromophore in holo PYP   +   CoASH

Figure 13.21 A different type of covalent chromophore in photoactive yellow protein: posttranslational introduction of a 4-hydroxycinnamoyl acyl moiety derived from deamination of tyrosine in thioester linkage to the active site cysteine.

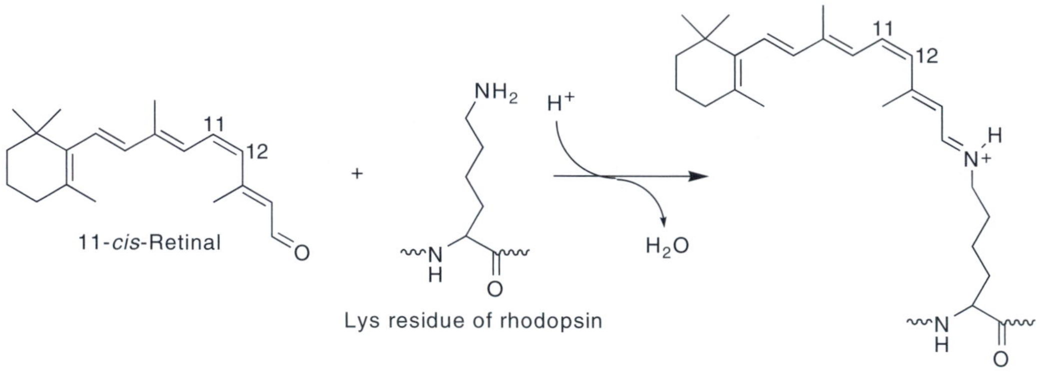

Figure 13.22 A different strategy for covalently tethering a chromophore to a protein: covalent imine linkage between a Lys side chain and an aldehyde for covalently tethering the retinal chromophore in the purple proteins bacteriorhodopsin and rhodospin.

philic purple bacteria. Neither the physics nor the biology of bacteriorhodopsin and eukaryotic rhodopsin are discussed here, other than to note that the chromophore is the polyene retinal, in imine linkage to the ϵ-NH_2 group of a lysine side chain (Figure 13.22), yet another posttranslational variation to amplify the inventory of chromophoric proteins. The imine linkage, in principle, is subject to hydrolysis and loss of the conjugated aldehyde, but in practice this occurs so infrequently that the imine linkage has sufficient lifetime to serve as a covalently tethered adduct.

Autocleavage Reactions of Peptide Bonds in Proteins

While proteolysis happens to all proteins inside and outside cells at the end of their individual life cycles, there are also many proteins converted from inactive precursor forms to active fragments by specific cleavages at one or more specific peptide bonds in the protein chain. In turn, most of these are effected *in trans* by proteases, such as the signal peptidases and proprotein convertases described in Chapter 8. A small fraction of proteins that undergo regiospecific peptide bond cleavage is autocatalytic, cleaving the peptide linkage by *in cis* mechanisms. These examples of intramolecular posttranslational modifications can arise in the late stages of protein folding or after the folded state has been achieved. Since peptide bonds in proteins are typically stable to hydrolysis under physiological conditions for many years, the mechanism of *regiospecific* autocleavage is of interest.

Four outcomes from the autocleavage of proteins are shown in Figure 13.23 (Paulus, 2000). First (Figure 13.23A) is the net hydrolysis to two peptide fragments, as happens in the production of the α and β subunits on activation of the catalytic form

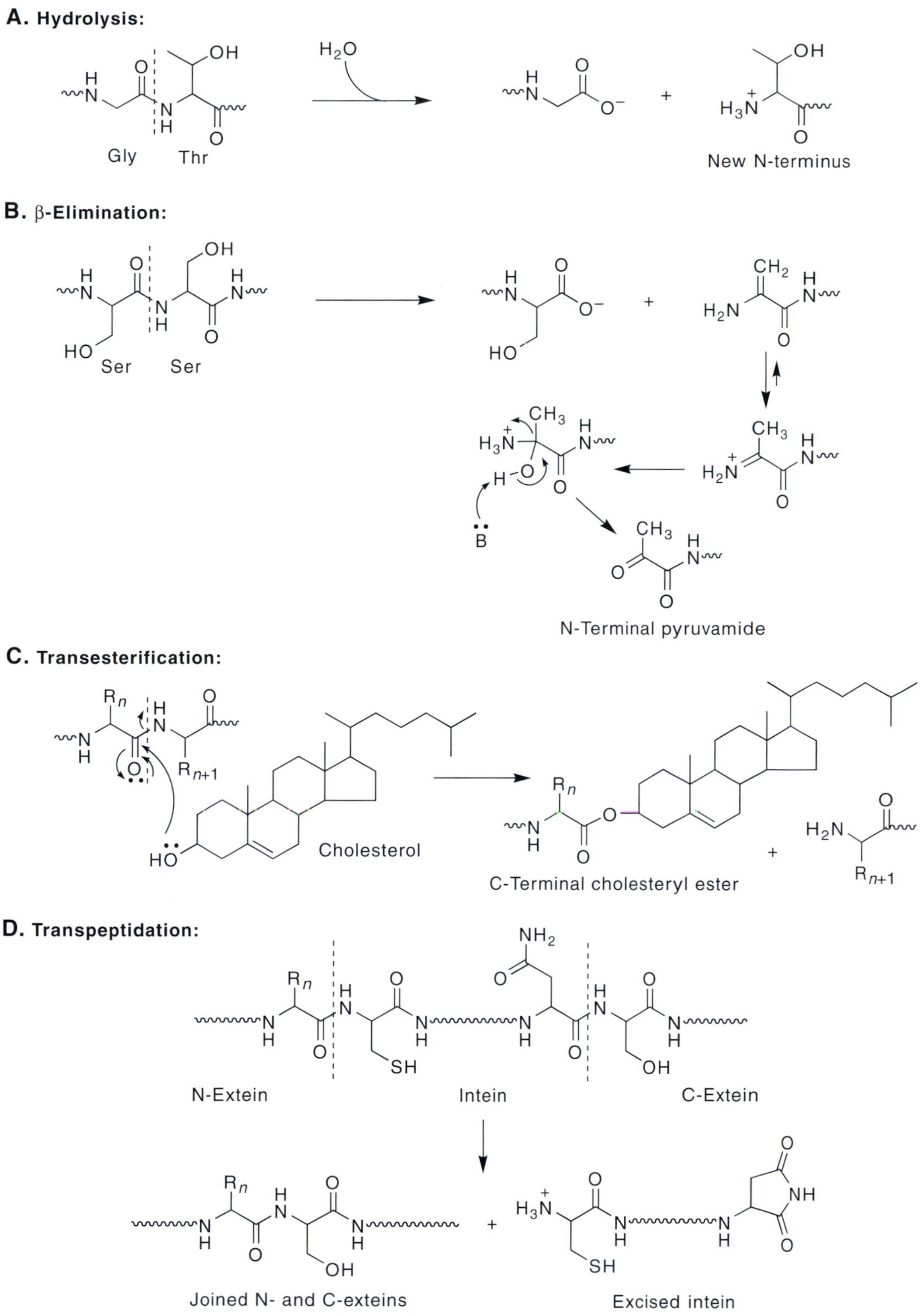

A. Hydrolysis:

Gly   Thr

New N-terminus

B. β-Elimination:

Ser   Ser

N-Terminal pyruvamide

C. Transesterification:

Cholesterol

C-Terminal cholesteryl ester

D. Transpeptidation:

N-Extein   Intein   C-Extein

Joined N- and C-exteins   Excised intein

Figure 13.23 Four outcomes from autocatalytic peptide bond cleavages in proteins.

of the enzyme GlcNAc-asparaginase (glycanase N), which is involved in the degradation of glycoproteins as noted in Chapters 8–10. A comparable cleavage at a GlyThr peptide bond is involved in autoactivation of the β subunits of the proteasome, releasing the Thr as the N-terminal nucleophile in proteasome catalytic cycles (Chapter 8). Second (Figure 13.23B) is the net cleavage at an internal SerSer or GlySer peptide bond in histidine or aspartate decarboxylases, respectively, to liberate the N-terminal Ser fragment that undergoes β-elimination and produces a pyruvoyl moiety that mediates substrate decarboxylation. A third variant (Figure 13.23C) is in the signaling proteins of the Hedgehog family where the C-terminal half of the protein carries out

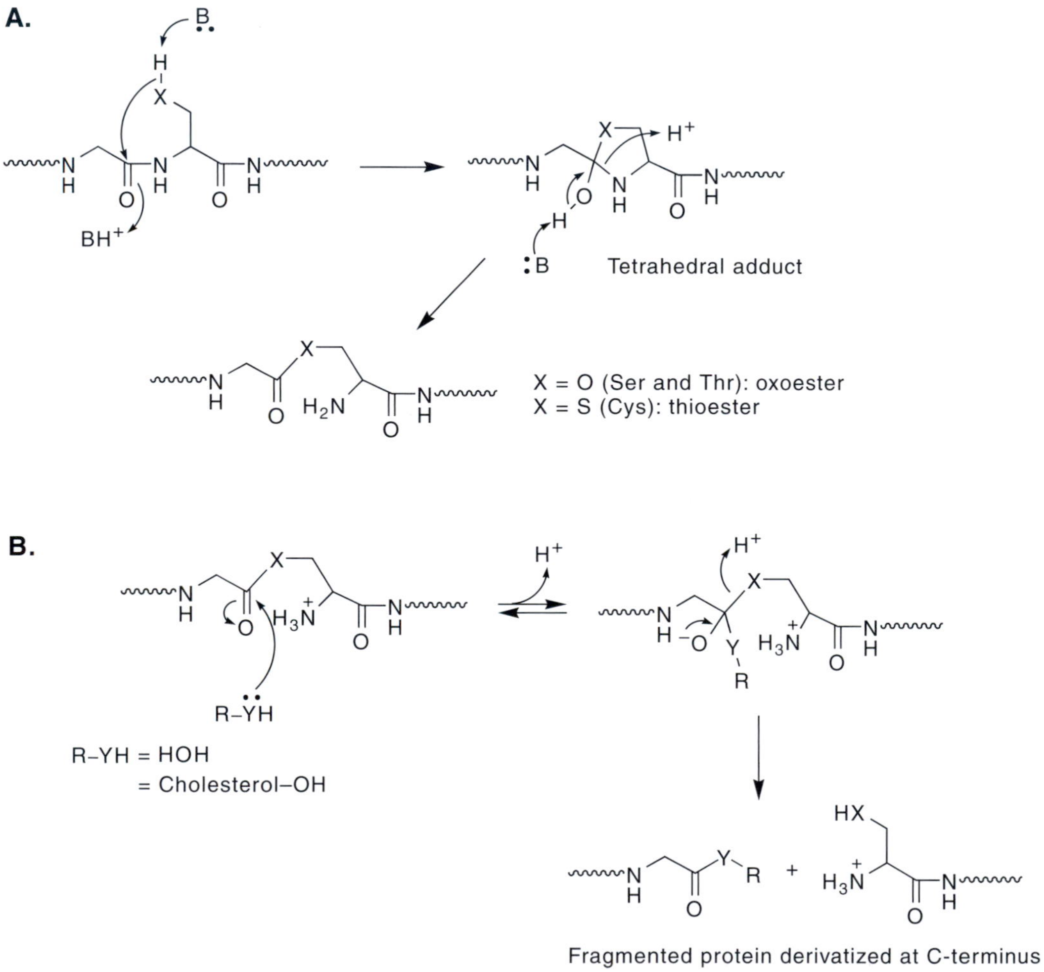

Figure 13.24 Autocleavage of peptide bonds in precursor forms of proteins. **A.** Intramolecular attack by a Ser or Cys side chain on the upstream amide carbonyl yields cyclic tetrahedral aminals or thiohemiaminals, followed by C–N bond cleavages to oxo- or thioesters. **B.** Transfer of the activated acyl moieties to external nucleophiles leads to net peptide bond cleavage and fragmentation of the single-chain protein precursor.

regiospecific fragmentation at a particular peptide bond, but transfers the N-terminal peptidyl fragment, not to H_2O for hydrolysis, but to the 3-OH of cholesterol to produce a protein-cholesterol ester localized to the cell membrane. The fourth outcome (Figure 13.23D) arises from net intramolecular transpeptidation, in which the transferring N-peptidyl fragment is transferred to an intraprotein nucleophile downstream (to a downstream serine side chain in Figure 13.23). This is the process of protein splicing, excising an intervening sequence (intein) and joining the adjacent upstream and downstream extein sequences back in frame.

All of these peptide bond autocleavages require that the protein carrying out the automodification induce a specific local conformation of the peptide region to be cleaved. With the loop in a high-energy state, it is set up for capture by the alcohol or thiolate side chain of the immediate downstream Ser, Thr, or Cys residue to produce a five-ring tetrahedral adduct at the peptide carbonyl under attack (Figure 13.24A). These adducts are resolved in the front direction by cleavage of the C–N single bond to produce an oxoester (Ser or Thr) or thioester (Cys) intermediate. This is the common intermediate in all four pathways. The *peptide bond has been cleaved.* The acyl moiety can then be transferred to water, transferred to the 3-OH group of cholesterol, or captured by a side chain Cys or Ser downstream in the protein on the way to protein splicing (Figure 13.24B).

Autocleavage of a Peptide Bond to Release the Catalytic N-Terminal Nucleophile

The N-acetylglucosaminyl asparaginase involved in cleaving the proximal sugar in N-glycoprotein degradations (Figure 13.25) (see Chapter 10 on N-glycoprotein biosynthesis and breakdown) is produced first in a single-chain, inactive proenzyme form

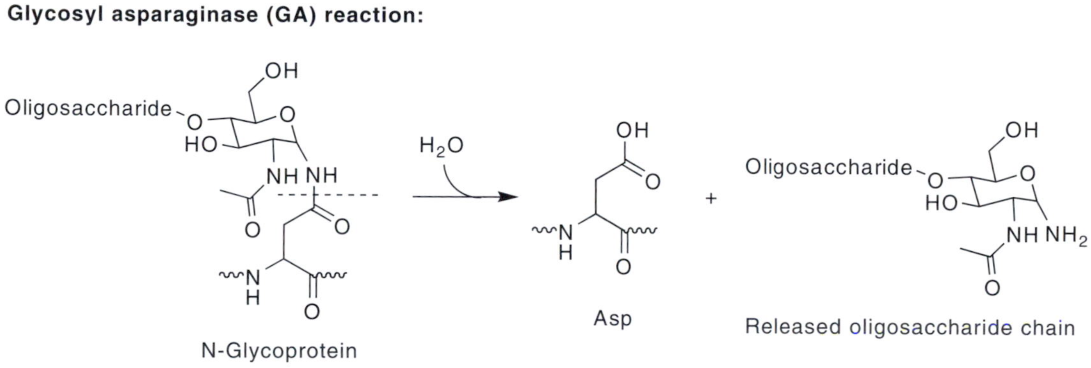

Glycosyl asparaginase (GA) reaction:

Figure 13.25 N-Glycosyl asparaginase (glycanase N) cleaves N-glycan chains at the GlcNAc-Asn bond to liberate the deglycosylated protein with an Asp residue and the free oligosaccharide chain.

(Figure 13.26A) (Tarentino et al., 1995). The enzyme autoactivates by self-cleavage at the $Asp_{151}Thr_{152}$ peptide bond to produce the two-chain αβ heterodimer with the Thr residue now N-terminal on the β subunit. This Thr_1 of the β subunit is the catalytic nucleophile in hydrolysis of the GlcNAc-Asn amide bond in all N-glycosylated

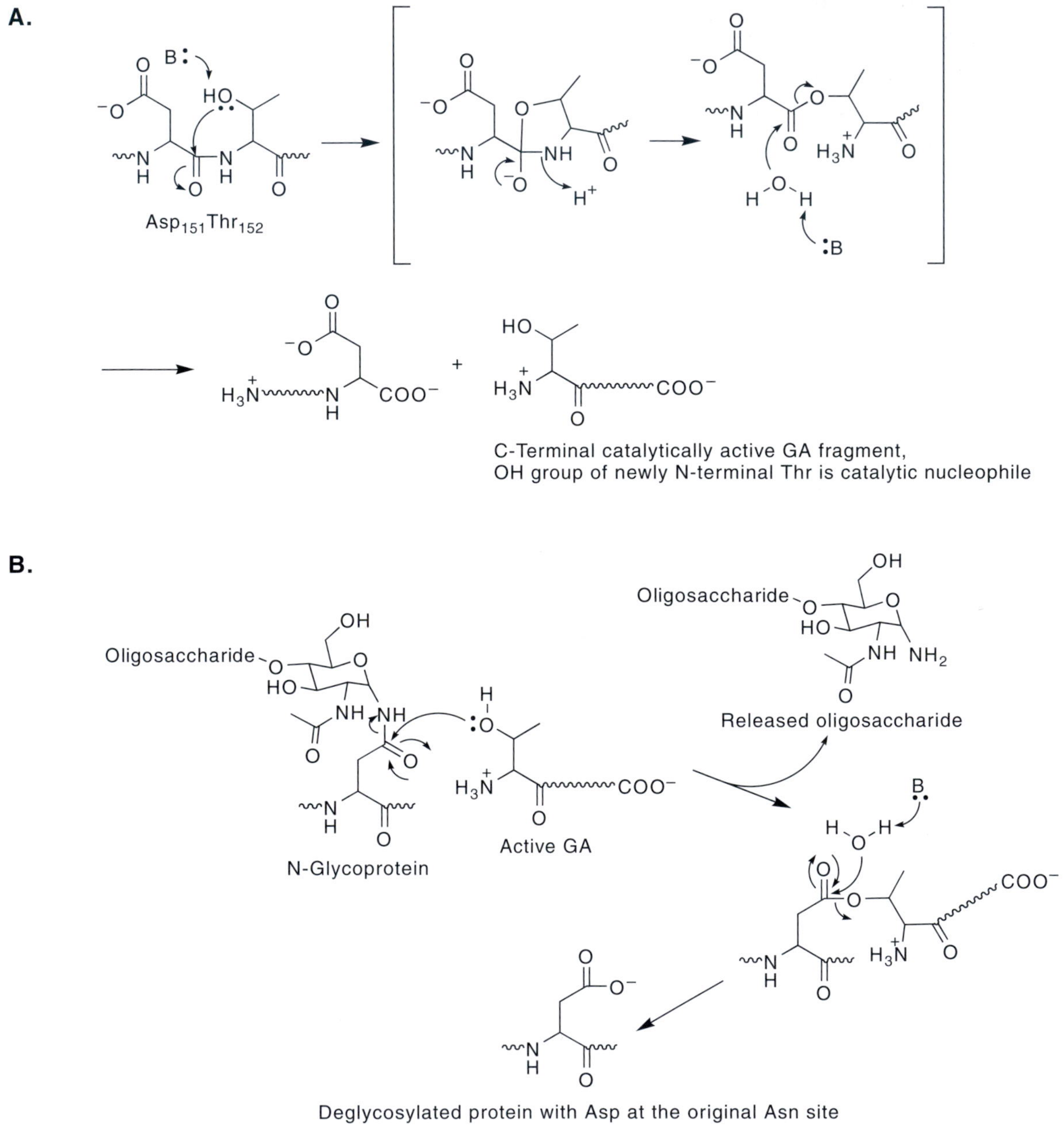
A.
$Asp_{151}Thr_{152}$
C-Terminal catalytically active GA fragment,
OH group of newly N-terminal Thr is catalytic nucleophile
B.
Oligosaccharide
N-Glycoprotein
Active GA
Released oligosaccharide
Deglycosylated protein with Asp at the original Asn site

Figure 13.26 N-Glycosyl asparaginase maturation. **A.** Autocleavage of the proenzyme form at $Asp_{151}Thr_{152}$ to generate the β subunit with Thr_1 as the active site nucleophile. **B.** Catalysis proceeds via a protein substrate acyl-O-Thr_1 covalent enzyme intermediate.

proteins. The enzyme uses the alcohol side chain of β-Thr$_1$ to add in to the GlcNAc-Asn amide bond, producing a tetrahedral adduct and then a covalent GlcNAc-O-Thr$_1$-acyl enzyme intermediate (Figure 13.26B). This is the typical covalent catalysis mechanism for the serine/threonine active site family of proteases (see Chapter 8). The oligosaccharide fragment is released as a 1-amino-GlcNAc and the acyl enzyme is then hydrolyzed in the second half-reaction, releasing the deglycosylated protein as an Asp side chain at the site of the original Asn and regenerating the Thr$_1$ in the resting form of the $\alpha\beta$ enzyme.

The x-ray structure of a Thr$_{152}$Cys mutant form of GlcNAc-asparaginase, unable to catalyze its autocleavage, has revealed that this region of the proenzyme is in a tight

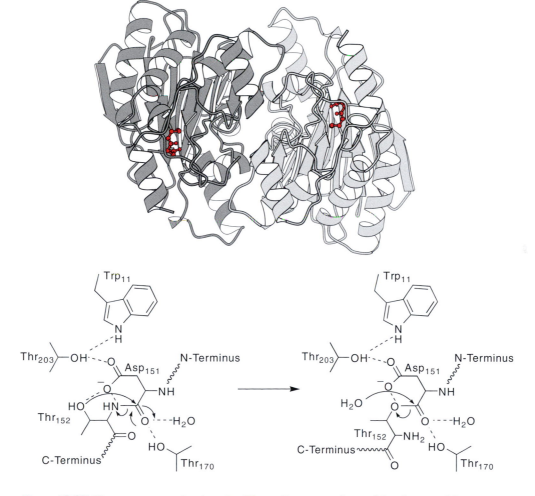

Figure 13.27 X-ray structure of an inactive Thr$_{152}$Cys mutant form of the glycanase N proenzyme dimer (figure made using PDB 9GAC) reveals a nonplanar conformer of the Asp$_{151}$Cys$_{152}$ peptide bond (shown in red ball and stick in both subunits) that would normally be cleaved. A schematic for the intramolecular attack of the OH of Thr$_{152}$ from that conformation to generate the intermediate in proenzyme cleavage is shown.

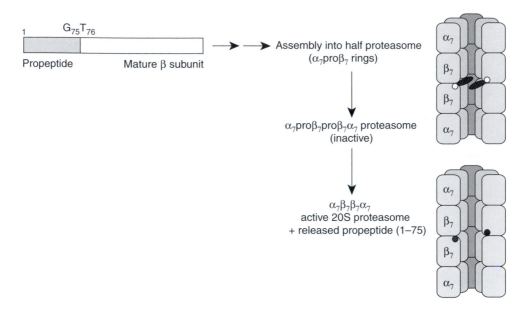

Figure 13.28 Autoproteolysis of the proenzyme single-chain form of the β subunit of the protea-some to liberate the N-terminal threonine nucleophile [adapted from Chen and Hochstrasser (1996)].

turn, imposed by the folded native structure of the proenzyme, and the $Asp_{151}Thr_{152}$ peptide bond deviates about 20° from planarity (Figure 13.27) (Xu et al., 1999). This nonplanar conformation decreases resonance overlap in the peptide bond and destabilizes it for attack by the side chain –OH of Thr_{152}, lowering the energy barrier for the formation of the oxyoxazolidinone tetrahedral adduct and favoring its pro-bability of accumulating to a reasonable fraction of the proenzyme population.

This type of autoactivation, self-cleavage to release a particular internal Thr, Ser, or Cys residue as the N-terminal residue of the fragmented protein, is seen in several other cases (Paulus, 2000), perhaps most notably in the formation of the active form of the proteasomes (Seemuller et al., 1996). In that case, once the double heptad ring of the proteasomal core has folded, this inactive precursor autoproteolyzes at the β subunits (Figure 13.28). The N-terminal Thr_1 is then capable of carrying out the pro-teolytic cleavage of protein subunits that have been threaded into the proteasome chamber by an analogous acyl enzyme mechanism as pictured above (see also Chapter 8). The nucleophilicity of the side chain β-OH of Thr_1 is thought to be enhanced by the free NH_2 group, perhaps acting as a general base to initiate the acyl enzyme for-mation. The autocleavage chemistry to activate enzymes with N-terminal nucle-ophiles may involve both the rearrangement of groups in the active site on proteolysis and the electronic and acid–base features of the newly generated NH_2 group in the N-terminal nucleophile.

Autocleavage of a Peptide Bond, β-Elimination, and Generation of a Pyruvoyl Prosthetic Group for Amino Acid Decarboxylation

In the three bacterial enzymes histidine decarboxylase, phosphatidyl serine decarboxylase, and aspartate decarboxylase, and in the mammalian enzyme S-adenosylmethionine decarboxylase (Figure 13.29), the anticipated coenzyme pyridoxal phosphate was not present in the active enzymes as isolated nor was it required for catalysis (van Poelje, 1990). Instead, an alternate electron sink, covalently tethered as a prosthetic group produced posttranslationally, was in use in the active sites. The active forms of

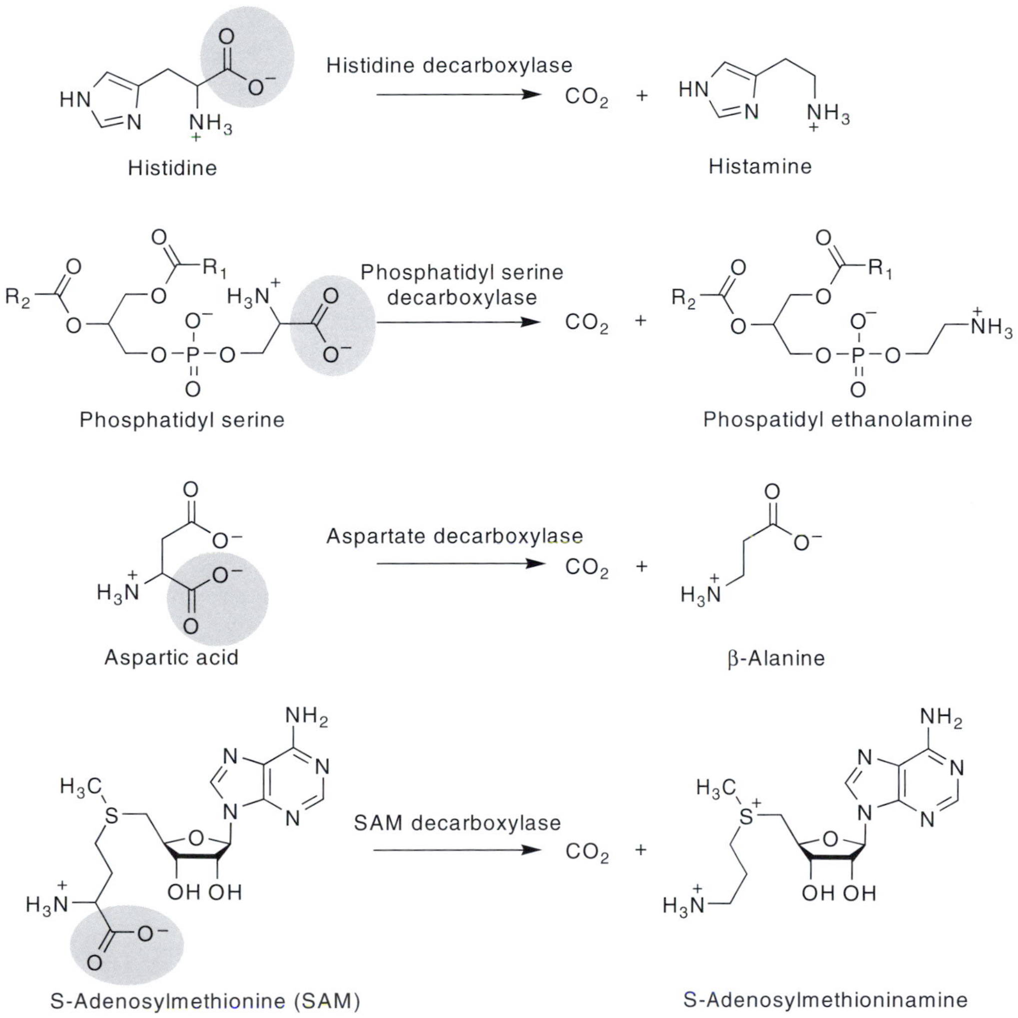

Figure 13.29 Reactions catalyzed by four amino acid decarboxylases that contain a covalently bound pyruvamide group as the electron sink in their active sites.

these enzymes are heterodimers. They are produced as inactive single-chain proenzymes that autocleave to produce active $(\alpha\beta)_n$ enzyme forms. The pyruvoyl group is on the N-terminus of one of the subunits and arises biosynthetically from serine (Figure 13.30). This pattern is fully reminiscent of the autoprocessing of the N-terminal nucleophile enzyme class discussed previously.

The histidine decarboxylase from *Lactobacillus* strain 30a autocleaves at the Ser_{81}-Ser_{82} peptide bond while aspartate decarboxylase cleaves at $Gly_{24}Ser_{25}$. The Ser_{82} and Ser_{25}, respectively, become the active site N-terminal pyruvamide moieties. X-ray structures indicate each autocleavage peptide site is likely to be in a strained loop, again destabilizing this site to intramolecular attack from the adjacent CH_2OH side chains (Gallagher et al., 1993). The oxyoxazolidinone tetrahedral adduct (Figure 13.32A) is the same as in the first class of examples mentioned previously and the C–N bond cleavage

Histidine decarboxylase from *Lactobacillus*:

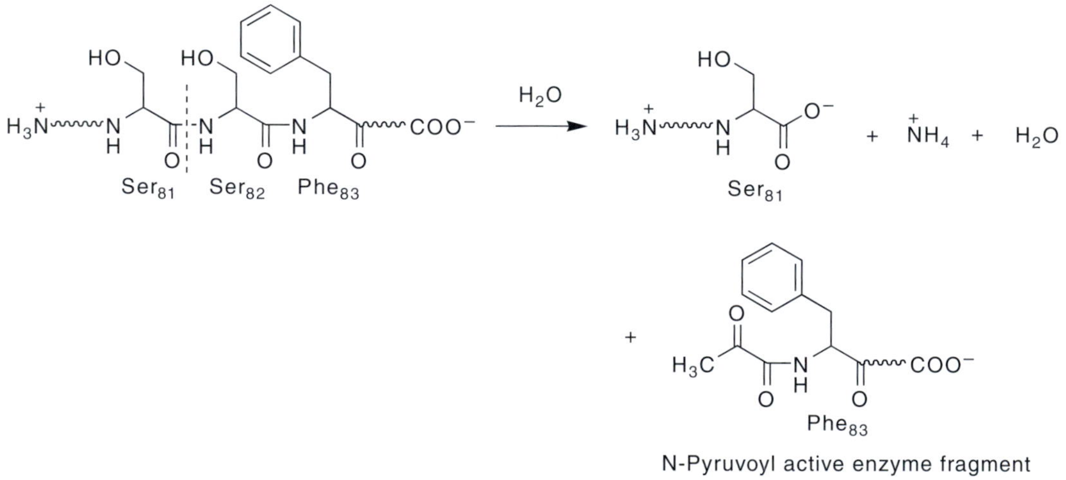

Aspartate decarboxylase from *E. coli*:

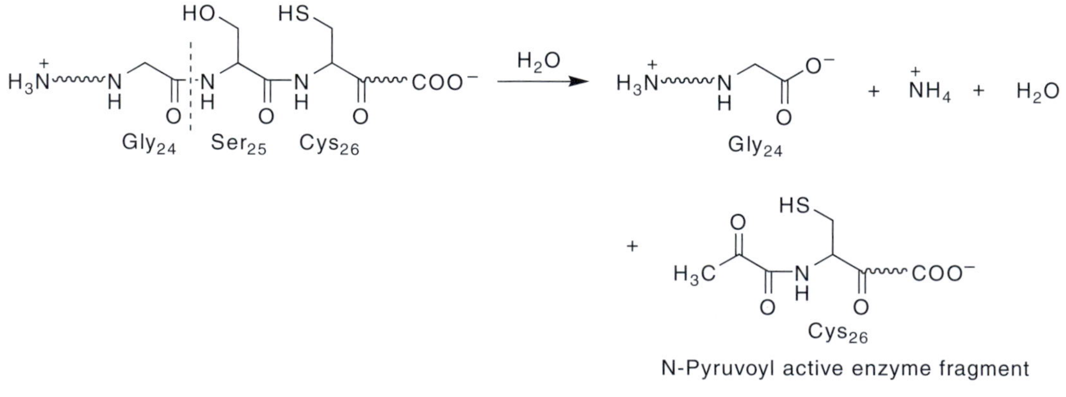

Figure 13.30 Generation of the N-terminal pyruvamide by autoproteolysis of the proenzyme forms of histidine decarboxylase (cleavage at $Ser_{81}Ser_{82}$) and aspartate decarboxylase ($Gly_{24}Ser_{25}$).

as it breaks down in the forward direction yields the two-subunit enzyme, which has the peptide bond broken while the two fragments are still tethered in oxoester linkage. Remarkably, x-ray analysis (Ramjee et al., 1997) of the tetrameric aspartate decarboxylase shows that three of the subunits have proceeded through the autoprocessing and bear N-terminal pyruvoyl groups while the fourth still has the α and β chains in oxoester linkage—a snapshot of a rearrangement in progress (Figure 13.31). The presumption is that the Ser_{82} (histidine decarboxylase) and Ser_{25} (aspartate decarboxylase), respectively, undergo α-hydrogen abstraction and β-hydroxyl elimination to generate the dehydroalanyl moiety. This could occur at the ester stage (when the α-hydrogen is activated) or after hydrolysis (Figure 13.32A). By either route the dehydroalanyl moiety, once it is N-terminal, is the unstable tautomer of the iminopyruvoyl group (Figure 13.32B). Hydrolysis of the imine will uncover the pyruvoyl moiety. If the dehydroalanine residue would be formed *without* internal peptide cleavage it would have reverse polarity and not be an electron sink for amine deaminations, a rationale for why autocleavage is required to uncover the new functionality in the enzyme.

The N-terminal pyruvamide thus uncovered acts as an electrophile because of its α-keto group. Ketones are not found in any of the 20 proteinogenic amino acids and this autocatalytic posttranslational modification clearly augments the catalytic inventory. The reactivity of the keto group in N-pyruvoyl posttranslationally activated enzymes is exemplified by aspartate α-decarboxylation, generating the β-alanine

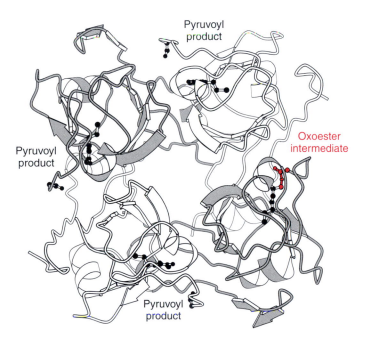

Figure 13.31 X-ray analysis of the tetrameric bacterial aspartate α-decarboxylase reveals three subunits with N-terminal pyruvoyl groups (shown in black and white ball and stick) and the fourth at the stage of the uncleaved oxoester intermediate (shown in red ball and stick) (figure made using PDB 1AW8).

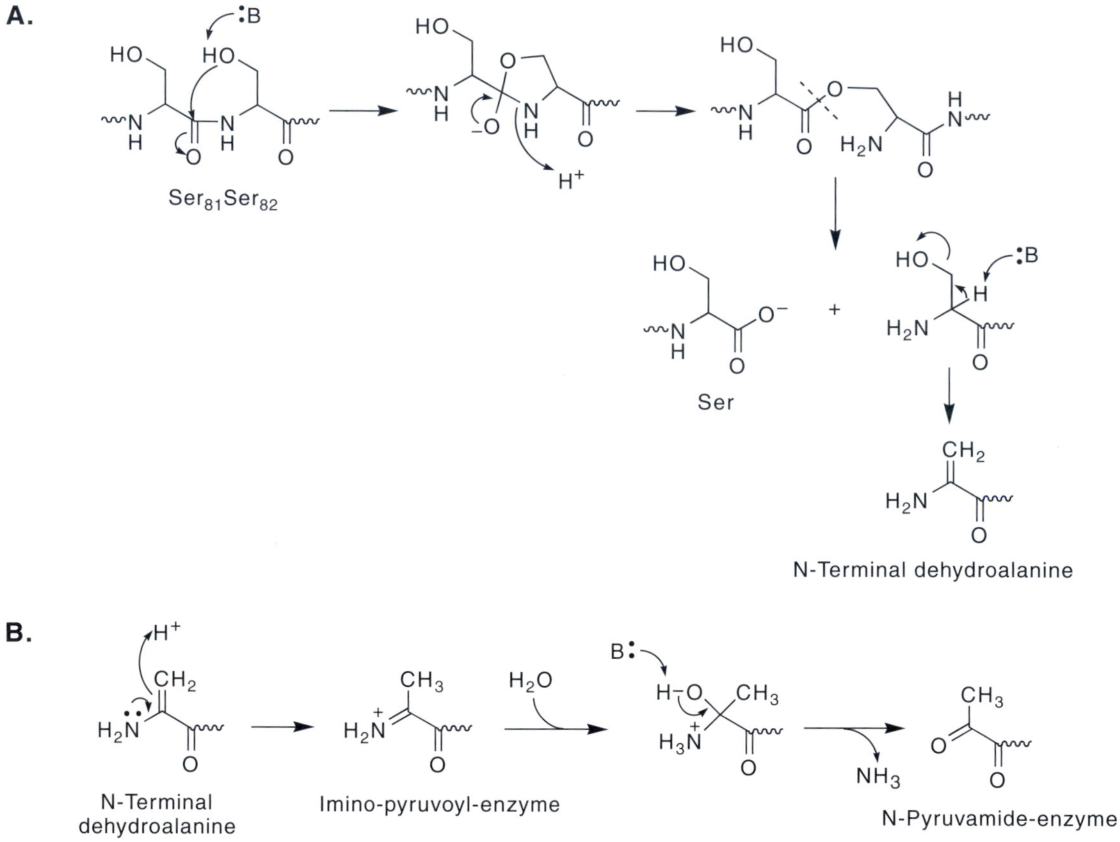

A.

B.

Figure 13.32 **A.** Dehydration of the Ser_{82} side chain of the proform of His decarboxylase to a dehydroalanyl moiety. **B.** This N-terminal dehydroalanine residue tautomerizes from the enamine to imine form and then hydrolyzes to release NH_3 and the ketone group of the N-pyruvamide of the active enzyme.

required for coenzyme A biosynthesis. The aspartate substrate bound in the active site can form a substrate aldimine with the pyruvoyl cofactor (Figure 13.33A). This sets up a low-energy path for decarboxylation since the resultant α-carbanion is delocalized throughout the extended conjugated system (Figure 13.33B). Reprotonation at Cα and product imine hydrolysis yields the β-alanine product (Figure 13.33C). The pyruvoyl group may be an evolutionary alternative to the aldehyde electron sink in the pyridoxal phosphate coenzyme (Walsh, 1979).

A third example of this class of enzymes is the enzyme D-proline reductase from the anaerobic bacterium *Clostridium sticklandii*, which ferments the cyclic imino acid to the linear 5-aminovalerate (Figure 13.34A), a reductive ring opening that is a different kind of transformation than the carbanionic decarboxylations of histidine or aspartate. Gene sequencing (Kabisch et al., 1999) validated that a single-chain 68-kDa

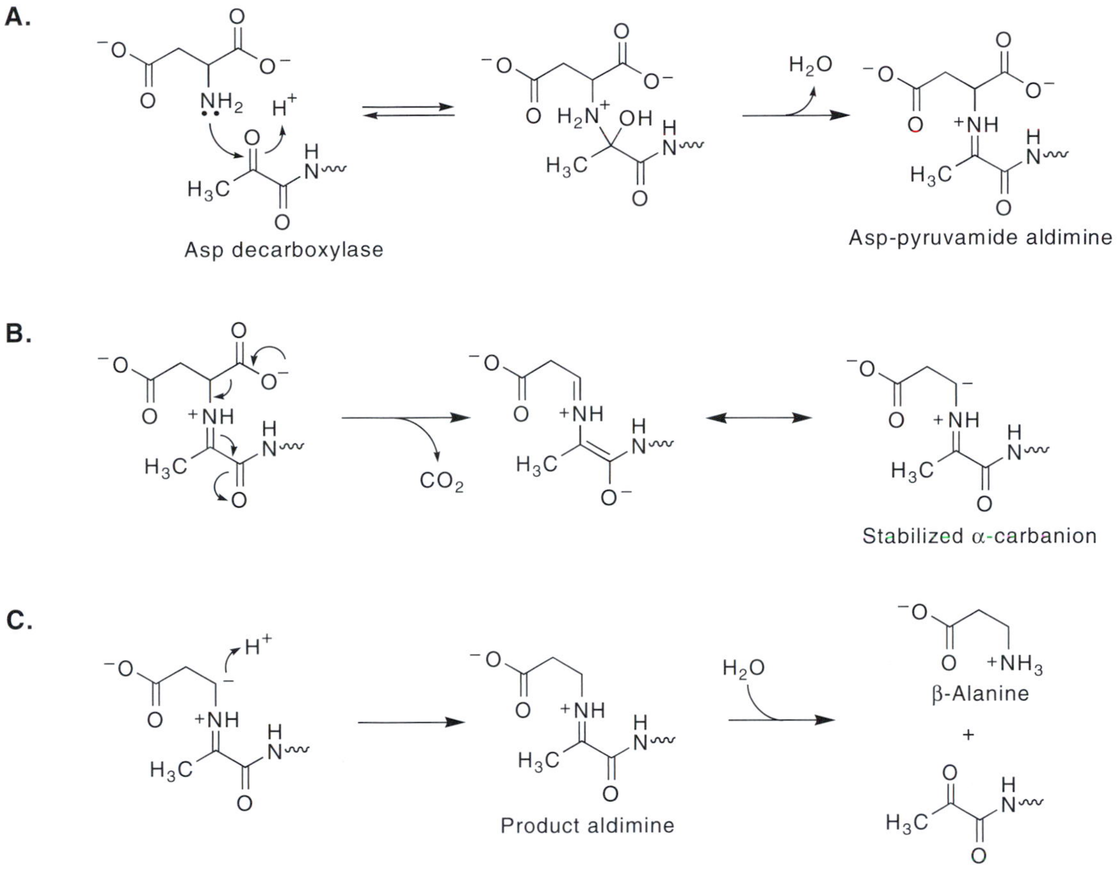

A.

Asp decarboxylase

Asp-pyruvamide aldimine

B.

Stabilized α-carbanion

C.

β-Alanine

Product aldimine

Figure 13.33 Three-step mechanism for decarboxylation of the substrate aspartate by the N-pyruvoyl form of active Asp decarboxylase. **A.** Substrate–pyruvamide aldimine formation. **B.** Decarboxylation to generate the stabilized carbanion. **C.** Reprotonation at Cα and hydrolysis of the product imine.

inactive precursor is cleaved to give 45-kDa and 23-kDa subunit fragments, the smaller of which bears an N-terminal pyruvoyl moiety. In this case the precursor amino acid at the peptide cleavage junction was a cysteine ($Thr_{425}Cys_{426}$), indicating that β-elimination of H_2S from the ester intermediate occurs to yield the β-alanyl precursor to the pyruvoyl group. An analogous internal Cys residue is the precursor to the N-terminal pyruvoyl group in the protein B subunit of glycine reductase.

The 23-kDa subunit of proline reductase also contains another prosthetic group, the 21st amino acid, selenocysteine, incorporated cotranslationally (see Chapters 1 and 4). The SeCys moiety is in the active site and the selenide side chain of $^-$SeCys is proposed to be the catalytic nucleophile on the iminium complex of the substrate D-proline that is tethered to the pyruvoyl group as shown in Figure 13.34B, and which cleaves the Cα–N bond as the Se–Cα bond forms. Proposed fragmentation of the

Se–Cα bond in the seleno adduct of the aminovaleryl-enzyme occurs by a net two-electron input and is initiated by mixed selenosulfide formation. The selenosulfide, akin to that proposed for glutathione peroxidase and thioredoxin reductase (Chapter 4), must undergo two-electron reduction back to the starting thiol/selenate oxidation state from an external reducing agent [e.g., thiols, perhaps via thioredoxin reductase (Chapter 4)] to set the stage for a subsequent catalytic cycle.

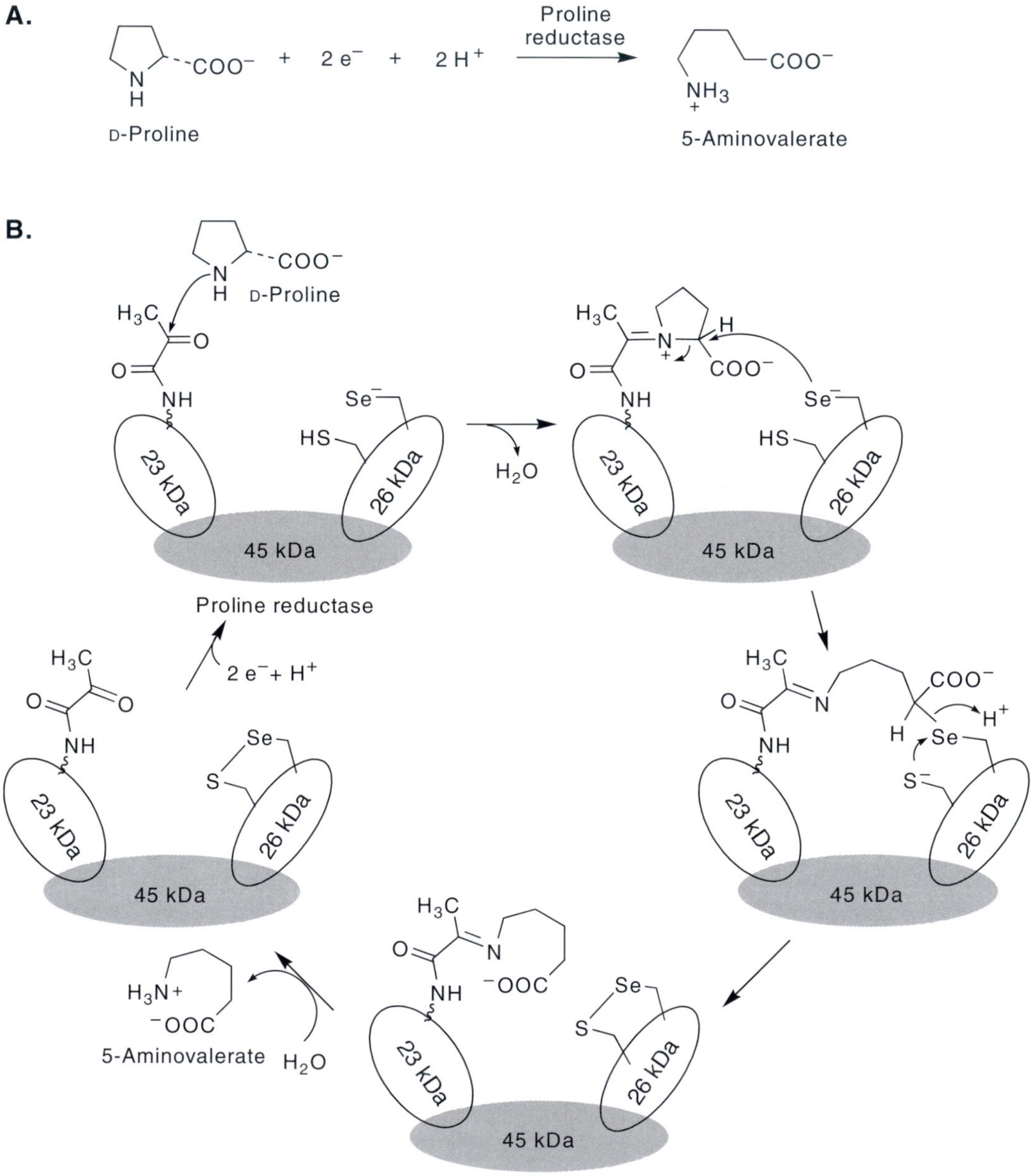

A.

B.

Figure 13.34 **A.** Reductive cleavage of D-proline to the linear 5-aminovalerate by an N-pyruvoyl enzyme. **B.** Role of selenoCys in the active site to enable Cα–Se bond formation concomitant with ring opening of the proline skeleton. Selenosulifde formation leads to Cα–SeCys bond cleavage followed by two-electron input to reduce the Se–S selenosulfide back to starting enzyme.

These first two classes of proteins use the downstream peptide fragment resulting from autocleavage as the business end (N-terminal nucleophile, N-terminal pyruvoyl group). The class to be described next uses the upstream acyl fragment after cleavage as the activated carbonyl fragment that gets captured by a specific alcohol cosubstrate.

Autocleavage of a Peptide Bond and Acyl Transfer to Cholesterol for Membrane Localization and Signaling

Members of the Hedgehog family of proteins are translocated through the secretory pathway to the surface of mammalian cells (Lee et al., 1994). During transit these 45-kDa proteins undergo autocleavage mediated by a specific cysteine side chain acting as the internal nucleophile to produce an oxythiazolidinone tetrahedral adduct and then the corresponding thioester intermediate (Figure 13.35), in strict analogy to the

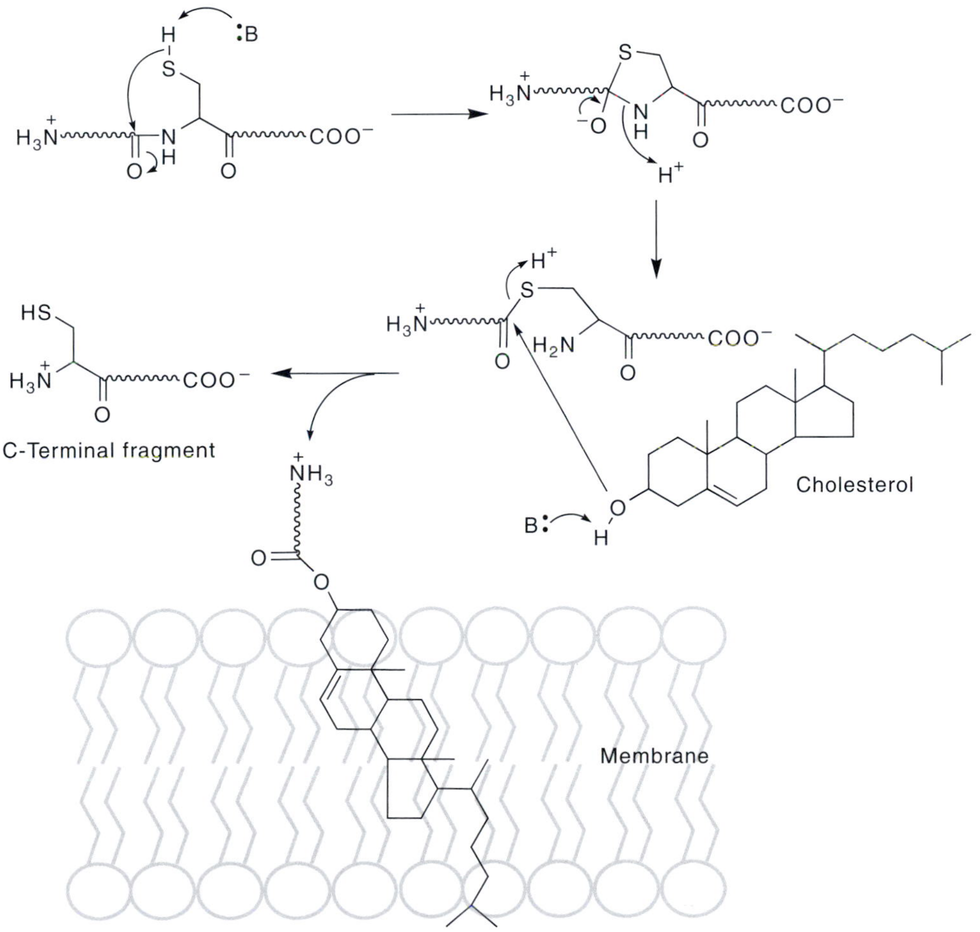

Figure 13.35 Hedgehog activation during cleavage of the single-chain precursor. Peptide bond fragmentation via oxythiazolidinone and thioester intermediates and capture of the 17-kDa N-terminal protein thioester by the C_3–OH of cholesterol to tether the active N-terminal fragment of Hedgehog as a cholesterol ester.

previous two classes of protein cleavage. This thioester formation is mediated by the C-terminal 25-kDa portion of the pro form of Hedgehog. The extreme C-terminal 63 residues contain a binding site for the membrane sterol cholesterol, as noted in Chapter 7. Cholesterol bound in this region is then used as the specific nucleophile to capture the N-terminal moiety of the thioester intermediate (Porter et al., 1996). Thus, the acyl transfer fate of the N-terminal portion of Hedgehog protein is not hydrolysis but alcoholysis, to the C_3-OH of cholesterol. The C-terminal fragment has done its work.

The N-terminal 20-kDa fragment of Hedgehog is now a protein-cholesterol ester. This lipid tail anchors the Hedgehog fragment in membranes, so it can diffuse two-dimensionally on cell surfaces to encounter the patched transmembrane receptor in *Drosophila* (Ye and Fortini, 2000) and initiate signal transduction via the cubitus interruptus pathway (see Chapter 8) and turn "on" gene expression of the morphogen protein wingless and the TGFβ family member decapentaplegic. The capture of a protein fragment by cholesterol for signaling purposes is the most novel biological role unearthed for this sterol in decades. It emphasizes the versatility conferred by the posttranslational modification of proteins and the different purposes to which autocatalytic peptide bond fragmentation via oxoester and thioester intermediates can be put.

Autocleavage of Peptide Bonds, Excision, and Religation: Protein Splicing

About a hundred examples of proteins predicted to be self-splicing have been identified on the basis of conserved sequences at the junctions of the splice sites (extein$_1$–intein and intein–extein$_2$) (Figure 13.36A) (Perler, 2002). About 25 of these have been verified experimentally to self-splice. Protein splicing occurs in single-cell eukarya such as yeast (nine examples), bacteria (24 species), and archaea (17 examples), consistent with splicing being an *ancient autocatalytic capacity* of folded proteins. Protein splicing comprises the excision of an intervening sequence, the **intein**, from a protein precursor, usually but not always a single polypeptide chain (Paulus, 2000; Perler, 2002), and the ligation of the flanking upstream (**extein$_1$**) and downstream (**extein$_2$**) protein sequences (Figure 13.23D). The mature protein consists of extein$_1$–extein$_2$ ligated through a normal peptide bond and the released intein. Inteins can be quite large: the intervening sequence in the inactive 109-kDa precursor form of the vacuolar H^+-ATPase of yeast is 454 residues. After splicing, the 69-kDa ATPase ligated subunit has gained catalytic activity (Figure 13.36B). Inteins are predicted to range from 134–608 residues, of which the N-terminal 100 and the C-terminal 40–50 residues are required for the splicing activity of the intein. The remaining sequence of inteins often encodes a homing endonuclease domain that helps intein DNA invade the host genes in which they are resident.

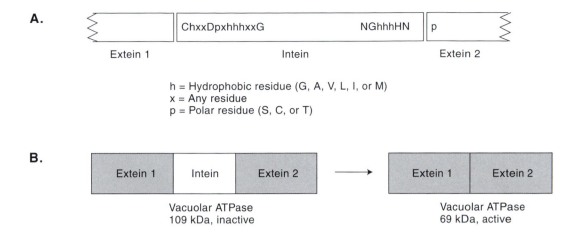

A. Conserved sequences at extein₁–intein and intein–extein₂ splice junctions. **B.** Splicing of the proenzyme form of the yeast vacuolar ATPase, making a single-chain 109-kDa inactive precursor into a single-chain 69-kDa active enzyme.

Figure 13.36

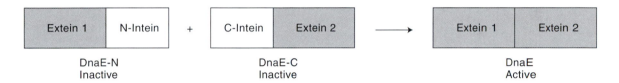

Figure 13.37 *In trans* splicing of two inactive subunits of *M. jannaschii* DNA polymerase (DnaE) to an active single-chain enzyme.

Inteins have been found to insert into 34 different protein families, including enzymes of nucleotide metabolism such as DNA helicases, the α and β subunits of ribonucleotide reductase, DNA polymerase, the A and B subunits of DNA gyrase, RecA, and the DNA polymerase III subunits (Perler, 2002). A protein can have more than one protein splicing junction. A recent compilation indicates 94 proteins with one intein inserted, 12 with two inteins, and three with three inteins inserted (Perler, 2002). In the DNA polymerase of *Methanococcus jannaschii* the N′ and C′ parts of an intein are on two protein precursors. Splicing occurs *in trans* as the two parts of the intein-self-assemble splice themselves out and create a single ligated Pol III α subunit (Figure 13.37). This involves ligation of the 774-residue fragment to the 36–420 fragment, as the 123-residue N′ and 36-residue C′ portions of the intein combine *in trans*, splice themselves out, and ligate the 774–420 fragments to yield the mature, active 1194-residue Pol III subunit (Paulus, 2000). This is a remarkable case of protein ligation, converting two nonfunctional protein substrates into one functional ligated protein product.

Inteins act stoichiometrically, not catalytically, on their extein substrates, acting intramolecularly at the E₁–I and I–E₂ junctions, first to cleave the peptide bonds at

A.

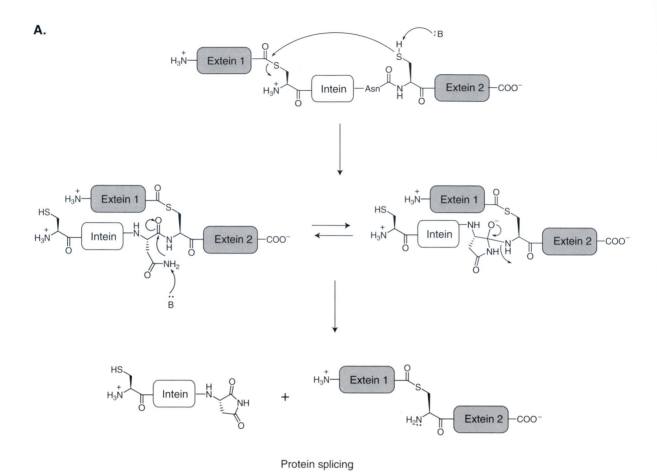

Protein splicing

Figure 13.38 Comparison of **A.** protein splicing and **B.** RNA splicing. Parallels include breakage of the covalent chain connection (peptide bond and phosphodiester bond, respectively), and formation of branched intermediates as a consequence of intramolecular attack by a downstream nucleophile within the chain (thiolate side chain of Cys or the 2′-OH of an adenosine moiety). Resolution of the branched intermediates involves excision of the intein/intron and religation of upstream and downstream exteins/exons [adapted from Kramer (1996)].

these junctions by distinct mechanisms, and then to religate them to the $E_1–E_2$ spliced product. The self-excised intein comes out in a chemically modified form, with its C-terminal asparagine cyclized as a succinimide, reflecting the mechanism of the $I–E_2$ excision step. As noted below, a branched or lariat intermediate is involved. In many aspects the logic of protein splicing parallels the logic of mRNA splicing, with the splicing out of the intervening RNA sequence (intron) and the ligation of the flanking sequence (exons) with *branched intermediates* (Figure 13.38). Unlike protein splicing though, mRNA splicing is not generally autocatalytic but instead requires the multicomponent ribonucleoprotein machinery of spliceosomes (Black, 2003).

The autocatalytic mechanism of protein splicing (Paulus, 2000) follows closely the first two steps of the N-terminal nucleophile amidase activation, the pyruvoyl enzyme

B.

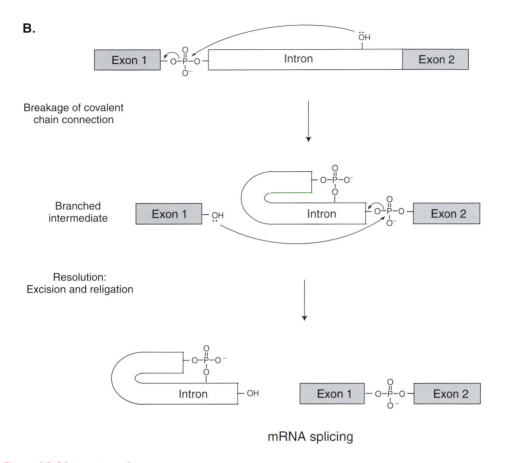

mRNA splicing

Figure 13.38 (continued)

activation, and Hedgehog family cleavage: intramolecular attack of a Cys, Ser, or Thr side chain at the E_1–I boundary to create the five-ring tetrahedral adduct (step 1) (Figure 13.39A), followed by resolution via C–N bond cleavage and generation of the common oxoester (Ser, Thr) or thioester (Cys) intermediate (step 2) (Figure 13.39B). The Ser, Thr, or Cys side chain initiating cleavage is the most N-terminal residue of the intein and the adjacent upstream peptide carbonyl is thus the E_1–I boundary. At the conclusion of these two steps the E_1–I peptide bond has been broken while the E_1 acyl chain retains thermodynamic activation (oxoester or thioester) for subsequent transfer. The catalytically competent nucleophile in protein splicing is the most N-terminal side chain of the E_2 domain, again a Ser, Thr, or Cys. The folded structure of the intein determines the positioning of the I–E_2 boundary and brings it in apposition to the E_1-ester–I boundary. Now, in step 3, a net transesterification occurs, where the E_1-acyl group is shuttled from one alcohol/thiol side chain (at the N-terminus of I) to the alcohol/thiol side chain at the N-terminus of E_2 (Figure 13.39C). This is likely an isoenergetic transfer when two Ser/Thr are involved or two Cys side chains and is downhill thermodynamically for a Ser/Thr and Cys pair of nucleophiles

A.

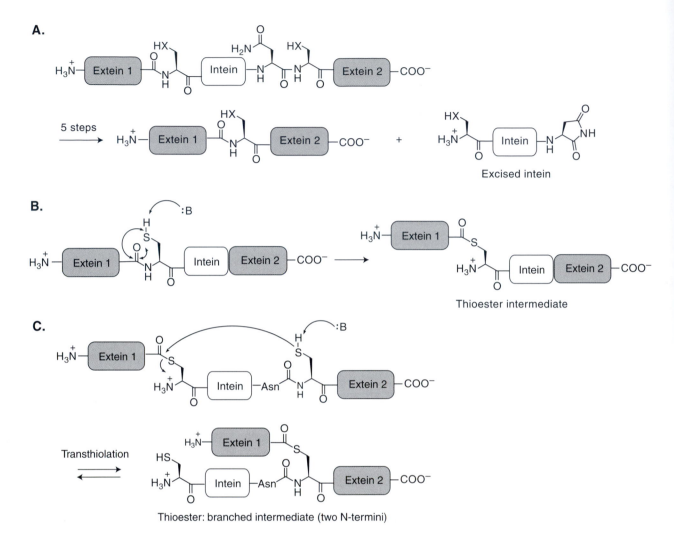

Figure 13.39 **A.** Five steps in protein splicing with a net cleavage of two peptide bonds and ligation of one new peptide bond. **B.** Cyclic adduct and subsequent oxoester/thioester formation. **C.** Transesterification by attack of the Ser, Thr, or Cys (example shown) at the N-terminus of E_2 with generation of a branched or lariat intermediate.

at I and E_2, respectively. Undoubtedly, the positioning of side chains in the intein active site controls their action as acid and base catalysts to facilitate the transacylation reaction. X-ray analysis of a catalytically incompetent mutant of the *M. xenopi* Gyrase A subunit 198-residue intein of the precursor protein (Klabunde et al., 1998) shows a *cis* peptide bond at the E_1–I junction (Figure 13.40). *Cis* peptide bonds can be 5 kcal/mol less stable than trans peptide bonds and thus do not generally accumulate. For protein splicing the pre-accumulation of a *cis* peptide conformer at the extein₁–intein splice junction sets up that peptide bond specifically for the N-to-S capture (by a Cys side chain) or the N-to-O capture (by a Ser/Thr side chain) that is

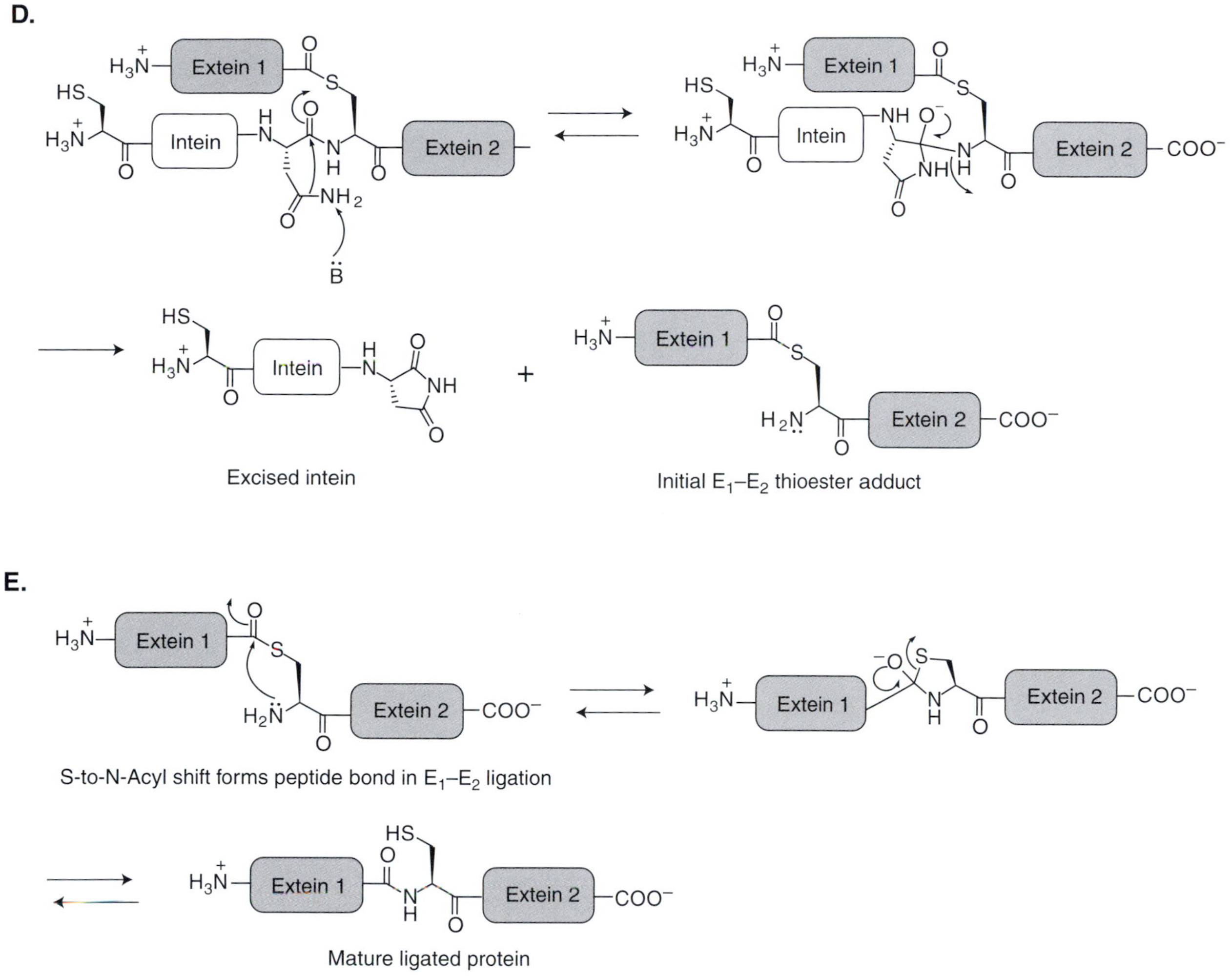

D.

Excised intein

Initial E_1–E_2 thioester adduct

E.

S-to-N-Acyl shift forms peptide bond in E_1–E_2 ligation

Mature ligated protein

Figure 13.39 **D.** Cleavage of the I-E_2 junction peptide bond by the Asn carboxamide at the C-terminus of the intein. This releases the intein with a cyclic succinimide terminus and yields E_1-E_2 joined in thioester linkage. **E.** S-to-N (or O-to-N) shift to reform the peptide bond at the E_1-E_2 junction to complete splicing and generate the mature ligated protein.

required to accumulate the otherwise unfavorable oxo/thioester intermediate and commit the precursor down the multistep splicing pathway.

The product of step 3 is the branched or lariat intermediate shown in Figure 13.39C. This lariat structure has E_1 joined to E_2 but in an oxo/thioester linkage, not a peptide bond yet. The amino group needed to engage in that peptide is still linked to the C-terminal residue, always a conserved asparagine, of the intein domain.

Step 4 of the splicing process requires cleavage of the I–E_2 junctional peptide bond (Figure 13.39D) (Asn-Cys is shown), releasing the amino group of the Cys from amide linkage, thus creating the nucleophilic amino group required for the last peptide

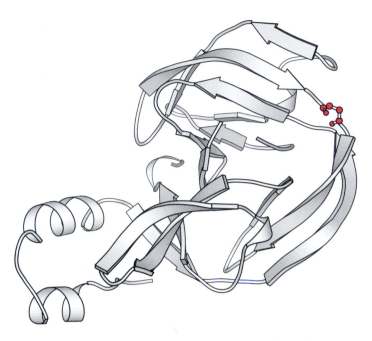

Figure 13.40 X-ray structure of a catalytically incompetent mutant form of *M. xenopi* gyrase A intein shows a cis peptide bond (red ball and stick) at the E1-I junction (figure made using PDB 1AM2).

ligation step. The cleavage of this second peptide bond (I–E$_2$) differs from the first (E$_1$–I) in the nature of the attacking nucleophile. Rather than a Thr/Ser or Cys side chain, the carboxamido-NH$_2$ of the C-terminal Asn residue of the intein attacks the adjacent peptide carbonyl. This generates a variant of the now familiar five-ring tetrahedral adduct (compare this I–E$_2$ tetrahedral adduct to the E$_1$–I tetrahedral adduct in the first peptide bond cleavage obtained in step 1). When it opens in the front direction by C–N bond cleavage, the intein has carved itself out, and the lariat structure has collapsed to a linear late stage intermediate. The form of the intein that has been sprung loose bears a cyclic five-membered aminosuccinimide in place of the Asn, which may hydrolyze with a half-life of a few minutes (Paulus, 2000) back to the ring-opened Asn. Protein splicing thus requires not only Ser, Thr, or Cys side chains as nucleophiles at the N-terminus of the intein and the N-terminus of extein$_2$, but also Asn at the C-terminus of the intein.

The last step of the protein splicing process is the thermodynamically favored reformation of a peptide bond at the E$_1$–E$_2$ junction by acyl transfer of the E$_1$-acyl moiety from oxo/thioester linkage to the now free amino group of the N-terminus of E$_2$ (Figure 13.39E). Such O-to-N (oxoester-to-peptide) or S-to-N (thioester-to-peptide) acyl migrations are not only thermodynamically favored but proceed spontaneously with low kinetic barrier under physiological conditions. The more thermodynamically activated peptidyl-thioester-to-peptide rearrangement is probably complete in

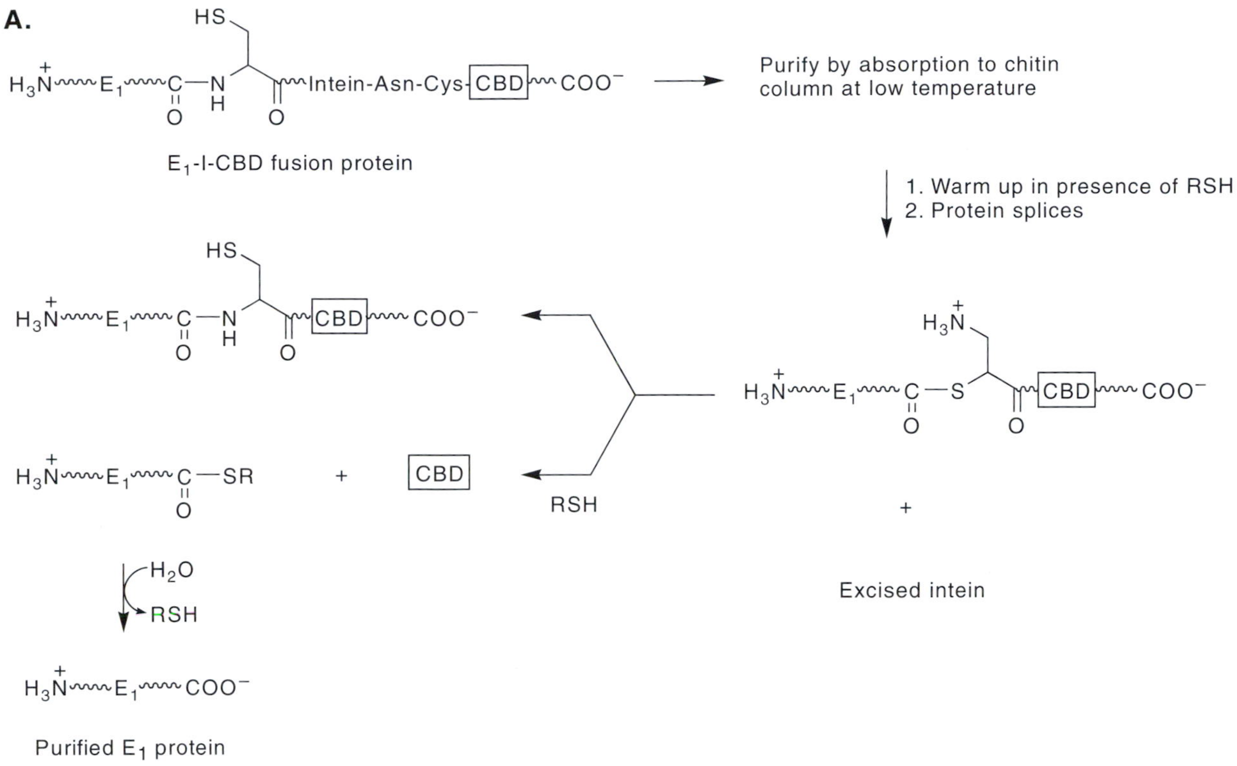

A.

Figure 13.41 Protein splicing during expressed protein ligations. **A.** Chitin-binding domains as affinity elements that are spliced out in the presence of thiols.

milliseconds, while the less activated oxoester-to-amide shift goes in less than one minute at 37 °C (Paulus, 2000).

The self-cleaving ability of inteins has found a number of biotechnology applications, first as self-cleaving tags that would permit affinity purification of fusion proteins, followed by excision of the intein. For example, use of the *S. cerevisiae* VMA intein with the C-terminal Asn mutated to an inactive Ala allowed fusion of the mutant intein to a chitin binding domain as E_2. The DNA sequence of the protein of interest was fused as E_1. On expression and purification in *E. coli*, the chitin binding domain served as an affinity tag on a chitin support at low temperature. Then, at higher temperature, the E_1–I N-to-S acyl transfer can occur and the thioester can be captured by free thiol in the buffer, such as dithiothreitol or benzene thiol. This releases the E_1 protein as the C-terminal DTT/aryl thioester, which rapidly hydrolyzes (Figure 13.41A). This approach has been extended in expressed protein ligation protocols (Muir et al., 1998) to capture the E_1-acyl fragment by peptides or proteins bearing N-terminal cysteines. The S-to-N shift goes rapidly and quantitatively to incorporate the added peptide/protein (Figure 13.41B). For example, one can ligate

B.

When RSH = Cys-peptide-COO⁻

Newly ligated C-terminally extended protein

Figure 13.41 **B.** Ligation of C-terminal thiol-containing peptides (thiol attack and S-to-N shift not shown).

C-terminal pY-peptides to protein kinases and protein phosphatases (Chapter 2) to create regulatory pY tails that modulate the kinase/phosphatase activity of the intramolecular, upstream catalytic domains (Lu et al., 2001 and 2003).

As a final comment on the four variants of protein autocleavage, x-ray structures of each of the protein domains responsible for the autocleavages have been solved (Paulus, 2000) and they indicate that the N-terminal nucleophile proteins (e.g., the glycosylasparaginase family) and the pyruvoyl forming amino acid decarboxylases represent the convergent evolution of catalysts that can destabilize particular peptide bonds intramolecularly for cleavage. On the other hand, the autoprocessing domains of Hedgehog and the inteins are clearly related folds. While the protein splicing autocatalysts are not found in multicellular eukaryotes, suggesting an earlier evolutionary origin, the Hedgehog family is fundamental in pattern formation in multicellular organisms, emphasizing the contemporary utility of this presumably ancient N-to-S and N-to-O intramolecular acyl transfer protein catalysis. It has been shown that the *trans* splicing of the DnaE protein precursors to create the 1994 residue DNA Pol III α catalytic subunit is essential in *Synechocystis sp*. Inteins may have initially evolved both for *in trans* and *in cis* splicing as regulatory elements to control functional protein formation (Paulus, 2000). Their subsequent invasion by homing endonuclease DNA may have created the possibility of horizontal transfer into conserved coding regions of genes.

References

Back, M.-c., P. M. Krosky, and D. C. Coen. "Relationship between autophosphorylation and phosphorylation of exogenous substrates by the human cytomegalovirus UL97 protein kinase," *J. Virol.* **76**:11943–11952 (2002).

Baedeker, M., and G. E. Schulz. "Autocatalytic peptide cyclization during chain folding of histidine ammonia-lyase," *Structure (Cambridge)* **10**:61–67 (2002a).

Baedeker, M., and G. E. Schulz. "Structures of two histidine ammonia-lyase modifications and implications for the catalytic mechanism," *Eur. J. Biochem.* **269**:1790–1797 (2002b).

Barondeau, D. P., C. D. Putnam, C. J. Kassmann, J. A. Tainer, and E. D. Getzoff. "Mechanism and energetics of green fluorescent protein chromophore synthesis revealed by trapped intermediate structures," *Proc. Natl. Acad. Sci. U.S.A.* **100**:12111–12116 (2003).

Bevis, B. J., and B. S. Glick. "Rapidly maturing variants of the Discosoma red fluorescent protein (DsRed)," *Nat. Biotechnol.* **20**:83–87 (2002).

Black, D. "Mechanisms of alternate pre-messengerRNA splicing," *Annu. Rev. Biochem.* **72**:291–316 (2003).

Borgstahl, G. E., D. R. Williams, and E. D. Getzoff. "1.4 Å structure of photoactive yellow protein, a cytosolic photoreceptor: Unusual fold, active site, and chromophore," *Biochemistry* **34**: 6278–6287 (1995).

Calabrese, J. C., D. B. Jordan, A. Boodhoo, S. Sariaslani, and T. Vannelli. "Crystal structure of phenylalanine ammonia lyase: Multiple helix dipoles implicated in catalysis," *Biochemistry* **43**:11403–11416 (2004).

Campbell, R. E., O. Tour, A. E. Palmer, P. A. Steinbach, G. S. Baird, D. A. Zacharias, and R. Y. Tsien. "A monomeric red fluorescent protein," *Proc. Natl. Acad. Sci. U.S.A.* **99**:7877–7882 (2002).

Chang, P., M. K. Jacobson, and T. J. Mitchison. "Poly(ADP-ribose) is required for spindle assembly and structure," *Nature* **432**:645–649 (2004).

Christenson, S. D., W. Liu, M. D. Toney, and B. Shen. "A novel 4-methylideneimidazole-5-one-containing tyrosine aminomutase in enediyne antitumor antibiotic C-1027 biosynthesis," *J. Am. Chem. Soc.* **125**:6062–6063 (2003).

Datta, S., Y. Mori, K. Takagi, K. Kawaguchi, Z. W. Chen, T. Okajima, S. Kuroda, T. Ikeda, K. Kano, K. Tanizawa, and F. S. Mathews. "Structure of a quinohemoprotein amine dehydrogenase with an uncommon redox cofactor and highly unusual crosslinking," *Proc. Natl. Acad. Sci. U.S.A.* **98**:14268–14273 (2001).

Duff, A. P., A. E. Cohen, P. J. Ellis, J. A. Kuchar, D. B. Langley, E. M. Shepard, D. M. Dooley, H. C. Freeman, and J. M. Guss. "The crystal structure of *Pichia pastoris* lysyl oxidase," *Biochemistry* **42**:15148–15157 (2003).

Gallagher, T., D. A. Rozwarski, S. R. Ernst, and M. L. Hackert. "Refined structure of the pyruvoyl-dependent histidine decarboxylase from *Lactobacillus 30a*," *J. Mol. Biol.* **230**:516–528 (1993).

Gomelsky, M., F. Biville, F. Gasser, and Y. D. Tsygankov. "Identification and characterization of the pqqDGC gene cluster involved in pyrroloquinoline quinone production in an obligate methylotroph *Methylobacillus flagellatum*," *FEMS Microbiol. Lett.* **141**:169–176 (1996).

Hyun Bae, J., M. Rubini, G. Jung, G. Wiegand, M. H. Seifert, M. K. Azim, J. S. Kim, A. Zumbusch, T. A. Holak, L. Moroder, R. Huber, and N. Budisa. "Expansion of the genetic code enables design of a novel 'gold' class of green fluorescent proteins," *J. Mol. Biol.* **328**:1071–1081 (2003).

Johnson, L., N., R. J. Lewis. "Structural basis for control by phosphorylation," *Chem. Rev.* **101**:2209–2242 (2001).

Ju, B. G., D. Solum, E. J. Song, K. J. Lee, D. W. Rose, C. K. Glass, and M. G. Rosenfeld. "Activating the PARP-1 sensor component of the Groucho/ TLE1 corepressor complex mediates a CaMKinase IIdelta-dependent neurogenic gene activation pathway," *Cell* **119:**815–829 (2004).

Kabisch, U. C., A. Grantzdorffer, A. Schierhorn, K. P. Rucknagel, J. R. Andreesen, and A. Pich. "Identification of D-proline reductase from *Clostridium sticklandii* as a selenoenzyme and indications for a catalytically active pyruvoyl group derived from a cysteine residue by cleavage of a proprotein," *J. Biol. Chem.* **274:**8445–8454 (1999).

Kim, M. Y., S. Mauro, N. Gevry, J. T. Lis, and W. L. Kraus. "NAD(+)-dependent modulation of chromatin structure and transcription by nucleosome binding properties of PARP-1," *Cell* **119:**803–814 (2004).

Klabunde, T., S. Sharma, A. Telenti, W. R. Jacobs, Jr., and J. C. Sacchettini. "Crystal structure of GyrA intein from *Mycobacterium xenopi* reveals structural basis of protein splicing," *Nat. Struct. Biol.* **5:**31–36 (1998).

Kobe, B., and B. E. Kemp. "Active site-directed protein regulation," *Nature* **402:**373–376 (1999).

Labas, Y. A., N. G. Gurskaya, Y. G. Yanushevich, A. F. Fradkov, K. A. Lukyanov, S. A. Lukyanov, and M. V. Matz. "Diversity and evolution of the green fluorescent protein family," *Proc. Natl. Acad. Sci. U.S.A.* **99:**4256–4261 (2002).

Langer, M., A. Lieber, and J. Retey. "Histidine ammonia-lyase mutant S143C is posttranslationally converted into fully active wild-type enzyme. Evidence for serine 143 to be the precursor of active site dehydroalanine," *Biochemistry* **33:**14034–14038 (1994).

Lee, J. J., S. C. Ekker, D. P. von Kessler, J. A. Porter, B. I. Sun, and P. A. Beachy. "Autoproteolysis in Hedgehog protein biogenesis," *Science* **266:**1528–1537 (1994).

Lu, W., D. Gong, D. Bar-Sagi, and P. A. Cole. "Site-specific incorporation of a phosphotyrosine mimetic reveals a role for tyrosine phosphorylation of SHP-2 in cell signaling," *Mol. Cell* **8:**759–769 (2001).

Lu, W., K. Shen, and P. A. Cole. "Chemical dissection of the effects of tyrosine phosphorylation of SHP-2," *Biochemistry* **42:**5461–5468 (2003).

Maehama, T., H. Nishina, S. Hoshino, Y. Kanaho, and T. Katada. "NAD(+)-dependent ADP-ribosylation of T lymphocyte alloantigen RT6.1 reversibly proceeding in intact rat lymphocytes," *J. Biol. Chem.* **270:**22747–22751 (1995).

Martynov, V. I., A. P. Savitsky, N. Y. Martynova, P. A. Savitsky, K. A. Lukyanov, and S. A. Lukyanov. "Alternative cyclization in GFP-like proteins family. The formation and structure of the chromophore of a purple chromoprotein from *Anemonia sulcata*," *J. Biol. Chem.* **276:**21012–21016 (2001).

Matz, M. V., A. F. Fradkov, Y. A. Labas, A. P. Savitsky, A. G. Zaraisky, M. L. Markelov, and S. A. Lukyanov. "Fluorescent proteins from nonbioluminescent *Anthozoa* species," *Nat. Biotechnol.* **17:**969–973 (1999).

Meyer, T. E. "Isolation and characterization of soluble cytochromes, ferredoxins, and other chromophoric proteins from the halophilic phototropic bacterium *Ectothiorhodospira halophila*," *Biochim. Biophys. Acta* **806:**175–183 (1985).

Muir, T. W., D. Sondhi, and P. A. Cole. "Expressed protein ligation: A general method for protein engineering," *Proc. Natl. Acad. Sci. U.S.A.* **95:**6705–6710 (1998).

Mure, M. "Tyrosine-derived quinone cofactors," *Acc. Chem. Res.* **37:**131–139 (2004).

Mure, M., S. A. Mills, and J. P. Klinman. "Catalytic mechanism of the topa quinone containing copper amine oxidases," *Biochemistry* **41:**9269–9278 (2002).

Okeley, N. M., and W. A. van der Donk. "Novel cofactors via post-translational modifications of enzyme active sites," *Chem. Biol.* **7:**R159–R171 (2000).

Paulus, H. "Protein splicing and related forms of protein autoprocessing," *Annu. Rev. Biochem.* **69:**447–496 (2000).

Pearson, A. R., T. De La Mora-Rey, M. E. Graichen, Y. Wang, L. H. Jones, S. Marimanikku-pam, S. A. Agger, P. A. Grimsrud, V. L. Davidson, and C. M. Wilmot. "Further insights into quinone cofactor biogenesis: Probing the role of mauG in methylamine dehydrogenase tryptophan tryptophylquinone formation," *Biochemistry* **43:**5494–5502 (2004).

Perler, F. B. "InBase: The Intein database," *Nucleic Acids Res.* **30:**383–384 (2002).

Porter, J. A., S. C. Ekker, W. J. Park, D. P. von Kessler, K. E. Young, C. H. Chen, Y. Ma, A. S. Woods, R. J. Cotter, E. V. Koonin, and P. A. Beachy. "Hedgehog patterning activity: Role of a lipophilic modification mediated by the carboxy-terminal autoprocessing domain," *Cell* **86:**21–34 (1996).

Ramjee, M. K., U. Genschel, C. Abell, and A. G. Smith. "*Escherichia coli* L-aspartate-alpha-decarboxylase: Preprotein processing and observation of reaction intermediates by electrospray mass spectrometry," *Biochem. J.* **323(3):**661–669 (1997).

Rosenow, M. A., H. A. Huffman, M. E. Phail, and R. M. Wachter. "The crystal structure of the Y66L variant of green fluorescent protein supports a cyclization-oxidation-dehydration mechanism for chromophore maturation," *Biochemistry* **43:**4464–4472 (2004).

Ruf, A., V. Rolli, G. de Murcia, and G. E. Schulz. "The mechanism of the elongation and branching reaction of poly(ADP-ribose) polymerase as derived from crystal structures and mutagenesis," *J. Mol. Biol.* **278:**57–65 (1998).

Schwede, T. F., J. Retey, and G. E. Schulz. "Crystal structure of histidine ammonia-lyase revealing a novel polypeptide modification as the catalytic electrophile," *Biochemistry* **38:**5355–5361 (1999).

Seemuller, E., A. Lupas, and W. Baumeister. "Autocatalytic processing of the 20S proteasome," *Nature* **382:**468–471 (1996).

Shagin, D. A., E. V. Barsova, Y. G. Yanushevich, A. F. Fradkov, K. A. Lukyanov, Y. A. Labas, T. N. Semenova, J. A. Ugalde, A. Meyer, J. M. Nunes, E. A. Widder, S. A. Lukyanov, and M. V. Matz. "GFP-like proteins as ubiquitous metazoan superfamily: Evolution of functional features and structural complexity," *Mol. Biol. Evol.* **21:**841–850 (2004).

Stites, T. E., A. E. Mitchell, and R. B. Rucker. "Physiological importance of quinoenzymes and the O-quinone family of cofactors," *J. Nutr.* **130:**719–727 (2000).

Tarentino, A. L., G. Quinones, C. R. Hauer, L. M. Changchien, and T. H. Plummer, Jr. "Molecular cloning and sequence analysis of *Flavobacterium meningosepticum* glycosyl-asparaginase: A single gene encodes the alpha and beta subunits," *Arch. Biochem. Biophys.* **316:**399–406 (1995).

Terskikh, A., A. Fradkov, G. Ermakova, A. Zaraisky, P. Tan, A. V. Kajava, X. Zhao, S. Lukya-nov, M. Matz, S. Kim, I. Weissman, and P. Siebert. "'Fluorescent timer': protein that changes color with time," *Science* **290:**1585–1588 (2000).

Tsien, R. Y. "The green fluorescent protein," *Annu. Rev. Biochem.* **67**:509–544 (1998).

van Poelje, P. D., and E. E. Snell. "Pyruvoyl-dependent enzymes," *Annu. Rev. Biochem.* **59**:29–59 (1990).

Vandenberghe, I., J. K. Kim, B. Devreese, A. Hacisalihoglu, H. Iwabuki, T. Okajima, S. Kuroda, O. Adachi, J. A. Jongejan, J. A. Duine, K. Tanizawa, and J. Van Beeumen. "The covalent structure of the small subunit from *Pseudomonas putida* amine dehydrogenase reveals the presence of three novel types of internal cross-linkages, all involving cysteine in a thioether bond," *J. Biol. Chem.* **276**:42923–42931 (2001).

Walsh, C. T. *Enzymatic Reaction Mechanisms*. Freeman: San Francisco (1979).

Wilce, M. C., D. M. Dooley, H. C. Freeman, J. M. Guss, H. Matsunami, W. S. McIntire, C. E. Ruggiero, K. Tanizawa, and H. Yamaguchi. "Crystal structures of the copper-containing amine oxidase from *Arthrobacter globiformis* in the holo and apo forms: Implications for the biogenesis of topaquinone," *Biochemistry* **36**:16116–16133 (1997).

Xu, Q., D. Buckley, C. Guan, and H. C. Guo. "Structural insights into the mechanism of intramolecular proteolysis," *Cell* **98**:651–661 (1999).

Yarbrough, D., R. M. Wachter, K. Kallio, M. V. Matz, and S. J. Remington. "Refined crystal structure of DsRed, a red fluorescent protein from coral, at 2.0-Å resolution," *Proc. Natl. Acad. Sci. U.S.A.* **98**:462–467 (2001).

Ye, Y., and M. E. Fortini. "Proteolysis and developmental signal transduction," *Semin. Cell Dev. Biol.* **11**:211–221 (2000).

CHAPTER 14

Swinging Arms for Biotin, Lipoate, and Phosphopantetheine Tethering to Proteins

The lipoamide redox and acylation cycle in α-keto acid dehydrogenase complexes.

405

Three distinct types of coenzymes are covalently tethered to their apo enzymes in posttranslational modifications that convert the inactive apo form of the protein into the active holo form with enzymatic activity. The coenzymes biotin, lipoate, and phosphopantetheine (Figure 14.1) have their functional groups (the ureido group of biotin, the disulfide of lipoate, and the thiolate of phosphopantetheine) at the terminus of a flexible tether that is covalently attached at the other end to the apo protein. Not covered in this chapter are the subset of flavoproteins, such as monoamine oxidases, where the flavin prosthetic group has been covalently captured by a protein side chain. Nor are the subset of cytochromes in which heme vinyl side chains have been similarly captured by protein side chains discussed.

For both biotin and lipoate the flexible tether is the five-carbon valerate side chain. Attachment of the carboxylate of this valeryl side chain to the ϵ-NH$_2$ of a lysyl residue by the posttranslational modifying enzymes creates the covalently attached biotinyl amide and lipoyl amide linkages, respectively. ATP is utilized by the posttranslational coenzyme ligases to activate the valeryl side chain carboxylates as AMP esters (biotinyl-AMP and lipoyl-AMP) and set up favorable thermodynamics for capture by the lysyl side chains (Figure 14.2A). The four methylene carbons of the lysyl moiety and the four methylene carbons of the valeryl side chain of the coenzymes create a flexible tether, which at full extension can stretch out to 20 Å, a reach that allows these tethered biotinyl and lipoyl coenzymes to gain access to distinct active sites in multisubunit enzyme complexes (Perham, 2000; Walsh, 1979).

Analogous posttranslational modification logic fragments CoASH during the creation of a covalently tethered phosphopantetheine. In this case the attacking nucleophiles are serine side chains of protein domains (Figure 14.2B). The phosphopantetheine (Ppant) attached to fatty acid synthases, polyketide synthases, and nonribosomal peptide synthetases (Walsh, 2003) is therefore a phosphoserine derivative. The phosphopantetheinyltransferases (PPTases) that modify the apo proteins utilize CoASH as the Ppant donor, releasing 3′,5′-ADP as coproduct (Figure 14.2B). The thiol of the cysteamine moiety of Ppant is the business end of the coenzyme and is also about 20 Å from the surface of the attached proteins at full extension, so it is able

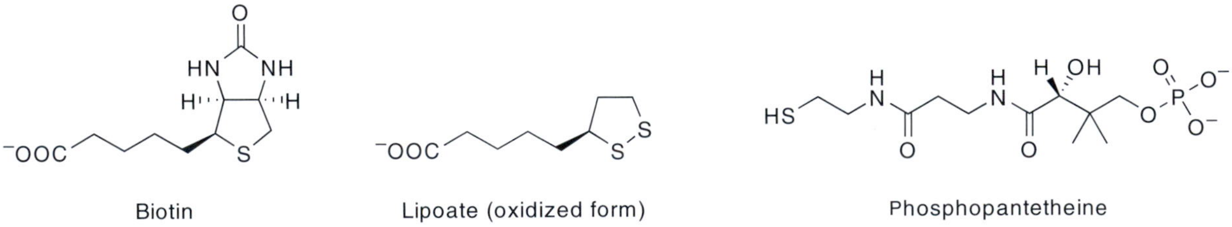

Figure 14.1 Coenzymes covalently tethered to apo proteins by swinging arms: biotin, lipoate, and phosphopantetheine prosthetic groups.

to service several modules in the multimodular fatty acid, polyketide, and peptide synthases and synthetases.

The covalently attached coenzymes are often termed prosthetic groups in such
proteins. While the chemistry enabled by the biotin, lipoate, and phosphopantetheine
prosthetic groups are quite distinct, as is elaborated below, all three carry portions of

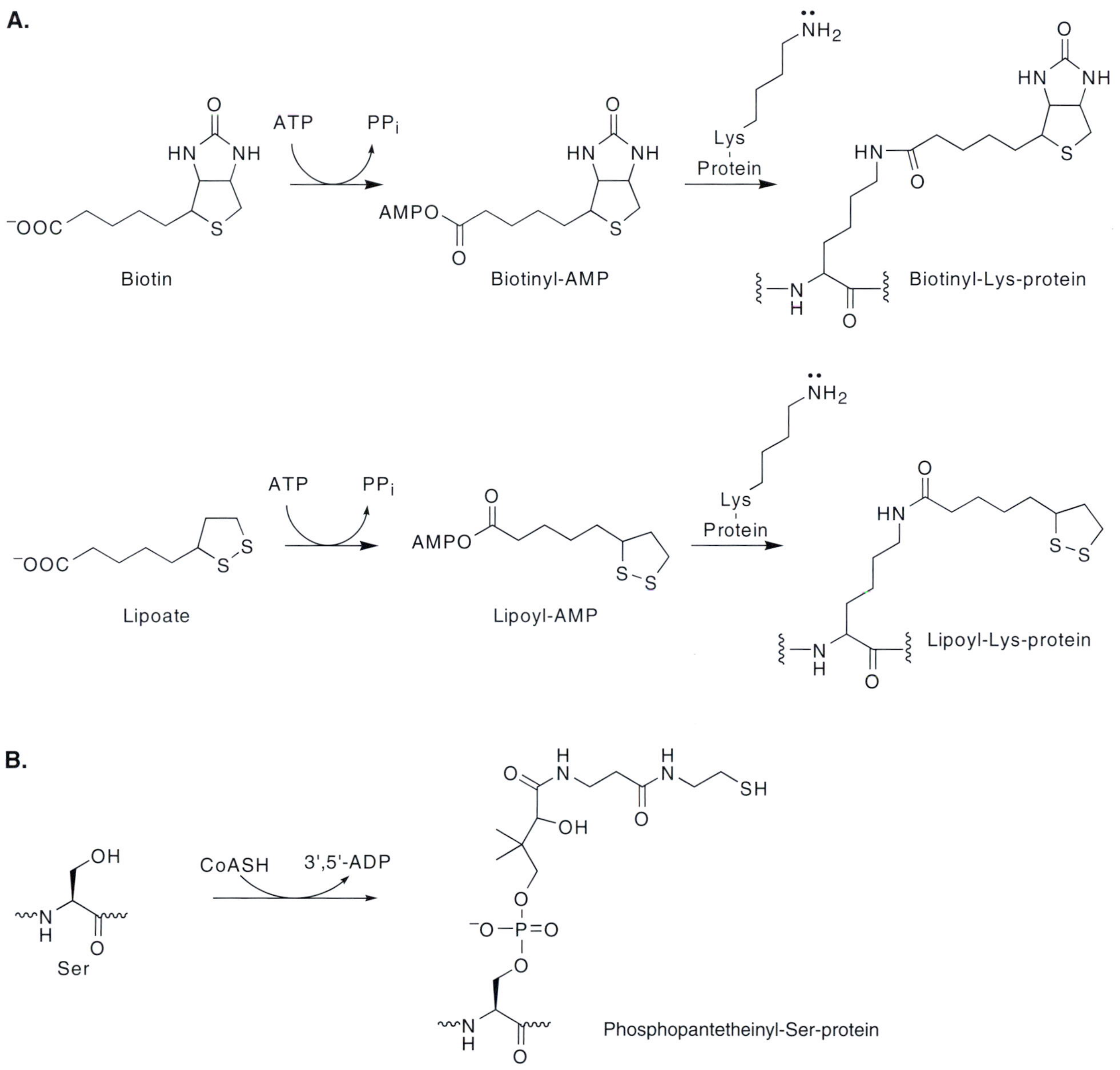

A.

B.

Figure 14.2 **A.** Activation of biotin and lipoate as acyl-AMP derivatives to enable capture by specific
lysine side chains to create attachment of biotin and lipoate in amide linkage. **B.** CoASH as phosphopantetheinyl donor to the side chain of a serine residue in carrier protein domains.

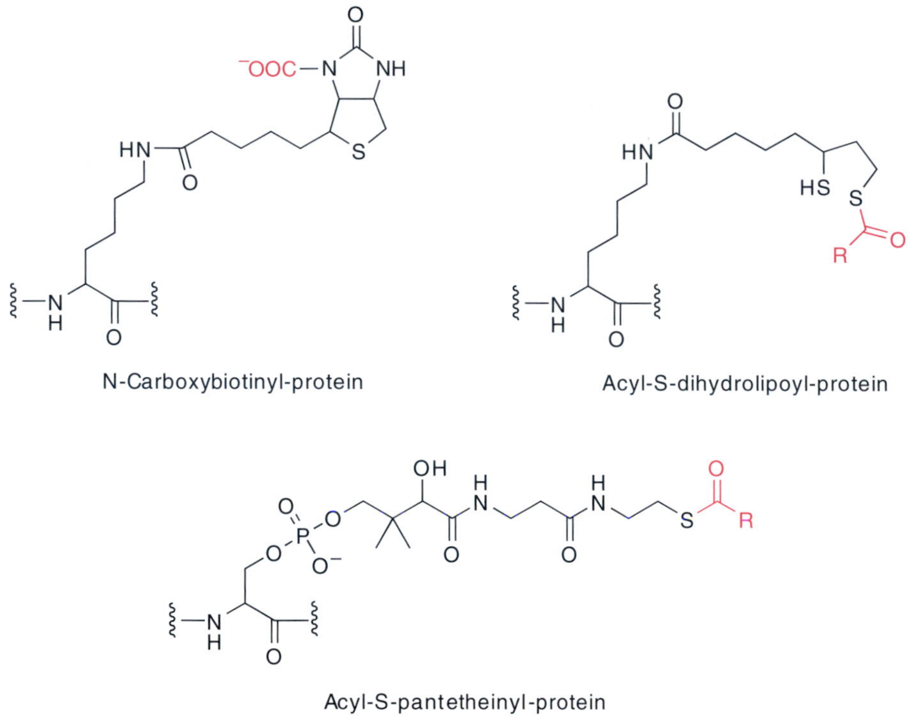

Figure 14.3 Substrate carboxyl or acyl groups carried by biotinyl, lipoyl, and phosphopantetheinyl prosthetic groups.

substrates as covalent adducts to and from enzyme active sites. Biotinyl enzymes carry around carboxyl groups; lipoyl enzymes carry around substrate-derived acyl groups, as do the Ppant prosthetic groups (Figure 14.3). The substrate–coenzyme adducts on the 20-Å arms visit different active sites in multisubunit/multienzyme complexes for specific chemical transformations. The classical formulation, first proposed for the lipoyl-enzymes, was that the prosthetic groups functioned as swinging arms, allowing the substrate fragments to be docked sequentially at different catalytic sites for progressive chemical transformation.

There are autonomously folded 80–100 amino acid domains that serve as biotinyl carrier proteins, lipoyl carrier proteins, and acyl-S-pantetheinyl carrier proteins. These domains are embedded in or associated with the catalytic domains of the multienzyme complexes and it is probably more appropriate to think of these domains plus the covalently tethered prosthetic groups as the structural elements that can move and dock in various active sites during carboxyl and acyl transfer cycles (Perham, 2000).

Biotin-Dependent Carboxylases

The fixation of CO_2 into carbon skeletons and the transfer of one-carbon fragments at the level of CO_2 constitute important steps in the biological assimilation of carbon.

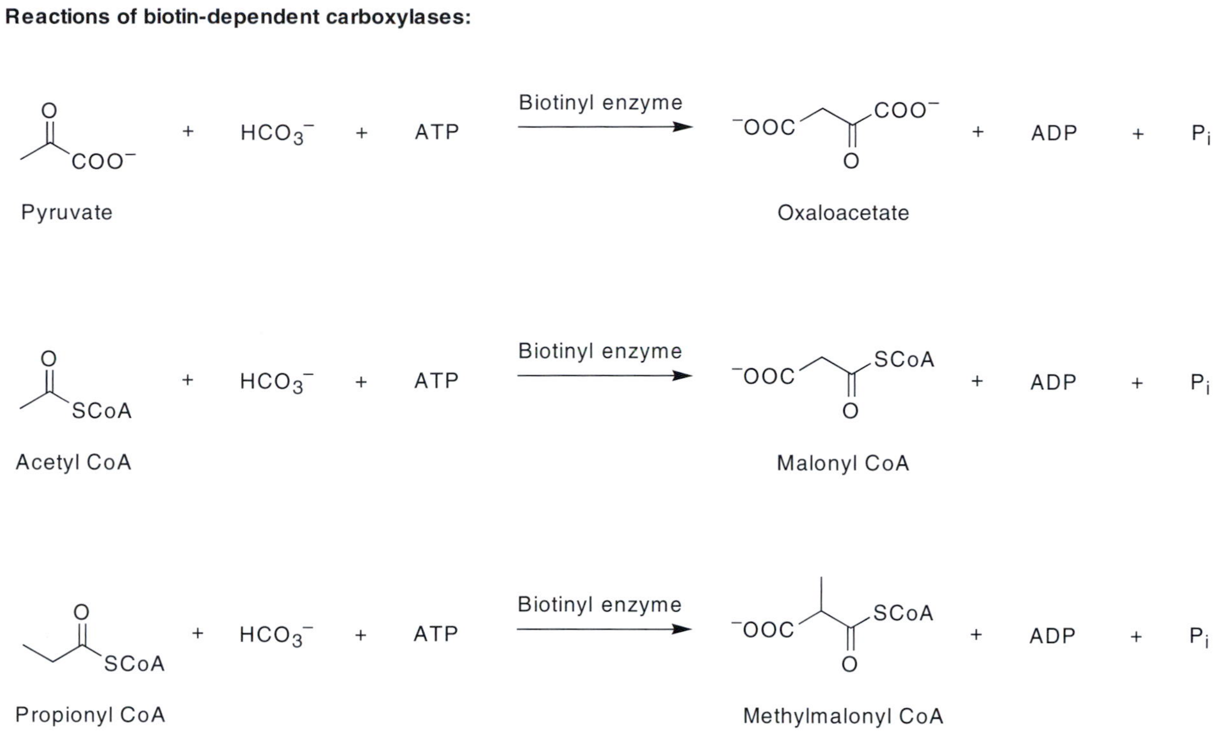

Reactions of biotin-dependent carboxylases:

Figure 14.4 Reactions catalyzed by biotin-dependent carboxylases.

Several kinds of carboxylases are known, including ribulose-1,5-bisphosphate carboxylase, which is responsible for the bulk of carbon assimilation in plants, and several enzymes utilizing phosphoenolpyruvate as latent C_3 carbanion for fixing CO_2 into oxaloacetate. The vitamin-K-dependent carboxylases that derivatize the γ-CH_2 of glutamyl residues in protein substrates to γ-carboxyGlu residues are discussed in Chapter 15. In addition, there are several biotin-dependent carboxylases, among them pyruvate carboxylase (which generates oxaloacetate), acetyl CoA carboxylase (which generates malonyl CoA as the initial step in fatty acid biosynthesis), and propionyl CoA carboxylase (which generates methylmalonyl CoA) (Figure 14.4).

These three biotin-dependent carboxylases require ATP for catalysis, utilizing it to activate cosubstrate bicarbonate, HCO_3^-, as the mixed carboxyphosphoric anhydride (Figure 14.5A). This thermodynamically activated carboxy derivative is formed in a separate subunit termed the ATP carboxylase domain. The carboxyphosphate held in that active site is then captured by the ureido nitrogen of the imidazolone ring of biotin tethered to the biotinyl carrier protein domain to produce N-carboxy-biotinyl enzyme (Figure 14.5B). The ureido moiety nitrogen of the imidazolone portion of biotin is a very weak nucleophile, but as previously noted, both in the green fluorescent protein autocyclization (Chapter 13) and in protein N-glycosylations (Chapter 10), enzymes can generate enough of the isoamide forms of these weak

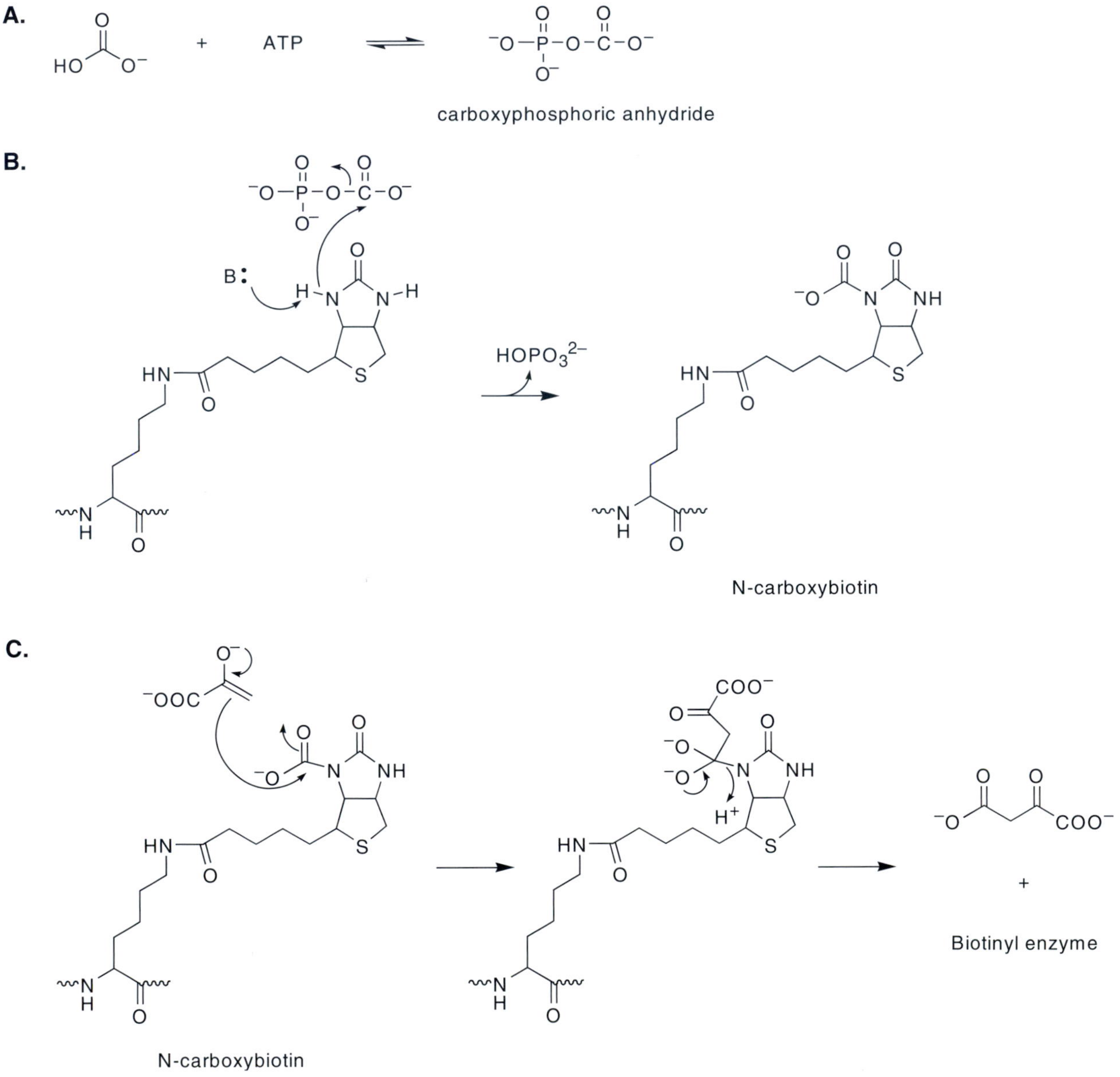

A.

carboxyphosphoric anhydride

B.

N-carboxybiotin

C.

N-carboxybiotin

Biotinyl enzyme

Figure 14.5 Component steps in biotin-dependent carboxylation enzymes. **A.** ATP-dependent formation of carboxyphosphate at the biotin carboxylase site. **B.** Transfer of the CO_2 to biotin as N-carboxybiotin. **C.** C–C bond formation as a stabilized substrate carbanion attacks the N-carboxybiotinyl enzyme.

bases to serve as effective nucleophiles in the catalytic site microenvironments (Knowles, 1989).

The N-carboxybiotinyl carrier protein domain then reorients and docks into a third domain or subunit that is substrate-specific (e.g., for acetyl CoA, propionyl

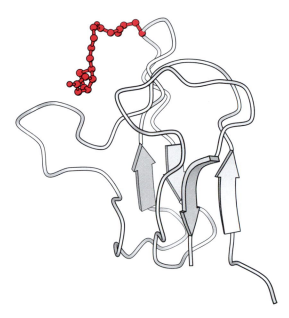

Figure 14.6 Architecture of the *E. coli* biotinyl carrier protein subunit with biotinyllysine chain as ball and stick (figure made using PDB 2BDO).

CoA, or pyruvate). In that active site the catalytic apparatus generates the carbon nucleophile, as transient C_2 carbanion for the two acyl CoA substrates or as the transient enolpyruvate, required for C–C bond formation by attack of the substrate carbanion on the carboxy group of the N-carboxybiotin (Figure 14.5C). The net reaction is ATP-driven C–C bond formation and the incorporation of bicarbonate as a carboxylate group into the carbon skeleton of primary metabolites.

The biotinyl carrier protein ferries the one-carbon carboxy fragment between the active site where bicarbonate is activated to carboxyphosphate and the substrate-specific sites where C–C bond formation is enabled. The covalent ferrying of a CO_2 equivalent tethered to biotin is the essence of the catalytic strategy.

The apo forms of these carboxylases have no activity for C–C bond formation. Thus, the posttranslational modification of the lysine side chain in the biotinyl carrier protein is the all or none switch. In *E. coli* the three domains of acetyl CoA carboxylase are present as separate protein subunits, which allowed early characterization of the biotinyl carrier protein (BCP) domain by x-ray crystallography (Figure 14.6). The biotinyl carrier protein is a small autonomous folded unit of 80 residues, in a capped β sandwich architecture. The lysine to be biotinylated is on a type I' β turn in one of the β sheets (Athappilly and Hendrickson, 1995). The ligase that activates biotin as biotinyl-AMP and then transfers it to the specific lysine of the BCP subunit requires the folded, native conformation and will not work on other lysine side chains (Wilson et al., 1992). This structure-based recognition of a folded domain rather than a local sequence motif by the posttranslational modification enzyme also holds for lipoyla-tion of apo forms of lipoyl carrier proteins (Dardel et al., 1993) and phosphopante-theinylation of apo forms of acyl carrier proteins by their respective modifying enzymes (Reuter et al., 1999).

Lipoyl-Dependent α-Keto Acid Dehydrogenases

Lipoate, 6,8-dithiooctanoate in the dihydro form and the 6,8-disulfide in the oxidized form (Figure 14.7), is a low-molecular-weight redox cofactor used for specialized roles in multicomponent α-keto acid dehydrogenase complexes. Pyruvate is the pri-mary α-keto acid oxidatively decarboxylated in cellular metabolism to produce acetyl CoA, but α-ketoglutarate from glutamate metabolism and branched chain α-keto acids from valine, leucine, and isoleucine catabolism are also substrates for com-parable enzymes in energy metabolism (Figure 14.8) (Perham, 2000).

Taking the transformation of pyruvate to acetyl CoA + CO_2 by pyruvate dehy-drogenase as the prototype, then energy released in the oxidation of the C_3 substrate to the C_2 product is captured in the chemical activation of the acetyl product as the thioester linkage in acetyl CoA. Closer inspection reveals that five coenzymes are required for this energy harvesting cellular machinery to run: CoASH, lipoate, thi-amine-PP (TPP), NAD, and FAD (Figure 14.9) (Walsh, 1979). The CoASH and NAD are cosubstrates, whereas the other three coenzymes are associated with the three enzymes of the pyruvate dehydrogenase complex. The first enzyme, the pyru-vate decarboxylase, E_1, has TPP bound noncovalently. The second enzyme, lipoate transacetylase, E_2, has the covalently attached lipoamide prosthetic group. The third enzyme, dihydrolipoate dehydrogenase, E_3, has bound FAD and uses NAD as the substrate.

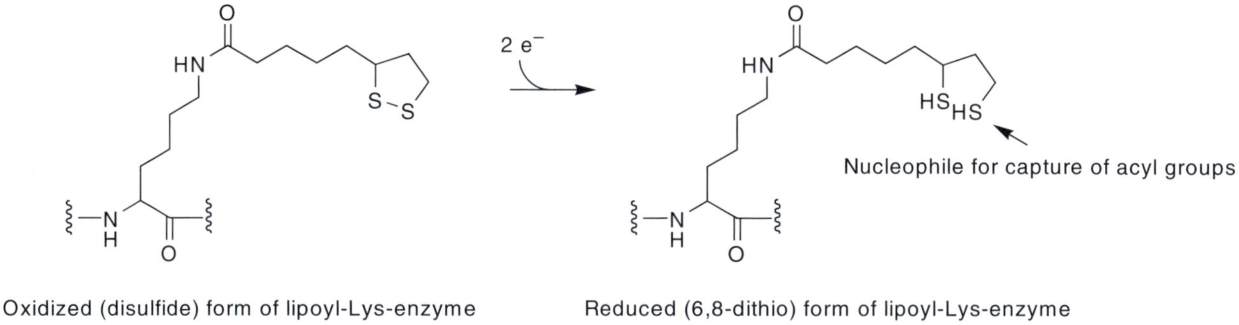

Oxidized (disulfide) form of lipoyl-Lys-enzyme   Reduced (6,8-dithio) form of lipoyl-Lys-enzyme

Figure 14.7 Redox interconversion of the lipoyl-ε-lysyl prosthetic group between the 6,8-dithio form and the two-electron oxidized cyclic disulfide.

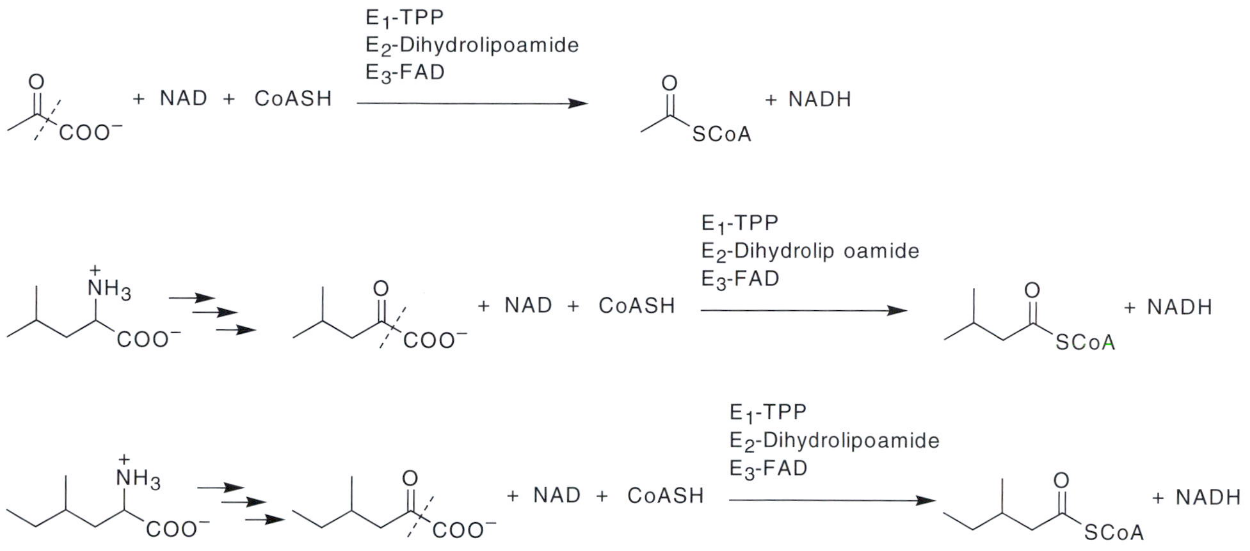

Figure 14.8 Oxidative decarboxylation of pyruvate and branched chain α-keto acids by multicomponent dehydrogenase complexes.

The pyruvate decarboxylase uses the kinetically acessible thiamine-PP thiazole C_2 carbanion generated from adduction to pyruvate, with subsequent loss of CO_2 and formation of the hydroxyethyl-TPP (HE-TPP) adduct in the E_1 active site (Figure 14.10A). If this adduct were to decompose at this step to regenerate the TPP carbanion, the C_2 product would be acetaldehyde (as in yeast fermentations). This would constitute a nonoxidative decarboxylation of pyruvate. To get the aldehyde center up to the oxidation state of acetate requires the special redox chemistry of the lipoate cofactor in E_2.

The resting form of E_2 in α-keto acid dehydrogenase complexes has the covalently attached lipoamide prosthetic group in the disulfide oxidation state. In every catalytic cycle that oxidized lipoamide gets reduced to the dihydro oxidation state as the transferring acetaldehyde fragment is oxidized to the acid state. The job of E_3 is to return the dihydrolipoamide back to the oxidized, disulfide form of lipoamide–E_2 for the next catalytic cycle to run. The acetaldehyde substrate fragment gets transferred from the active site of E_1 to the lipoamide swinging arm of E_2 by formation of the carbanion of HE-TPP (stabilized by delocalization into the thiazole ring of the TPP). The HE-TPP carbanion attacks the electrophilic sulfur of the oxidized lipoamide, opening the ring, forming a thioester between the transferring C_2 substrate fragment and S_8 of the lipoamide (Figure 14.10B). This is the first energy-capturing step in pyruvate dehydrogenase action, because while the 6,8-disulfide of lipoamide is being reductively cleaved, the transferring fragment is oxidized to the acyl level and captured as a high-

Five coenzymes in α-keto acid dehydrogenase-mediated oxidative decarboxylation:

Lipoamide-Lys-E$_2$

CoASH

FAD

NAD$^+$

Thiamine pyrophosphate

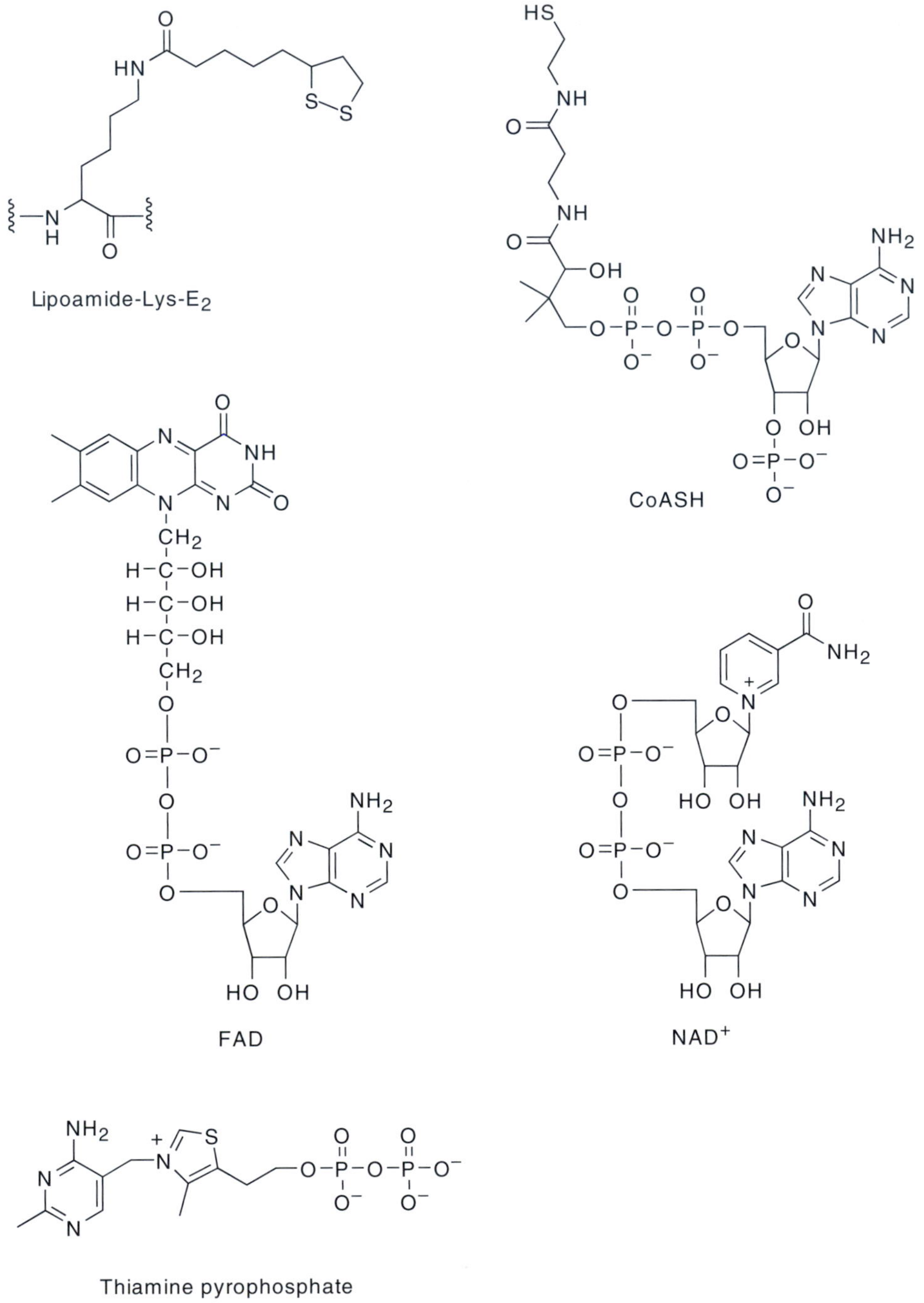

Figure 14.9 Five coenzymes required for the enzymatic chemistry of oxidative decarboxylation of α-keto acid substrates.

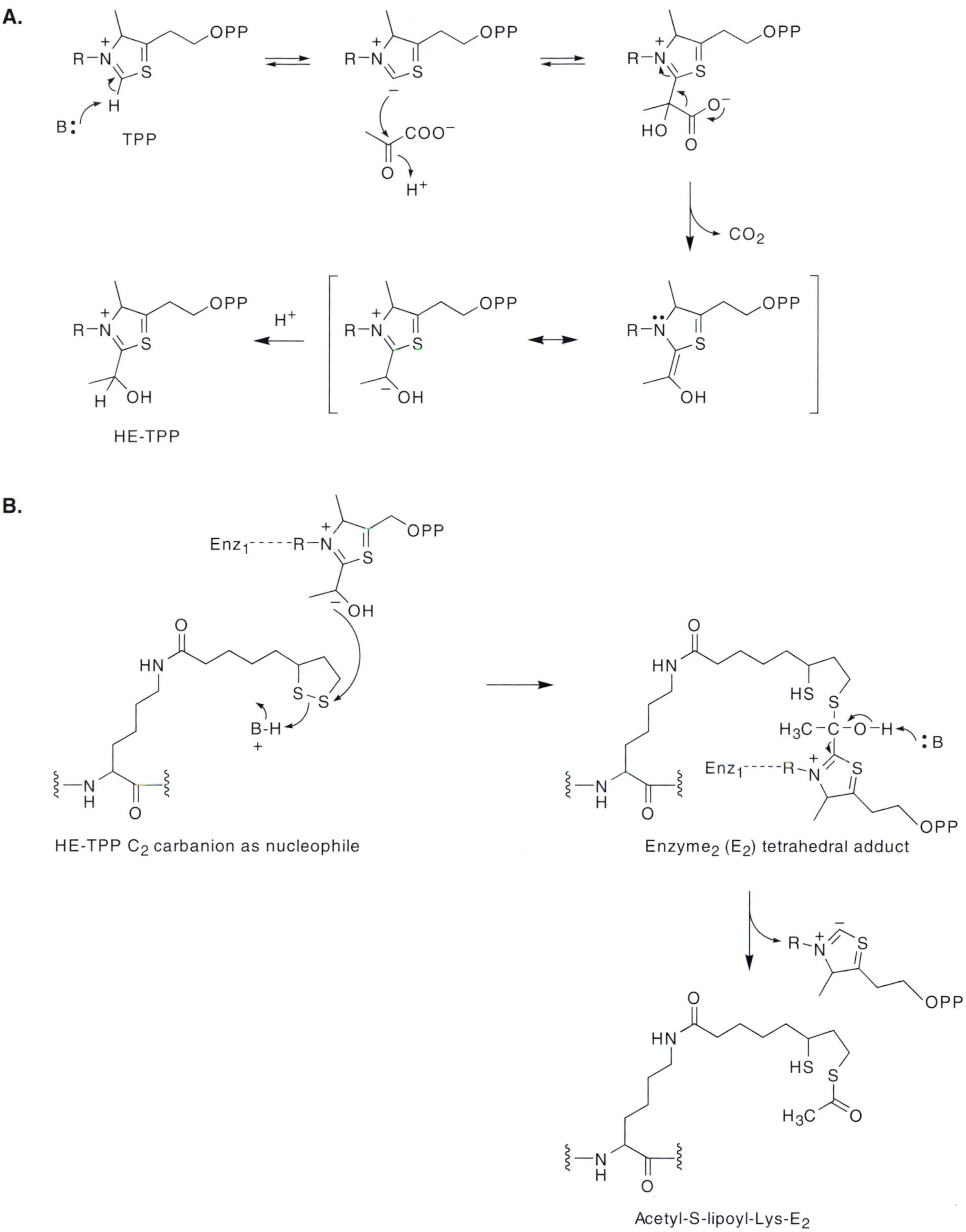

Figure 14.10 **A.** Adduct formation between pyruvate and the TPP C_2 thiazolium anion to form hydroxyethyl-TPP (HE-TPP). **B.** Attack of the HE-TPP C_2 carbanion on the lipoyl disulfide.

energy thioester. All subsequent operations of the multienzyme complex are clean up. In Chapter 4 the versatility of the dithiol-disulfide redox cycles in proteins was noted. Here is the same chemistry, on the intramolecular disulfide of lipoamide, being used in an energy-harvesting maneuver crucial to cellular energy metabolism.

The second half of the E_2–lipoamide transacetylase reaction is the transfer of the now-activated acetyl fragment to cosubstrate CoASH. This yields acetyl CoA, the

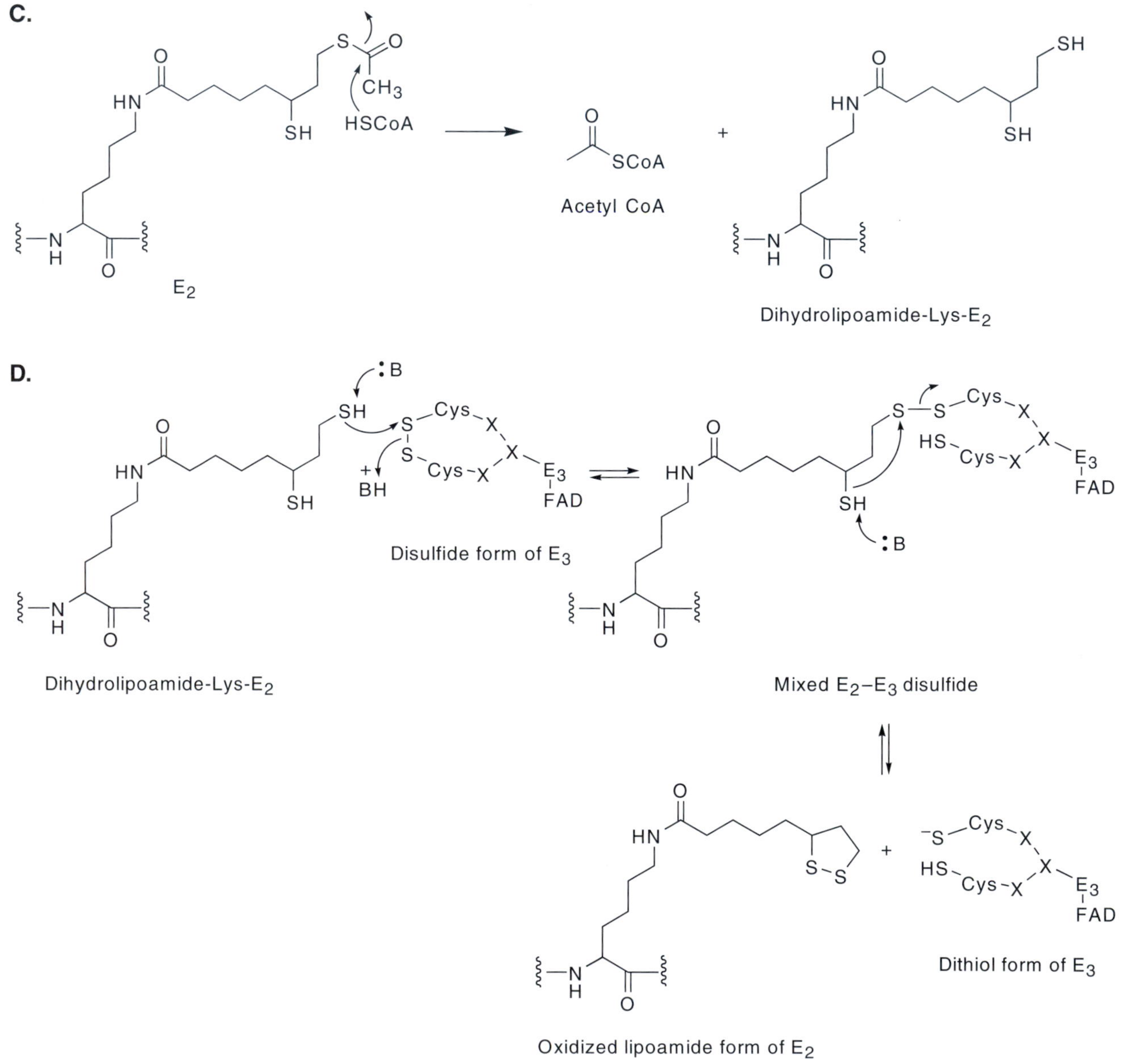

C.

D.

Figure 14.10 **C.** Transacetylation between acetyl-S-lipoyl E_2 and CoASH to yield acetyl-S-CoA. **D.** Reoxidation of the dihydrolipoyl-E_2 active site disulfide in E_3.

central acyl energy currency in cells, by an energy-neutral transthiolation, and produces the dithiol form of the lipoamide–E_2 (Figure 14.10C). The function of E_3 is to reoxidize the dithiol of dihydrolipoamide–E_2 back to the starting disulfide form. This is entirely analogous to the glutathione reductase and thioredoxin reductase reactions, run in the opposite direction, discussed in Chapter 4. Those enzymes have an active site CXXXC disulfide and an FAD coenzyme. NADPH is used to reduce the FAD and the resulting $FADH_2$ reduces the CXXXC disulfide to yield the dithiol form that can participate in thiol-disulfide exchange reactions (Chapter 4). The

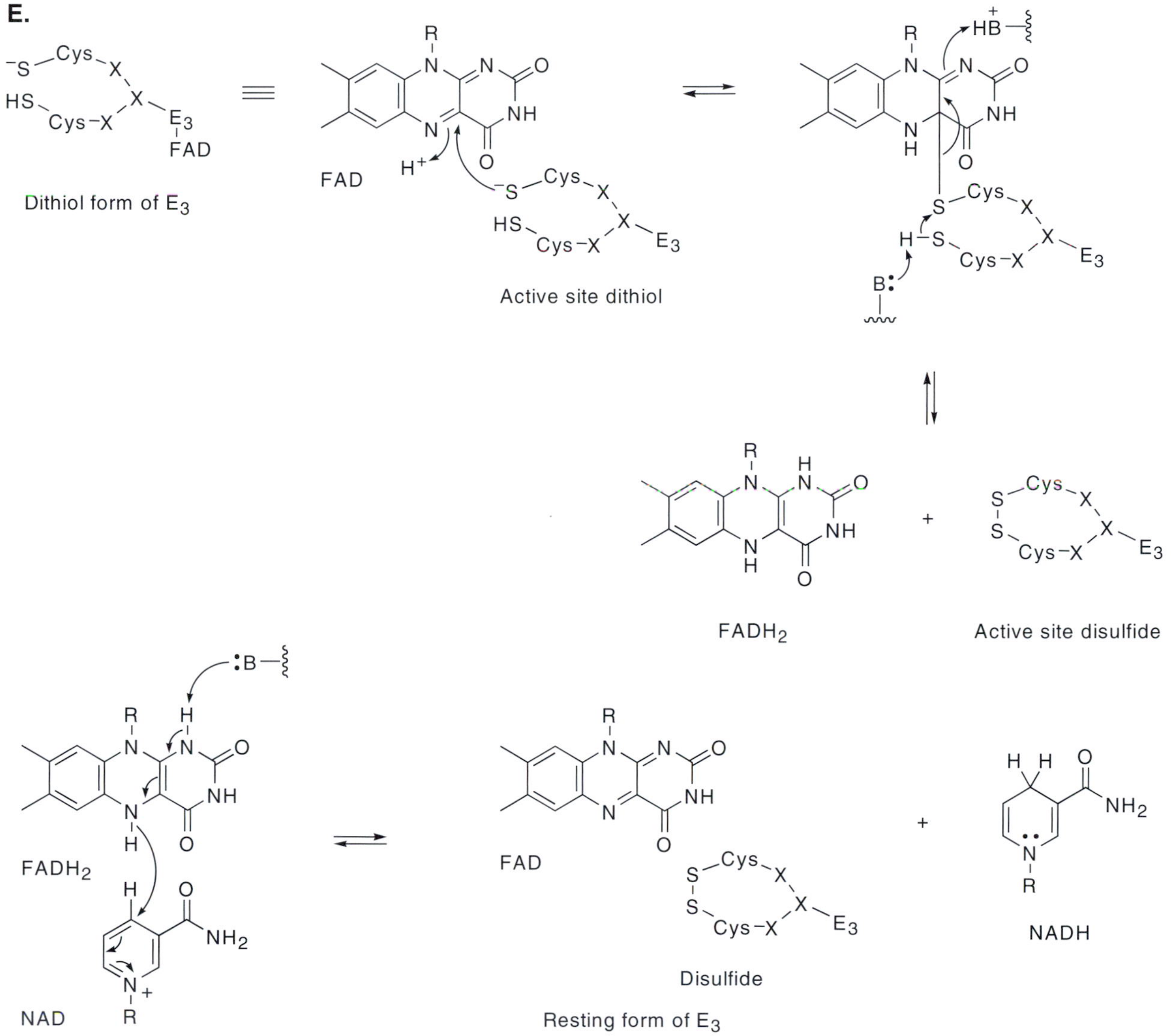

E.

Figure 14.10 **E.** Reoxidation of the dihydro form of E_3 by NAD, regenerating the starting state of the enzyme and producing NADH.

equivalent logic holds for dihydrolipoamide dehydrogenase, an FAD enzyme with an active site disulfide. The dithiol form of the dihydrolipoamide–E_2 attacks the active site disulfide of lipoamide dehydrogenase, and engages in thiol–disulfide interchange to create the disulfide form of E_2, readying it for a subsequent round of catalysis, and the dithiol form of E_3 (Figure 14.10D). The dithiol of E_3 passes two electrons to the FAD in the active site and the resulting $FADH_2$ then passes electrons to NAD (Figure 14.10E). This is the second energy-generating step in pyruvate dehydrogenase catalysis. NADH formation represents the second aliquot of energy captured in a chemically useful form for cells as a consequence of the acetaldehyde-to-acetate oxidation state change effected by E_2.

The covalently tethered lipoamide prosthetic group on E_2 has to swing between the E_1 active site to pick up the transferring hydroxyethyl carbanion, transfer it to the E_2 active site for acetyl transfer to CoASH, and then swing to the E_3 active site to dock at the E_3 active site disulfide to engage in thiol–disulfide exchange of the dihydrolipoamide moiety. The lipoamide cofactor is not just a chaperone or ferrying agent, but is the key player in the redox chemistry that captures energy in chemically useful forms for the cell. It is probably an ancient molecule in cellular energy metabolism.

The lipoyl carrier domains, like the biotinyl carrier protein domains, are about 80-residue, autonomously folding domains (Figure 14.11) that are largely β sheet

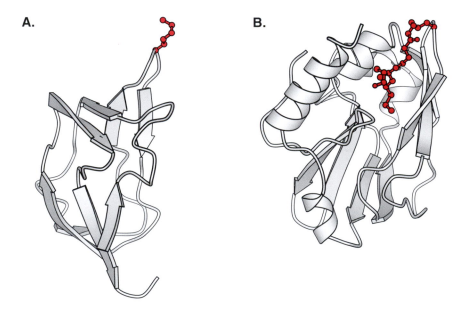

Figure 14.11 **A.** An 80-residue lipoyl domain is the apo protein that is modified on a specific lysine (red ball and stick) during covalent lipoylation in the E_2 subunit of pyruvate dehydrogenase (figure made using PDB 1LAB). **B.** The aminomethylated holo form of the H protein component of the glycine cleavage enzyme with aminomethyl lipoylLys in red ball and stick (figure made using PDB 1HTP).

structures (Dardel et al., 1993). Again the lysine side chain to be modified is in a loop region, in this case between strands β_4 and β_5 (Figure 14.11A). The apo form of the lipoyl carrier protein domains can be lipoylated by the LipA enzyme using lipoic acid and ATP, via the lipoyl-AMP intermediate, in strict analogy to the biotinylation logic (Figure 14.11B), or a second enzyme LipB can use octanoyl-S-acyl carrier protein as substrate (Perham, 2000). The LipA route utilizes an octanoyl-NH-Lys form of E_2, generated from octanoyl-S-ACP. The insertion of the two sulfurs and the disulfide ring formation then occurs (Zhao et al., 2003). Studies with lipoylated carrier proteins show that E_1 recognition involves the carrier domain as well as the swinging arm itself. This suggests that the actual participants in the docking at E_1 and probably E_3 (Perham, 2000) are a composite surface of lipoamide and its presenting carrier protein domain. X-ray analysis of the branched chain keto acid dehydrogenase of *P. putida* shows a hydrophobic tunnel for the lipoyllysyl arm reaching into the E_1 active site (Figure 14.12). The glycine cleavage system in microbial metabolism is also a well-studied case of comparable lipoamide function (Perham, 2000).

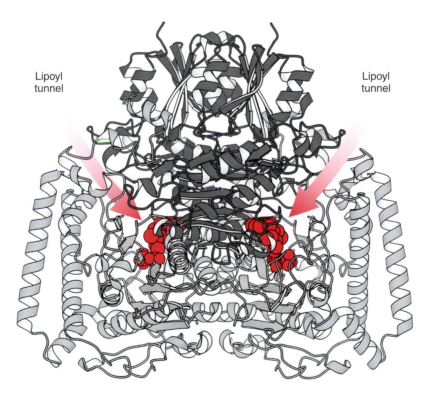

Figure 14.12 A tunnel for the lipoyllysyl arm of E_2 reaching to the active site of E_1, where the HE-TPP carbanion is generated. The thiamine diphosphate cofactor is shown in red space-filling representation (figure made using PDB 1QS0). One E_1 subunit in dark gray, one subunit in light gray.

Phosphopantetheinylated Carrier Proteins in Acyl and Peptidyl Transfers

Fatty Acid Synthases (FAS)

The detection of phosphopantetheinyl prosthetic groups covalently attached to fatty acid synthases was the first example of Ppant posttranslational modification (Wakil, 1989). Fatty acid synthases are multimodular, existing as separate subunits in prokaryotes such as *E. coli* (Type II FAS) and as single polypeptide chains containing multiple domains in yeast and in higher eukaryotes, including humans (Type I FAS) (Figure 14.13A). The acyl carrier protein (ACP) in *E. coli* was easy to characterize because it is a small (10 kDa) heat stable protein, and its phosphopantetheinylation was crucial to the activity of fatty acid synthase. In animal fatty acid synthases the ACP is embedded in a 2500-residue polypeptide containing multiple domains (Smith et al., 2003; Wakil, 1989). As its name implies, the ACP domain is noncatalytic, functioning to carry the elongating fatty acyl chain between the active sites in the other domains of the enzyme. The other domains or subunits of fatty acid synthases are catalytic. The acyl-transferase (AT) domain adds a malonyl group to the holo form of the ACP domain (Figure 14.13B). Other domains carry out the C–C bond-forming chain elongation step [ketosynthase (KS)], the β-ketoacyl thioester reduction [ketoreductase (KR)], the dehydration of the β-OH acyl thioester [dehydratase (DH)], and the enoyl reduction of the enoyl thioester to the saturated thioester [enoylreductase (ER)] (Figures 14.13C–F).

In each cycle of acyl chain elongation, the growing chain is attached in thioester linkage to the terminal thiol of the phosphopantetheine prosthetic group in the ACP domain. All chemical steps, from C–C bond formation through the four-electron reduction of the β-keto to the β-CH_2 acyl thioester, occur while the acyl chain remains on the ACP domain, so the acyl-S-ACP visits four catalytic sites (KS, KR, DH, and ER) in every catalytic cycle. For elongation of malonyl CoA substrates to the C_{16} palmitoyl CoA that is the predominant product released by fatty acid synthases, this means seven elongation cycles for a total of 28 sequential docking events of the growing acyl-S-ACP to the FAS catalytic sites. In principle, the growing acyl chain could have been attached to a cysteinyl thiol side chain with equivalent thermodynamic activation. Presumably the 20-Å reach of the Ppant-SH and the Ppant-S-acyl arm is crucial for the efficient visitation to the catalytic domains of FAS, both in the type I and type II FAS (Figure 14.14).

The posttranslational modification enzymes converting the apo ACP to the holo ACP domains are known generically as phosphopantetheinyltransferases (PPTases)

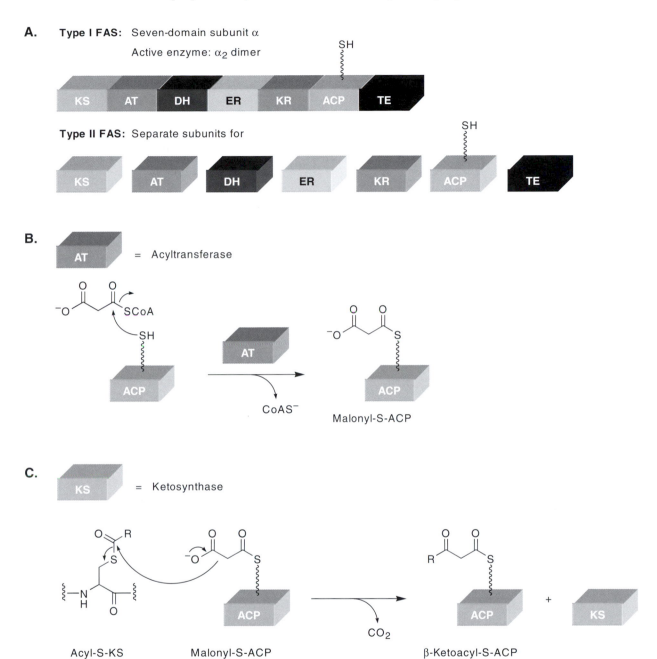

A. **Type I FAS:** Seven-domain subunit α

Active enzyme: α_2 dimer

KS | AT | DH | ER | KR | ACP | TE

Type II FAS: Separate subunits for

KS | AT | DH | ER | KR | ACP | TE

B.

AT = Acyltransferase

CoAS⁻

Malonyl-S-ACP

C.

KS = Ketosynthase

Acyl-S-KS   Malonyl-S-ACP CO_2 β-Ketoacyl-S-ACP   KS

Figure 14.13 **A.** Domains in fatty acid synthase assembly line enzymes. The domains can be collected together in one multidomainal polypeptide chain (Type I) or as separate subunits (Type II). Action of acyltransferase (AT), ketosynthase (KS), ketoreductase (KR), dehydratase (DH), and enoylreductase (ER) domains (14.13**B–F**).

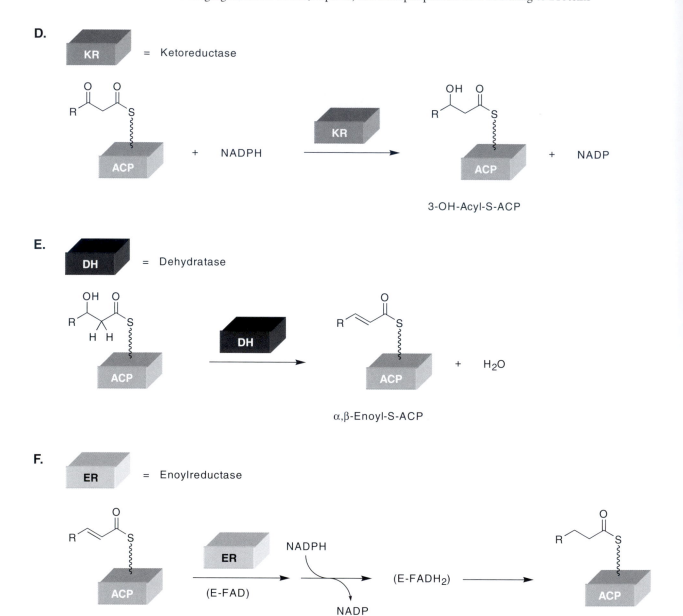

Figure 14.13 (continued)

(Lambalot et al., 1996). There is one in the human genome, two in yeast, and multiple forms in some bacteria that make antibiotics, as discussed below. PPTases, like their biotinylating and lipoylating PTM attachment enzyme counterparts, work only on folded apo carrier protein domains. In this case the coenzyme fragment is covalently attached, not to a lysine, but to a specific serine–CH_2OH side chain (Figure 14.15).

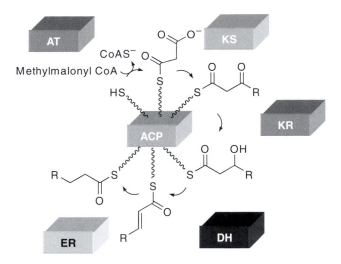

Figure 14.14 Cycles of condensation, reduction, dehydration, and saturation during acyl chain growth attached to the pantetheinyl tether on the ACP domain of fatty acid synthases.

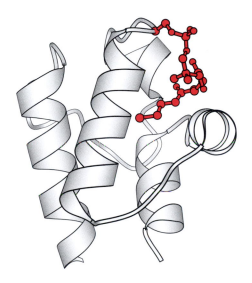

Figure 14.15 Covalent attachment of the Ppant tether (red ball and stick) via phosphodiester linkage to a serine side chain in the ACP domain (figure made using PDB 1F80).

The serine side chain attacks the CoASH cosubstrate at the diphosphate linkage, releasing 3′,5′-ADP while the Ppant moiety forms a phosphodiester to the serine side chain. This phosphorylation of serine has formal analogy to the phosphorylation by serine/threonine protein kinases (Figure 14.16). The HS-Ppant-ACP, the holo form of the acyl carrier protein, now provides the nucleophilic thiol on which acyl chain growth can occur.

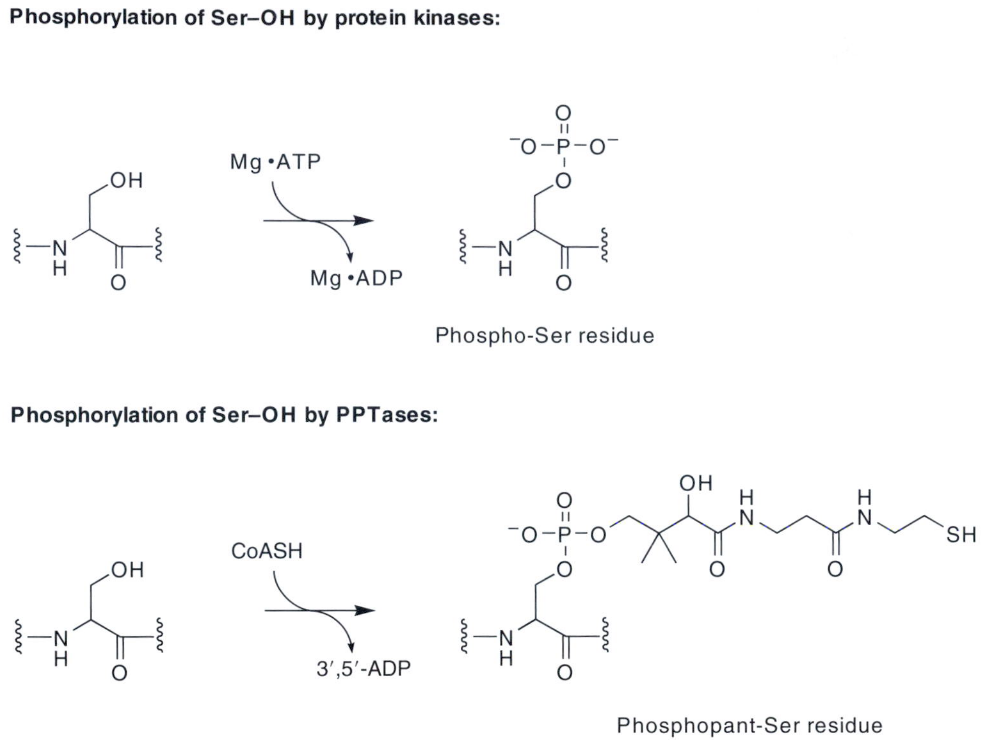

Phosphorylation of Ser–OH by protein kinases:

Phosphorylation of Ser–OH by PPTases:

Figure 14.16 Analogy between serine side chain phosphorylation by protein kinases and by phosphopantetheinyltransferases.

Polyketide Synthases (PKS)

Polyketide synthases, found in bacteria and fungi, are multimodular catalysts related to fatty acid synthases. They are also found in multimodular type I forms and separate subunit type II forms (Rawlings, 2001a and 2001b). The antibiotic erythromycin is made on type I PKS assembly lines while the antitumor agent doxorubicin is made by distributed type II PKS (Figure 14.17). Parallel logic of using HS-Ppant-ACP domains as carriers and tethering way stations for acyl chain growth applies in PKS catalysis.

In the type II systems, such as for daunomycin formation, the KR catalytic function is disabled, so in each cycle a β-keto acyl thioester is produced and then subject to elongation. After nine cycles a decaketidyl-S-ACP would be formed. The nine keto groups allow for directed carbanion chemistry to construct the tetracyclic skeleton of the antitumor drug. Analogous logic is used to generate a polyketidyl-S-PCP to form the antibiotic tetracycline (Figure 14.18) (Shen, 2000; Walsh, 2003).

In the erythromycin type I PKS assembly line there are six modules, distributed two at a time in three protein subunits (deoxyerythronolide B synthase = DEBS1–DEBS3). There are seven ACP domains (one in the initial loading module),

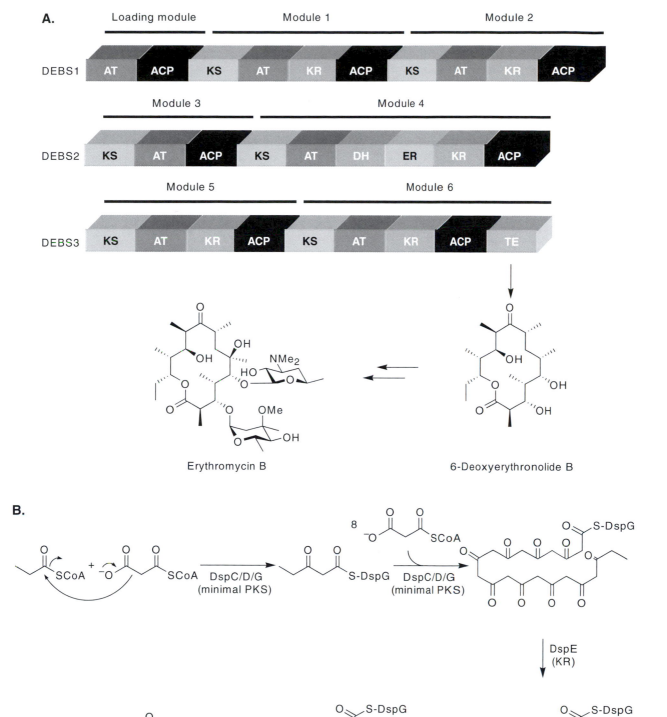

Figure 14.17 **A.** Biogenesis of the carbon skeleton of erythromycin on a modular type I PKS. **B.** Generation of the tetracyclic skeleton of doxorubicin on type II PKS machinery.

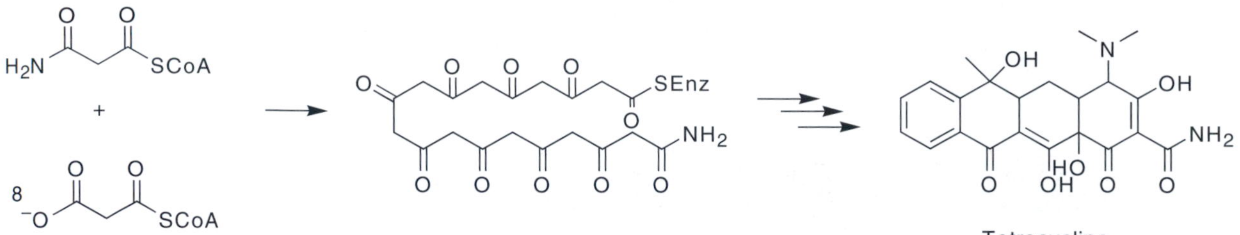

Figure 14.18 The nonaketidyl-S-pantetheinyl-P-enzyme intermediate in tetracycline biosynthesis.

each of which must be primed with HS-Ppant for the assembly line to run (Figure 14.19). The acyl chain growth starts on the first ACP. When all catalytic domains in the first module of DEBS1 have been visited by the acyl-S-ACP, the acyl chain is transferred to the next downstream HS-Ppant-ACP by attack of the downstream pantetheinyl thiol. Now the acyl chain can be presented to each of the catalytic domain active sites in module 2. When all chemical operations are completed, the acyl chain is transferred to HS-Ppant-ACP$_3$, now an inter-subunit transfer to DEBS2. Operation on modules 3 and 4 occur in DEBS2 before chain transfer to the third protein, DEBS3, with ACP$_6$ and ACP$_7$ as the last two way stations for the growing acyl chain. Within a given module, some catalytic domains may be inactivated or even absent, so incomplete processing of the acyl chain will occur and functionality will persist as the chain moves to the next downstream ACP way station. When the acyl chain has reached the last ACP domain in DEBS3, and six chain elongations have occurred, concomitant with chain transfer to the downstream ACPs, the 15-carbon backbone of the acyl chain has six methyl substituents (from the six methylmalonyl CoAs used in the elongation steps) and five functional groups (four hydroxyls and one ketone). These five functional groups reflect the incomplete processing of the chain at modules 1, 2, 3, 5, and 6. Only module 4 (in DEBS2) has the DH and ER domains typical of FAS modules, allowing full processing of the β-keto and β-OH groups to the β-CH$_2$ substituents.

The chain termination step on the FAS and PKS assembly lines occurs by enzymatic disconnection of the covalent attachment of the full length acyl chain to the last ACP. The thioester linkage between the acyl chain and the S-Ppant prosthetic group is cleaved by the thioesterase (TE) domain, the most C-terminal domain in the protein assembly line, usually just downstream of the last ACP. TE domains have nucleophilic active site serine side chains that attack the acyl chain and translocate it to the TE active site (Figure 14. 20A). The acyl-O-TE intermediate can now suffer hydrolysis (capture by water), as in the release of the daunomycin acid. It can suffer thiolysis, as in capture by CoASH to release the fatty acyl CoA in fatty acid synthesis. Or, an intramolecular nucleophile could capture the thioester and result in macrocyclization (Figure 14.20B).

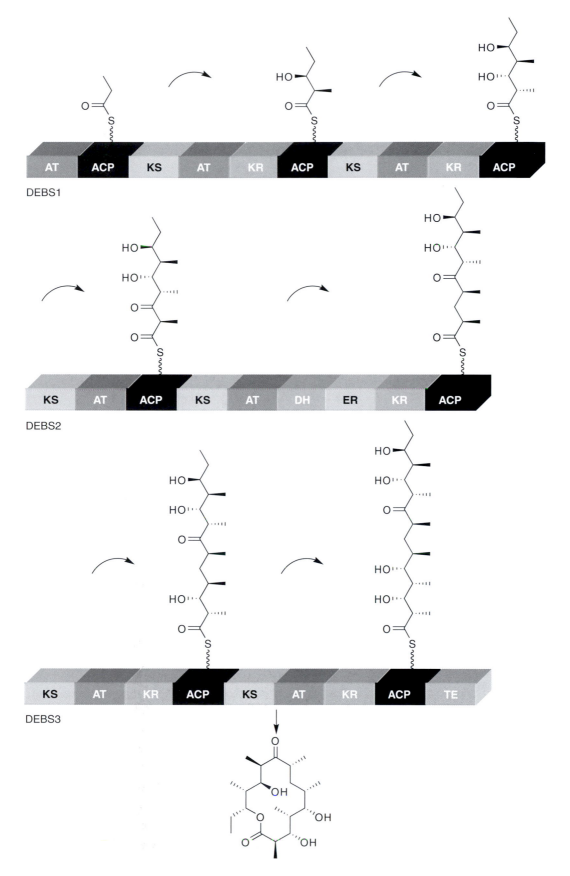

Figure 14.19 The elongating chains of the precursors of deoxyerythronolide B on the seven carrier protein domains of the DEBS assembly line.

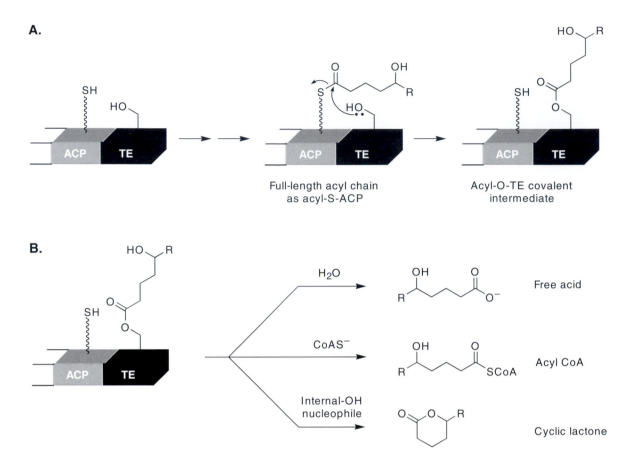

Figure 14.20 Chain termination by fatty acid synthases and polyketide synthases. **A.** Chain transfer from the last ACP to the active site serine of the TE domain. **B.** Thiolysis versus macrocyclization to yield acyl CoA or macrolactone products.

In the acyl chains built up by fatty acid synthase, all the carbons have been reduced to CH_2 groups and bear no potential nucleophiles. For polyketide acyl chains, the incomplete reduction leaves –OH groups, as in the erythromycin assembly line. Indeed, the DEBS TE acts as a regiospecific macrolactonization catalyst, catalyzing the attack of the C_{13}–OH group on the oxoester in the TE active site to produce the 14-membered macrolide, deoxyerythronolide B (DEB). The incomplete reduction during chain elongation by PKS assembly lines thus sets up the possibility of regiospecific macrolactonization in this natural product class.

The distinguishing feature of the type I PKS assembly lines is the presence of multiple modules, with one HS-Ppant-ACP per module, such that chain growth and translocation down the modules is a kind of bucket brigade action, with the HS-Ppant in each downstream module the hand that covalently grasps the transferring acyl group (the bucket).

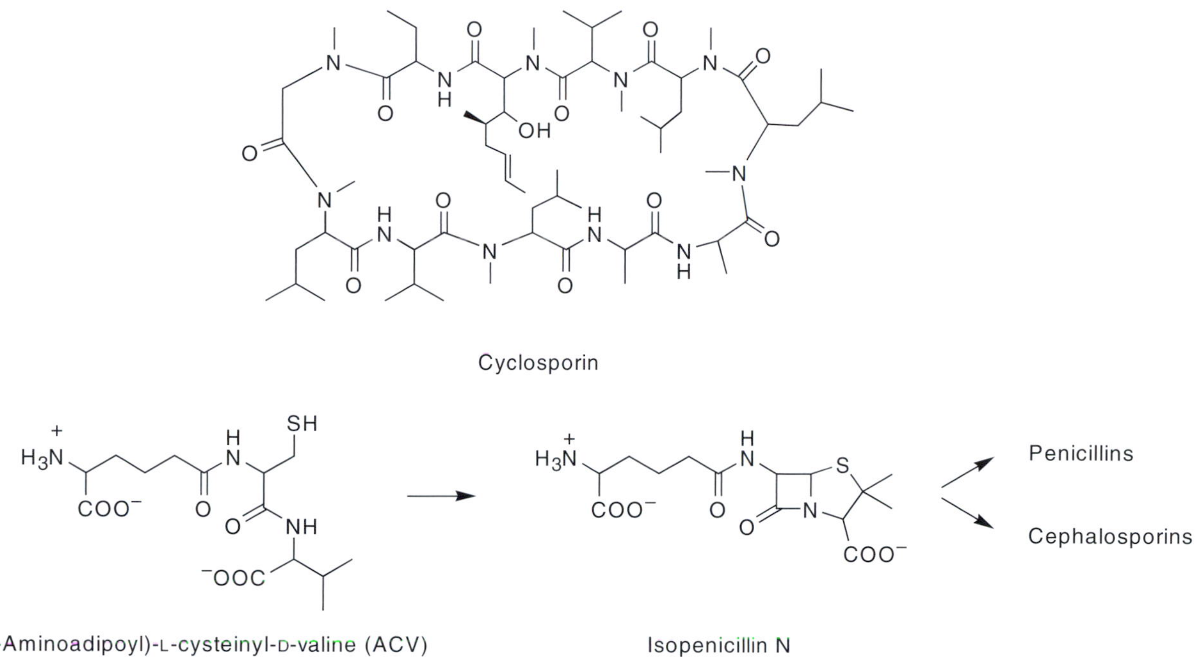

Cyclosporin

L-δ-(α-Aminoadipoyl)-L-cysteinyl-D-valine (ACV)

Isopenicillin N

Penicillins

Cephalosporins

Figure 14.21 Nonribosomal peptides, cyclosporine and ACV, generated on Ppant-containing multimodular enzymatic assembly lines.

Nonribosomal Peptide Synthetase (NRPS) Assembly Lines

The third place where HS-Ppant prosthetic groups are embedded in multimodular enzymatic assembly lines is in the biosynthesis of nonribosomal peptides by bacteria and fungi (Marahiel, 1997). These include cyclosporin, the immunosuppressive drug used to fight the rejection of organ transplants, and the tripeptide aminoadipoylcysteinylvaline (ACV), the precursor of penicillins and cephalosporins (Figure 14.21). The logic of NRPS assembly lines is analogous to type I PKS assembly lines: one module for each amino acid activated and incorporated into the growing peptide chain. Each module has an HS-Ppant carrier protein (the peptidyl carrier protein = PCP), and catalytic domains (Doekel, 2001; Marahiel, 1997). One catalytic domain is for the selection of the amino acid to be activated (as the aminoacyl-AMP) and incorporated via aminoacyl-S-PCP formation. This is called the adenylation (A) domain. The second core catalytic domain in each NRPS elongation module is the condensation (C) domain, which catalyzes peptide bond condensation between the growing peptidyl-S-PCP upstream and the aminoacyl-S-PCP downstream (Figure 14.22). Thus, each NRPS elongation module should have C-A-PCP domains.

The ACV synthetase, involved in penicillin biosynthesis, is fashioned according to this logic (Walsh, 2003). The enzyme is a 450-kDa polypeptide with three modules, one for each amino acid, with a total of 10 domains (Figure 14.23). The first module

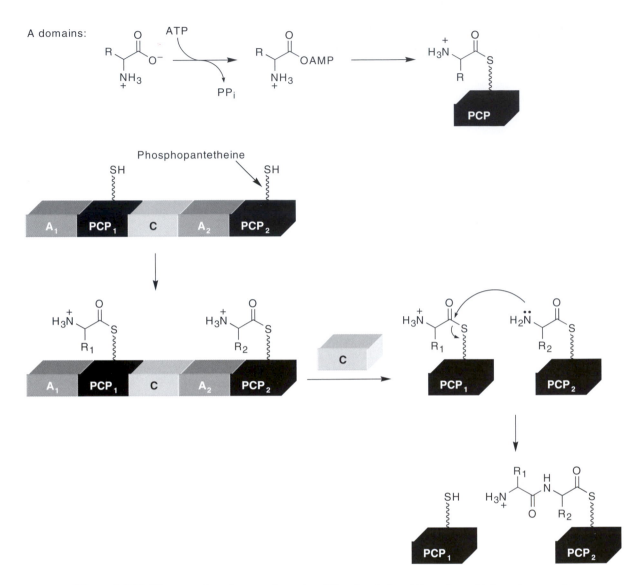

Figure 14.22 The core domains, condensation (C), adenylation (A), and peptidyl carrier protein (PCP), of NRPS modules.

has an A_1-PCP_1 for selection and loading of the nonproteinogenic amino acid ε-aminoadipate onto HS-Ppant-PCP_1. The A_1 domain actually activates the ε-carboxylate and makes an ε-aminoadipoyl thioester linkage to the Ppant prosthetic group on PCP_1. The second module has to activate cysteine (A_2), load it onto HS-Ppant-PCP_2, and make the first peptide bond (C_1) with chain translocation to create ε-aminoadipoyl-Cys-S-PCP_2. The third module needs to activate Val (A_3), load it onto HS-Ppant-PCP_3, and make the second peptide bond (C_2). There are two other domains in the third module for two remaining operations. The first is epimerization of L-Val

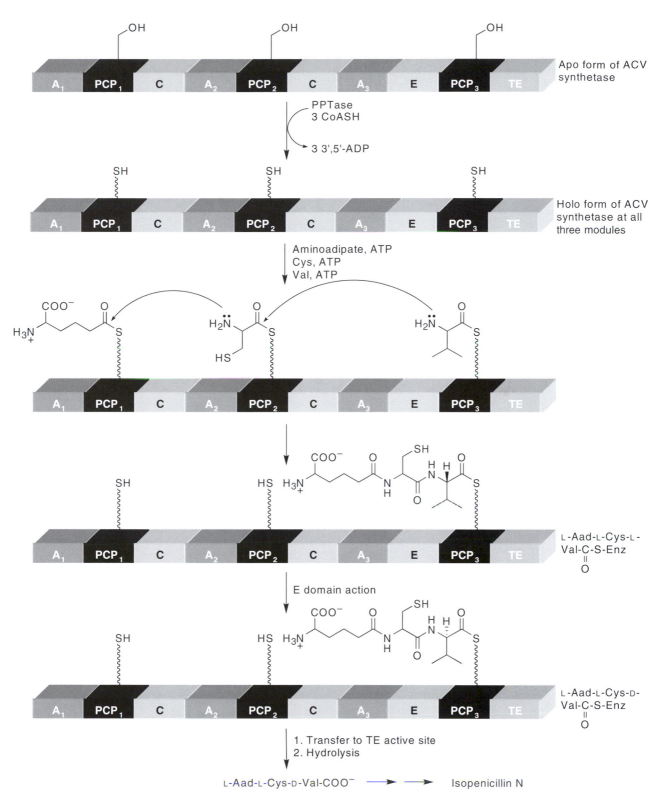

Figure 14.23 The three modules of the 450-kDa ACV synthetase involved in generation of the ACV tripeptide for penicillin and cephalosporin biosynthesis.

to D-Val, achieved by the epimerase (E) domain, so the tripeptidyl-S-enzyme accumulating on PCP$_3$ is ε-aminoadipoyl-L-Cys-D-Val-S-PCP$_3$. This is the substrate for the 10th and most C-terminal domain of ACV synthetase, the TE domain. This TE domain acts as a hydrolase, making the tripeptidyl-O-TE acyl enzyme intermediate and then hydrolyzing it to release the free tripeptide. Subsequent metabolism by isopenicillin N-synthase converts ACV by double cyclization into the 4,5-bicyclic ring system of the β-lactam antibiotics.

In the NRPS assembly lines, the peptide bonds are formed via activated thioester intermediates. The A domains can incorporate nonproteinogenic amino acids and produce nonstandard peptide bonds (as in ACV), creating useful diversity for biological activity (Doekel, 2001). The PCP domains and the HS-Ppant arm again serve the key functions of chaperoning the growing chain, preserving thermodynamic activation in thioester linkages, and allowing local catalytic domains in a given module to carry out their chemical operations (e.g., epimerization) before the chain is passed downstream to the next aminoacyl-S-Ppant-PCP.

Summary

The biotin, lipoic acid, and phosphopantetheine cofactors have been captured as covalent prosthetic groups by their respective apo proteins. The three coenzymes are introduced posttranslationally onto prefolded 8- to 10-kDa protein domains that perform carrier functions, allowing the business end of the coenzymes to visit or dock into the active sites of catalytic components of multienzyme complexes. The carrier protein domains serve as recognition platforms for protein–protein interactions to provide affinity for the docking operations.

Each of the three coenzymes in turn carries portions of the substrates in covalent linkage to ensure adequate lifetime and concentration of the substrate-derived intermediates as they traverse between the different catalytic sites of the multienzyme, multimodule complexes.

Biotin carries CO_2 for primary metabolic carboxylations, while lipoate is involved in the reverse reaction, decarboxylation of α-keto acids. Lipoate combines dithiol–disulfide redox chemistry with thiamine-PP mediated carbanion chemistry to harvest energy from the oxidative decarboxylations that generate acyl CoA substrates for the Krebs cycle. The terminal thiol of phosphopantetheine in reaction with substrate acyl groups enables acyl thioester chemistry and provides thermodynamic activation that permits Claisen condensation chemistry for C–C bond forming steps in fatty acid and polyketide assemblies. The acyl-S-pantetheine intermediates are also central to nonribosomal peptide bond formations in the biogenesis of antibacterial and antitumor natural products.

References

Athappilly, F. K., and W. A. Hendrickson. "Structure of the biotinyl domain of acetyl-coenzyme A carboxylase determined by MAD phasing," *Structure* **3**:1407–1419 (1995).

Dardel, F., A. L. Davis, E. D. Laue, and R. N. Perham. "Three-dimensional structure of the lipoyl domain from *Bacillus stearothermophilus* pyruvate dehydrogenase multienzyme complex," *J. Mol. Biol.* **229**:1037–1048 (1993).

Doekel, S., and M. Marahiel. "Biosynthesis of natural products on modular peptide synthetases," *Metab. Eng.* **3**:64–77 (2001).

Knowles, J. R. "The mechanism of biotin-dependent enzymes," *Annu. Rev. Biochem.* **58**:195–221 (1989).

Lambalot, R. H., A. M. Gehring, R. S. Flugel, P. Zuber, M. LaCelle, M. A. Marahiel, R. Reid, C. Khosla, and C. T. Walsh. "A new enzyme superfamily—the phosphopantetheinyl transferases," *Chem. Biol.* **3**:923–936 (1996).

Marahiel, M., H. Mootz, and T. Stachelhaus. "Modular peptide synthetases involved in nonribosomal peptide synthesis," *Chem. Rev.* **97**:2651–2674 (1997).

Perham, R. N. "Swinging arms and swinging domains in multifunctional enzymes: Catalytic machines for multistep reactions," *Annu. Rev. Biochem.* **69**:961–1004 (2000).

Rawlings, B. J. "Type I polyketide biosynthesis in bacteria (part A—erythromycin biosynthesis)," *Nat. Prod. Rep.* **18**:190–227 (2001a).

Rawlings, B. J. "Type I polyketide biosynthesis in bacteria (part B)," *Nat. Prod. Rep.* **18**:231–281 (2001b).

Reuter, K., M. R. Mofid, M. A. Marahiel, and R. Ficner. "Crystal structure of the surfactin synthetase-activating enzyme sfp: A prototype of the 4′-phosphopantetheinyl transferase superfamily," *EMBO J.* **18**:6823–6831 (1999).

Shen, B. "Biosynthesis of aromatic polyketides," *Top. Curr. Chem.* **209**:1–51 (2000).

Smith, S., A. Witkowski, and A. K. Joshi. "Structural and functional organization of the animal fatty acid synthase," *Prog. Lipid Res.* **42**:289–317 (2003).

Wakil, S. J. "Fatty acid synthase, a proficient multifunctional enzyme," *Biochemistry* **28**:4523–4530 (1989).

Walsh, C. T. *Enzymatic Reaction Mechanisms*. Freeman: San Francisco (1979).

Walsh, C. T. *Antibiotics: Actions, Origins, Resistance*. ASM Press: Washington (2003).

Wilson, K. P., L. M. Shewchuk, R. G. Brennan, A. J. Otsuka, and B. W. Matthews. "*Escherichia coli* biotin holoenzyme synthetase/bio repressor crystal structure delineates the biotin- and DNA-binding domains," *Proc. Natl. Acad. Sci. U.S.A.* **89**:9257–9261 (1992).

Zhao, X., J. R. Miller, Y. Jiang, M. A. Marletta, and J. E. Cronan. "Assembly of the covalent linkage between lipoic acid and its cognate enzymes," *Chem. Biol.* **10**:1293–1302 (2003).

15

Protein Carboxylation and Amidation

Calcium ions organize the Gla domain of factor IX by multidentate coordination: $Gla_{7,17,21}$ as ligands to one Ca^{2+}.

The side chains of glutamyl and glutaminyl residues in proteins can be sites for several types of posttranslational modification. For example, recall the enzymatic methylation of glutamyl carboxylate side chains in bacterial chemotaxis transmembrane receptor proteins as a titrant for signal duration and intensity in Chapter 5. Analogously, at the end of Chapter 11 the ability of the *E. coli* cytotoxic necrotizing protein to inactivate Rho family GTPase activity by enzymatic deamidation of Gln_{63} to Glu_{63} was mentioned. The homologous toxin from *Bordetella bronchiseptica* deamidates Gln_{61} at the active site of Ras to block its GTPase activity.

When Gln is present as the N-terminal residue, having been generated after methionyl aminopeptidase action to remove Met_1 (Chapter 6), there are some proteins where cyclization of this Gln_1 occurs by intramolecular attack of the free α-NH_2

435

group on the δ-carboxamido group. This intramolecular cyclization and loss of the δ-NH$_2$ as ammonia creates a pyroglutamate residue (also known as 5-oxoproline) at the N-terminus. Such a derivatization at the N-terminus can stabilize a protein towards aminopeptidase action since there is no longer a free N-terminal amino group as discussed later in this chapter. There are 5-oxoprolinase enzymes that can specifically recognize and hydrolyze the 5-oxoPro amide linkage in such modified proteins.

In this chapter several distinct types of posttranslational protein modifications at Glu, Gln, and Gly side chains, respectively (Figure 15.1), are discussed, with some

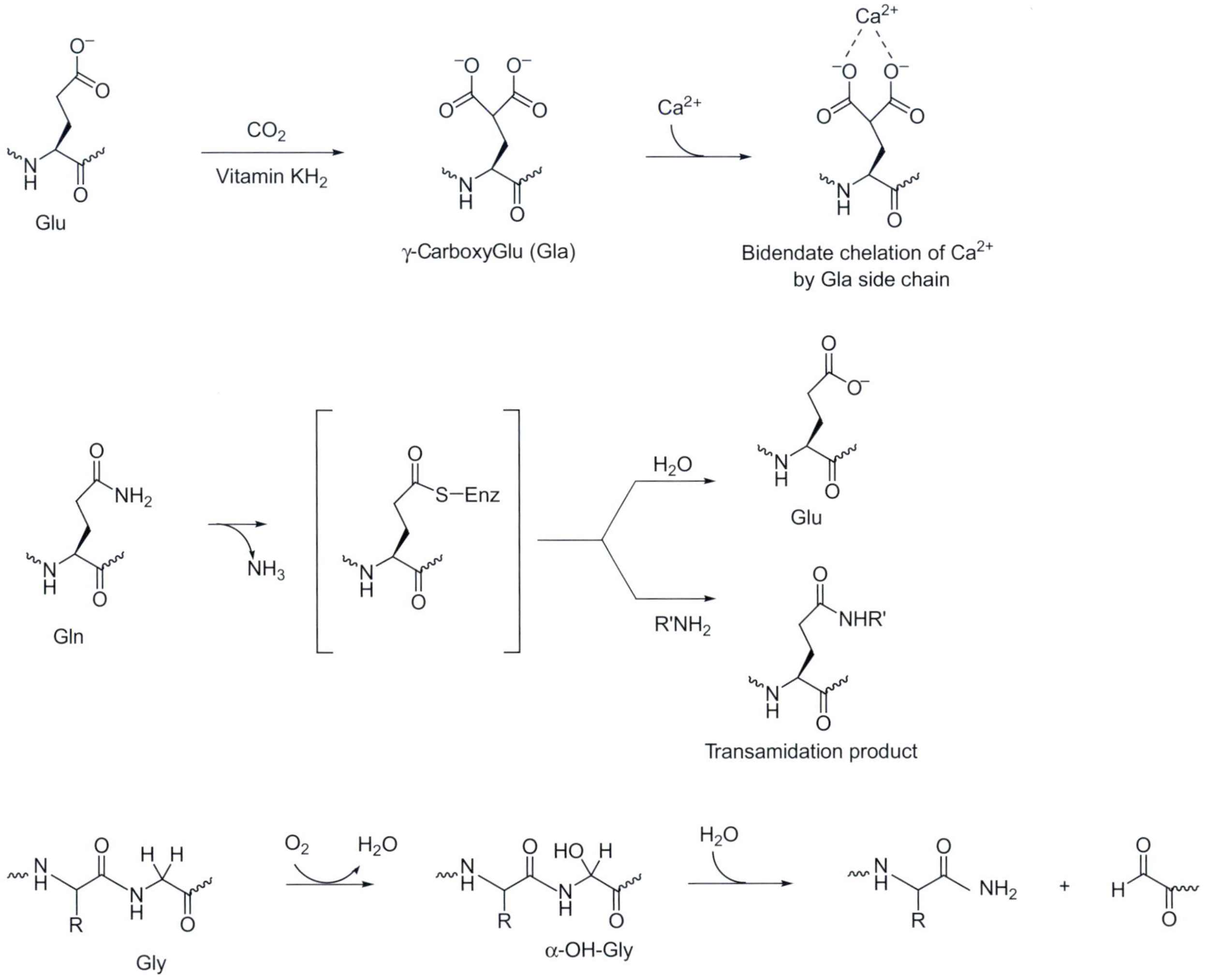

Figure 15.1 Posttranslational modifications of Glu, Gln, and Gly: C–C bond formation at the γ-carbon during Gla formation; glutaminase action of Gln side chains leading to net hydrolysis or transglutamination; and α-hydroxylation of Gly sets up peptide bond cleavage to the C-terminal amide.

functional complementarity. The Glu modification is fixation of CO_2 via the creation of new C–C bonds as γ-carboxy-Glu residues are formed, most famously in mammalian zymogen forms of proteases functioning in blood coagulation cascades. The Gln modification occurs in proteins at the γ-carboxamido nitrogen locus, and can be a net hydrolysis (Gln to Glu, as in the above toxin actions) or a net transamidation. Transamidation of Gln side chains to Lys side chains occurs at the later stages of clot formation to provide cross-links that add mechanical strength to fibrin clots or the extracellular matrix.

Also grouped in this chapter is the C-terminal cleavage of glycine residues in proteins that are precursors of peptide hormones in neural and neuroendocrine tissues. In Chapter 9 the posttranslational activation of the C-terminal glycine residue of ubiquitin and ubiquitin-like protein tags to create isopeptide linkages was examined in detail. Here, the oxidative removal of C-terminal glycine residues, leaving the amino group of the glycine moiety behind as an amide linkage in the newly liberated C-terminus, is discussed. Finally, both polyglycine and polyglutamate chain formation at the C-terminal tails of tubulin subunits is covered.

Protein Carboxylations

Two of the swinging-arm coenzymes noted in Chapter 14, biotin and lipoate, act while tethered to enzymes that carry out carboxylations and decarboxylations of small-molecule substrates, respectively. The lipoate cofactor on E_2 of the pyruvate dehydrogenase complex accepts the two-carbon acyl fragment *after* decarboxylation at the thiamine-PP center of E_1, so the covalently attached lipoamide does not participate directly with CO_2 as the substrate or product. By contrast, the biotinyl-lysine cofactors of carboxylases for acetyl CoA, propionyl CoA, and pyruvate combine with carboxyphosphate, derived from substrate CO_2, at the active site of the biotin carboxylase subunits (Figure 15.2A). The resultant N-carboxybiotionyl adduct is the form in which CO_2 is transferred between the subunit active sites of those carboxylases.

N-Carboxy Protein Adducts

Given the millimolar concentrations of bicarbonate in intracellular fluids and the hydration–dehydration equilibrium between CO_2 and bicarbonate, the reversible addition of CO_2 to lysine side chains can occur in proteins to form the N-carbamates (Figure 15.2B), in some analogy to the N-carboxyureido moiety in the biotin case above. These are generally of low fractional stoichiometry, but N-carbamates can be stabilized in enzyme active sites by electrostatic interactions with cationic side chains or by complexation with metal cations. A well-documented example where the N-

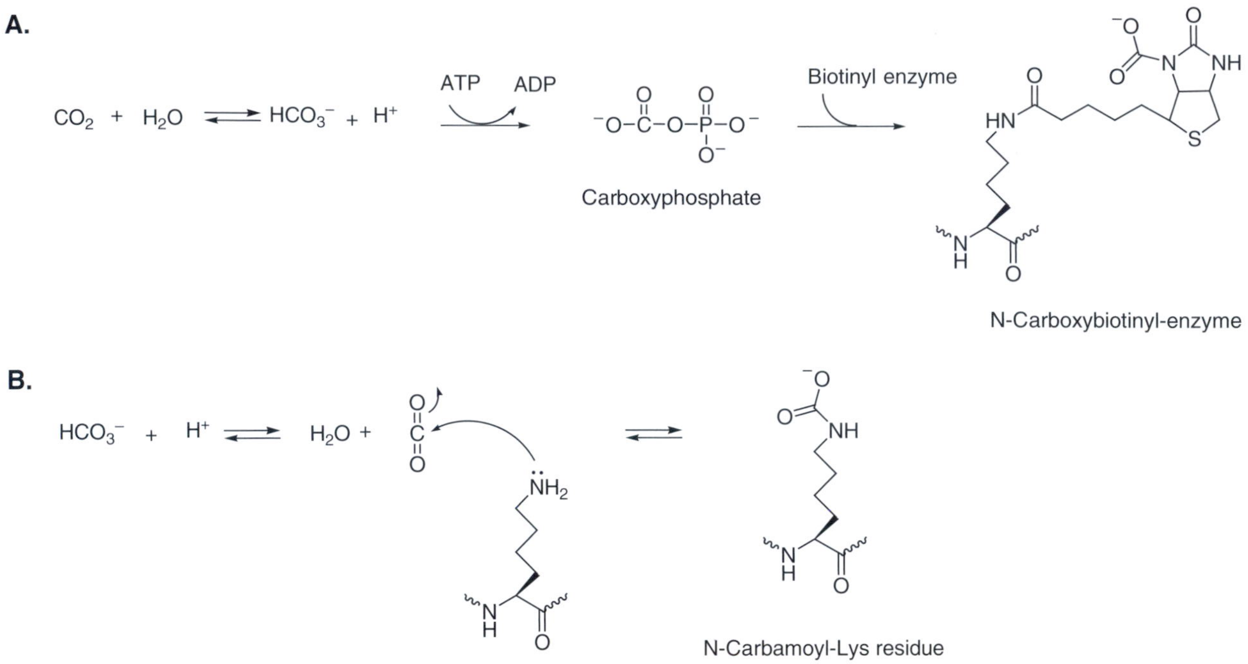

A.

N-Carboxybiotinyl-enzyme

B.

N-Carbamoyl-Lys residue

Figure 15.2 N-Carboxylation of proteins. **A.** N-Carboxybiotinyl enzyme intermediates. **B.** N-Carbamoyl-Lys enzyme species.

carbamate accumulates and is an integral part of the active site catalytic machinery is the class D β-lactamase OXA10 from *Pseudomonas* bacteria (Golemi et al., 2001). The enzyme is active only when Lys_{70} is N-carbamoylated and is then poised to activate a water molecule in the active site for substrate β-lactam hydrolysis (Figure 15.3). The K_D value for CO_2 addition to Lys_{70} was measured to be 0.2 μM. The periplasmic enzyme was found at >15 μM concentration, the bicarbonate levels were estimated at mM concentrations, consistent with the high fractional occupancy of the enzyme as the carbamate. This catalytically requisite N-carbamate appears to accumulate without metal ion coordination and stabilization, but with electrostatic stabilization by three residues in the enzyme.

C-Carboxy Protein Adducts via Vitamin KH_2 and O_2

The main type of covalent protein carboxylation reactions that happen directly on protein side chains, rather than via tethered biotinyl prosthetic groups, are not the reversible N-carbamate formations, but instead are irreversible $C–COO^-$ formations. These are irreversible carboxylations at the γ-CH_2 loci of glutamate side chains of a small set of proteins in invertebrate and vertebrate cells during the transit of those

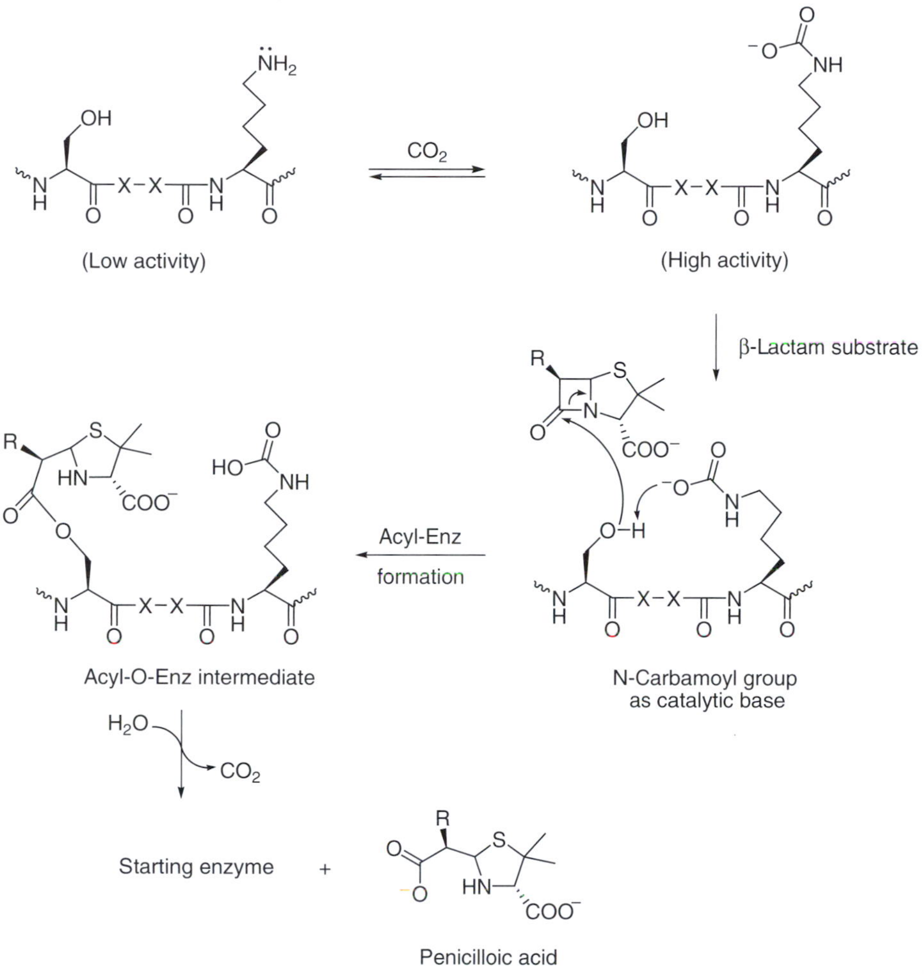

Figure 15.3 Catalytic function of the N-carboxyLys in the hydrolysis of lactam antibiotics by class D penicillinases.

proteins through the ER compartment of cell secretory systems. The product of such enzymatic posttranslational fixation of CO_2 is the γ-carboxyglutamyl side chain, known as Gla (Furie et al., 1999). Gla is actually a γ-dicarboxylate, more typically known as a malonate side chain. The vicinal carboxylates in such malonyl side chains are high-affinity bidentate ligands for cations, particularly for calcium ions (Figure 15.1). This metal coordination is thought to be the underlying logic for posttranslational Gla residue formation, the generation of a network of high affinity Ca^{2+} binding sites in a cluster of Gla residues.

The C–C bond formations in the enzymatic C-carboxylation of specific glutamyl residues in particular eukaryotic proteins require regio- and stereoselective activation

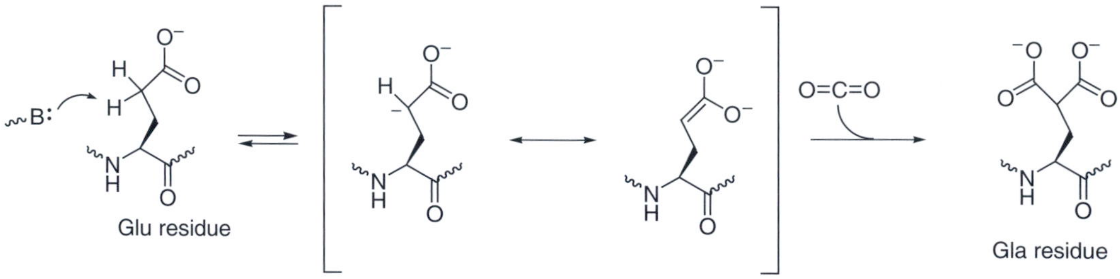

Figure 15.4 γ-Carboxylation of Glu side chains in certain proteins to yield Gla residues substituted malonates that are bidentate ligands for Ca^{2+} ions.

of one of the prochiral γ-CH_2 hydrogens at each methylene locus carboxylated. This is achieved by the generation of a transient carbanionic species (Figure 15.4) to capture the electrophilic CO_2. Although the γ-CH_2 group is adjacent to the side chain carboxylate in glutamyl residues, the C–H hydrogens are still relatively nonacidic and require powerful catalysis by the posttranslational modification enzyme to generate the γ-carbanion nucleophile to be captured by CO_2.

The glutamyl side chain activation is achieved by the chemistry of the other two cosubstrates, the dihydro form of the naphthoquinone vitamin K, known as KH_2, and molecular oxygen. In its dihydro oxidation state the naphthoquinol ring of KH_2 (Figure 15.5A) is reactive with O_2 by low-barrier one-electron pathways to generate the one-electron-oxidized KH• semiquinone radical and one-electron-reduced O_2^-•. Rapid recombination of the semiquinone radical and superoxide yields the peroxyanion (Figure 15.5B). Dowd and colleagues (Dowd et al., 1994) hypothesized that the vitamin K hydroperoxy anion might not be a strong enough base in the active site of the carboxylase posttranslational modification catalyst to abstract one of the glutamyl residue γ-CH_2 hydrogens as a proton and yield the transient carbanion species. Instead, intramolecular conversion of the peroxy anion to the epoxyquinone could yield the indicated alkoxy anion as proposed active site base.

The vitamin-KH_2-dependent protein carboxylase, a 94-kDa transmembtrane ER protein, oxidizes one molecule of KH_2 and reduces one molecule of O_2 for every protein glutamyl side chain it carboxylates (Figure 15.6A). The O_2 is reduced and fragmented, with one atom ending up as water and the other installed as the epoxide oxygen in the coproduct vitamin K-2,3-epoxide, validated by $^{18}O_2$ enzymatic and model transfer studies (Dowd et al., 1995; Kuliopulos et al., 1992). The K-2,3-epoxide product is incompetent for utilization in subsequent cycles of carboxylation until it undergoes epoxide ring opening, water elimination, and quinone reduction, catalyzed by a poorly characterized ER membrane enzyme vitamin K-2,3-epoxide reductase (Figure 15.6B). This enzyme is the target of the anticoagulant warfarin, and

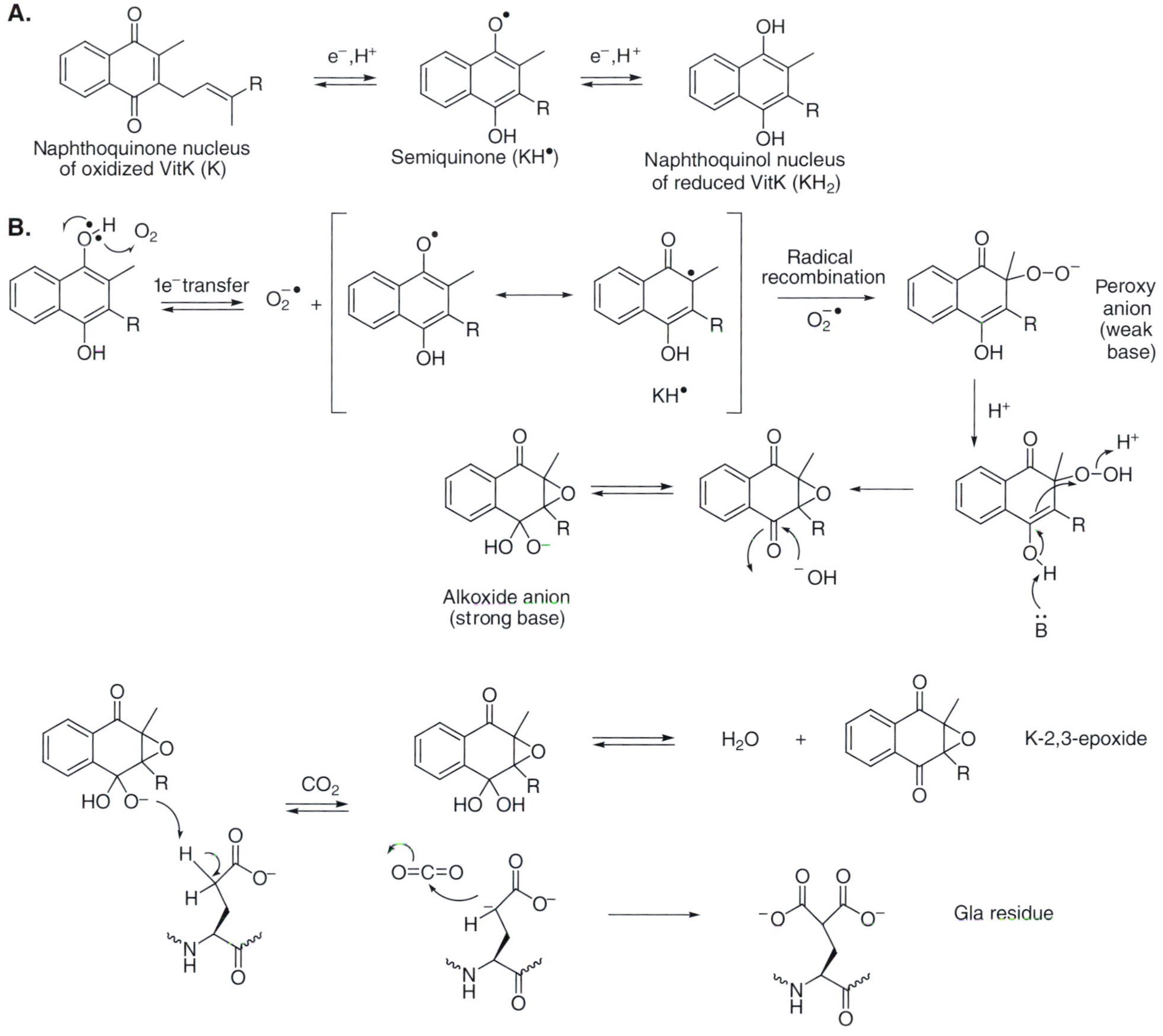

Figure 15.5 **A.** Redox form of vitamin K. **B.** Reaction of vitamin KH_2 with oxygen to generate the alkoxide anion proposed to be the base that abstracts a proton from the Glu γ-CH_2 site.

inhibition of the regeneration of KH_2 leads to generalized undercarboxylation of protein glutamyl residues and coagulation defects, as noted below.

Given the location of the KH_2-utilizing protein carboxylase in the membranes of the endoplasmic reticulum, it is reasonable that some proteins transiting through the ER would be substrates for modification. To date about a dozen proteins are known to be modified at 3–12 glutamyl side chains, clustered generally in the N-terminal region of the mature protein (Berkner, 2000; Furie et al., 1999). Functional classes of carboxylase substrates include the proenzyme forms of four serine proteases secreted

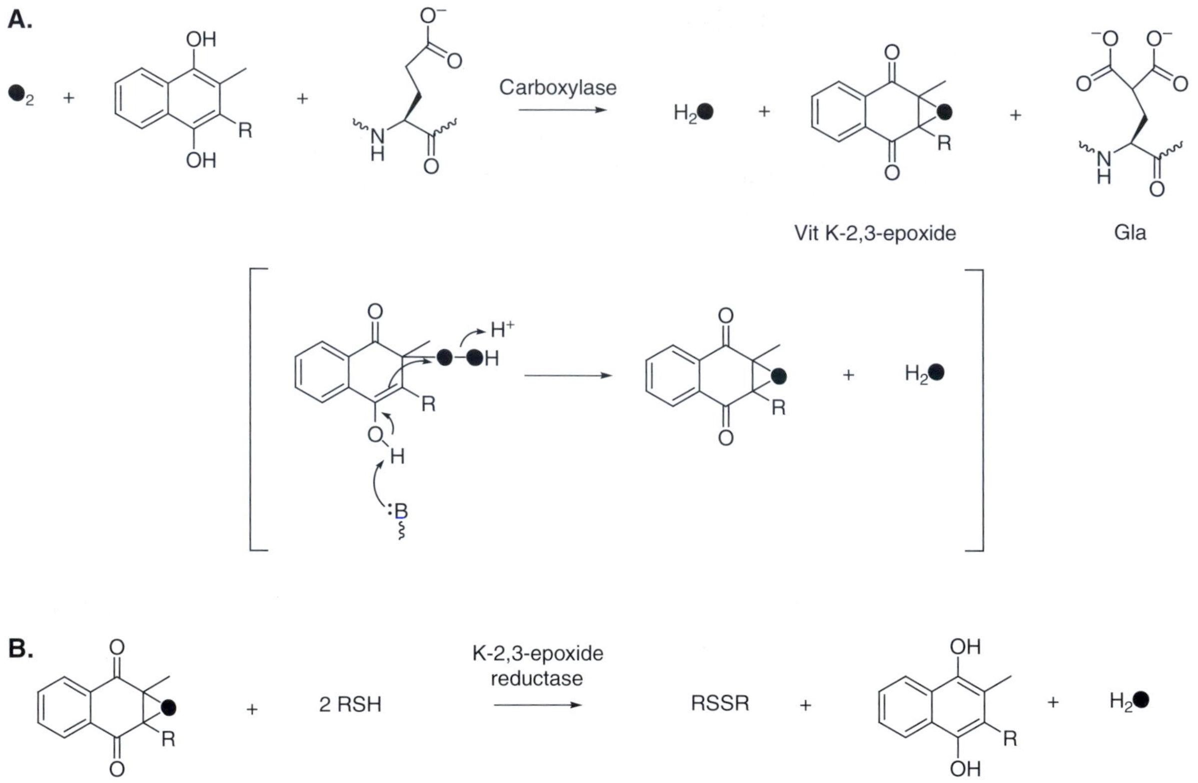

A.

Vit K-2,3-epoxide   Gla

B.

Figure 15.6 Stoichiometry and redox adjustment in the carboxylase/epoxide reductase cycle. **A.** Generation of one vitamin K-2,3-epoxide coupled to every Gla formed. **B.** Reductive cleavage of the K-2,3-epoxide back to KH_2.

by the liver and involved in extracellular phases of blood coagulation: factors VII, IX, and X, and prothrombin. They also include two coagulation cascade activator proteins, proteins C and S, and the bone Gla protein (osteocalcin) and matrix Gla protein in mineralized tissues. One protein, Gas6, is secreted by smooth muscle cells and serves as a protein receptor for receptor tyrosine kinases on other cells. Finally, some neurotoxic peptides in venoms of cone snails can be carboxylated to contain multiple Gla residues.

The selectivity for recognition of partner protein substrates of the KH_2-linked protein carboxylase appears to depend on a docking region, usually found in a propeptide just in front of the Gla domain, shown as residues –18 to –1 of preprothrombin in Figure 15.7A (Furie et al., 1999). For example, in the maturation of the serine proteases such as prothrombin, carboxylation in the ER is thought to follow the cleavage of the 19-residue signal peptide by signal peptidase, but precede action of the proprotein convertases. Thus, the nascent chain of preprothrombin is processed twice by proteases in the secretory pathway to release the mature prothrombin. During coagulation cascades the single-chain prothrombin gets cleaved by activated factor X (Xa) to yield the two-chain active protease thrombin.

After signal peptidase action the N-terminal 18 residues serve as a γ-carboxylase recognition sequence (γ-CRS). The Phe at −16, the Ala at −10, and the Leu at −6 are three crucial hydrophobic side chains that comprise part of the docking information for the immature forms of prothrombin and proFIX to be recognized by the ER-associated carboxylase (Furie et al., 1999). Once docked to carboxylase via the γ-CRS, 10 glutamates in the downstream 37 residues (Figure 15.7B) are carboxylated to Gla (in Figure 15.7B Gla residues are shown as γ symbols). For proFIX, 10 glutamates in a comparable 40-residue stretch of the N-terminal region are also carboxylated to Gla residues. The steric, architectural, and/or mechanistic features that limit carboxylase reach to glutamates (and not aspartates) only 35–40 residues downstream from a CRS sequence are unknown. Nor are the rules for the recognition and affinity of the different γ-CRS sequences in the 15 different protein substrates. Also unanswered is the exact degree to which the carboxylase works processively on a bound protein substrate (e.g., to fully carboxylate all 10 Glu residues to Gla before release and so avoid the production of highly heterogeneous undercarboxylated proteins). There is one report (Berkner and Pudota, 1998) that the carboxylase will automodify itself on up to three glutamyl residues, generating Gla side chains both *in vitro* and *in vivo*. The functional consequences of this for specificity, catalytic efficiency, or regulation are not yet evaluated, but this may be a further example of the kinds of covalent autocatalysis discussed in Chapter 13.

This Gla_{10}-containing prothrombin or these proFIX molecules then pass into the Golgi and trans Golgi networks where the 18-residue N-terminal γ-CRS peptide is cleaved by a proprotein convertase (see Chapter 10). The Gla forms of the proenzymes accumulate in secretory granules.

The Gla domain is at the N-terminus of the mature zymogen form of proFIX, where Glas at positions 7, 8, 15, 17, 20, 21, 26, 27, 30, 33, 36, and 40 can form a high-density microenvironment for coordination of 10 calcium ions. Structures of the N-terminal fragments of prothrombin (Huang et al., 2004) and factor IX (Huang et al., 2004) show that the Gla residues are in bidentate ligation to divalent Ca^{2+} ions, creating a highly organized network (Figure 15.8A). This string of 10 Gla residues, going from the absence to the presence of Ca^{2+}, can drive protein conformational change from a disorganized to a highly organized local structure (Figure 15.8B). This reorganization exposes a hydrophobic patch on these Gla-containing proteins that can interact with the anionic phospholipid surfaces on the external sides of cell membranes. In turn, this drives the assembly of these proteins into cell-associated complexes to initiate and propagate protease cascades in coagulation (Schenone et al., 2004). The biologic rationale for the clustered Glu-to-Gla modifications is to create a

A.   Maturation of preprothrombin:

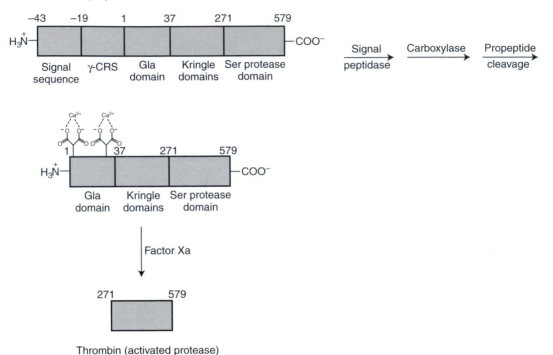

B.   Gla domain of human prothrombin:

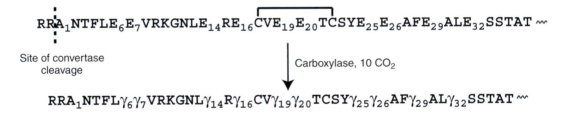

Figure 15.7 **A**. Posttranslational processing steps in the maturation of the zymogen form of pre-prothrombin to thrombin. **B**. Ten Glu residues converted to Gla residues, shown as γ, in the 37-residue Gla domain of prothrombin.

string of dianionic side chains that can act cooperatively and rapidly to structure a stretch of extracellular protein in the presence of a signal, Ca^{2+} ions, to drive platelet membrane associations for defensive response to bleeding. All of this happens on the zymogen forms of the proteases (e.g., factor IX, factor X, and prothrombin), and they must subsequently be activated by the third proteolytic fragmentation in their life cycles.

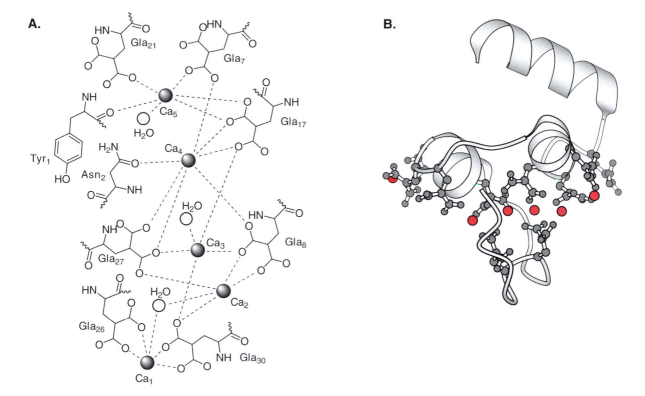

Figure 15.8 Seven Gla residues in the N-terminal 30 residues of FIX. **A.** Bidentate ligation of the 7 Gla residues of FIX to Ca^{2+} ions organizes the Gla domain into a constrained architecture that exposes the membrane binding surface of FIX. **B.** Fold of the Gla domain of factor IX with Gla residues in gray ball and stick; Ca^{2+} ions in red (figure made using PDB 1NL0).

Enzymatic Posttranslational Transamidations and Competing Hydrolyses

A variety of posttranslational modifications result in the construction of amide bonds in proteins undergoing modification and a number have been cited in previous chapters. These all fall in the general category of acylation of a protein amine, almost always one or more $\epsilon\text{-}NH_2$ groups of lysine residues (Figure 15.9A). Thus, there were discussions of protein N-acetylation in Chapter 6 and protein N-myristoylation in Chapter 7. Also noted in Chapters 6 and 7 were the sortase-mediated protein acylation of peptidoglycan chains in Gram-positive bacteria and GPI anchor attachment to protein acyl chains. The *isopeptide* bonds formed to ubiquitin, SUMO, and other Ubl protein modifiers were detailed in Chapter 9, from the C-terminus of the ubiquitin or Ubl to the $\epsilon\text{-}NH_2$ of lysine residues in the acceptor proteins, including ubiquitins in polyubiquitin chain buildup.

A.

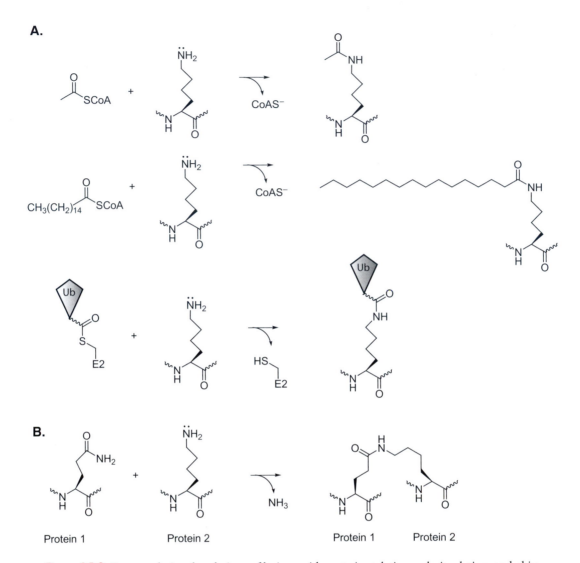

B.

Protein 1   Protein 2   Protein 1   Protein 2

Figure 15.9 Posttranslational acylations of lysine residues. **A.** Acetylation, palmitoylation, and ubiquitylation of Lys residues. **B.** Transamidation of Lys and Glu residues by transglutaminase.

There is one other category of isopeptide bond formation, which links two proteins via an amide linkage to an ϵ-NH_2 of a Lys residue that occurs when transglutaminase enzymes are in action. As the name implies, this family of posttranslational modifying enzymes uses glutamine side chains as the acyl donors and typically lysine side chains as the nucleophilic acceptors (Figure 15.9B). The net result is the conversion of a glutamine side chain to a γ-glutamyllysine amide bond. If the glutamine donor and lysine acceptor are in the same protein, this constitutes an intramolecular cross-link. If the two residues are on distinct partner proteins, an intermolecular cross-linking isopeptide bond has been generated.

The enzyme chemistry of glutaminases acting on free glutamine as substrate was established some decades ago, with the reaction proceeding through a covalent γ-glutamyl-S-enzyme intermediate, involving an active site cysteine thiolate in the glutaminase active sites (Figure 15.10A). The thioester intermediate is decomposed hydrolytically in a water addition step. The glutaminases are members of the papain superfamily of protein hydrolases (Chapter 8) that use an active site Cys as the nucleophile, and a conserved His residue as the general base in the first step of acyl-S-enzyme formation.

Transglutaminases fall into the same superfamily. Like papain they use proteins as substrates, and like glutaminase they engage in covalent glutamyl-S-enzyme and NH_3 release from the glutaminyl donor in the first half-reaction (Figure 15.10B). The protein-glutamyl-S-transaminase acyl enzyme intermediate can decompose by two kinetic routes. The preferred pathway is capture of the protein-glutamyl moiety by the deprotonated ϵ-NH_2 of a lysine side chain in a protein cosubstrate. Sometimes soluble polyamines such as spermidine are kinetically competent and lead to isopeptide bonds that can react at the other end for protein cross-linking via such polyamide bridges. In certain cell types, such as platelets, the concentration of the low molecular-weight neurotransmitter amine serotonin, 5-hydroxytryptamine, builds up to sufficient concentrations (e.g., in α-granules) that transglutaminase-mediated serotonylation of proteins occurs. This happens to small GTPases such that serotonylated RhoA and Rab GTPases are generated (Figure 15.10C) and these modified GTPases are constitutively activated during platelet activation and aggregation (Walther et al., 2003).

Alternatively, in the absence of kinetically competent concentrations of lysine side chains or polyamines, the long-lived acyl-S-transglutaminase intermediates can decompose hydrolytically, a net glutaminase reaction, to yield the substrate protein that has undergone deamination at one or more glutamine side chains. Pathological sequelae to this hydrolysis pathway are noted in the example of celiac sprue below.

There are multiple genes for transglutaminases in mammalian tissues, some with restricted expression, appropriate to specialized tissue function (Griffin et al., 2002; Lorand and Graham, 2003). For example, factor XIII in the blood coagulation cascade is secreted as a zymogen form, proteolytically activatable to FXIIIa, with much elevated transglutaminase (TGase) activity. Its physiological substrates are the subunit of fibrin molecules that have been activated in coagulation cascades by thrombin and have assembled to form the "soft" clots. FXIIIa shows selectivity for cross-linking Q_{398} on one fibrin α chain to K_{406} on a facing antiparallel α chain. The Q_{398}-to-K_{406} isopeptide bonds happen on both chains from FXIIIa action to produce mechanically stable clots (Figure 15.11). Lorand and Graham (2003) have described this posttranslational cross-linking enzymology as "spot welding" of one protein to another

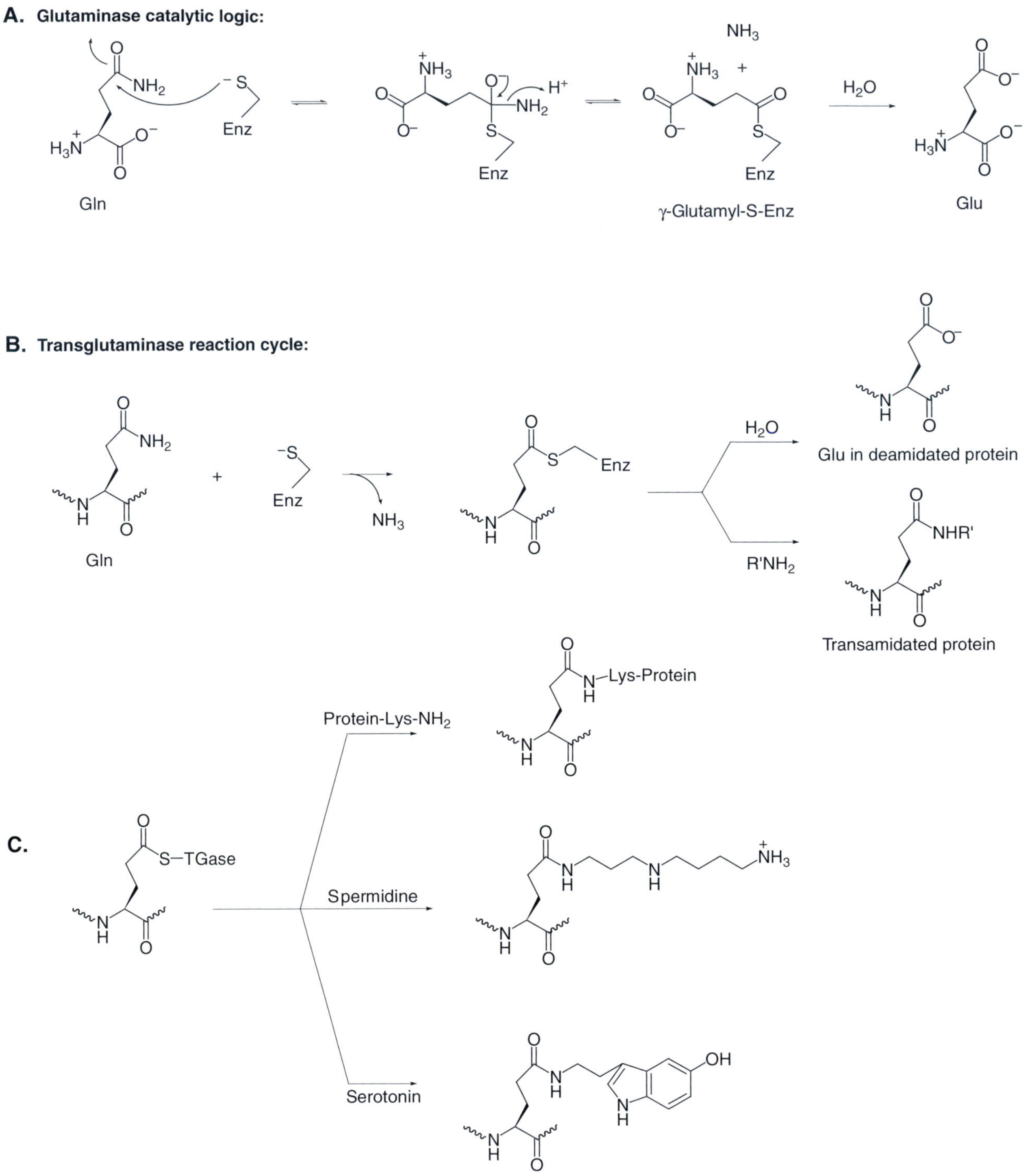

A. Glutaminase catalytic logic:

B. Transglutaminase reaction cycle:

C.

Figure 15.10 **A.** Acyl-S-enzyme formation in glutaminase catalysis. **B.** Transglutaminase cross-links via polyamine-bis-amide bridges. **C.** The glutamyl-S-transglutaminase intermediate can be captured by low molecular-weight amines such as spermidine or serotonin.

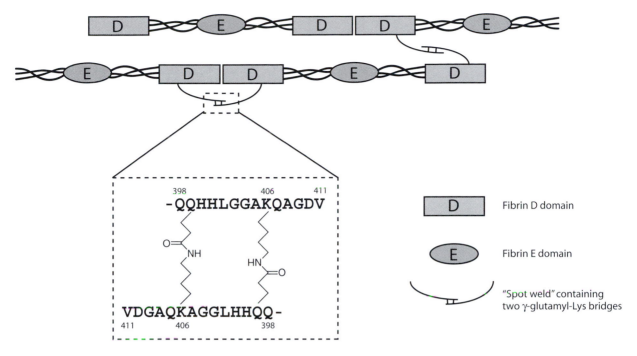

Figure 15.11 Double cross-link formation between pairs of Q_{398} and K_{406} residues in fibrin α chains during the factor XIIIa transglutaminase-mediated cross-linking of soft fibrin clots.

for creating the molecular connections that impart tensile strength to such protein scaffolds.

Comparable protein intermolecular cross-linking occurs in transglutaminase action in skin and hair and in wound healing processes to provide stable forms of assembled proteins. Keratinocyte transglutaminase activity is a central part of the terminal differentiation program in keratinocytes to build cornified epithelial tissue layers that provide impermenant barriers.

The glutamyllysine isopeptide linkages generated by the different transglutaminases create protease-resistant, irreversible bonds. It follows that transglutaminase activity needs to be carefully regulated in time and place to avoid disregulated activity and pathophysiologic consequences to inappropriate action. The proteolytic activation of FXIII only during coagulation cascades is one such example. The x-ray structure of FXIII indicates a core catalytic/regulatory domain (residues 140–454) downstream from an N-terminal β sandwich and upstream from a tandem pair of β barrels at the C-terminus (Figure 15.12). The TGases, unlike many members of the papain superfamily, are dramatically activated by Ca^{2+}, which serves as an additional governor of activity. Intracellular Ca^{2+} is generally below the μM range required for TGase activation, but extracellular Ca^{2+} is present in many extracellular spaces above the EC_{50}, thus turning on TGase activity in extracellular milieus. Analysis of the

α,β Catalytic core

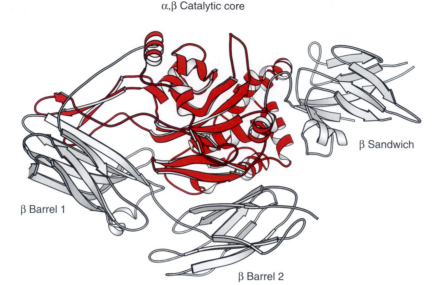

β Sandwich

β Barrel 1

β Barrel 2

Figure 15.12 X-ray structure of TG2. The α,β catalytic core domain is shown in red, and the β barrel and β sandwich domains are shown in grey (figure made using PDB 1KV3).

effects of Ca^{2+} addition on TGase conformation indicate substantial movement in the active site residues with Ca^{2+}-driven unfolding of the helical regions and increased access to protein substrates (e.g., the glutaminyl donor side chains) as the C-terminal β barrel domains are melted out.

The TGase2 isoform, also known as tissue TGase, has been enigmatic in function. It is found intracellularly but is not presumed to work until displayed on the surfaces of cells, perhaps controlled by extracellular $[Ca^{2+}]$. This TGase isoform has been implicated both as a catalyst and an antigen in the immunologic response in the GI tract to certain glutamine-rich proteins, gliadins, in wheat by individuals who have celiac sprue. Antibodies to both TGase2 and gliadin-derived peptides are observed as well as circulating T cells that mount responses to the peptide antigens. It is known that enzymatic conversion of particular glutamine side chains, presumably by TGase2 on epithelial cell surfaces, during gliadin digestion increases the immunogenicity of the peptides. Mass-spectrometry-based analyses of gliadin peptides indicated that TGase action can convert 19 of 95 glutamine side chains to glutamates. Furthermore, Piper et al. (2002) showed that a peptide fragment that was produced during gliadin digestion by GI proteases was deaminated at the two Q residues in the sequence PQPQLPY by recombinant TGase2 to the PEPELPY peptide products. When the TGase2 was assayed in the absence of amine cosubstrates, it was detected as a long-lived peptidyl-S-TGase2 acyl enzyme, consistent with the mechanism of Figure 15.10B. A long lifetime for this acyl enzyme species might also account for the recognition of TGase2 as a foreign antigen in patients. Inhibitors of TGase and/or prior digestion of the gliadin immunodominant peptide epitopes in food might alleviate symptoms (Piper et al., 2002).

Other examples of deamidations and related reactions on Asn and Arg side chains have been discussed in Chapter 5, such as the deamidation of Asn in connection with the repair of isoaspartyl linkages and the deimination of Arg to citrulline residues.

Amidations

Some proteins and peptides are found with C-terminal amides instead of the normal C-terminal residue as a free carboxylate. These amides typically arise from a two-step posttranslational modification. The C-terminal carboxamido NH_2 substituent in eukaryotic peptide hormones such as oxytocin, vasopressin, gastrin calcitonin, and cholycystokinin (Eipper et al., 1993) comprises the residuum of what had been a C-terminal glycine residue. The peptides themselves have originated by proprotein convertase action on protein precursors. The immediate precursor of the hormonal peptides, with the C-terminal glycine intact, is typically not biologically active in the neural or neuroendocrine function of the mature peptide.

The glycine was carved out by posttranslational enzymatic oxygenation that generates an α-OH glycine residue as the initial product. This is a hemiaminal and can be decomposed in a lyase step to release the two carbons as glyoxalate and leave the NH_2 moiety derived from glycine as the C-terminal amide on what had been the $(n-1)$th amino acid residue, but is now the newly liberated C-terminus.

As shown in Figure 15.13, the production of the nonapeptide hormone oxytocin occurs by the processing of a 95-residue precursor of both oxytocin (OT) and neurophysin. The OT nonapeptide sequence is just downstream of the signal sequence, which is removed during transit into the ER (Chapter 8). The precursor has a Gly_9-Gly_{10} sequence that is cleaved, leaving the amino group from Gly_{10} as the C-terminal amide functional group of active oxytocin.

The posttranslational enzyme responsible for the Gly_9Gly_{10} cleavage and generation of the C-terminal amide in this and other peptide amide forming systems is known as peptidylglycine α-hydroxylating monooxygenase (Eipper et al., 1993) (PHM). It is a bidomainal enzyme that is upregulated in neural and neuroendocrine cells and can localize into secretory granules where the C-terminal oxidative trimming step occurs on the peptide hormone precursors. The PHM enzyme itself is produced as a precursor protein sent into the secretory pathway via an N-terminal signal peptide removed by signal peptidase in the ER. Then the first catalytic domain, the hydroxylase, precedes the second domain, the peptidylamidoglycolate lyase, before a transmembrane domain that anchors the enzyme into the secretory vesicle membranes (Figure 15.14).

Oxytocin maturation:

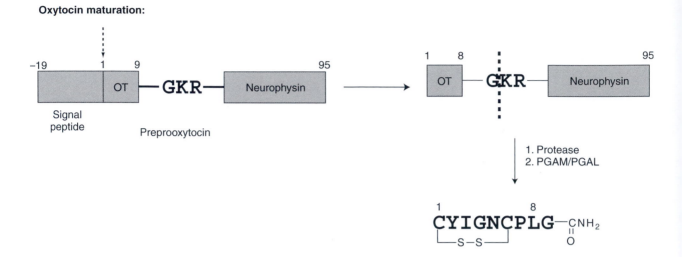

Figure 15.13 Proteolytic processing of preprooxytocin to the nonapeptide hormone oxytocin.

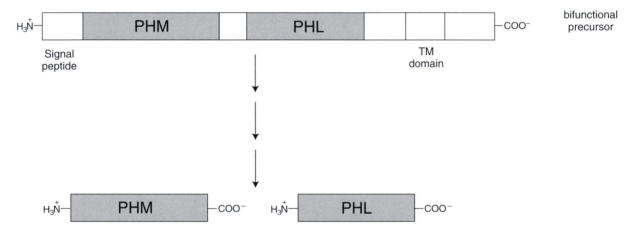

Figure 15.14 Organization of the bidomainal peptidylglycine α-hydroxylating monooxygenase.

The x-ray structure of the enzyme has been solved (Prigge et al., 1997 and 1999), revealing that the enzyme is a bicopper enzyme. The Cu ion pair undergoes redox chemistry (e.g., via dihydroascorbate as the electron donor) to enable reaction with both the cosubstrate O_2 and the C-terminal glycine residue of the protein substrate by one-electron pathways (Figure 15.15). The pro-(S) hydrogen of the glycine CH_2 group is stereospecifically abstracted as H• and reacted with a high-valent oxocopper species derived from the reductive activation and cleavage of the O_2 bound to copper to yield the α-OH glycyl residue at the C-terminus of the protein. Inspection of this product

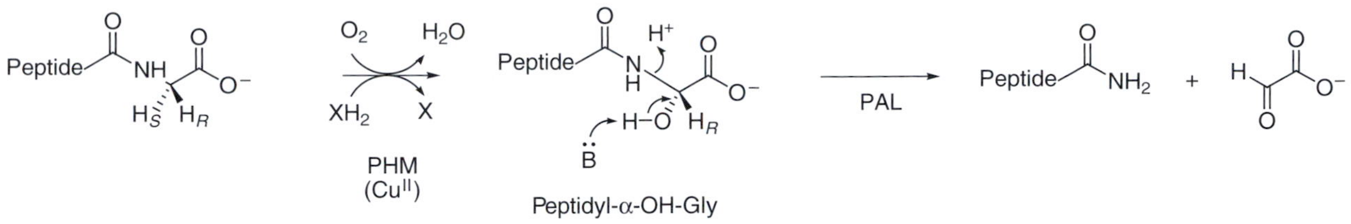

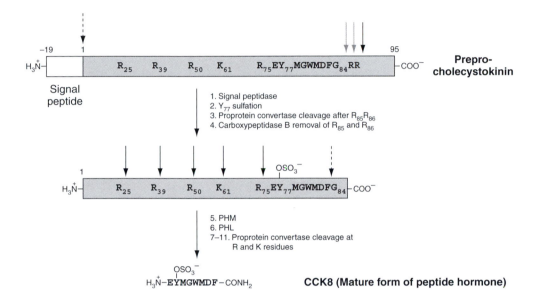

Figure 15.15 Pathway for PHM and PAL catalysis.

Figure 15.16 Maturation of the 115-residue preprocholecystokinin to CCK8: seven proteolytic steps, one tyrosine sulfation, and one Gly α-hydroxylation, which occurs at the CCK83 intermediate stage.

shows that the C-terminal α-OH Gly is the adduct of glyoxalate and the amide of the preceding amino acid residue. The accelerated decomposition into these two fragments is catalyzed by the peptidylamidoglycolate lyase domain of the PHM enzyme.

These two-step oxidative tailorings of terminal glycines occur as the endgame in the multistep posttranslational processing of proproteins that give rise to all the peptides that end in C-terminal amides. One example is in the generation of the octapeptide cholecystokinin 8 (CCK8) from a 115-residue proprotein, preprocholecystokinin (Figure 15.16). There are an estimated nine steps of posttranslational processing to convert the 115-residue primary translation product to the octapeptide hormone CCK8 [NH_3-Asp-Tyr (OSO_3^-)-Met-Gly-Trp-Met-Asp-Phe-$CONH_2$]. Seven of these are proteolytic clips, one involves sulfation of the tyrosyl residue (see Chapter

3), and the ninth is the C-terminal oxidative amidation under discussion (Eberlein et al., 1992).

The first modification step is the removal of the N-terminal 20-residue signal sequence, followed by regiospecific Tyr_{77} residue O-sulfation by a protein O-sulfo-transferase. Then a furin-type proprotein convertase (see Chapter 8) cleaves at the $Arg_{85}Arg_{86}$ dibasic site. Next, the trimming of these two C-terminal Arg residues occurs by exopeptidic removal by carboxypeptidase B to expose Gly_{84} as the new C-terminus. At this stage it appears the oxygenation of Gly_{84} to α-OH-Gly_{84} is carried out by the two-domain PHM enzyme, yielding CCK83, an amidated intermediate in CCK maturation (Eberlein et al., 1992). Subsequent endoproteolytic clips occur by proteases recognizing double-basic and single-basic residues during CCK transit through the secretory pathway to generate CCK58, CCK39, CCK38, and CCK22 on the way to CCK8. One assumes this is representative of all the coordinated posttranslational enzymatic processings that yield other neuroendocrine peptide C-terminal amides from proproteins.

Tandem Modifications of N-Glutaminyl and C-Glycyl Termini During Maturation of a Peptide Hormone

Another relevant example of processing of a preprohormone to a modified, biologically active small peptide is the maturation of TRH, the thyrotropin releasing hormone. TRH is a modified tripeptide, pyroGlu-His-Pro-amide (pEHP-NH_2, where pE represents pyroGlu; Figure 15.17), generated from a 255-residue precursor by hypothalamic neurons. The processed TRH is secreted and acts on a G protein coupled receptor in thyroid tissue to stimulate thyroid hormone processing and secretion. The preproTRH is cleaved by suites of proteases described in chapter 8, first to remove the signal peptide and then by propeptide convertases that cleave at basic residues to liberate five copies of the tetrapeptide QHPG.

We have just discussed how terminal Gly residues are hydroxylated and then deconvoluted to release glyoxylate and leave the NH_2 moiety from Gly as the terminal amide—in this case the QHP-NH_2) by the two component peptidyl amide hydroxylase. The amino terminus of this tripeptide hormone is also protected from proteolytic degradation, by the enzymatic action of glutaminyl cyclase, a relative of zinc aminopeptidases, that uses the newly released α-NH_2 of the Gln residue as nucleophile on the γ-carboxamido nitrogen. Loss of NH_3 from the tetrahedral adduct cyclizes the original Gln side chain to the five membered ring of pyroGlu (pE) (also known as 5-oxoPro as well as pyrrolidone carboxylate). PyroGlu residues are resistant to hydrolytic cleavage by most aminopeptidases although there are specific pyrrolidone peptidases known.

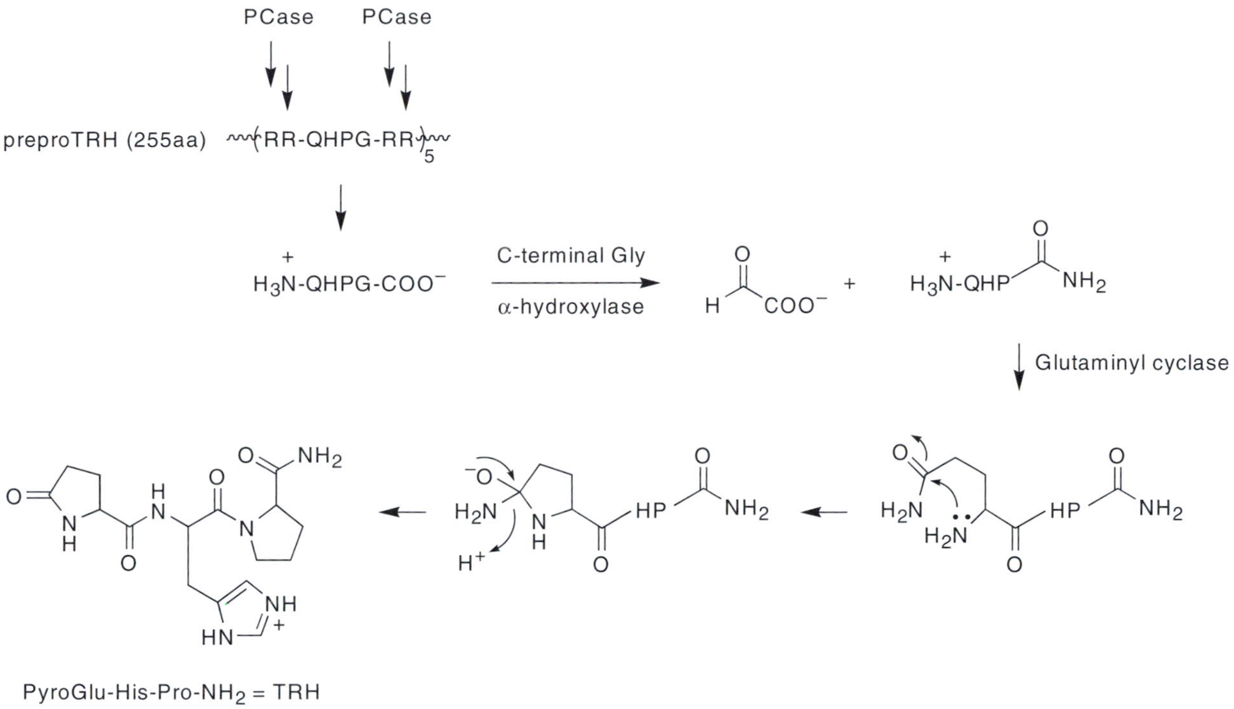

Figure 15.17. Processing of preproTRH by Proprotein Convertases (PCases), C-terminal Gly hydroxylation and N-terminal glutaminyl cyclase action to yield the tripeptide hormone TRH blocked at the N- and C-terminus.

The tandem action of the C-terminal peptidyl glycine hydroxylase and the N-terminal glutaminyl cyclase caps both ends of the tripeptide. Glutaminyl cyclase acts at the N-termini of some full length proteins but most often on other peptides such as gastrin and neurotensin released from precursors during analogous maturations from prepro forms. The 40 residue amyloid peptide Aβ (3-42) implicated in the pathogenesis of Alzheimer's disease is a mixture of peptides forms, with about 25% containing an N-terminal pyroglutamyl moiety instead of the Q_3 encoded in the gene (Harigaya et al, 2000).

Protein Circularizations

One last variant of the posttranslational proteolysis of proproteins coupled to amide bond formation at the C-terminus of the mature product occurs in the biosynthesis of circular proteins, with head-to-tail joining of the amino and the carboxy terminus in peptide bond formation (Figure 15.18). The largest cyclic peptide/protein known is a 70-residue protein AS-48 from *Enterococcus faecalis*, which is used as a lytic toxin

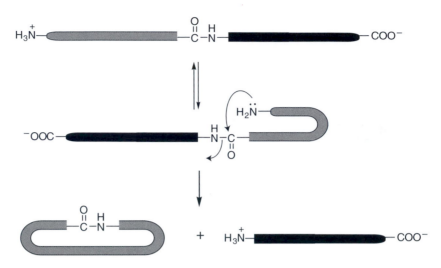

Figure 15.18 Protein circularization: head-to-tail peptide bond formation between the amino and carboxy termini.

against sensitive bacterial strains [see Trabi and Craik (2002) for review]. There is a family of about 45 cyclic peptides from plants, termed cyclotides (Trabi and Craik, 2002), that range in size from 28–37 residues. Finally, the antibiotic defensin peptide from rhesus macaques, known as RTD-1 (rhesus theta defensin 1), is a head-to-tail cyclic 18-residue peptide. The RTD-1 has three disulfides arranged in a ladder-like fashion (Cys_3–SS–Cys_{16}, Cys_5–SS–Cys_{14}, and Cys_7–SS–Cys_{12}), while the plant cyclotide proteins have three disulfides in a knotted topology (Trabi et al., 2001; Trabi and Craik, 2002) (Figure 15.19).

The circular nature aids in resistance to protease and the trio of disulfide bonds in both RTD-1 and the cyclotides confer rigid architectures, allowing the population of a small number of restricted conformers that contribute to diverse biological activities. These are ribosomally synthesized peptides, from protein precursors that undergo typical proprotein convertase-mediated proteolytic clipping. For example, the genes for some of the plant cyclotides have been sequenced and encode precursors with typical N-terminal signal sequences, propeptide regions, and then 1–3 cyclotide repeats before a hydrophobic C-terminus. In some analogy to the CCK maturation noted in the previous section, the cyclotide segments must be excised by specific proteolytic cuts upstream and downstream during transit through the secretory pathway in plant cells. As yet unknown is the mechanism and driving force for cyclization. It is likely that the precursor, hydrolytically excised at one or both ends, undergoes regiospecific disulfide bond formations. The trisulfide containing linear forms may be constrained with N- and C-termini in proximity, such that the high local concentration of ends leads to favored attack of the amine on the carboxylate with subsequent dehydrative capture.

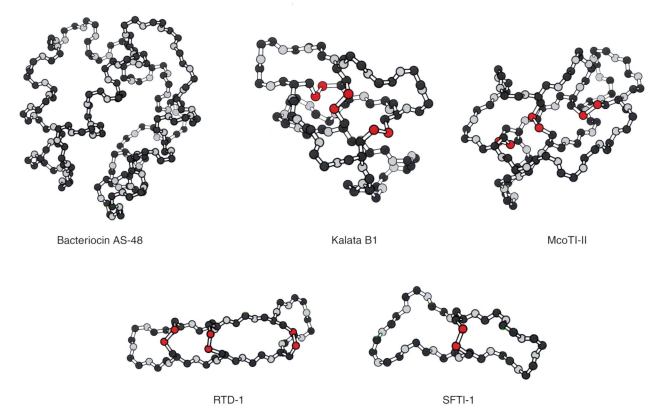

Bacteriocin AS-48   Kalata B1   McoTI-II

RTD-1   SFTI-1

Figure 15.19 Backbone structures of naturally occurring circular proteins. Disulfide bonds are shown in red [figure made using PDBs 1E68 (bacteriocin AS-48), 1KAL (kalata B1), 1HA9 (McoTI-II), 1HVZ (RTD-1), and 1JBL (SFTI-1)].

Posttranslational Modifications at the C-Terminal Tails of α- and β-Tubulin Subunits

Microtubules have a diverse array of functions in cells with the dynamic addition and release of α- and β-tubulin subunits, modulated by microtubule-associated protein components, contributing to the lifetime and function of microtubular structures. Tubulin subunits are known to undergo distinct types of covalent enzymatic modifications. One, mentioned in Chapter 6, is acetylation on lysine side chains, which tends to stabilize microtubules. Typically this occurs at Lys_{40} near the N-terminus of the tubulin subunit and occurs after assembly of the monomers into microtubules [see Westermann and Weber (2003) for review]. This is reversed by the action of two types of deacetylases also discussed in Chapter 6 for their activity on acetylated histones. HDAC6 and the NAD-dependent SIRT2 each have been validated to deacetylate modified microtubules.

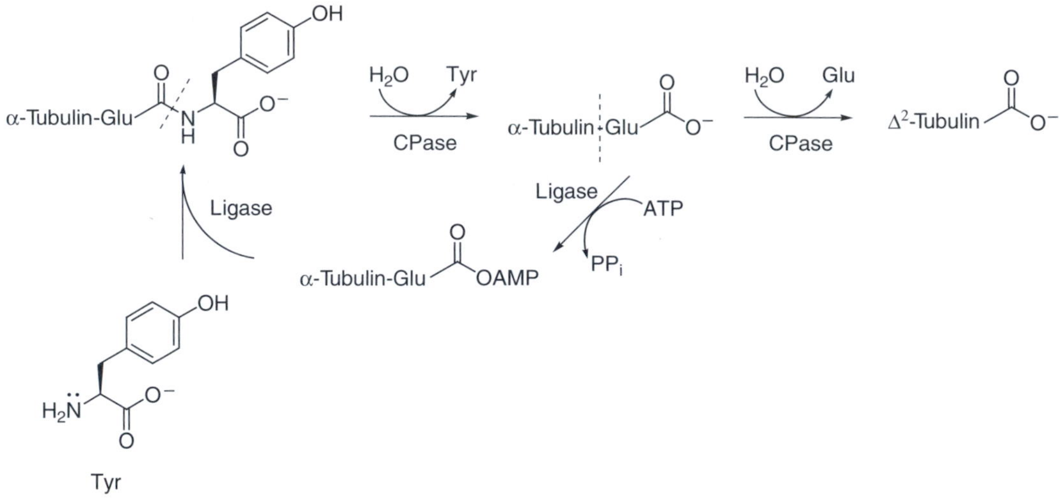

Figure 15.20 Posttranslational modification at the C-terminus of α-tubulin by removal and religation of the C-terminal tyrosine.

Three other distinctive posttranslational modifications occur near the C-termini of tubulin subunits in regions that extend as tails outside the microtubules, in some analogy to the N-terminal tails of histones (Westermann and Weber, 2003). One modification is the specific removal of the C-terminal tyrosyl residue of α-tubulin by a specific carboxypeptidase (tubulin tyrosine carboxypeptidase). This tyrosine can be added back by a separate amide ligase activity that cleaves ATP to ADP, presumably making the tyrosyl phosphate intermediate (Figure 15.20) as the required carboxyl activated form of tyrosine. An as yet uncharacterized carboxypeptidase can act on the detyrosinated tubulin to remove the new terminal glutamate and yield Δ^2-tubulin, which is not a substrate for religation with tyrosine and accumulates in substantial amounts in brain microtubules.

The more curious modifications occur at glutamyl side chains clustered around E_{445} near the C-terminus and comprise either enzymatic polyglycylation or enzymatic polyglutamylation (Figure 15.21). The Gly chain starts by the formation of an isopeptide bond to the first Gly as a Glu_{445}-γ-Gly amide linkage is produced. Subsequent additions of Gly residues to the first Gly make standard peptide bonds. The $(Gly)_n$ chain can be as short as 1–2 residues or as long as 30–40 residues (Westermann and Weber, 2003). This modification is predominant in mammalian sperm tubulin.

The corresponding assembly of a polyglutamyl chain at E_{445} or neighboring E residues at the C-terminus of tubulins also starts with the construction of a Glu_{445}-γ-Glu-isopeptide bond. Subsequent additions of Glu residues one at a time to build up 20- to 30-long $(Glu)_n$ chains that apparently can be a mix of α- or γ-peptide linkages. The Glu_n chains differ from the Gly_n chains in that the Glu_n chains have n negative charges. Polyglutamylated tubulins are found in tubulin fractions from mammalian

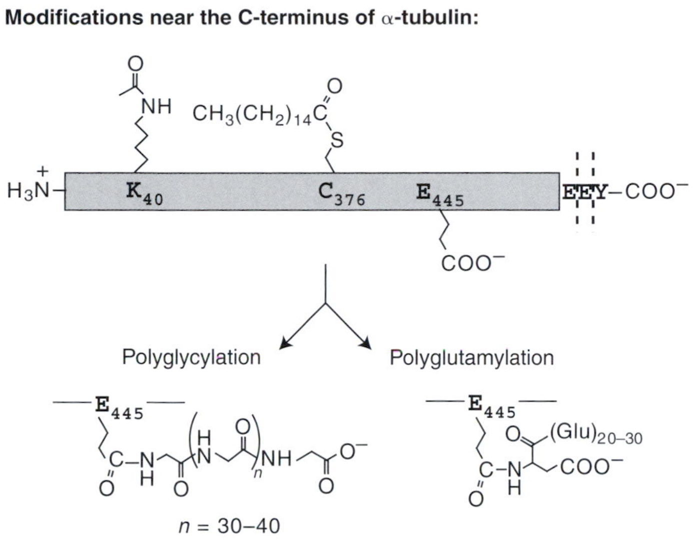

Modifications near the C-terminus of α-tubulin:

Figure 15.21 Polyglycylation and polyglutamylation of Glu$_{445}$ near the C-terminus of tubulin.

brain. It is anticipated that distinct physical properties are imparted to the two modified C-termini of tubulin subunits and the polymerized microtubules, including selective recruitment of partner proteins. The enzymes building the Gly$_n$ and Glu$_n$ chains are as yet incompletely characterized but they do not involve amino acids activated on tRNAs, but rather have analogous logic to the assembly of poly-γ-glutamate chains on folate cofactors. The 30- to 40-long Gly or 20- to 30-long Glu tails on tubulin may be handles for the modulation of structure or function on different types of microtubular structures, including those in flagella and cilia (Westermann and Weber, 2003).

References

Berkner, K. L "The vitamin K-dependent carboxylase," *J. Nutr.* **130:**1877–1880 (2000).

Berkner, K. L., and B. N. Pudota. "Vitamin K-dependent carboxylation of the carboxylase," *Proc. Natl. Acad. Sci. U.S.A.* 95: 466–471 (1998).

Dowd, P., S. W. Ham, S. Naganathan, and R. Hershline. "The mechanism of action of vitamin K," *Annu. Rev. Nutr.* **15:**419–440 (1995).

Dowd, P., R. Hershline, S. W. Ham, and S. Naganathan. "Mechanism of action of vitamin K," *Nat. Prod. Rep.* **11:**251–264 (1994).

Eberlein, G. A., V. E. Eysselein, M. T. Davis, T. D. Lee, J. E. Shively, D. Grandt, W. Niebel, R. Williams, J. Moessner, J. Zeeh, and et al. "Patterns of prohormone processing. Order revealed by a new procholecystokinin-derived peptide," *J. Biol. Chem.* **267:**1517–1521 (1992).

Eipper, B. A., S. L. Milgram, E. J. Husten, H. Y. Yun, and R. E. Mains. "Peptidylglycine alpha-amidating monooxygenase: A multifunctional protein with catalytic, processing, and routing domains," *Protein Sci.* **2:**489–497 (1993).

Furie, B., B. A. Bouchard, and B. C. Furie. "Vitamin K-dependent biosynthesis of gamma-carboxyglutamic acid," *Blood* **93:**1798–1808 (1999).

Golemi, D., L. Maveyraud, S. Vakulenko, J. P. Samama, and S. Mobashery. "Critical involvement of a carbamylated lysine in catalytic function of class D beta-lactamases," *Proc. Natl. Acad. Sci. U.S.A.* **98:**14280–14285 (2001).

Griffin, M., R. Casadio, and C. M. Bergamini. "Transglutaminases: Nature's biological glues," *Biochem. J.* **368:**377–396.

Harigaya, Y., T.C. Saido, C.B. Eckman, C-M. Prada, M.Shoji, and S.G. Younkin. "Amyloid β protein starting pyroglutamate at position 3 is a major component of the amyloid deposits in the Alzheimer's disease brain," *Biochem. Biophys. Res. Commun.* **276:** 422–427 (2000).

Huang, M., B. C. Furie, and B. Furie. "Crystal structure of the calcium-stabilized human factor IX Gla domain bound to a conformation-specific anti-factor IX antibody," *J. Biol. Chem.* **279:**14338–14346 (2004).

Kuliopulos, A., B. R. Hubbard, Z. Lam, I. J. Koski, B. Furie, B. C. Furie, and C. T. Walsh. "Dioxygen transfer during vitamin K dependent carboxylase catalysis," *Biochemistry* **31:**7722–7728 (1992).

Lorand, L., and R. M. Graham. "Transglutaminases: Crosslinking enzymes with pleiotropic functions," *Nat. Rev. Mol. Cell. Biol.* **4:**140–156 (2003).

Piper, J. L., G. M. Gray, and C. Khosla. "High selectivity of human tissue transglutaminase for immunoactive gliadin peptides: Implications for celiac sprue," *Biochemistry* **41:** 386–393 (2002).

Prigge, S. T., A. S. Kolhekar, B. A. Eipper, R. E. Mains, and L. M. Amzel. "Amidation of bioactive peptides: The structure of peptidylglycine alpha-hydroxylating monooxygenase," *Science* **278:**1300–1305 (1997).

Prigge, S. T., A. S. Kolhekar, B. A. Eipper, R. E. Mains, and L. M. Amzel. "Substrate-mediated electron transfer in peptidylglycine alpha-hydroxylating monooxygenase," *Nat. Struct. Biol.* **6:**976–983 (1999).

Schenone, M., B. C. Furie, and B. Furie. "The blood coagulation cascade," *Curr. Opin. Hematol.* **11:**272–277 (2004).

Trabi, M., and D. J. Craik. "Circular proteins—no end in sight," *Trends Biochem. Sci.* **27:**132–138 (2002).

Trabi, M., H. J. Schirra, and D. J. Craik. "Three-dimensional structure of RTD-1, a cyclic antimicrobial defensin from Rhesus macaque leukocytes," *Biochemistry* **40:**4211–4221 (2001).

Walther, D. J., J. U. Peter, S. Winter, M. Holtje, N. Paulmann, M. Grohmann, J. Vowinckel, V. Alamo-Bethencourt, C. S. Wilhelm, G. Ahnert-Hilger, and M. Bader. "Serotonylation of small GTPases is a signal transduction pathway that triggers platelet alpha-granule release," *Cell* **115:**851–862 (2003).

Westermann, S., and K. Weber. "Post-translational modifications regulate microtubule function," *Nat. Rev. Mol. Cell Biol.* **4:**938–947 (2003).

CHAPTER 16

Diversification of Proteomes

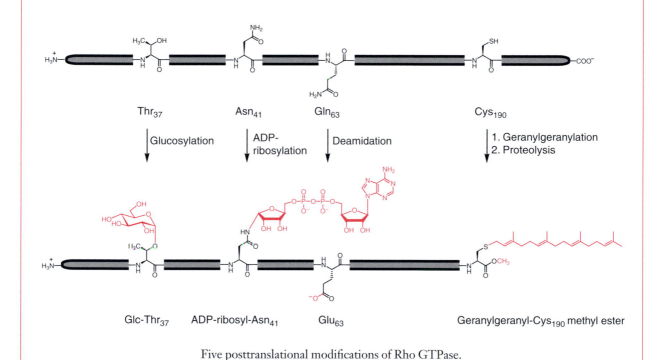

Five posttranslational modifications of Rho GTPase.

The Range of Covalent Posttranslational Modifications

The preceding chapters in this volume have catalogued the major classes of posttranslational modifications introduced by enzymes that diversify the proteomes. As noted in Chapter 1, 15 of the 20 common proteinogenic amino acid side chains undergo covalent modifications. Some side chain nucleophiles, such as the thiolate of cysteine, the ϵ-NH$_2$ of lysine, the β-OH of serine (and threonine), and, surprisingly, the carboxa-

461

mido nitrogen ($CONH_2$) of asparagine residues, are used repeatedly for posttranslational decoration by electrophilic groups derived from small-molecule substrates and coenzymes in cells (Tables 1.1 and 1.2). The result is dozens of different modifications built on the limited diversity of the 20-amino-acid-derived protein scaffolds.

The covalent groups stably introduced vary from as small as one carbon (at the methyl oxidation state or the CO_2 oxidation state) to the 76-carbon-containing small protein ubiquitin and its congeners. Each modification creates a new handle that enables the derivatized protein to gain a specific function, exemplified by histone-tail methylations to effect transcription factor binding to DNA, γ-carboxyGlu (Gla) binding of bidentate Ca^{2+} ions in blood coagulation, and the information-rich protein tag ubiquitin to control the location and/or lifetime of tagged proteins.

The covalent groups introduced can be devoid of carbon and contain only heteroatoms, as in protein kinase action to introduce $-PO_3^{2-}$ groups and sulfuryl transferases to introduce $-SO_3^-$ groups. These charged tetrahedral inorganic functionalities create multiple architectural possibilities in the local region where introduced, for partner protein recognition and both intra- and inter-chain domain reorganizations that drive conformer population redistributions in proteins. The utility of the dianionic phosphate group to drive protein conformational change and partner recognition is apparent by the dominant role this class of protein posttranslational modifications has assumed in the eukaryotic evolution of signal transduction pathways in the 500-member human kinome.

One set of modifications, which introduces the coenzyme forms of the vitamins, biotin, lipoate, and pantetheine at the end of 20-Å tethers, has extended the physical reach of protein side chains. The swinging arm coenzymes, tethered to their host proteins by amide attachement to lysine (biotin and lipoamide) or phosphodiester attachment to serine (phosphopantetheine), are free at the other end to reach to the active sites of domains in intra- and inter-protein contexts for the carboxylation and acyl transfer reactions carried out by enzymes of primary metabolism.

Two types of posttranslational modifications result in covalent cross-linking of proteins, one by a nonoxidative route, the other oxidative. The nonoxidative route is carried out by transglutaminases, substituting the $-NH_2$ of glutamine residues by $-NH-Lys$-proteins in transamidation reactions, such as in cornifying epithelial layers. The oxidative cross-linking route is the predominant one for linking protein subunits and fragments, such as the thiol-to-disulfide transformations via protein sulfenic acid/thiol capture mechanisms. The stabilizing disulfide linkages build up as proteins move from the reducing cytosolic compartment to the more oxidizing compartments of the ER, Golgi complex, and post-Golgi complex vesicles in the protein secretory pathways of eukaryotic cells.

Covalent modifications of protein side chains can introduce hydrophobic or hydrophilic groups. Hydrophobic groups can be added by posttranslational enzymatic catalysts to Lys-ϵ-NH$_2$ and Cys–S$^-$ groups by alkylation or by acylation routes. The alkyl donors are SAM for the C$_1$ methyl fragment and farnesyl-PP and geranyl-geranyl-PP for the C$_{15}$ and C$_{20}$ prenyl fragments, respectively. The acyl groups covalently added span a corresponding range from the C$_2$ acetyl group (on N-terminal amino groups as well as lysine side chains) to the C$_{14}$ myristoyl and the C$_{16}$ palmitoyl groups, added to the N-terminal glycine residues and internal cysteine residues, respectively. Both the long-chain alkylation and the acylation reactions redirect proteins from cytoplasm/nucleoplasm to membrane interfaces, altering their intracellular addresses. The differential stability and degree of hydrolytic reversibility of palmitoyl-S-Cys linkages compared to myristoyl-N-amide linkages to lysine residues and the distinct biological consequences have been noted. The C$_1$-methylations and C$_2$-acetylations do not generally impart sufficient hydrophobicity to lysine side chains of modified proteins to partition them into membranes, but do alter the lysine hydrophobicity and/or charge sufficiently to dramatically affect recognition by partner proteins in the multicomponent complexes for gene transcription.

The most common hydrophilic organic modifications put in by posttranslational modification enzymes are sugar residues in both O- and N-linkages. The O-glycosylations can be a single O-GlcNAc, O-GalGlu disaccharides in collagen processing, O-glucosyl-trisaccharides or O-fucosyl-tetrasaccharides in Notch maturation, or O-mannosyl-oligosaccharide chains. The N-glycan chains linked to Asn carboxamido nitrogens in eukaryotes start as branched tetradecasaccharyl chains before being trimmed back to Glc$_1$Man$_9$GlcNAc$_2$ in the ER. This dodecasaccharide group is the lever for chaperone-mediated refolding cycles, which increases the yield of native glycoproteins that have passed into the ER compartment. Further trimming back to a core Man$_5$GlcNAc$_2$ heptasaccharide in the Golgi complex sets up multitudinous variations of glycosyltransferase decoration to yield bi-, tri-, and tetraantennary N-glycan chains in mature glycoproteins in the late stages of eukaryotic cell secretory pathways. Partial occupancy and site heterogeneity are typical, such that mixtures of N-glycoproteins are generated with a wide variety of structural and functional roles of the glycan chains in the mature proteins.

More limited examples of hydrophilic posttranslational modifications are the γ-carboxylations of zymogen forms of secreted serine proteases and hydroxylations at carbon sites on the side chains of proteins. These include the classical cases of the 4-OH-prolines and 5-OH-lysines essential for collagen triple-helix formation and secretion, respectively. The contemporary case of the molecular basis of oxygen sensing in cells by hydroxylation of the transcription factor HIF-1α at a specific proline and a specific asparagine residue has reinforced the biological importance of such protein monooxygenases.

Oxidative posttranslational transformations of proteins are not limited to sulfur oxygenation (Cys and Met) or the carbon hydroxylations just noted. The generation of orthoquinone forms of tyrosine and tryptophan in amine oxidase maturation creates the electron sinks required for low-energy pathways for the deamination of amine substrates. The orthoquinone group is a superactivated ketone not readily available in the side chains of the 20 proteinogenic amino acids. Two other posttranslational modifications lead to ketone groups in enzyme active sites. One is found in a distinct class of amine deaminases for the amino acids phenylalanine and histidine, where the MIO group is created as the active site electrophile. The other is the fragmentive conversion of internal serine residues to N-terminal pyruvamides.

MIO group formation is representative also of a series of other transformations that are autocatalytic in the active sites of folded proteins. This one involves the rearrangement of the $Ala_{142}Ser_{143}Gly_{144}$ tripeptide to the cyclic imidazolone MIO electron sink. Comparable logic in autocatalytic rerrangement of a $Ser_{65}Tyr_{66}Gly_{67}$ tripeptide loop to an aryl imidazolone that autoxidizes to the conjugated fluorophore generates green fluorescent proteins and congeners. These tripeptide loop rearrangements are rare and require finely tuned folded protein conformational states to set in motion the intramolecular attack of peptide bond nitrogen on upstream peptide carbonyl.

Cleavages, Autocleavages, and Religations

There are several mechanistically related examples of autocatalysis where internal peptide bonds in natively folded proteins are cleaved by intramolecular attack of Ser, Thr, and Cys side chains on the upstream peptide carbonyls. The resultant tetrahedral adducts can eliminate the peptide nitrogen and generate oxoester (from Ser or Thr) or thioester (from Cys) intermediates.

A variety of fates are seen for such rearranged protein intermediates. One is cleavage of the protein into two fragments, generating an N-terminal pyruvamide group on the downstream protein fragment. This represents autocleavage of an inactive proform of a protein to an active fragment in which the active site electrophile, the ketone group of the pyruvamide moiety, has been formed during the autocleavage. The C-terminus of the upstream peptide fragment can be captured by water or by other ROH substrates in such autocleavages. During autocleavage of the precursor form of the Hedgehog family of signaling proteins, the specific ROH cosubstrate is the 3-OH of cholesterol, producing the cholesterol ester of the N-terminal 17-kDa fragment of Hedgehog that is the active form of the protein for signaling.

Recall, too, the hundred or so examples of protein autocleavage to the peptidyl-oxoester/thioester intermediates and then religation during protein splicing reactions. The net reaction is excision of the intervening peptide (intein) and joining of the upstream and downstream exteins. The reformation of the peptide bond, as extein$_1$ and extein$_2$ are ligated, is driven by favorable S-to-N or O-to-N acyl shift chemistry.

Protein cleavages *without religation* are the fates of proteins that have been cleaved posttranslationally by proteases. The molecular logic of the main families of proteases for the mechanism of hydrolytic peptide bond cleavage and specificity were examined. Also noted were the different subcellular locations and the timing of controlled proteolytic cuts in proteins as they make their way through the secretory pathways of eukaryotic cells. Signal peptidases remove the N-terminal 20–30 residues during passage into the ER. Furin and related proprotein convertases act later, in the Golgi and trans Golgi network of vesicles, to effect specific cuts, such as cleaving Notch from one polypeptide to two associated protein fragments. A third stage of trimming proteolysis in the life cycle of a protein displayed at the cell surface is carried out by proteases acting as sheddases, liberating some or all of the extracellular domain of transmembrane protein substrates. A fourth stage of specific, limited proteolysis can then occur by regulated intramembrane proteolysis (RIP), for example on Notch, the sterol response binding protein, and the amyloid precursor peptides. The cytoplasmic stubs of such a protein, four cuts after the initial translation of the protein, can then move to the nucleus and act as a transcriptional activator of specific subsets of genes.

Specific peptide bond cleavages at a site near the carboxy terminus occur in proteins that get transferred to GPI anchors in eukaryotic cells and to peptidoglycan chains in Gram-positive bacteria. These are net transamidations. In a real sense these are cleavages and religations in parallel to the steps in splicing noted above, but the incoming amine nucleophile is not part of the same, or even another protein. Rather, in GPI anchor covalent attachment it is an ethanolamine moiety on the oligosaccharylphosphatidyl inositiol lipid. In the bacterial sortase transamidations the attacking nucleophile is the terminal amine of a peptidyl cross-bridge from Lys$_3$ of the peptidoglycan chains.

Covalent Protein Tags

Perhaps the most widespread example in eukaryotic cells of C-terminal derivatization of proteins occurs in the activation and covalent tethering of the 8-kDa ubiquitin (and its ten or so ubiquitin-like homologs) as thioesters to various members of the families of E$_2$ ubiquitin ligases. These ubiquityl-S-enzyme adducts are thermodynamically activated for transfer of the 76-residue protein tag to the ϵ-NH$_2$ of lysine residues in hundreds of cellular proteins, generating isopeptide linkages in the proteins marked by ubiquitin. These isopeptide linkages can be cleaved hydrolytically, not by the normal complement of proteases but rather by a family of isopeptidases termed deubiquitylating enzymes.

The covalent ubiquitin tag is architecturally and informationally rich and presents several faces for recognition by the partner proteins that usher the marked proteins to different subcellular locales. There is a sharp distinction in fate between proteins covalently marked with one ubiquitin and those tandemly marked with four or more ubiquitins, oligomerized as a multiubiquitin chain. The Ub_1-ϵ-Lys-proteins get chaperoned into the trans Golgi vesicular network as cargo and can be deposited in lysosomes where they are proteolyzed as part of the down-regulation and removal of transmembrane receptor proteins. The Ub_n-ϵ-Lys-proteins get shepherded by distinct protein machinery to the proteasomes for degradation to limit peptides. In either fate the tagged proteins, by Ub_1 or by Ub_n, have their lifetimes shortened by subsequent proteolytic destruction.

Tandem Modifications

The tandem, processive addition of four or more ubiquitins to create the recognition signal for chaperoning to proteasomes sets up the ensuing removal of the Ub_n tags by deubiquitylases (DUBs), unfolding and threading of the protein into the proteasome chambers, and its irreversible cleavage into 8- to 12-residue peptide fragments. There are two general attributes of posttranslational modification reactions thus exemplified. One is the enumeration of the large array of distinct proteins that undergo that type of chemical modification. The second is the multiplicity and regioselectivity of modifications within that protein subset.

The enumeration of the family of proteins subjected to a given type of modification has engendered proteomics approaches, heavily dependent on mass spectrometric methodologies. Thus, ongoing efforts to delineate the phosphoprotein proteomes in particular cell types at given times and conditions have been noted, as well as recent studies to define the ubiquitylated proteome in yeast, with many more studies to come.

In the second facet, several examples of the multiple phosphorylation of proteins have been noted, with distinct consequences, from the pT and pY double phosphorylation requirement for the activation of MAP kinases to the 11 phosphorylation sites spread over the several domains of the Abl tyrosine kinase. The competition between acetylation and ubiquitylation for five lysines at the C-terminus of the transcription factor p53 has been examined. The four distinct types of covalent processing (S-palmitoylation, S-farnesylation, proteolysis, and C-terminal COOH methylation) that occur in the maturation of Ras GTPase have been delineated.

As a paradigmatic example of both of the above concerns, the decipherment of the histone code on histone tails continues to be a challenging technical and intellectual problem for analysis of the combinatorial logic of the covalent control of selective gene transcription. The complexity of the problem and a way forward has been proposed with the prescriptive annotation of histone modifications (Pesavento, 2004). Using histone H4 as a starting point and seven known sites of modification (Ser_1 phosphorylation, Arg_3 mono- or dimethylation, and $Lys_{5,8,12,16,20}$ mono- or dimethylation or acetylation), along with N-acetylation and the cotranslation proteolytic removal of the initiator Met as input, 46,875 possible combinations were catalogued in a relational database (Pesavento, 2004). Histone H4 extracted from cells was subjected to ESI/Q-FTMS and a peak at +112 Da in the MS–MS spectrum from electron capture dissociation was unambiguously assigned to N-terminal acetylation, Lys_{16} acetylation, and Lys_{20} dimethylation. Analogously in the presence of the HDAC inhibitor butyrate, using the high mass accuracy enabled by this approach, a +238-Da mass peak was assigned to histone H4 having seven posttranslational modifications. The precise regiospecificity was N-terminal K_5-acetyl, K_8-acetyl, K_{12}-acetyl, K_{16}-acetyl, and K_{20}-dimethyl (Pesavento, 2004) (Figure 6.5). This approach illustrates the combination of bioinformatics and high-resolution mass spectrometry that should be applicable not just to one of the histones, but to all four histone tails in nucleosomes from different physiological states at specific genes in particular cell types, and on to more complex slices of the covalently modified proteome.

The Rho family of GTPases, involved in the organization of the cytoskeleton by nucleation of stress fibers and focal adhesion, undergo at least five different chemical types of posttranslational modifications as noted in Chapters 7, 10, and 11, and schematized in the chapter-opener figure to this chapter. There are 10 Rho family members, all acting as protein switches, driven from "on" to "off" states by the hydrolysis of bound GTP to GDP (Bishop and Hall, 2000). The switching cycle is controlled by Rho partner proteins acting as GTPases (GAPs) and guanine nucleotide exchange factors (GEFs).

Three of the five modifications are catalyzed by bacterial toxins that disable Rho catalytic activity and thus disrupt cytoskeletal organization. Two of these are glycosylation reactions, the O-glucosylation of Thr_{35} and the N-glycosylation of Asn_{41} by ADP-ribose (from NAD), that affect residues in and around the active site. A third toxin-mediated modification is the deamidation of Gln_{63} required for GTP hydrolysis. A variant of the hydrolysis of glutamine is transglutaminase-mediated capture of the glutamyl-S-enzyme intermediate (Chapter 15), not by water but by the amino group of serotonin. This has been reported for Rho in platelets (Walther et al., 2003). The other two posttranslational modifications happen near the C-terminus of the protein. One involves lipid-mediated anchoring of Rho at a Cys side chain via geranylgeranylation to create a nonreversible covalent S-prenyl linkage that helps direct Rho to membrane–cytoplasm interfaces. The prenylated Cys can be cleaved by a C-terminal carboxypeptidase activity that produces Rho truncated at Gly_{189}.

Another remarkable example of tandem sets of posttranslational modifications of a protein during maturation is the conversion of thyroglobulin to the thyroid hormone thyroxine (tetraiodo- and triiodothyronine). In the thyroid gland a 300-kDa protein thyroglobulin, stored as a 600-kDa dimer in the thyroid follicular lumen, serves as a platform for posttranslational diiodination of some of its tyrosine residues, followed by radical coupling of two such residues to create the tetraiodinated aryl-ether derivative T_4 and the triiodo-T_3, still embedded in the protein backbone (Dunn, 2000). The protein framework is then proteolyzed to release the free T_4 and T_3 forms of thyroxine (Figure 16.1A). The 2768-residue thyroglobulin is translated, directed into the ER by its 19-residue N-terminal signal sequence whose cleavage by signal peptidase is an early posttranslational processing step. Thyroglobulin is also cotranslationally subjected to N-glycosylation, and transits through the secretory pathway to the plasma membrane of the producing thyroid cells and into the lumen of thyroid follicles.

At the membrane there are NADH oxidase enzymes, generating H_2O_2 from O_2 reduction, and also the heme-dependent thyroperoxidase that is the iodination catlyst for the halogenation of Tyr residues of thyroglobulin (Kimura et al., 1987). The

A.

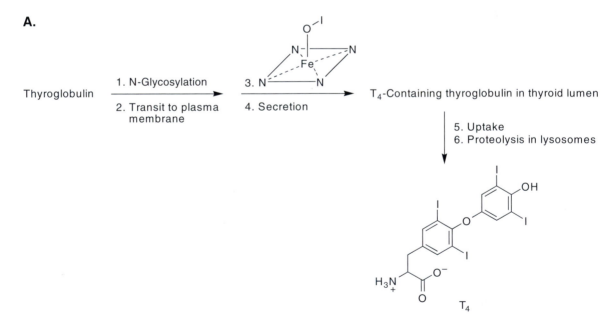

Figure 16.1 Posttranslational modification of thyroglobulin. **A.** Multiple steps for conversion of thyroglobulin residues to T_4.

iodoperoxidase is thought to use H_2O_2 to generate a typical high-valent oxoiron species ($Fe^V=O$) that activates iodide ions as Fe^{III}–OI, which is a source of an "I^+" equivalent for the reaction with Tyr residues at positions 3 and 5 in the aryl ring, ortho to the activating phenol-OH (Figure 16.1B). Multiple Tyr residues in thyroglobulin are

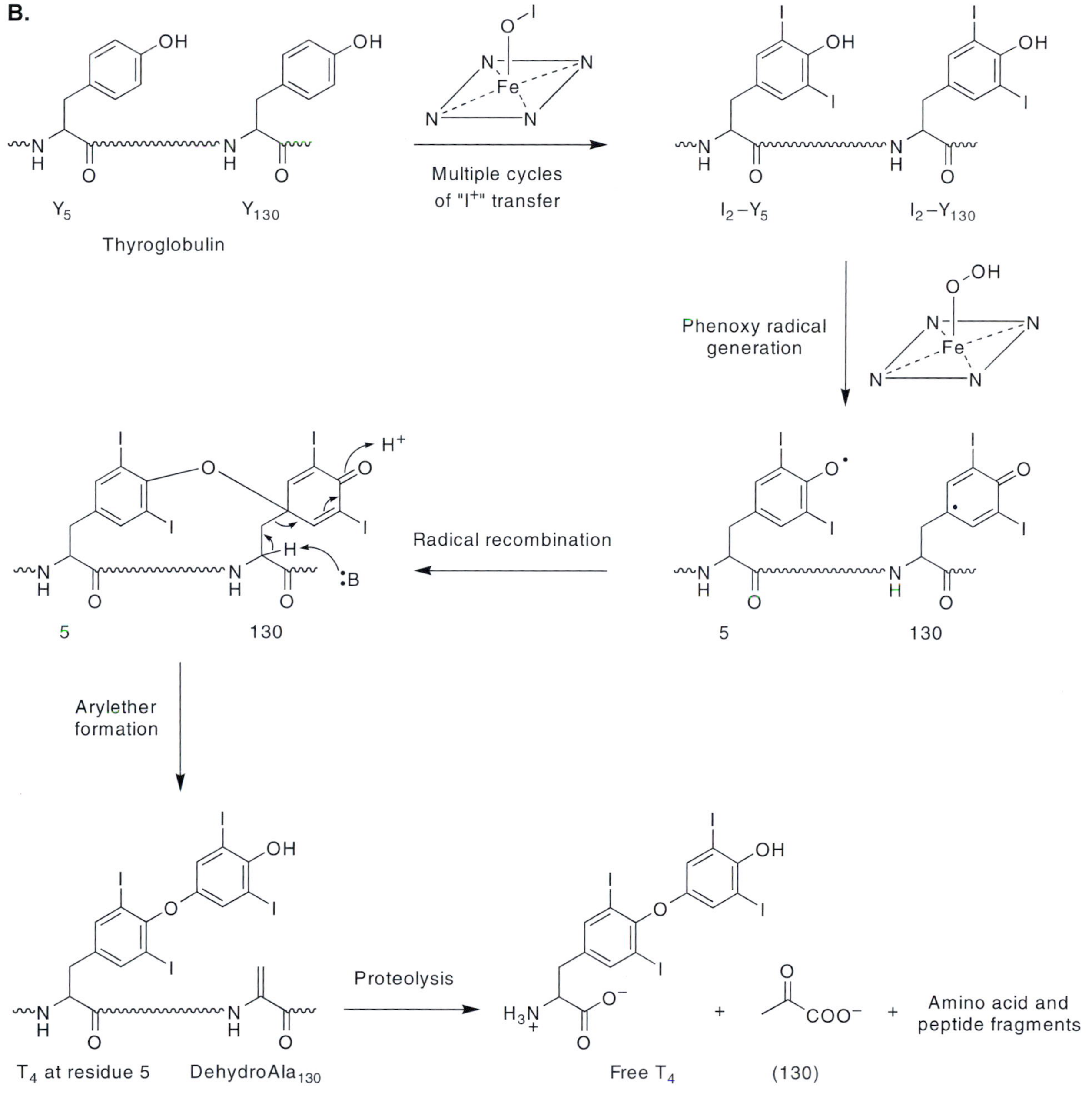

B.

Figure 16.1 **B.** Mechanism for iodination and oxidative coupling of Y_5 and Y_{130} of thyroglobulin to T_4.

iodinated by the peroxidase with up to five monoiodo-Tyr, yielding five diiodo-Tyr. Of these 10 residues, Y_5 near the N-terminus and Y_{2554} near the C-terminus are most often converted to embedded T_3 and T_4.

In the arylether coupling reaction at the N-terminus, I_2-Y_5 can act as the acceptor and I_2-Y_{130} as the donor of the aryl moiety. Thyroperoxidase also catalyzes this coupling step in posttranslational maturation—that is, in the fragmentation of the diiodo-Tyr_{130} side chain by its transfer to the phenolic OH of diiodo-Tyr_5. This requires that I_2-Y_5 and I_2-Y_{130} be proximal in the folded structure of thyroglobulin at the plasma membrane or in the lumen adjacent to the peroxidase externally directed active site. Arylether coupling could proceed via phenoxyradical species on Y_5 and Y_{130} as depicted in Figure 16.1B. The fragmentation step is written as a β-elimination, generating a dehydroAla residue at residue 130, that deconvolutes to pyruvate (Dunn et al., 1998) on subsequent proteolysis, akin to the transformations noted in Chapter 13.

The T_4-containing thyroglobulin is stored in the lumen of the thyroid gland and turns over at about 1% per day in normal humans to release free T_3 and T_4. This occurs by receptor-mediated re-uptake from the lumen by polarized thyroid epithelial cells, transport to lysosomes, and proteolysis. The N-terminal and C-terminal T_4 residues (from Y_5 and Y_{2554}) are released early by endo- and exoprotease action followed by complete digestion of the 300-kDa thyroglobulin scaffold. Mono- and diiodo-Tyr residues released are deiodinated enzymatically to recycle iodide ions.

The global strategy in thyroid hormone generation is akin to the production of small peptide hormones by a series of proteolytic cleavages to release small peptide fragments. In this case a 2800-residue protein precursor is proteolytically cleaved to about two molecules of T_4 in each thyroglobulin life cycle. Presumably the microenvironment of the folded thyroglobulin is essential for both multiple iodinations and radical recombination to make T_4. This is a very expensive way to make the tetraiodo arylether derivative of tyrosine, but it is presumably balanced by the essential and widespread activity of this hormone that is central to energy balance and homeostasis. Recent evidence suggests that free T_4 is processed, deiodinated enzymatically to T_1, and decarboxylated to the T_1-thyronamine that acts as a ligand for the G-protein-coupled receptors in distinct signaling pathways (Scanlan et al., 2004) that counterbalance the T_4-derived signals.

Cotranslational Modifications and the 21st and 22nd Amino Acids

In addition to posttranslational modification reactions of proteins, a few covalent modifications occur cotranslationally. The N-glycosylation of proteins noted above and discussed in detail in Chapter 10 is one such example, with the nascent chain emerging from the ribosome tunnel once the chain has reached about 40 residues and then subject to the oligosaccharyltransferase machinery in the ER membrane. The deformylation of the N-formyl-Met$_1$ residue of bacterial proteins is a widespread prokaryotic cotranslational example. The corresponding removal of Met$_1$ from eukaryotic proteins emerging from the ribosomal tunnel by one of two methionine amino-peptidases is another (Lowther and Matthews, 2000). The N-acetylation of the new N-terminus by acetyl CoA and the NAT enzyme or the special subset of Gly$_1$-residue acylation by myristoyl CoA via protein myristoyltransferase action are also cotranslational and happen before substantial folding of the emerging nascent protein.

Recall the occurrence of selenocysteine as the 21st proteinogenic amino acid in a small number of proteins in eukaryotes and prokaryotes. Special emphasis in Chapter 4 was devoted to the active site selenoCys selenate anion in thioredoxin reductase catalytic cycles, where the enhanced nucleophilicity of the selenate anion provides extra driving force. The logic of utilization of the 21st amino acid and also of the 22nd, pyrrolysine, so far detected only in a methanogen methylamine methyltransferase, is to conscript a transfer RNA that normally recognizes one of the stop codons in the triplet genetic code and so would normally function as a blank (a termination signal). During SeCys insertion into proteins this termination tRNA has been resculpted to accept cysteine in the presence of an isoform of cysteinyl-tRNA synthetase specific for that tRNA (Figure 1.3A). The generation of Ser-tRNA$_{term}$ provides the scaffold for enzymatic conversion of the seryl moiety, via a dehydroalanyl intermediate, to selenoCys-tRNA$_{term}$. Now the second phase of incorporation of the 21st amino acid requires that the ribosome recognize an AUG stop codon in the middle of a protein coding sequence with the CUA anticodon of the charged SeCys-tRNA$_{term}$. This behavior is called termination/stop codon supression and results in the insertion of SeCys at the location where a stop codon would normally be read. The same logic is utilized by the methanogenic archaebacteria for incorporation of pyrrolysine (Figure 1.3B). A Lys-tRNA$_{CUA}$, acting as a suppressor tRNA for an AUG stop codon, is loaded with Lys, then converted to the pyrroLys moiety in pyrroLys-tRNA$_{CUA}$. That is base paired at the AUG codon in the peptidyltransferase center of the ribosome to insert pyrroLys at a particular residue of the nascent methylamine methyltransferase protein. This is the logic that is being appropriated for efforts to expand the genetic code noted below.

Prospects for Expanding the Genetic Code: 23rd to *n*th Proteinogenic Amino Acid

In principle the genetic code could be expanded to a 23rd and 24th amino acid and further if one could engineer the following protein synthetic machinery in cells. First, one would need a $tRNA_{term}$ that could suppress a given triplet stop codon that showed orthogonal specificity to an amino acid distinct from any of the existing 22 proteinogenic amino acids. One would need additionally an aminoacyl-tRNA synthetase also doubly orthogonal from the other aa-tRNA synthetases in a cell, which recognized only the suppressor tRNA and transferred a particular nonproteinogenic amino acid. Schultz and colleagues have had some remarkable successes in the design and implementation of this strategy (Wang and Schultz, 2002). They started with a methanogen-suppressor tRNA and a tyrosyl-tRNA synthetase and co-evolved them to recognize, activate, and transfer particular tyrosine analogs to create the aminoacyl-$tRNA_{term}$ that would insert the novel amino acid by suppression of a stop codon in proteins expressed in *E. coli* (Santoro et al., 2002; Wang and Schultz, 2001) (Figure 16.2A). Over a dozen novel amino acids have been site-specifically incorporated into proteins whose gene contained a suppressable stop codon at a particular site in the protein (Mehl et al., 2003). Some of these are shown in Figure 16.2A (Wang et al., 2002), including ketone-containing amino acids (Wang et al., 2003) such as *p*-acetyl-Phe, whose utility could be shown by imine formation to fluorescamine, allowing the location of the fluorescent protein in cells. A ketone moiety embedded in benzoyl-Phe creates a photoactivatable side chain for cross-link studies (Chin et al., 2002a). All of these derivatives could be placed at any residue at will by the creation of the gene variant with a stop codon placed at any desired residue in the sequence of any target protein that can be expressed in *E. coli*. A 27th and 28th amino acid incorporated are the azidoPhe (Chin et al., 2002b) and the alkyne ether derivative of tyrosine. When placed in two separate proteins biosynthetically, the azide and alkyne can be activated by copper catalysis in aqueous solutions under physiological conditions to the triazene (Figure 16.2B). This represents a novel nonoxidative cross-link for proteins. Each component amino acid could in principle be placed at any locus in a protein pair to map the surfaces that come close enough to form the triazene cross-link. Homoglutamine has been incorporated, as has the L-DOPA analog of tyrosine, setting the engineered protein up for redox chemistry via orthoquinone formation, in analogy to the natural TOPA quinone cofactors noted in Chapter 13.

The final example of what is sure to be an expanding list is the first example of incorporation of an O-GlcNAc-serine at residue 4 of myoglobin expressed in *E. coli* (Zhang et al., 2004). The monoglycosyl myoglobin was characterized by mass spectrometry and the GlcNAc chain could be enzymatically elongated by a galactosyltransferase and UDP-Gal to create a Gal-GlcNAc-O-Ser_4 variant of myoglobin (Figure 16.3). This sets the stage for site-specific formation of O-glycoproteins at any residue and, furthermore, would allow stoichiometric and homogenous O-neoglycoprotein production.

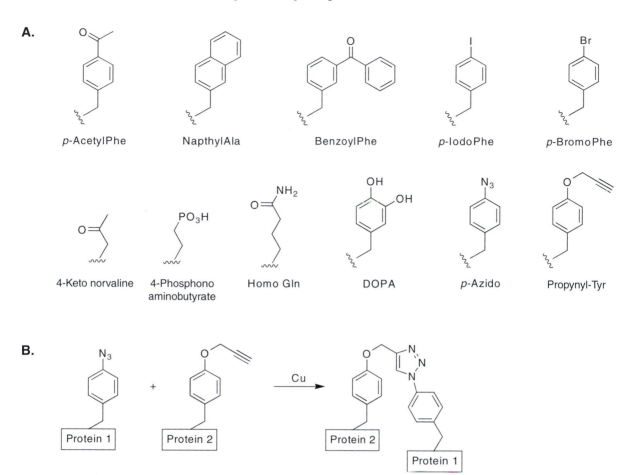

Figure 16.2 **A.** Examples of nonproteinogenic amino acids incorporated site-specifically into proteins by orthogonal pairs of Tyr-tRNA and the synthetase that insert the amino acids at stop codons. **B.** Interprotein cross-linking by Cu-mediated click chemistry between an azido amino acid incorporated on one protein and an alkynyl amino acid residue on a second protein.

Progress has been made in the corresponding evolution of an orthogonal pair of suppressor tRNA and paired aminoacyl-tRNA synthetase in yeast such that site-specifically engineered proteins can be biosynthesized via an expanded genetic code in eukaryotes (Chin et al., 2003), incorporating p-acetylPhe, p-iodoPhe, methoxy-Tyr, p-azidoPhe, and benzoylPhe.

Challenges that lie ahead for routine expansion of the genetic code to the nth amino acids in prokaryotic and eukaryotic protein expression systems include the coevolution of orthogonal pairs of suppressor tRNA and aminoacyl-tRNA synthetases that will take novel nonaromatic amino acids (i.e., generalization beyond the evolution of Tyr-tRNA synthetases). The efficiency of incorporation of the unnatural

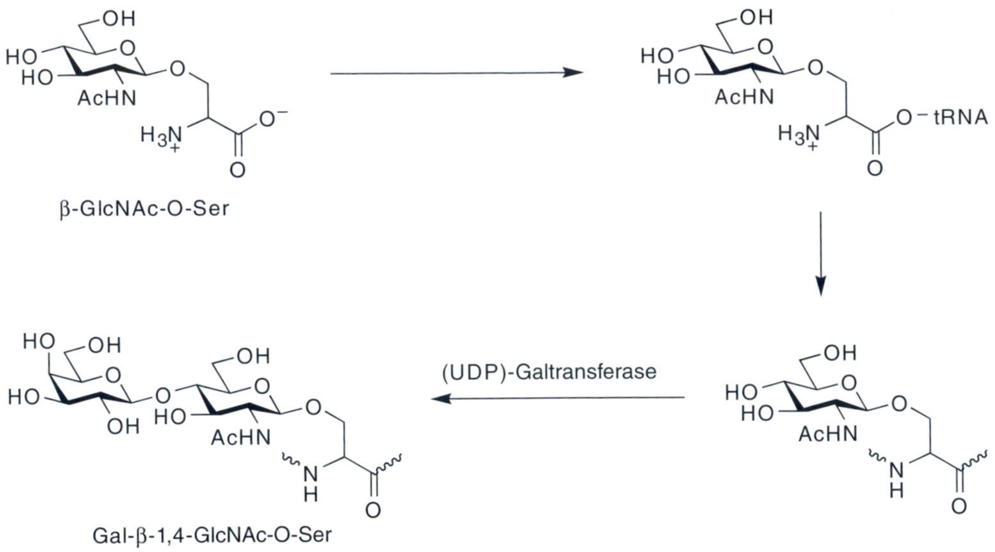

Figure 16.3 Activation and incorporation of GlcNAc-O-Ser at codon 4 of myoglobin via recognition of GlcNAc-O-Ser by an orthogonal tRNA–tRNA synthetase pair and nonsense codon suppression.

amino acids will depend on the successful uptake of the desired amino acids by amino acid transport proteins when added exogenously to producer microbial/eukaryotic cells. In a few cases it may be possible to construct a metabolic pathway in the producer organism to generate the variant amino acid, as demonstrated for the *in situ* generation of *p*-aminoPhe and its site-specific incorporation by an orthogonal tRNA-tRNA synthetase pair (Mehl et al., 2003), but in general this will not be possible for unnatural amino acids. Nonetheless, the demonstration that a Tyr-tRNA synthetase can be evolved to recognize and load GlcNAc-O-serine as substrate augurs well for expanding the genetic code.

References

Bishop, A. L., and A. Hall. "Rho GTPases and their effector proteins," *Biochem. J.* **348(2)**:241–255 (2000).

Chin, J. W., A. B. Martin, D. S. King, L. Wang, and P. G. Schultz. "Addition of a photocrosslinking amino acid to the genetic code of *Escherichia coli*," *Proc. Natl. Acad. Sci. U.S.A.* **99**:11020–11024 (2002a).

Chin, J. W., S. W. Santoro, A. B. Martin, D. S. King, L. Wang, and P. G. Schultz. "Addition of p-azido-ʟ-phenylalanine to the genetic code of *Escherichia coli*," *J. Am. Chem. Soc.* **124**:9026–9027 (2002b).

Chin, J. W., T. A. Cropp, J. C. Anderson, M. Mukherji, Z Zhang, P. Schultz. "An expanded eukaryotic genetic code," *Science* **301**:964–967 (2003)

Dunn, A. D., C. M. Corsi, H. E. Myers, and J. T. Dunn. "Tyrosine 130 is an important outer ring donor for thyroxine formation in thyroglobulin," *J. Biol. Chem.* **273**:25223–25229 (1998).

Dunn, J., and A. D. Dunn. *Thyroglobulin: Chemistry, Biochemistry and Proteolysis*. Lippincott, Williams and Wilkins: Philadelphia (2000).

Kimura, S., T. Kotani, O. W. McBride, K. Umeki, K. Hirai, T. Nakayama, and S. Ohtaki. "Human thyroid peroxidase: Complete cDNA and protein sequence, chromosome mapping, and identification of two alternately spliced mRNAs," *Proc. Natl. Acad. Sci. U.S.A.* **84**:5555–5559 (1987).

Lowther, W. T., and B. W. Matthews. "Structure and function of the methionine aminopeptidases," *Biochim. Biophys. Acta* **1477**:157–167 (2000).

Mehl, R. A., J. C. Anderson, S. W. Santoro, L. Wang, A. B. Martin, D. S. King, D. M. Horn, and P. G. Schultz. "Generation of a bacterium with a 21 amino acid genetic code," *J. Am. Chem. Soc.* **125**:935–939 (2003).

Pesavento, J. J., Y-b. Kim, G. K. Taylor, and N. L. Kelleher. "Prescriptive annotation of histone modifications: A new approach for streamlined characterization of proteins by top down mass spectrometry," *J. Am. Chem. Soc.* **126**:3386–3387 (2004).

Santoro, S. W., L. Wang, B. Herberich, D. S. King, and P. G. Schultz. "An efficient system for the evolution of aminoacyl-tRNA synthetase specificity," *Nat. Biotechnol.* **20**:1044–1048 (2002).

Scanlan, T. S., K. L. Suchland, M. E. Hart, G. Chiellini, Y. Huang, P. J. Kruzich, S. Frascarelli, D. A. Crossley, J. R. Bunzow, S. Ronca-Testoni, E. T. Lin, D. Hatton, R. Zucchi, and D. K. Grandy. "3-Iodothyronamine is an endogenous and rapid-acting derivative of thyroid hormone," *Nat. Med.* **10**:638–642 (2004).

Walther, D. J., J. U. Peter, S. Winter, M. Holtje, N. Paulmann, M. Grohmann, J. Vowinckel, V. Alamo-Bethencourt, C. S. Wilhelm, G. Ahnert-Hilger, and M. Bader. "Serotonylation of small GTPases is a signal transduction pathway that triggers platelet alpha-granule release," *Cell* **115**:851–862 (2003).

Wang, L., A. Brock, and P. G. Schultz. "Adding L-3-(2-Naphthyl)alanine to the genetic code of *E. coli*," *J. Am. Chem. Soc.* **124**:1836–1837 (2002).

Wang, L., and P. G. Schultz. "A general approach for the generation of orthogonal tRNAs," *Chem. Biol.* **8**:883–890 (2001).

Wang, L., and P. G. Schultz. "Expanding the genetic code," *Chem. Commun. (Cambridge)*:1–11 (2002).

Wang, L., Z. Zhang, A. Brock, and P. G. Schultz. "Addition of the keto functional group to the genetic code of *Escherichia coli*," *Proc. Natl. Acad. Sci. U.S.A.* **100**:56–61 (2003).

Zhang, Z., J. Gildersleeve, Y. Y. Yang, R. Xu, J. A. Loo, S. Uryu, C. H. Wong, and P. G. Schultz. "A new strategy for the synthesis of glycoproteins," *Science* **303**:371–373 (2004).

Index

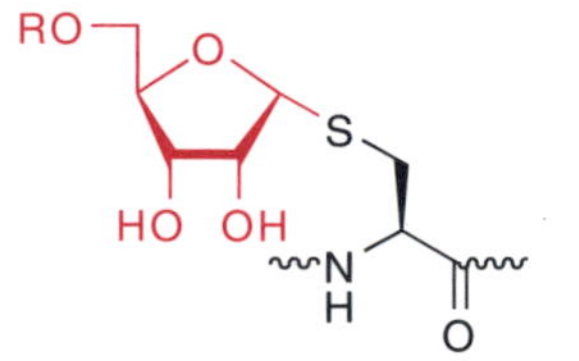

ADP-ribosyl-Cys-protein

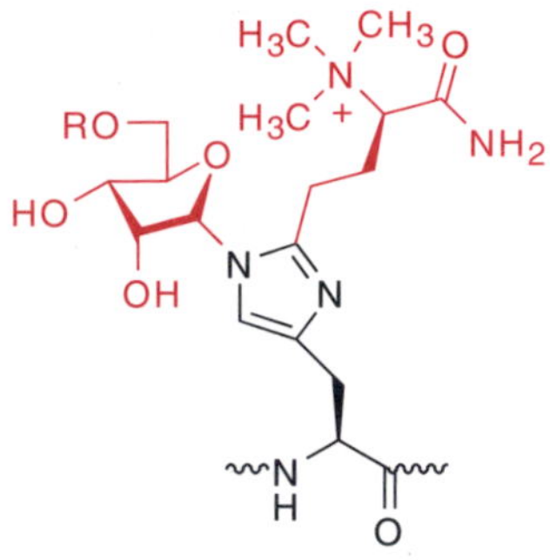

ADP-ribosylated diphthamide

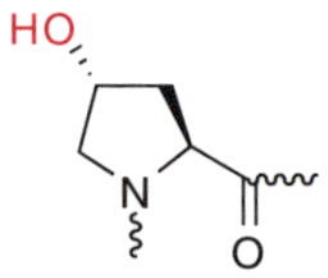

4-OH-Pro

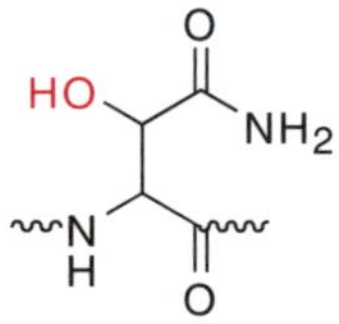

3-OH-Asn

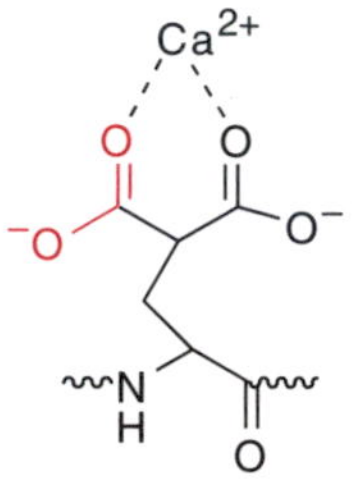

Ca^{2+}-Gla

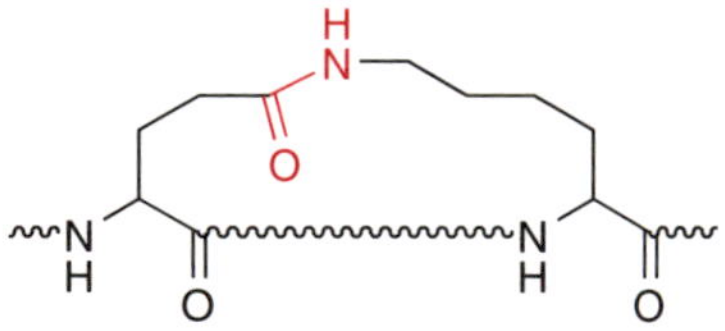

Gln-Lys-cross-link

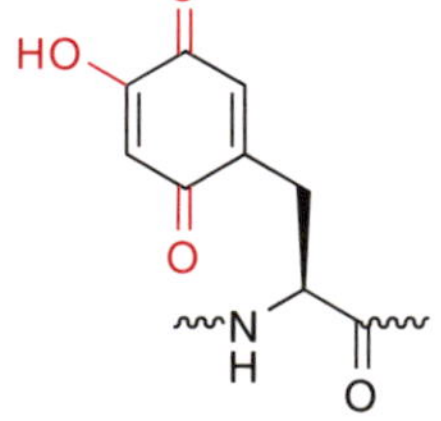

TOPA quinone